Trends in Mathematics is a series devoted to the publication of volumes arising from conferences and lecture series focusing on a particular topic from any area of mathematics. Its aim is to make current developments available to the community as rapidly as possible without compromise to quality and to archive these for reference.

Proposals for volumes can be sent to the Mathematics Editor at either

Birkhäuser Verlag
P.O. Box 133
CH-4010 Basel
Switzerland

or

Birkhäuser Boston Inc.
675 Massachusetts Avenue
Cambridge, MA 02139
USA

Material submitted for publication must be screened and prepared as follows:

All contributions should undergo a reviewing process similar to that carried out by journals and be checked for correct use of language which, as a rule, is English. Articles without proofs, or which do not contain any significantly new results, should be rejected. High quality survey papers, however, are welcome.

We expect the organizers to deliver manuscripts in a form that is essentially ready for direct reproduction. Any version of TeX is acceptable, but the entire collection of files must be in one particular dialect of TeX and unified according to simple instructions available from Birkhäuser.

Furthermore, in order to guarantee the timely appearance of the proceedings it is essential that the final version of the entire material be submitted no later than one year after the conference. The total number of pages should not exceed 350. The first-mentioned author of each article will receive 25 free offprints. To the participants of the congress the book will be offered at a special rate.

Modules
and Comodules

Tomasz Brzeziński
José Luis Gómez Pardo
Ivan Shestakov
Patrick F. Smith
Editors

Birkhäuser
Basel · Boston · Berlin

Editors:

Tomasz Brzeziński
Department of Mathematics
Swansea University
Singleton Park
Swansea, SA2 8PP
Wales, UK
e-mail: t.brzezinski@swansea.ac.uk

Ivan Shestakov
Instituto de Matemática e Estatística da
Universidade de São Paulo (IME-USP)
Rua do Matão, 1010
Cidade Universitária
CEP 05508-090 São Paulo – SP
Brasil
e-mail: shestak@ime.usp.br

José Luis Gómez Pardo
Departamento de Alxebra
Universidade de Santiago
15782 Santiago de Compostela
Spain
e-mail: alpardo@usc.es

Patrick F. Smith
Department of Mathematics
University of Glasgow
University Gardens
Glasgow G12 8QW
Scotland, UK
e-mail: p.smith@maths.gla.ac.uk

2000 Mathematical Subject Classification: 16Dxx, 16Wxx, 17Dxx

Library of Congress Control Number: 2008925376

Bibliographic information published by Die Deutsche Bibliothek. Die Deutsche Bibliothek lists
this publication in the Deutsche Nationalbibliografie; detailed bibliographic data is available in
the Internet at http://dnb.ddb.de

ISBN 978-3-7643-8741-9 Birkhäuser Verlag AG, Basel - Boston - Berlin

© 2008 Birkhäuser Verlag AG
Basel · Boston · Berlin
P.O. Box 133, CH-4010 Basel, Switzerland
Part of Springer Science+Business Media
Printed on acid-free paper produced from chlorine-free pulp. TCF ∞
Cover Design: Alexander Faust, CH-4051 Basel, Switzerland
Printed in Germany
ISBN 978-3-7643-8741-9

e-ISBN 978-3-7643-8742-6

9 8 7 6 5 4 3 2 1

www.birkhauser.ch

Contents

Preface vii

List of Participants viii

Acknowledgment ix

Robert Wisbauer *by J. Clark* xi

J.Y. Abuhlail and S.K. Nauman
Injective Morita Contexts (Revisited) 1

A. Ardizzoni and C. Menini
A Categorical Proof of a Useful Result 31

N. Andruskiewitsch and I.E. Angiono
On Nichols Algebras with Generic Braiding 47

V.A. Artamonov and I.A. Chubarov
Dual Algebras of Some Semisimple Finite-dimensional
Hopf Algebras 65

M. Beattie, D. Bulacu and Ş. Raianu
Balanced Bilinear Forms for Corings 87

G.F. Birkenmeier, J.K. Park and S.T. Rizvi
Ring Hulls of Semiprime Homomorphic Images 101

T. Brzeziński
Notes on Formal Smoothness 113

E. Sanchez Campos and P.F. Smith
Certain Chain Conditions in Modules over Dedekind
Domains and Related Rings 125

S. Charalambides and J. Clark
τ-Injective Modules 143

vi   Contents

M.A. Chebotar, W.-F. Ke, P.-H. Lee and E.R. Puczyłowski
  A Note on Polynomial Rings over Nil Rings 169

J. Dauns and Y. Zhou
  QI-modules 173

L. El Kaoutit and J. Gómez-Torrecillas
  Corings with Exact Rational Functors and Injective Objects 185

M.L. Galvão
  Preradicals of Associative Algebras and Their Connections
  with Preradicals of Modules 203

P.A. Guil Asensio, M.C. Izurdiaga and B. Torrecillas
  On the Construction of Separable Modules 227

S.K. Jain and A.K. Srivastava
  Essential Extensions of a Direct Sum of Simple Modules-II 243

L. Kadison
  Pseudo-Galois Extensions and Hopf Algebroids 247

D.K. Tütüncü, N.O. Ertaş and R. Tribak
  Cohereditary Modules in $\sigma[M]$ 265

F. Kourki
  When Maximal Linearly Independent Subsets of a
  Free Module Have the Same Cardinality? 281

P.A. Linnell
  Embedding Group Algebras into Finite von Neumann
  Regular Rings 295

M. Mathieu
  The Local Multiplier Algebra: Blending Noncommutative
  Ring Theory and Functional Analysis 301

A.Ç. Özcan, D.K. Tütüncü and M.F. Yousif
  On Some Injective Modules In $\sigma[M]$ 313

D.E. Radford and H.J. Schneider
  Biproducts and Two-cocycle Twists of Hopf Algebras 331

Preface

This volume consists of articles contributed to the Proceedings of the International Conference on Modules and Comodules, held in Portugal from September 6 to September 8, 2006, and dedicated to Robert Wisbauer, on the occasion of his 65th birthday. The conference was attended by 70 mathematicians from 26 countries representing 6 continents, and was hosted by the University of Porto.

The first article in this volume is by John Clark, in which he reflects on Robert's career and his influence on Modern Algebra, in particular Module Theory. The articles that follow reflect Robert's wide interests. These include topics in the formal Theory of Modules bordering on Category Theory, in Ring Theory, in Hopf Algebras and Quantum Groups, and in Corings and Comodules. Some of these fields have a long established tradition, whereas others have emerged in recent years. To all of them Robert has made significant contributions and proved to be their untiring ambassador.

Many of the contributing articles have been written by Robert's students and collaborators. The enthusiasm with which they embraced the idea of organising a conference for Robert and wrote articles dedicated specially to him, bears testimony to Robert's role in the development of Modern Algebra. They also testify to Robert's extreme friendliness, openness and helpfulness to all.

We would like to take this opportunity to thank all the local organisers of the meeting, who created such a stimulating working atmosphere and an exciting social programme in the beautiful surroundings of Porto. Special thanks go to Christian Lomp for his personal involvement in organising the conference and in editing the current volume. Without his enthusiasm, hard work and dedication, the conference and this volume would not have been possible.

Finally we would like to wish Robert many happy returns. We hope he will derive great pleasure in reading the articles contained in this volume and dedicated to him.

Editors:
Tomasz Brzeziński (Swansea)
José Luis Gómez Pardo (Santiago de Compostela)
Ivan Shestakov (São Paulo)
Patrick Smith (Glasgow)

List of Participants

of the International Conference on Modules and Comodules dedicated to Robert Wisbauer's 65th birthday.

Albuquerque, Helena
(Coimbra, Portugal)
Alev, Jacques (Reims, France)
Antunes Simões, Maria Elisa
(Lisbon, Portugal)
Ardizzoni, Alessandro (Ferrara, Italy)
Artamonov, Viatcheslav
(Moscow, Russia)
Bagheri, Saeid (Düsseldorf, Germany)
Beattie, Margaret
(Sackville NB, Canada)
Borges, Inês (Coimbra, Portugal)
Bresar, Matej (Maribor, Slovenia)
Brzeziński, Tomasz (Swansea, Wales)
Carvalho, Paula (Porto, Portugal)
Catoiu, Stefan (Chicago, IL, USA)
Clark, John (Otago, New Zealand)
Coelho, Flavio (São Paulo, Brasil)
Crivei, Septimiu (Cluj-Napoca, Romenia)
EL Kaoutit, Laiachi (Granada, Spain)
Estrada, Sergio (Murcia, Spain)
Facchini, Alberto (Padova, Italy)
Ferrero, Miguel (Porto Alegre, Brasil)
Galvão, Maria Luisa (Lisbon, Portugal)
García Hernandez, José Luis
(Murcia, Spain)
González Peláez, Marcela
(México, Mexico)
González Rodríguez, Ramón
(Vigo, Spain)
Guil Asensio, Pedro (Murcia, Spain)
Guroglu, Tugba (Izmir, Turkey)
Gómez Pardo, José Luis
(Santiago Comp., Spain)
Gómez Torrecillas, José (Granada, Spain)
Herbera, Dolors (Barcelona, Spain)
Herzog, Ivo (Lima OH, USA)
Iyudu, Natalia (London, England)
Jain, Surender (Athens OH, USA)
Jesus, Bruno (Porto, Portugal)
Kaucikas, Algirdas (Vilnius, Lithuania)
Kharchenko, Vladislav (Mexico, Mexico)

Kourki, Farid (Tangier, Morocco)
Lomp, Christian (Porto, Portugal)
Lopatin, Artem (Omsk, Russia)
Lopes, Samuel (Porto, Portugal)
Machiavelo, Antonio (Porto, Portugal)
Marín, Leandro (Murcia, Spain)
Mathieu, Martin
(Belfast, Northern Ireland)
Mazurek, Ryszard (Białystok, Poland)
Menini, Claudia (Ferrara, Italy)
Narayanaswami, Vanaja (Mumbai, India)
Nogueira, Maria Teresa
(Lisbon, Portugal)
Oliveira, Ana Cristina (Porto, Portugal)
Özcan, Ayse Çigdem (Ankara, Turkey)
Pozhidaev, Alexander
(Novosibirsk, Russia)
Prest, Mike (Manchester, England)
Puczyłowski, Edmund (Warsaw, Poland)
Pusat-Yilmaz, Dilek (Izmir, Turkey)
Raggi, Federico Francisco
(México, Mexico)
Rios, Jose (Mexico D.F., Mexico)
Rodríguez González, Nieves
(Santiago Comp., Spain)
Sant'Ana, Alveri (Porto Alegre, Brasil)
Santa-Clara, Catarina (Lisboa, Portugal)
Santos, Jose Carlos (Porto, Portugal)
Saraiva, Paulo (Coimbra, Portugal)
Shestakov, Ivan (São Paulo, Brasil)
Simões, Raquel (Lisboa, Portugal)
Smith, Patrick (Glasgow, Scotland)
Solotar, Andrea
(Buenos Aires, Argentina)
Sow, Djiby (Dakar, Senegal)
Toksoy, Sultan Eylem (Izmir, Turkey)
Tribak, Rachid (Tanger, Morocco)
van den Berg, John
(Pietermaritzburg, S. Africa)
Wisbauer, Robert (Düsseldorf, Germany)
Zhukavets, Natalia
(Prague, Czech Republic)

Acknowledgment

This volume was partially financed by
Centro de Matemática da Universidade do Porto
through the
Programa Operacional Ciência e Inovação 2010
(POCI2010)

The organizers of the conference would like to thank the following institutions for their support:

Centro de Matemática da Universidade do Porto
Faculdade de Ciências da Universidade do Porto
Fundação para a Ciência e a Tecnologia
Deutscher Akademischer Austauschdienst
Câmera Municipal do Porto

As well as the companies:

Caixa Geral de Depôsitos,
Chagas,
ASA and
Comissão de Viticultura da região dos vinhos verdes.

Robert Wisbauer

John Clark

Robert Wisbauer retires this year after a long, distinguished and energetic career in algebra. Born and raised in Bavaria, close to the border with Czechoslovakia, after his undergraduate years in Würzburg Robert moved to the University of Düsseldorf and obtained his doctorate there in 1971. He has been with the Mathematics Institute in Düsseldorf ever since. Well, not quite "ever since" because in the last 25 years Robert has certainly done a lot of travelling. Indeed, during that time he has acted as a roving ambassador for module theory and non-associative rings, spreading their gospel to the four corners of the world and, in particular, developing collaborations and encouraging young up-and-coming algebraists around the globe. Of course, he was only able to do this so effectively because of his considerable language skills and, more importantly, his vast knowledge of and enthusiasm for his subject.

Perhaps his enthusiasm and love of travel began with a year of study at the University of Edinburgh during his student days. Then, with his PhD completed, he visited the University of Moscow as an exchange scientist in 1973 and from this grew a lasting, working friendship with algebraists from the USSR. In 1978, he spent some time at the University of Nantes and this encouraged him to develop bilateral programmes which later included Spanish mathematicians. We could continue listing his visits abroad and the beneficial effects these had, but instead we emphasize that the traffic was by no means one way – he has proved a generous, ever-friendly host in Düsseldorf to a large number of algebraists, both already established and those in the early stages of their careers (including several doctoral students).

We have already alluded to the enviable knowledge he has of his subject. This is most evidenced by his 1991 text *"Foundations of Modules and Ring Theory"*. Its subtitle *"A Handbook for Study and Research"* describes it aptly, with over 250 citations listed in the Science Citation Index as evidence of how useful the *"Foundations"* has been to the module-theoretic community. Of course, *"Foundations"* also brought to the fore Robert's beloved $\sigma[M]$, the category of R-modules subgenerated by a fixed module M over a given ring R. A substantial part of Robert's legacy will be due to his tenet that much can be done, with more effect, if one can recast a problem or property in $\sigma[M]$ from its setting in the (usually) larger category of all R-modules.

But of course, there's much more to his legacy than *"Foundations"*. Indeed he has nigh on 100 publications, including six monographs, with more than 40 coauthors (from all continents). Consequently, his retirement certainly deserves attention and it is indeed fitting to have a conference to acknowledge his outstanding career. It is hoped that his retirement will not curb his travel, (but it has been said that the moon may soon be his only travel destination left, perhaps a lunar modules mission).

John Clark
Department of Mathematics and Statistics
University of Otago
Dunedin, PO Box 56
New Zealand
e-mail: `jclark@maths.otago.ac.nz`

Modules and Comodules
Trends in Mathematics, 1–30
© 2008 Birkhäuser Verlag Basel/Switzerland

Injective Morita Contexts (Revisited)

J.Y. Abuhlail and S.K. Nauman

Dedicated to Prof. Robert Wisbauer

Abstract. This paper is an exposition of the so-called *injective Morita contexts* (in which the connecting bimodule morphisms are injective) and *Morita α-contexts* (in which the connecting bimodules enjoy some local projectivity in the sense of Zimmermann-Huisgen). Motivated by situations in which only one trace ideal is in action, or the compatibility between the bimodule morphisms is not needed, we introduce the notions of Morita *semi-contexts* and *Morita data*, and investigate them. Injective Morita data will be used (with the help of *static* and *adstatic modules*) to establish equivalences between some *intersecting subcategories* related to subcategories of categories of modules that are localized or colocalized by trace ideals of a Morita datum. We end up with applications of Morita α-contexts to ∗-*modules* and *injective right wide Morita contexts*.

1. Introduction

Morita contexts, in general, and (*semi-*)*strict Morita* contexts (with surjective connecting bilinear morphisms), in particular, were extensively studied and developed exponentially during the last few decades (e.g., [AGH-Z1997]). However, we sincerely feel that there is a gap in the literature on *injective Morita contexts* (i.e., those with injective connecting bilinear morphisms). Apart from the results in [Nau1994-a], [Nau1994-b] (where the second author initially explored this notion) and from an application to Grothendieck groups in the recent paper ([Nau2004]), it seems that injective Morita contexts were not studied *systematically* at all.

We noticed that in several results of ([Nau1993], [Nau1994-a] and [Nau1994-b]) that are related to Morita contexts, only one trace ideal is used. Observing this fact, we introduce the notions of *Morita semi-contexts* and *Morita data* and investigate them. Several results are proved then for *injective* Morita semi contexts and/or injective Morita data.

Consider a Morita datum $\mathcal{M} = (T, S, P, Q, \langle, \rangle_T, \langle, \rangle_S)$, with not necessarily compatible bimodule morphisms $\langle, \rangle_T : P \otimes_S Q \to T$ and $\langle, \rangle_S : Q \otimes_T P \to S$. We say that $\mathcal{M}$ is *injective,* iff $\langle, \rangle_T$ and $\langle, \rangle_S$ are injective, and to be a *Morita α-datum,* iff the associated dual pairings $\mathbf{P}_l := (Q, {}_TP)$, $\mathbf{P}_r := (Q, P_S)$, $\mathbf{Q}_l := (P, {}_SQ)$ and $\mathbf{Q}_r := (P, Q_T)$ satisfy the α-condition (which is closely related to the notion of local projectivity in the sense of Zimmermann-Huisgen [Z-H1976]). The α-condition was introduced in [AG-TL2001] and further investigated by the first author in [Abu2005].

While (semi-)strict unital Morita contexts induce equivalences between the whole module categories of the rings under consideration, we show in this paper how injective Morita (semi-)contexts and injective Morita data play an important role in establishing equivalences between suitable *intersecting subcategories* of module categories (e.g., intersections of subcategories that are localized/colocalized by trace ideals of a Morita datum with subcategories of static/ adstatic modules, etc.). Our main applications, in addition to equivalences related to the Kato-Ohtake-Müller *localization-colocalization theory* (developed in [Kat1978], [KO1979] and [Mül1974]), will be to *∗-modules* (introduced by Menini and Orsatti [MO1989]) and to *right wide Morita contexts* (introduced by F. Castaño Iglesias and J. Gómez-Torrecillas [C-IG-T1995]).

Most of our results will be stated for *left modules,* while deriving the "dual" versions for right modules is left to the interested reader. Moreover, for Morita contexts, some results are stated/proved for only one of the Morita semi-contexts, as the ones corresponding to the second semi-context can be obtained analogously. For the convenience of the reader, we tried to make the paper self-contained, so that it can serve as a reference on injective *Morita (semi-)contexts* and their applications. In this respect, and for the sake of completeness, we have included some previous results of the authors that are (in most cases) either provided with new shorter proofs, or are obtained under weaker conditions.

This paper is organized as follows: After this brief introduction, we give in Section 2 some preliminaries including the basic properties of *dual α-pairings,* which play a central role in rest of the work. The notions of *Morita semi-contexts* and *Morita data* are introduced in Section 3, where we clarify their relations with the *dual pairings* and the so-called *elementary rngs. Injective Morita (semi-)contexts* appear in Section 4, where we study their interplay with dual α-pairings and provide some examples and a counter-example. In Section 5 we include some observations regarding *static* and *adstatic* modules and use them to obtain equivalences among suitable *intersecting subcategories* of modules related to a Morita (semi-)context. In the last section, more applications are presented, mainly to subcategories of modules that are *localized* or *colocalized* by a trace ideal of an injective Morita (semi-)context, to *∗-modules* and to *injective right wide Morita contexts.*

2. Preliminaries

Throughout, R denotes a commutative ring with $1_R \neq 0_R$ and A, A', B, B' are unital R-algebras. We have reserved the term "*ring*" for an associative ring with a multiplicative unity, and we will use the term "*rng*" for a general associative ring (not necessarily with unity). All modules over rings are assumed to be unitary, and ring morphisms are assumed to respect multiplicative unities. If $\mathfrak{T}$ and $\mathfrak{S}$ are categories, then we write $\mathfrak{T} < \mathfrak{S}$ ($\mathfrak{T} \leq \mathfrak{S}$) to mean that $\mathfrak{T}$ is a (full) subcategory of $\mathfrak{S}$, and $\mathfrak{T} \approx \mathfrak{S}$ to indicate that $\mathfrak{T}$ and $\mathfrak{S}$ are equivalent.

Rngs and their modules

2.1. By an A-**rng** (T, μ_T), we mean an (A, A)-bimodule T with an (A, A)-bilinear morphism $\mu_T : T \otimes_A T \to T$, such that $\mu_T \circ (\mu_T \otimes_A id_T) = \mu_T \circ (id_T \otimes_A \mu_T)$. We call an A-rng (T, μ_T) an A-**ring**, iff there exists in addition an (A, A)-bilinear morphism $\eta_T : A \to T$, called the **unity map**, such that $\mu_T \circ (\eta_T \otimes_A id_T) = \vartheta_T^l$ and $\mu_T \circ (id_T \otimes_A \eta_T) = \vartheta_T^r$ (where $A \otimes_A T \overset{\vartheta_T^l}{\simeq} T$ and $T \otimes_A A \overset{\vartheta_T^r}{\simeq} T$ are the canonical isomorphisms). So, an A-ring is a unital A-rng; and an A-rng is (roughly speaking) an A-ring not necessarily with unity.

2.2. A morphism of rngs $(\psi : \delta) : (T : A) \to (T' : A')$ consists of a morphism of R-algebras $\delta : A \to A'$ and an (A, A)-bilinear morphism $\psi : T \to T'$, such that $\mu_{T'} \circ \chi_{(T',T')}^{(A,A')} \circ (\psi \otimes_A \psi) = \psi \circ \mu_T$ (where $\chi_{(T',T')}^{(A,A')} : T' \otimes_A T' \to T' \otimes_{A'} T'$ is the canonical map induced by δ). By $\mathbb{RNG}$ we denote the category of associative rngs with morphisms being rng morphisms, and by $\mathbb{URNG} < \mathbb{RNG}$ the (non-full) subcategory of *unital* rings with morphisms being the morphisms in $\mathbb{RNG}$ which respect multiplicative unities.

2.3. Let (T, μ_T) be an A-rng. By a **left T-module** we mean a left A-module N with a left A-linear morphism $\phi_T^N : T \otimes_A N \to N$, such that $\phi_T^N \circ (\mu_T \otimes_A id_N) = \phi_T^N \circ (id_T \otimes_A \phi_T^N)$. For left T-modules M, N, we call a left A-linear morphism $f : M \to N$ a T-**linear morphism**, iff $f(tm) = tf(m)$ for all $t \in T$. The category of left T-modules and left T-linear morphisms is denoted by ${}_T\mathbb{M}$. The category $\mathbb{M}_T$ of right T-modules is defined analogously. Let $(T : A)$ and $(T' : A')$ be rngs. We call an (A, A')-bimodule N a (T, T')-**bimodule**, iff (N, ϕ_T^N) is a left T-module and $(N, \phi_{T'}^N)$ is a right T'-module, such that $\phi_{T'}^N \circ (\phi_T^N \otimes_{A'} id_{T'}) = \phi_T^N \circ (id_T \otimes_A \phi_{T'}^N)$. For (T, T')-bimodules M, N, we call an (A, A')-bilinear morphism $f : M \to N$ (T, T')-**bilinear**, provided f is left T-linear and right T'-linear. The category of (T, T')-bimodules is denoted by ${}_T\mathbb{M}_{T'}$. In particular, for any A-rng T, a left (right) T-module M has a canonical structure of a *unitary* right (left) S-module, where $S := End({}_TM)^{op}$ ($S := End(M_T)$); and moreover, with this structure M becomes a (T, S)-bimodule (an (S, T)-bimodule).

Remark 2.1. Similarly, one can define rngs over arbitrary (not-necessarily unital) ground rngs and rng morphisms between them. Moreover, one can define (bi)modules over such rngs and (bi)linear morphisms between them.

Notation 2.4. *Let* T *be an* A-*rng. We write* $_TU$ *(*U_T*) to denote that* U *is a left (right)* T-*module. For a left (right)* T-*module* $_TU$, *we consider the set* $^*U :=$ $\mathrm{Hom}_{T-}(U,T)$ *(*$U^* := \mathrm{Hom}_{-T}(U,T)$*) of all left (right)* T-*linear morphisms from* U *to* T *with the canonical right (left)* T-*module structure.*

Generators and cogenerators

Definition 2.1. *Let* T *be an* A-*rng. For a left* T-*module* $_TU$ *consider the following subclasses of* $_T\mathbb{M}$:

$$\begin{aligned}
\mathrm{Gen}(_TU) &:= \{_TV \mid \exists \text{ a set } \Lambda \text{ and an exact sequence } U^{(\Lambda)} \to V \to 0\};\\
\mathrm{Cogen}(_TU) &:= \{_TW \mid \exists \text{ a set } \Lambda \text{ and an exact sequence } 0 \to W \to U^{\Lambda}\};\\
\mathrm{Pres}(_TU) &:= \{_TV \mid \exists \text{ sets } \Lambda_1, \Lambda_2 \text{ and}\\
&\qquad\qquad \text{an exact sequence } U^{(\Lambda_2)} \to U^{(\Lambda_1)} \to V \to 0\};\\
\mathrm{Copres}(_TU) &:= \{_TW \mid \exists \text{ sets } \Lambda_1, \Lambda_2 \text{ and}\\
&\qquad\qquad \text{an exact sequence } 0 \to W \to U^{\Lambda_1} \to U^{\Lambda_2}\};
\end{aligned}$$

A left T-*module in* $\mathrm{Gen}(_TU)$ *(respectively* $\mathrm{Cogen}(_TU)$, $\mathrm{Pres}(_TU)$, $\mathrm{Copres}(_TU)$*) is said to be* U-**generated** *(respectively* U-**cogenerated**, U-**presented**, U-**copresented**). *Moreover, we say that* $_TU$ *is a* **generator** *(respectively* **cogenerator**, **presentor**, **copresentor**), *iff* $\mathrm{Gen}(_TU) = {}_T\mathbb{M}$ *(respectively* $\mathrm{Cogen}(_TU) = {}_T\mathbb{M}$, $\mathrm{Pres}(_TU) =$ $_T\mathbb{M}$, $\mathrm{Copres}(_TU) = {}_T\mathbb{M}$*).*

Dual α-pairings

In what follows we recall the definition and properties of dual α-pairings introduced in [AG-TL2001, Definition 2.3.] and studied further in [Abu2005].

2.5. Let T be an A-rng. A **dual left** T-**pairing** $\mathbf{P}_l = (V, {}_TW)$ consists of a left T-module W and a right T-module V with a right T-linear morphism $\kappa_{\mathbf{P}_l} : V \to {}^*W$ (equivalently a left T-linear morphism $\chi_{\mathbf{P}_l} : W \to V^*$). For dual left pairings $\mathbf{P}_l = (V, {}_TW)$, $\mathbf{P}_l' = (V', {}_{T'}W')$, a morphism of dual left pairings $(\xi, \theta) : (V', W') \to (V, W)$ consists of a triple

$$(\xi, \theta : \varsigma) : (V, {}_TW) \to (V', {}_{T'}W'),$$

where $\xi : V \to V'$ and $\theta : W' \to W$ are T-linear and $\varsigma : T \to T'$ is a morphism of rngs, such that considering the induced maps $\langle, \rangle_T : V \times W \to T$ and $\langle, \rangle_{T'} :$ $V' \times W' \to T'$ we have

$$\langle \xi(v), w' \rangle_{T'} = \varsigma(\langle v, \theta(w') \rangle_T) \text{ for all } v \in V \quad \text{and} \quad w' \in W'. \tag{1}$$

The dual left pairings with the morphisms defined above build a category, which we denote by $\mathcal{P}_l$. With $\mathcal{P}_l(T) \leq \mathcal{P}_l$ we denote the full subcategory of dual T-pairings. The category $\mathcal{P}_r$ of dual right pairings and its full subcategory $\mathcal{P}_r(T) \leq \mathcal{P}_r$ of dual right T-pairings are defined analogously.

Remark 2.2. The reader should be warned that (in general) for a non-commutative rng T and a dual left T-pairing $\mathbf{P}_l = (V, {}_TW)$, the following map induced by the right T-linear morphism $\kappa_{\mathbf{P}_l} : V \to {}^*W$:

$$\langle, \rangle_T : V \times W \to T, \ \langle v, w \rangle_T := \kappa_{\mathbf{P}_l}(v)(w)$$

is not necessarily T-*balanced*, and so does not induce (in general) a map $V \otimes_T W \to T$. In fact, for all $v \in V$, $w \in W$ and $t \in T$ we have

$$\langle vt, w \rangle = \kappa_{\mathcal{P}_l}(vt)(w) = [\kappa_{\mathcal{P}_l}(v)t](w) = [\kappa_{\mathcal{P}_l}(v)(w)]t = \langle v, w \rangle_T t;$$
$$\langle v, tw \rangle = \kappa_{\mathcal{P}_l}(v)(tw) = t[\kappa_{\mathcal{P}_l}(v)(w)] = t\langle v, w \rangle_T.$$

2.6. Let T be an A-rng, N, W be left T-modules and identify N^W with the set of all mappings from W to N. Considering N with the *discrete topology* and N^W with the product topology, the induced *relative topology* on $\mathrm{Hom}_{T-}(W, N) \hookrightarrow N^W$ is a linear topology (called the **finite topology**), for which the *basis of neighborhoods of* 0 is given by the set of annihilator submodules:

$$\mathcal{B}_f(0) := \{ F^{\perp(\mathrm{Hom}_{T-}(W,N))} \mid F = \{w_1, \ldots, w_k\} \subset W \text{ is a finite subset}\},$$

where

$$F^{\perp(\mathrm{Hom}_{T-}(W,N))} := \{ f \in \mathrm{Hom}_{T-}(W, N)) \mid f(W) = 0 \}.$$

2.7. Let T be an A-rng, $\mathbf{P}_l = (V, {}_TW)$ a dual left T-pairing and consider for every right T-module U_T the following canonical map

$$\alpha_U^{\mathbf{P}_l} : U \otimes_T W \to \mathrm{Hom}_{-T}(V, U), \quad \sum u_i \otimes_T w_i \mapsto [v \mapsto \sum u_i \langle v, w_i \rangle_T]. \qquad (2)$$

We say that $\mathbf{P}_l = (V, {}_TW) \in \mathcal{P}_l(T)$ **satisfies the left α-condition** (or is a **dual left α-pairing**), iff $\alpha_U^{\mathbf{P}_l}$ is injective for every right T-module U_T. By $\mathcal{P}_l^{\alpha}(T) \leq \mathcal{P}_l(T)$ we denote the *full* subcategory of dual left T-pairings satisfying the left α-condition. The full subcategory of **dual right α-pairings** $\mathcal{P}_r^{\alpha}(T) \leq \mathcal{P}_r(T)$ is defined analogously.

Definition 2.2. *Let T be an A-rng, $\mathbf{P}_l = (V, {}_TW)$ be a dual left T-pairing and consider*

$$\kappa_{\mathbf{P}_l} : V \to {}^*W \quad and \quad \alpha_V^{\mathbf{P}_l} : V \otimes_T W \to \mathrm{End}(V_T).$$

We say $\mathbf{P}_l \in \mathcal{P}_l(T)$ is

dense, *iff $\kappa_{\mathbf{P}_l}(V) \subseteq {}^*W$ is dense (w.r.t. the* finite topology *on ${}^*W \hookrightarrow T^W$);*

injective *(resp. **semi-strict, strict**), iff $\alpha_V^{\mathbf{P}_l}$ is injective (resp. surjective, bijective);*

non-degenerate, *iff $V \overset{\kappa_{\mathbf{P}_l}}{\hookrightarrow} {}^*W$ and $W \overset{\chi_{\mathbf{P}_l}}{\hookrightarrow} V^*$ canonically.*

2.8. Let T be an A-rng. We call a T-module W **locally projective** (in the sense of B. Zimmermann-Huisgen [Z-H1976]), iff for every diagram of T-modules

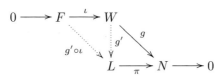

with exact rows and finitely generated T-submodule $F \subseteq W$: for every T-linear morphism $g : W \to N$, there exists a T-linear morphism $g' : W \to L$, such that $g \circ \iota = \pi \circ g' \circ \iota$.

For proofs of the following basic properties of *locally projective modules* and *dual α-pairings* see [Abu2005] and [Z-H1976]:

Proposition 2.1. *Let T be an A-ring and $\mathbf{P}_l = (V, {}_TW) \in \mathcal{P}_l(T)$.*

1) *The left T-module ${}_TW$ is locally projective if and only if $({}^*W, W)$ is an α-pairing.*

2) *The left T-module ${}_TW$ is locally projective, iff for any finite subset $\{w_1, \ldots, w_k\} \subseteq W$, there exists $\{(f_i, \widetilde{w}_i)\}_{i=1}^k \subset {}^*W \times W$ such that $w_j = \sum_{i=1}^k f_i(w_j)\widetilde{w}_i$ for all $j = 1, \ldots, k$.*

3) *If ${}_TW$ is locally projective, then ${}_TW$ is flat and T-cogenerated.*

4) *If $\mathbf{P}_l \in \mathcal{P}_l^\alpha(T)$, then ${}_TW$ is locally projective.*

5) *If ${}_TW$ is locally projective and $\kappa_P(V) \subseteq {}^*W$ is dense, then $\mathbf{P}_l \in \mathcal{P}_l^\alpha(T)$.*

6) *Assume T_T is an injective cogenerator. Then $\mathbf{P}_l \in \mathcal{P}_l^\alpha(T)$ if and only if ${}_TW$ is locally projective and $\kappa_{\mathbf{P}_l}(V) \subseteq {}^*W$ is dense.*

7) *If T is a QF ring, then $\mathbf{P}_l \in \mathcal{P}_l^\alpha(T)$ if and only if ${}_TW$ is projective and $W \overset{\chi_{\mathbf{P}_l}}{\hookrightarrow} V^*$.*

The following result completes the nice observation [BW2003, 42.13.] about locally projective modules:

Proposition 2.2. *Let T be a ring, ${}_TW$ a left T-module, $S := \mathrm{End}({}_TW)^{op}$ and consider the canonical (S, S)-bilinear morphism*

$$[,]_W : {}^*W \otimes_T W \to \mathrm{End}({}_TW), \ f \otimes_T w \mapsto [\widetilde{w} \mapsto f(\widetilde{w})w].$$

1) *${}_TW$ is finitely generated projective if and only if $[,]_W$ is surjective.*

2) *${}_TW$ is locally projective if and only if $\mathrm{Im}([,]_W) \subseteq \mathrm{End}({}_TW)$ is dense.*

Proof. 1) This follows by [Fai1981, 12.8.].

2) Assume ${}_TW$ is locally projective and consider for every left T-module N the canonical mapping

$$[,]_N^W :{}^* W \otimes_T N \to \mathrm{Hom}_T(W, N), \ f \otimes_T n \mapsto [\widetilde{w} \mapsto f(\widetilde{w})n].$$

It follows then by [BW2003, 42.13.], that $\mathrm{Im}([,]_N^W) \subseteq \mathrm{Hom}_T(W, N)$ is dense. In particular, setting $N = W$ we conclude that $\mathrm{Im}([,]_W) \subseteq \mathrm{End}({}_TW)$ is dense. On the other hand, assume $\mathrm{Im}([,]_W) \subseteq \mathrm{End}({}_TW)$ is dense. Then for every finite subset $\{w_1, \ldots, w_k\} \subseteq W$, there exists $\sum_{i=1}^n \widetilde{g}_i \otimes_T \widetilde{w}_i \in {}^*W \otimes_T W$ with

$$w_j = id_W(w_j) = [,]_W(\sum_{i=1}^n \widetilde{g}_i \otimes_T \widetilde{w}_i)(w_j) = \sum_{i=1}^n \widetilde{g}_i(w_j)\widetilde{w}_i \text{ for } j = 1, \ldots, k.$$

It follows then by Proposition 2.1 2) that ${}_TW$ is locally projective. $\square$

3. Morita (semi-)contexts

We noticed, in the proofs of some results on equivalences between subcategories of module categories associated to a given Morita context, that no use is made of the *compatibility* between the connecting bimodule morphisms (or even that only one trace ideal is used and so only one of the two bilinear morphisms is really in action). Some results of this type appeared, for example, in [Nau1993], [Nau1994-a] and [Nau1994-b]. Moreover, in our considerations some Morita contexts will be formed for arbitrary associative rngs (i.e., not necessarily unital rings). These considerations motivate us to make the following general definitions:

3.1. By a **Morita semi-context** we mean a tuple

$$\mathbf{m}_T = ((T : A), (S : B), P, Q, \langle,\rangle_T, I), \tag{3}$$

where T is an A-rng, S is a B-rng, P is a (T, S)-bimodule, Q is an (S, T)-bimodule, $\langle,\rangle_T : P \otimes_S Q \to T$ is a (T, T)-bilinear morphism and $I := \mathrm{Im}(\langle,\rangle_T) \lhd T$ (called the **trace ideal associated to $\mathbf{m}_T$**). We drop the ground rings A, B and the trace ideal $I \lhd T$, if they are not explicitly in action. If $\mathbf{m}_T$ (3) is a Morita semi-context and T, S are unital rings, then we call $\mathbf{m}_T$ a **unital Morita semi-context**.

3.2. Let $\mathbf{m}_T = ((T:A),(S:B),P,Q,\langle,\rangle_T)$, $\mathbf{m}_{T'} = ((T':A'),(S':B'),P',Q',\langle,\rangle_{T'})$ be Morita semi-contexts. By a **morphism of Morita semi-contexts** from $\mathbf{m}_T$ to $\mathbf{m}_{T'}$ we mean a four fold set of morphisms

$$((\beta : \delta),(\gamma : \sigma),\phi,\psi) : ((T : A), (S : B), P, Q) \to ((T' : A'), (S' : B'), P', Q'),$$

where $(\beta : \delta) : (T : A) \to (T' : A')$ and $(\gamma : \sigma) : (S : B) \to (S' : B')$ are rng morphisms, $\phi : P \to P'$ is (T, S)-bilinear and $\psi : Q \to Q'$ is (S, T)-bilinear, such that

$$\beta(\langle p, q \rangle_T) = \langle \phi(p), \psi(q) \rangle_{T'} \text{ for all } p \in P, q \in Q .$$

Notice that we consider P' as a (T, S)-bimodule and Q' as an (S, T)-bimodule with actions induced by the morphism of rngs $(\beta : \delta)$ and $(\gamma : \sigma)$. By $\mathbb{MSC}$ we denote the *category of Morita semi-contexts* with morphisms defined as above, and by $\mathbb{UMSC} < \mathbb{MSC}$ the (non-full) subcategory of *unital Morita semi-contexts*.

Morita semi-contexts are closely related to *dual pairings* in the sense of [Abu2005]:

3.3. Let $(T, S, P, Q, <, >_T) \in \mathbb{MSC}$ and consider the canonical isomorphisms of Abelian groups

$$\mathrm{Hom}_{(S,T)}(Q, {}^*P) \overset{\xi}{\simeq} \mathrm{Hom}_{(T,T)}(P \otimes_S Q, T) \overset{\zeta}{\simeq} \mathrm{Hom}_{(T,S)}(P, Q^*).$$

This means that we have two dual T-pairings $\mathbf{P}_l := (Q, {}_TP) \in \mathcal{P}_l(T)$ and $\mathbf{Q}_r := (P, Q_T) \in \mathcal{P}_r(T)$, induced by the canonical T-linear morphisms

$$\kappa_{\mathbf{P}_l} := \xi^{-1}(\langle,\rangle_T) : Q \to {}^*P \quad \text{and} \quad \kappa_{\mathbf{Q}_r} := \zeta(\langle,\rangle_T) : P \to Q^*.$$

On the other hand, let $(S, T, Q, P, \langle, \rangle_S) \in \mathbb{MSC}$ and consider the canonical isomorphisms of Abelian groups

$$\mathrm{Hom}_{(S,T)}(Q, P^*) \overset{\xi'}{\simeq} \mathrm{Hom}_{(S,S)}(Q \otimes_T P, S) \overset{\zeta'}{\simeq} \mathrm{Hom}_{(T,S)}(P, {}^*Q).$$

Then we have two dual S-pairings $\mathbf{P}_r := (Q, P_S) \in \mathcal{P}_r(S)$ and $\mathbf{Q}_l := (P, {}_SQ) \in \mathcal{P}_l(S)$, induced by the canonical morphisms

$$\kappa_{\mathbf{P}_r} := \xi'^{-1}(\langle, \rangle_S) : Q \to P^* \quad \text{and} \quad \kappa_{\mathbf{Q}_r} := \zeta'(\langle, \rangle_S) : P \to {}^*Q.$$

3.4. By a **Morita datum** we mean a tuple

$$\mathcal{M} = ((T : A), (S : B), P, Q, \langle, \rangle_T, \langle, \rangle_S, I, J), \tag{4}$$

where the following are Morita semi-contexts:

$$\mathcal{M}_T := ((T:A), (S:B), P, Q, \langle, \rangle_T, I) \quad \text{and} \quad \mathcal{M}_S := ((S:B), (T:A), Q, P, \langle, \rangle_S, J) \tag{5}$$

If, moreover, the bilinear morphisms $\langle, \rangle_T : P \otimes_S Q \to T$ and $\langle -, \rangle_S : Q \otimes_T P \to S$ are *compatible,* in the sense that

$$\langle q, p \rangle_S q' = q \langle p, q' \rangle_T \quad \text{and} \quad p \langle q, p' \rangle_S = \langle p, q \rangle_T p' \; \forall \, p, p' \in P, \; q, q' \in Q, \tag{6}$$

then we call $\mathcal{M}$ a **Morita context.** If T, S in a Morita datum (context) $\mathcal{M}$ are unital, then we call $\mathcal{M}$ a **unital Morita datum (context).**

3.5. Let $\mathcal{M} = ((T : A), (S : B), P, Q, \langle, \rangle_T, \langle, \rangle_S)$ and $\mathcal{M}' = ((T' : A'), (S' : B'), P', Q', \langle, \rangle_{T'}, \langle, \rangle_{S'})$ be Morita contexts. Extending [Ami1971, Page 275], we mean by a **morphism of Morita contexts** from $\mathcal{M}$ to $\mathcal{M}'$ a four fold set of maps

$$((\beta : \delta), (\gamma : \sigma), \phi, \psi) : ((T : A), (S : B), P, Q) \to ((T' : A'), (S' : B'), P', Q'),$$

where $(\beta : \delta) : (T : A) \to (T' : A')$, $(\gamma : \sigma) : (S : B) \to (S' : B')$ are rng morphisms, $\phi : P \to P'$ is (T, S)-bilinear and $\psi : Q \to Q'$ is (S, T)-bilinear, such that

$$\beta(\langle p, q \rangle_T) = \langle \phi(p), \psi(q) \rangle_{T'} \quad \text{and} \quad \gamma(\langle q, p \rangle_S) = \langle \psi(q), \phi(p) \rangle_{S'} \; \forall \, p \in P, q \in Q.$$

By $\mathbb{MC}$ we denote the *category of Morita contexts* with morphisms defined as above, and by $\mathbb{UMC} < \mathbb{MC}$ the (non-full) subcategory of *unital Morita contexts.*

Example 3.1. If R is commutative, then any Morita semi-context $(R, R, P, Q, \langle, \rangle_R)$ yields a Morita context

$$(R, R, P, Q, \langle, \rangle_R, [,]_R),$$

where $[,]_R := Q \otimes_R P \simeq P \otimes_R Q \xrightarrow{\langle, \rangle_R} R.$ □

3.6. We call a Morita semi-context $\mathbf{m}_T = (T, S, P, Q, \langle, \rangle_T)$ **semi-derived (derived)**, iff $S := \mathrm{End}(_T P)^{op}$ (and $Q = {}^*P$). We call a Morita datum, or a Morita context, $\mathcal{M} = (T, S, P, Q, \langle, \rangle_T, \langle, \rangle_S)$ **semi-derived (derived)**, iff $S = \mathrm{End}(_T P)^{op}$, or $T = \mathrm{End}(P_S)$ ($S = \mathrm{End}(_T P)^{op}$ and $Q = {}^*P$, or $T = \mathrm{End}(P_S)$ and $Q = P^*$).

Remark 3.1. Following [Cae1998, 1.2.] (however, dropping the condition that the bilinear map $\langle,\rangle_T : P \otimes_S Q \to T$ is surjective), Morita semi-contexts $(T,S,P,Q,\langle\rangle_T)$ in our sense were called *dual pairs* in [Ver2006]. However, we think the terminology we are using is more informative and avoids confusion with other notions of dual pairings in the literature (e.g., the ones studied by the first author in [Abu2005]). The reason for this specific terminology (i.e., Morita semi-contexts) is that every Morita context contains two Morita semi-contexts as clear from the definition; and that any Morita semi-context can be *extended* to a (not necessarily unital) Morita context in a natural way as explained below.

Elementary rngs

In what follows we demonstrate how to build new Morita (semi-)contexts from a given Morita semi-context. These constructions are inspired by the notion of *elementary rngs* in [Cae1998, 1.2.] (and [Ver2006, Remark 3.8.]):

Lemma 3.1. *Let* $\mathbf{m}_T := ((T : A), (S : B), P, Q, \langle,\rangle_T) \in \mathbb{MSC}$.

1) *The* (T,T)-*bimodule* $\mathbb{T} := P \otimes_S Q$ *has a structure of a* T-*rng* (A-*rng*) *with multiplication*

$$(p \otimes_S q) \cdot_\mathbb{T} (p' \otimes_S q') := \langle p, q \rangle_T p' \otimes_S q' \; \forall \, p, p' \in P, \; q, q' \in Q,$$

such that $\langle,\rangle_T : \mathbb{T} \to T$ *is a morphism of* A-*rngs,* P *is a* $(\mathbb{T}, S)$-*bimodule and* Q *is an* $(S, \mathbb{T})$-*bimodule, where*

$$(p \otimes_S q) \rightharpoonup \widetilde{p} := \langle p, q \rangle_T \widetilde{p} \quad and \quad \widetilde{q} \leftharpoonup (p \otimes_S q) := \widetilde{q}\langle p, q \rangle_T.$$

Moreover, we have morphisms of T-*rngs* (A-*rngs*)

$$\psi : \mathbb{T} \to \operatorname{End}(P_S), \quad p \otimes_S q \mapsto [\widetilde{p} \mapsto \langle p, q \rangle_T \widetilde{p}];$$
$$\phi : \mathbb{T} \to \operatorname{End}(_S Q)^{op}, \, p \otimes_S q \mapsto [\widetilde{q} \mapsto \widetilde{q}\langle p, q \rangle_T],$$

$((\mathbb{T} : A), (S : B), P, Q, id_\mathbb{T}) \in \mathbb{MSC}$ *and we have a morphism of Morita semi-contexts*

$$(\langle,\rangle_T, id_S, , id_P, id_Q) : (\mathbb{T}, S, P, Q, id_\mathbb{T}) \to (T, S, P, Q, \langle,\rangle_T).$$

2) *The* (S,S)-*bimodule* $\mathbf{S} := Q \otimes_T P$ *has a structure of an* S-*rng* (B-*rng*) *with multiplication*

$$(q \otimes_T p) \cdot_\mathbf{S} (q' \otimes_T p') := q\langle p, q' \rangle_T \otimes_T \; p'$$
$$= q \otimes_T \langle p, q' \rangle_T p' \; \forall \, p, p' \in P, \; q, q' \in Q,$$

such that $\langle,\rangle_S : \mathbf{S} \to S$ *is a morphism of* B-*rngs,* P *is a* $(T, \mathbf{S})$-*bimodule and* Q *is an* $(\mathbf{S}, T)$-*bimodule, where*

$$\widetilde{p} \leftharpoonup (q \otimes_T p) := \langle \widetilde{p}, q \rangle_T p \quad and \quad (q \otimes_T p) \rightharpoonup \widetilde{q} := q\langle p, \widetilde{q} \rangle_T.$$

Moreover, we have morphisms of S-*rngs* (B-*rngs*)

$$\Psi : \mathbf{S} \to \operatorname{End}(_T P)^{op}, \, q \otimes_T p \mapsto [\widetilde{p} \mapsto \langle \widetilde{p}, q \rangle_T p],$$
$$\Phi : \mathbf{S} \to \operatorname{End}(Q_T), \quad q \otimes_T p \mapsto [\widetilde{q} \mapsto q\langle p, \widetilde{q} \rangle_T],$$

and $\mathcal{M} := ((T : A), (\mathbf{S} : B), P, Q, \langle,\rangle_T, id_\mathbf{S})$ *is a Morita context.*

Remarks 3.2.

1. Given $((S : B), (T : A), Q, P, \langle, \rangle_S) \in \mathbb{MSC}$, the (S, S)-bimodule $\mathbb{S} := Q \otimes_T P$ becomes an S-rng with multiplication

$$(q \otimes_T p) \cdot_{\mathbb{S}} (q' \otimes_T p') := \langle q, p \rangle_S q' \otimes_T p' \ \forall \ p, p' \in P, \ q, q' \in Q;$$

and the (T, T)-bimodule $\mathbf{T} := P \otimes_S Q$ becomes a T-rng with multiplication

$$(p \otimes_S q) \cdot_{\mathbf{T}} (p' \otimes_S q') := p \langle q, p' \rangle_S \otimes_S q'$$
$$= p \otimes_S \langle q, p' \rangle_S q' \ \forall \ p, p' \in P, \ q, q' \in Q.$$

Analogous results to those in Lemma 3.1 can be obtained for the S-rng $\mathbb{S}$ and the T-rng $\mathbf{T}$.

2. Given a Morita semi-context $(T, S, P, Q, \langle, \rangle_T)$ several equivalent conditions for the T-rng $\mathbf{T} := P \otimes_S Q$ to be unital and the modules $_T P, Q_\mathbf{T}$ to be *firm* can be found in [Ver2006, Theorem 3.3.]. Analogous results can be formulated for the S-rng $Q \otimes_T P$ and the S-modules $P_\mathbf{S}, {}_\mathbf{S} Q$ corresponding to any $(S, T, Q, P, \langle, \rangle_S) \in \mathbb{MSC}$.

Proposition 3.1.

1) *Let* $\mathbf{m}_T = (T, S, P, Q, \langle, \rangle_T) \in \mathbb{UMSC}$ *and assume the A-rng* $\mathbb{T} := P \otimes_S Q$ *to be unital. If* $\langle, \rangle_T : \mathbb{T} \to T$ *respects unities (and* $\mathbf{m}_T$ *is injective), then* $\langle, \rangle_T$ *is surjective* ($\mathbb{T} \overset{\langle,\rangle_T}{\simeq} T$ *as A-rings*).

2) *Let* $\mathbf{m}_S = (S, T, Q, P, \langle, \rangle_S) \in \mathbb{UMSC}$ *and assume the B-rng* $\mathbb{S} := Q \otimes_S P$ *to be unital. If* $\langle, \rangle_S : \mathbb{S} \to S$ *respects unities (and* $\mathbf{m}_S$ *is injective), then* $\langle, \rangle_S$ *is surjective* ($\mathbb{S} \overset{\langle,\rangle_S}{\simeq} S$ *as B-rings*).

3) *Let* $\mathcal{M} = (T, S, P, Q, \langle, \rangle_T, \langle, \rangle_S) \in \mathbb{UMC}$ *and assume the rngs* $\mathbb{T} := P \otimes_S Q$, T, $\mathbb{S} := Q \otimes_S P$ *to be unital. If* $\langle, \rangle_T : P \otimes_S Q \to T$ *and* $\langle, \rangle_S : \mathbb{S} \to S$ *respect unities, then* $\mathbb{T} \overset{\langle,\rangle_T}{\simeq} T$ *as A-ring,* $\mathbb{S} \overset{\langle,\rangle_S}{\simeq} S$ *as B-rings and we have equivalences of categories* $_\mathbb{T}\mathbb{M} \approx {}_\mathbb{S}\mathbb{M}$ *(and* $\mathbb{M}_\mathbb{T} \approx \mathbb{M}_\mathbb{S}$*).*

Proof. Assume $\mathbb{T}$ is unital with $1_\mathbb{T} = \sum_{i=1}^n p_i \otimes_S q_i$. If $\langle, \rangle_T$ respects unities, then we have $\sum_{i=1}^n \langle p_i, q_i \rangle_T = 1_T$, and so for any $t \in T$ we get $t = t1_T = \sum_{i=1}^n t \langle p_i, q_i \rangle_T = \sum_{i=1}^n \langle tp_i, q_i \rangle_T \in \mathrm{Im}(\langle, \rangle_T)$. One can prove 2) analogously. As for 3), it is well known that a unital Morita context with surjective connecting bimodule morphisms is strict (e.g., [Fai1981, 12.7.]), hence $\mathbb{T} \overset{\langle,\rangle_T}{\simeq} T$, $\mathbb{S} \overset{\langle,\rangle_S}{\simeq} S$. The equivalences of categories $_\mathbb{T}\mathbb{M} \simeq {}_T\mathbb{M} \approx {}_S\mathbb{M} \simeq {}_\mathbb{S}\mathbb{M}$ (and $\mathbb{M}_\mathbb{T} \simeq \mathbb{M}_T \approx \mathbb{M}_S \simeq \mathbb{M}_\mathbb{S}$) follow then by classical Morita Theory (e.g., [Fai1981, Chapter 12]). $\square$

Definition 3.1. *Let T be an A-rng, V_T a right T-module and consider for every left T-module $_T L$ the annihilator*

$$\mathrm{ann}_L^\otimes(V_T) := \{l \in L \mid V \otimes_T l = 0\}.$$

Following [AF1974, Exercises 19], *we say V_T is L-faithful, iff* $\operatorname{ann}_L^{\otimes}(V_T) = 0$; *and to be* **completely faithful**, *iff V_T is L-faithful for every left T-module ${}_S L$. Similarly, we can define completely faithful left T-modules.*

Under suitable conditions, the following result characterizes the Morita data, which are Morita contexts:

Proposition 3.2. *Let $\mathcal{M} = (T, S, P, Q, \langle,\rangle_T, \langle,\rangle_S)$ be a Morita datum.*

1) *If $\mathcal{M} \in \mathbb{MC}$, then $\mathbf{S} \overset{id}{\simeq} \mathbb{S}$ and $\mathbf{T} \overset{id}{\simeq} \mathbb{T}$ as rngs.*
2) *Assume ${}_T P$ is Q-faithful and Q_T is P-faithful. Then $\mathcal{M} \in \mathbb{MC}$ if and only if $\mathbf{S} \overset{id}{\simeq} \mathbb{S}$ and $\mathbf{T} \overset{id}{\simeq} \mathbb{T}$ as rngs.*

Proof. 1) Obvious.

2) Assume $\mathbf{S} \overset{id}{\simeq} \mathbb{S}$ and $\mathbf{T} \overset{id}{\simeq} \mathbb{T}$ as rngs. If $p \in P$ and $q, q' \in Q$ are arbitrary, then we have for any $\widetilde{p} \in P$:

$$\langle q, p\rangle_S q' \otimes_T \widetilde{p} = (q \otimes_T p) \cdot_S (q' \otimes_T \widetilde{p}) = (q \otimes_T p) \cdot_S (q' \otimes_T \widetilde{p}) = q\langle p, q'\rangle_T \otimes_T \widetilde{p},$$

hence $\langle q, p\rangle_S q' - q\langle p, q'\rangle_T \in \operatorname{ann}_Q^{\otimes}(P) = 0$ (since ${}_T P$ is Q-faithful), i.e., $\langle q, p\rangle_S q' = q\langle p, q'\rangle_T$ for all $p \in P$ and $q, q' \in Q$. Assuming Q_T is P-faithful, one can prove analogously that $\langle p, q\rangle_T p' = p\langle q, p'\rangle_S$ for all $p, p' \in P$ and $q \in Q$. Consequently, $\mathcal{M}$ is a Morita context. $\qquad\qquad\square$

4. Injective Morita (semi-)contexts

Definition 4.1. *We call a Morita semi-context $\mathbf{m}_T = (T, S, P, Q, \langle,\rangle_T, I)$:*

- **injective** *(resp.* **semi-strict, strict**)*, iff $\langle,\rangle_T : P \otimes_S Q \to T$ is injective (resp. surjective, bijective);*
- **non-degenerate**, *iff $Q \hookrightarrow {}^*P$ and $P \hookrightarrow Q^*$ canonically;*
- **Morita α-semi-context**, *iff $\mathbf{P}_l := (Q, {}_T P) \in \mathcal{P}_l^\alpha(T)$ and $\mathbf{Q}_r := (P, Q_T) \in \mathcal{P}_r^\alpha(T)$.*

Notation 4.1. *By $\mathbb{MSC}^\alpha \leq \mathbb{MSC}$ ($\mathbb{UMSC}^\alpha \leq \mathbb{UMSC}$) we denote the full subcategory of (unital) Morita semi-contexts satisfying the α-condition. Moreover, we denote by $\mathbb{IMSC} \leq \mathbb{MSC}$ ($\mathbb{IUMSC} \leq \mathbb{UMSC}$) the full subcategory of injective (unital) Morita semi-contexts.*

Definition 4.2. *We say a Morita datum (context) $\mathcal{M} = (T, S, P, Q, \langle,\rangle_T, \langle,\rangle_S, I, J)$:*

- *is* **injective** *(resp.* **semi-strict, strict**)*, iff $\langle,\rangle_T : P \otimes_S Q \to T$ and $\langle,\rangle_S : Q \otimes_T P \to S$ are injective (resp. surjective, bijective);*
- *is* **non-degenerate**, *iff $Q \hookrightarrow {}^*P$, $P \hookrightarrow Q^*$, $Q \hookrightarrow P^*$ and $P \hookrightarrow {}^*Q$ canonically;*
- **satisfies the left α-condition**, *iff $\mathbf{P}_l := (Q, {}_T P) \in \mathcal{P}_l^\alpha(T)$ and $\mathbf{Q}_l := (P, {}_S Q) \in \mathcal{P}_l^\alpha(S)$;*
- **satisfies the right α-condition**, *iff $\mathbf{Q}_r := (P, Q_T) \in \mathcal{P}_r^\alpha(T)$ and $\mathbf{P}_r := (Q, P_S) \in \mathcal{P}_r^\alpha(S)$;*

   – **satisfies the α-condition,** *or* $\mathcal{M}$ *is a* **Morita α-datum** (**Morita α-context**), *iff* $\mathcal{M}$ *satisfies both the left and the right α-conditions.*

Notation 4.2. *By* $\mathrm{MC}_l^\alpha \leq \mathrm{MC}$ $(\mathrm{UMC}_l^\alpha \leq \mathrm{UMC})$ *we denote the full subcategory of Morita contexts satisfying the left α-condition, and by* $\mathrm{MC}_r^\alpha \leq \mathrm{MC}$ $(\mathrm{UMC}_r^\alpha \leq \mathrm{UMC})$ *the full subcategory of (unital) Morita contexts satisfying the right α-condition. Moreover, we set* $\mathrm{MC}^\alpha := \mathrm{MC}_l^\alpha \cap \mathrm{MC}_r^\alpha$ *and* $\mathrm{UMC}^\alpha := \mathrm{UMC}_l^\alpha \cap \mathrm{UMC}_r^\alpha.$

Lemma 4.1. *Let* $\mathcal{M} = (T, S, P, Q, \langle,\rangle_T, \langle,\rangle_S, I, J) \in \mathrm{MC}.$ *Consider the Morita semi-context* $\mathcal{M}_S := (S, T, Q, P, \langle,\rangle_S),$ *the dual pairings* $\mathbf{P}_l := (Q, {}_T P) \in \mathcal{P}_l(T),$ $\mathbf{Q}_r := (P, Q_T) \in \mathcal{P}_r(T)$ *and the canonical morphisms of rings*

$$\rho_P : S \to \mathrm{End}({}_T P)^{op} \quad and \quad \lambda_Q : S \to \mathrm{End}(Q_T).$$

1) *If* $\mathbf{Q}_r$ *is injective (semi-strict), then* $\mathcal{M}_S$ *is injective* $(\rho_P : S \to \mathrm{End}({}_T P)^{op}$ *is a surjective morphism of B-rngs).*
2) *Assume* P_S *is faithful and let* $\mathbf{Q}_r$ *be semi-strict. Then* $S \simeq \mathrm{End}({}_T P)^{op}$ *(an isomorphism of unital B-rings) and* $\mathcal{M}_S$ *is strict.*
3) *If* $\mathbf{P}_l$ *is injective (semi-strict), then* $\mathcal{M}_S$ *is injective* $(\lambda_Q : S \to \mathrm{End}(Q_T)$ *is a surjective morphism of B-rngs).*
4) *Assume* ${}_S Q$ *is faithful and let* $\mathbf{P}_l$ *be semi-strict. Then* $S \simeq \mathrm{End}(Q_T)$ *(an isomorphism of unital B-rings) and* $\mathcal{M}_S$ *is strict.*

Proof. We prove only 1) and 2), as 3) and 4) can be proved analogously.

Consider the following butterfly diagram with canonical morphisms

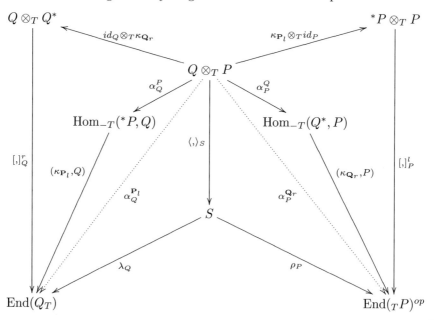

$$\tag{7}$$

Let $\sum q_i \otimes_T p_i \in Q \otimes_T P$ be arbitrary. For every $\widetilde{p} \in P$ we have

$$\left[\left(\kappa_{\mathbf{Q}_r}, P\right) \circ \alpha_P^Q\right) \left(\sum q_i \otimes_T p_i\right)\right] (\widetilde{p}) = \sum \langle \widetilde{p}, q_i \rangle_T p_i$$

$$= \sum \widetilde{p}\langle q_i, p_i \rangle_S = \rho_P \left(\sum \langle q_i, p_i \rangle_S\right) (\widetilde{p}) = (\rho_P \circ \langle, \rangle_S) \left(\sum q_i \otimes_T p_i\right) (\widetilde{p}),$$

i.e., $\alpha_P^{\mathbf{Q}_r} := (\kappa_{\mathbf{Q}_r}, P) \circ \alpha_P^Q = \rho_P \circ \langle, \rangle_S$; and

$$[,]_P^l \circ (\kappa_{\mathbf{P}_l} \otimes_T id_P)) \left(\sum q_i \otimes_T p_i\right)\right] (\widetilde{p}) = \sum \kappa_{\mathbf{P}_l}(q_i)(\widetilde{p})p_i = \sum \langle \widetilde{p}, q_i \rangle_T p_i$$

$$= \sum \widetilde{p}\langle q_i, p_i \rangle_S = \rho_P \left(\sum \langle q_i, p_i \rangle_S\right) (\widetilde{p}) = \left[(\rho_P \circ \langle, \rangle_S) \left(\sum q_i \otimes_T p_i\right)\right] (\widetilde{p}),$$

i.e., $[,]_P^l \circ (\kappa_{\mathbf{P}_l} \otimes_T id_P) = \rho_P \circ \langle, \rangle_S$. On the other hand, for every $\widetilde{q} \in Q$ we have

$$\left((\kappa_{\mathbf{P}_l}, Q) \circ \alpha_Q^{\mathbf{P}_l}\right) \left(\sum q_i \otimes_T p_i\right) (\widetilde{q}) = \sum q_i \langle p_i, \widetilde{q} \rangle_T$$

$$= \left(\sum \langle q_i, p_i \rangle_S\right) \widetilde{q} = \lambda_Q \left(\sum \langle q_i, p_i \rangle_S\right) (\widetilde{q}) = (\lambda_Q \circ \langle, \rangle_S) \left(\sum q_i \otimes_T p_i\right),$$

i.e., $\alpha_Q^{\mathbf{P}_l} := (\kappa_{\mathbf{P}_l}, Q) \circ \alpha_Q^{\mathbf{P}_l} = \lambda_Q \circ \langle, \rangle_S$ and

$$([,]_Q^r \circ (id_Q \otimes_T \kappa_{\mathbf{Q}_r})) \left(\sum q_i \otimes_T p_i\right)\right] (\widetilde{q}) = \sum q_i \kappa_{\mathbf{Q}_r}(p_i)(\widetilde{q}) = \sum q_i \langle p_i, \widetilde{q} \rangle_T$$

$$= \sum \langle q_i, p_i \rangle_S \widetilde{q} = \lambda_Q \left(\sum \langle q_i, p_i \rangle_S\right) (\widetilde{q}) = \left[(\lambda_Q \circ \langle, \rangle_S) \left(\sum q_i \otimes_T p_i\right)\right] (\widetilde{q}),$$

i.e., $[,]_Q^r \circ (id_Q \otimes_T \kappa_{\mathbf{Q}_r}) = \lambda_Q \circ \langle, \rangle_S$. Hence Diagram (7) is commutative.

(1) Follows directly from the assumptions and the equality $\alpha_P^{\mathbf{Q}_r} = \rho_P \circ \langle, \rangle_S$.

(2) Let P_S be faithful, so that the canonical left S-linear map $\rho_P : S \to \mathrm{End}(_T P)^{op}$ is injective. Assume now that $\mathbf{Q}_r$ is semi-strict. Then ρ_P is surjective by 1), whence bijective. Since rings of endomorphisms are unital, we conclude that $S \simeq \mathrm{End}(_T P)^{op}$ is a unital B-ring as well (with unity $\rho_P^{-1}(id_P)$). Moreover, the surjectivity of $\alpha_P^{\mathbf{Q}_r} = \rho_P \circ \langle, \rangle_S$ implies that $\langle, \rangle_S$ is surjective (since ρ_P is injective), say $1_S = \sum_j \langle \widetilde{q}_j, \widetilde{p}_j \rangle_S$ for some $\{(\widetilde{q}_j, \widetilde{p}_j)\}_J \subseteq Q \times P$. For any $\sum_i q_i \otimes_T p_i \in \mathrm{Ker}(\langle, \rangle_S)$, we have then

$$\sum_i q_i \otimes_T p_i = \left(\sum_i q_i \otimes_T p_i\right) \cdot 1_S \qquad = \sum_i (q_i \otimes_T p_i) \cdot \left(\sum_j \langle \widetilde{q}_j, \widetilde{p}_j \rangle_S\right)$$

$$= \sum_{i,j} q_i \otimes_T p_i \langle \widetilde{q}_j, \widetilde{p}_j \rangle_S \qquad = \sum_{i,j} q_i \otimes_T \langle p_i, \widetilde{q}_j \rangle_T \widetilde{p}_j$$

$$= \sum_{i,j} q_i \langle p_i, \widetilde{q}_j \rangle_T \otimes_T \widetilde{p}_j \qquad = \sum_{i,j} \langle q_i, p_i \rangle_S \widetilde{q}_j \otimes_T \widetilde{p}_j$$

$$= \sum_j (\sum_i \langle q_i, p_i \rangle_S) \widetilde{q}_j \otimes_T \widetilde{p}_j = 0,$$

i.e., $\langle, \rangle_S$ is injective, whence an isomorphism. $\qquad\square$

The following result shows that Morita α-contexts are injective:

Corollary 4.1. $\mathrm{MC}_l^\alpha \cup \mathrm{MC}_r^\alpha \leq \mathrm{IMC}$.

Example 4.1. Let $\mathbf{m}_T = (T, S, P, Q, \langle, \rangle_T)$ be a non-degenerate Morita semi-context. If T is a QF ring and the T-modules $_T P$, Q_T are projective, then by Proposition 2.1 7) $\mathbf{P}_l := (Q, {}_T P) \in \mathcal{P}_l^\alpha(T)$ and $\mathbf{Q}_r := (P, Q_T) \in \mathcal{P}_r^\alpha(T)$ (i.e., $\mathbf{m}_T$ is a Morita α-semi-context, whence injective). On the other hand, let $\mathcal{M} = (T, S, P, Q, \langle, \rangle_T, \langle, \rangle_S)$ be a non-degenerate Morita datum. If T, S are QF rings and the modules $_T P$, Q_T, P_S, $_S Q$ are projective, then $\mathcal{M}$ is a Morita α-datum (whence injective).   □

Every semi-strict *unital* Morita context is injective (whence strict, e.g., [Fai1981, 12.7.]). The following example, which is a modification of [Lam1999, Example 18.30]), shows that the converse is not necessarily true:

Example 4.2. Let $T = M_2(\mathbb{Z}_2)$ be the ring of 2×2 matrices with entries in $\mathbb{Z}_2$. Notice that $e = \begin{bmatrix} 1 & 0 \\ 0 & 0 \end{bmatrix} \in T$ is an idempotent, and that $eTe \simeq \mathbb{Z}_2$ as rings. Set

$$P := Te = \left\{ \begin{bmatrix} a' & 0 \\ c' & 0 \end{bmatrix} \mid a', c' \in \mathbb{Z}_2 \right\} \quad \text{and} \quad Q := eT = \left\{ \begin{bmatrix} a & b \\ 0 & 0 \end{bmatrix} \mid a, b \in \mathbb{Z}_2 \right\}.$$

Then $P = Te$ is a (T, eTe)-bimodule and $Q = eT$ is an (eTe, T)-bimodule. Moreover, we have a Morita context

$$\mathcal{M}_e = (T, eTe, Te, eT, \langle, \rangle_T, \langle . \rangle_{eTe}),$$

where the connecting bilinear maps are

$$\langle, \rangle_T \quad : Te \otimes_{eTe} eT \qquad \to T,$$

$$\begin{bmatrix} a' & 0 \\ c' & 0 \end{bmatrix} \otimes_{eTe} \begin{bmatrix} a & b \\ 0 & 0 \end{bmatrix} \mapsto \begin{bmatrix} a'a & a'b \\ c'a & c'b \end{bmatrix}$$

$$\langle, \rangle_{eTe} : eT \otimes_T Te \qquad \to eTe$$

$$\begin{bmatrix} a & b \\ 0 & 0 \end{bmatrix} \otimes_T \begin{bmatrix} a' & 0 \\ c' & 0 \end{bmatrix} \mapsto \begin{bmatrix} aa' + bc' & 0 \\ 0 & 0 \end{bmatrix}.$$

Straightforward computations show that $\langle, \rangle_T$ is injective but not surjective (as $\begin{bmatrix} 1 & 1 \\ 1 & 0 \end{bmatrix} \notin \mathrm{Im}(\langle, \rangle_T)$) and that $\langle, \rangle_{eTe}$ is in fact an isomorphism. This means that $\mathcal{M}_e$ is an injective Morita context that is not semi-strict (whence not strict).   □

Definition 4.3. *Let T be a rng and $I \lhd T$ an ideal. For every left T-module $_T V$ consider the canonical T-linear map*

$$\zeta_{I,V} : V \to \mathrm{Hom}_T(I, V), \quad v \mapsto [t \mapsto tv].$$

*We say $_T I$ is **strongly V-faithful**, iff $\mathrm{ann}_V(I) := \mathrm{Ker}(\zeta_{I,V}) := 0$. Moreover, we say I is **strongly faithful**, if $_T I$ is V-faithful for every left T-module $_T V$. Strong faithfulness of I w.r.t. right T-modules can be defined analogously.*

Remark 4.1. Let T be a rng, $I \lhd T$ an ideal and $_T U$ a left ideal. It's clear that $\mathrm{ann}_U^\otimes(I_T) \subseteq \mathrm{ann}_U(I) := \mathrm{Ker}(\zeta_{I,U})$. Hence, if $_T I$ is *strongly U-faithful*, then I_T is *U-faithful* (which justifies our terminology). In particular, if $_T I$ is *strongly faithful*, then I_T is *completely faithful*.

Morita α-contexts are injective by Corollary 4.1. The following result gives a partial converse:

Lemma 4.2. *Let* $\mathcal{M} = (T, S, P, Q, \langle,\rangle_T, \langle,\rangle_S, I, J) \in \mathbb{MC}$ *and assume that the Morita semi-context* $\mathcal{M}_S := (S, T, Q, P, \langle,\rangle_S, J)$ *is injective.*

1) *If* $_S J$ *is strongly faithful, then* $\mathbf{Q}_r := (P, Q_T) \in \mathcal{P}_r^\alpha(T)$.
2) *If* J_S *is strongly faithful, then* $\mathbf{P}_l := (Q, {}_T P) \in \mathcal{P}_l^\alpha(T)$.

Proof. We prove only 1), since 2) can be proved similarly. Assume $\mathcal{M}_S$ is injective and consider for every left T-module U the following diagram

$$Q \otimes_T U \xrightarrow{\quad\quad\quad\quad \alpha_U^{\mathbf{Q}_r} \quad\quad\quad\quad} \mathrm{Hom}_{T-}(P, U) \qquad (8)$$

$$\searrow{\zeta_{J,Q\otimes_T U}} \qquad \swarrow{\psi_{Q,U}}$$

$$\mathrm{Hom}_{S-}(J, Q \otimes_T U)$$

where for all $f \in \mathrm{Hom}_{T-}(P, U)$ and $\sum \langle q_j, p_j \rangle_S \in J$ we define

$$\psi_{Q,U}(f)(\sum \langle q_j, p_j \rangle_S) := \sum q_j \otimes_T f(p_j).$$

Then we have for every $\sum \tilde{q}_i \otimes_T \tilde{u}_i \in Q \otimes_T U$ and $s = \sum_j \langle q_j, p_j \rangle_S \in J$:

$$\left(\psi_{Q,U} \circ \alpha_U^{\mathbf{Q}_r}\right)\left(\sum_i \tilde{q}_i \otimes_T \tilde{u}_i\right)(s) = \sum_j q_j \otimes_T \left[\alpha_U^{\mathbf{Q}_r}\left(\sum_i \tilde{q}_i \otimes_T \tilde{u}_i\right)\right](p_j)$$

$$= \sum_j q_j \otimes_T \sum_i \langle p_j, \tilde{q}_i \rangle_T \tilde{u}_i] = \sum_{i,j} q_j \otimes_T \langle p_j, \tilde{q}_i \rangle_T \tilde{u}_i$$

$$= \sum_{i,j} q_j \langle p_j, \tilde{q}_i \rangle_T \otimes_T \tilde{u}_i = \sum_{i,j} \langle q_j, p_j \rangle_S \tilde{q}_i \otimes_T \tilde{u}_i$$

$$= \zeta_{J,Q\otimes_T U}\left(\sum_i \tilde{q}_i \otimes_T \tilde{u}_i\right)(s),$$

i.e., diagram (8) is commutative. If $_S J$ is strongly faithful, then $\mathrm{Ker}(\zeta_{J,Q\otimes_T U}) = \mathrm{ann}_{Q\otimes_T U}(J) = 0$, hence $\zeta_{J,Q\otimes_T U}$ is injective and it follows then that $\alpha_U^{\mathbf{Q}_r}$ is injective. $\qquad\square$

Proposition 4.1. *Let* $\mathcal{M} = (T, S, P, Q, \langle,\rangle_T, \langle,\rangle_S, I, J) \in \mathbb{IMC}$. *If* $_T I$, I_T, $_S J$ *and* J_S *are strongly faithful, then* $\mathcal{M} \in \mathbb{MC}^\alpha$.

5. Equivalences of categories

In this section we give some applications of *injective Morita (semi-)contexts* and *injective Morita data* to equivalences between suitable subcategories of modules arising in the Kato-Müller-Ohtake localization-colocalization theory (as developed in (e.g., [Kat1978], [KO1979], [Mül1974]). All rings, hence all Morita (semi-)contexts and data, in this section are unital.

Static and adstatic modules

5.1. ([C-IG-TW2003]) Let $\mathcal{A}$ and $\mathcal{B}$ be two complete cocomplete Abelian categories, $\mathbf{R} : \mathcal{A} \to \mathcal{B}$ an additive covariant functor with left adjoint $\mathbf{L} : \mathcal{B} \to \mathcal{A}$ and let

$$\omega : \mathbf{LR} \to 1_{\mathcal{A}} \quad \text{and} \quad \eta : 1_{\mathcal{B}} \to \mathbf{RL}$$

be the induced natural transformations (called the *counit* and the *unit* of the adjunction, respectively). Related to the adjoint pair $(\mathbf{L}, \mathbf{R})$ are two *full* subcategories of $\mathcal{A}$ and $\mathcal{B}$:

$$\text{Stat}(\mathbf{R}) := \{X \in \mathcal{A} \mid \mathbf{LR}(X) \overset{\omega_X}{\simeq} X\} \quad \text{and} \quad \text{Adstat}(\mathbf{R}) := \{Y \in \mathcal{B} \mid Y \overset{\eta_Y}{\simeq} \mathbf{RL}(Y)\},$$

whose members are called **R-static objects** and **R-adstatic objects,** respectively. It is evident (from definition) that we have equivalence of categories $\text{Stat}(\mathbf{R}) \approx \text{Adstat}(\mathbf{R})$.

A typical situation, in which static and adstatic objects arise naturally is the following:

5.2. Let T, S be rings, $_T U_S$ a (T, S)-bimodule and consider the covariant functors

$$\mathbf{H}_U^l := \text{Hom}_T(U, -) : \; _T\mathbb{M} \to \; _S\mathbb{M} \quad \text{and} \quad \mathbf{T}_U^l := U \otimes_S - : \; _S\mathbb{M} \to \; _T\mathbb{M}.$$

It is well known that $(\mathbf{T}_U^l, \mathbf{H}_U^l)$ is an adjoint pair of covariant functors via the *natural isomorphism*

$$\text{Hom}_T(U \otimes_S M, N) \simeq \text{Hom}_S(M, \text{Hom}_T(U, N)) \text{ for all } M \in \; _S\mathbb{M} \quad \text{and} \quad N \in \; _T\mathbb{M}$$

and the natural transformations

$$\omega_U^l : U \otimes_S \text{Hom}_T(U, -) \to 1_{_T\mathbb{M}} \quad \text{and} \quad \eta_U^l : 1_{_S\mathbb{M}} \to \text{Hom}_T(U, U \otimes_S -)$$

yield for every $_T K$ and $_S L$ the canonical morphisms

$$\omega_{U,K}^l : U \otimes_S \text{Hom}_T(U, K) \to K \quad \text{and} \quad \eta_{U,L}^l : L \to \text{Hom}_T(U, U \otimes_S L). \quad (9)$$

We call the $\mathbf{H}_U^l$-*static* modules U-**static w.r.t.** S and set

$$\text{Stat}^l(_T U_S) := \text{Stat}(\mathbf{H}_U^l) = \{_T K \mid U \otimes_S \text{Hom}_{T-}(U, K) \overset{\omega_{U,K}^l}{\simeq} K\};$$

and the $\mathbf{H}_U^l$-*adstatic* modules U-**adstatic w.r.t.** S and set

$$\text{Adstat}^l(_T U_S) := \text{Adstat}(\mathbf{H}_U^l) = \{_S L \mid L \overset{\eta_{U,L}^l}{\simeq} \text{Hom}_{T-}(U, U \otimes_S L)\}.$$

By [Nau1990a] and [Nau1990b], there are equivalences of categories

$$\text{Stat}^l(_T U_S) \approx \text{Adstat}^l(_T U_S). \quad (10)$$

On the other hand, one can define the full subcategories $\mathrm{Stat}^r({}_TU_S) \approx \mathrm{Adstat}^r({}_TU_S)$:

$$\mathrm{Stat}^r({}_TU_S) \quad := \{K_S \mid \mathrm{Hom}_{-S}(U, K) \otimes_T U \simeq K\};$$
$$\mathrm{Adstat}^r({}_TU_S) := \{L_T \mid L \simeq \mathrm{Hom}_{-S}(U, L \otimes_T U)\}.$$

In particular, setting

$$\mathrm{Stat}({}_TU) := \mathrm{Stat}^l({}_TU_{\mathrm{End}({}_TU)^{op}}); \quad \mathrm{Adstat}({}_TU) := \mathrm{Adstat}^l({}_TU_{\mathrm{End}({}_TU)^{op}});$$
$$\mathrm{Stat}(U_S) := \mathrm{Stat}^r({}_{\mathrm{End}({}_SU)}U_S); \quad \mathrm{Adstat}(U_S) := \mathrm{Adstat}^r({}_{\mathrm{End}({}_SU)}U_S),$$

there are equivalences of categories:

$$\mathrm{Stat}({}_TU) \simeq \mathrm{Adstat}({}_TU) \quad \text{and} \quad \mathrm{Stat}(U_S) \simeq \mathrm{Adstat}(U_S). \tag{11}$$

Remark 5.1. The theory of static and adstatic modules was developed in a series of papers by the second author (see the references). They were also considered by several other authors (e.g., [Alp1990], [CF2004]). For other terminologies used by different authors, the interested reader may refer to a comprehensive treatment of the subject by R. Wisbauer in [Wis2000].

Intersecting subcategories

Several intersecting subcategories related to Morita contexts were introduced in the literature (e.g., [Nau1993], [Nau1994-b]). In what follows we introduce more and we show that many of these coincide, if one starts with an injective Morita semi-context. Moreover, other results on equivalences between some intersecting subcategories related to an injective Morita context will be reframed for arbitrary (not necessarily compatible) injective Morita data.

Definition 5.1.

1) *For a right T-module X, a T-submodule $X' \subseteq X$ is called K-**pure** for some left T-module ${}_TK$, iff the following sequence of Abelian groups is exact*

$$0 \to X' \otimes_T K \to X \otimes_T K \to X/X' \otimes_T K \to 0;$$

2) *For a left T-module Y, a T-submodule $Y' \subseteq Y$ is called L-**copure** for some left T-module ${}_TL$, iff the following sequence of Abelian groups is exact*

$$0 \to \mathrm{Hom}_T(Y/Y', L) \to \mathrm{Hom}_T(Y, L) \to \mathrm{Hom}_T(Y', L) \to 0.$$

Definition 5.2. (*Compare* [KO1979, Theorems 1.3., 2.3.]) *Let T be a ring, $I \lhd T$ an ideal, U a left T-module and consider the canonical T-linear morphisms*

$$\zeta_{I,U} : U \to \mathrm{Hom}_T(I, U) \quad \text{and} \quad \xi_{I,U} : I \otimes_T U \to U.$$

1) *We say ${}_TU$ is I-**divisible**, iff $\xi_{I,U}$ is surjective (equivalently, iff $IU = U$).*

2) *We say ${}_TU$ is I-**localized**, iff $U \overset{\zeta_{I,U}}{\simeq} \mathrm{Hom}_T(I, U)$ canonically (equivalently, iff ${}_TI$ is strongly U-faithful and ${}_TI \subseteq T$ is U-copure).*

3) *We say a left T-module U is I-**colocalized**, iff $I \otimes_T U \overset{\xi_{I,U}}{\simeq} U$ canonically (equivalently, iff ${}_TU$ is I-divisible and $I_T \subseteq T$ is U-pure).*

Notation 5.3. *For a ring T, an ideal $I \lhd T$, and with morphisms being the canonical ones, we set*

$$
\begin{aligned}
{}_I\mathfrak{D} &:= \{{}_T U \mid IU = U\}; & {}_I\mathfrak{F} &:= \{{}_T U \mid U \hookrightarrow \mathrm{Hom}_{T-}(I,U)\}; \\
{}_I\mathcal{L} &:= \{{}_T U \mid U \simeq \mathrm{Hom}_T(I,U)\}; & {}_I\mathcal{C} &:= \{{}_T U \mid I \otimes_T U \simeq U\}; \\
\mathfrak{D}_I &:= \{U_T \mid UI = U\}; & \mathfrak{F}_I &:= \{U_T \mid U \hookrightarrow \mathrm{Hom}_{-T}(I,U)\}; \\
\mathcal{L}_I &:= \{U_T \mid U \simeq \mathrm{Hom}_T(I,U)\}; & \mathcal{C}_I &:= \{U_T \mid U \otimes_T I \simeq U\};.
\end{aligned}
$$

The following result is due to T. Kato, K. Ohtake and B. Müller (e.g., [Mül1974], [Kat1978], [KO1979]):

Proposition 5.1. *Let $\mathcal{M} = (T,S,P,Q,\langle,\rangle_T,\langle,\rangle_S,I,J) \in \mathbb{UMC}$. Then there are equivalences of categories*

$$
{}_I\mathcal{C} \approx {}_J\mathcal{C}, \quad \mathcal{C}_I \approx \mathcal{C}_J, \quad {}_I\mathcal{L} \approx {}_J\mathcal{L} \quad and \quad \mathcal{L}_I \approx \mathcal{L}_J.
$$

5.4. Let $\mathbf{m}_T = (T,S,P,Q,\langle,\rangle_T,I) \in \mathbb{UMSC}$ and consider the dual pairings $\mathbf{P}_l := (Q, {}_TP) \in \mathcal{P}_l(T)$ and $\mathbf{Q}_r := (P, Q_T) \in \mathcal{P}_r(T)$. For every left (right) T-module U consider the canonical S-linear morphism induced by $\langle,\rangle_T$:

$$
\alpha_U^{\mathbf{Q}_r} : Q \otimes_T U \to \mathrm{Hom}_{T-}(P,U) \ (\alpha_U^{\mathbf{P}_l} : U \otimes_T P \to \mathrm{Hom}_{-T}(Q,U)).
$$

We define

$$
\mathcal{D}_l(\mathbf{m}_T) := \{{}_T U \mid Q \otimes_T U \overset{\alpha_U^{\mathbf{Q}_r}}{\simeq} \mathrm{Hom}_{T-}(P,U)\};
$$

$$
\mathcal{D}_r(\mathbf{m}_T) := \{U_T \mid U \otimes_T P \overset{\alpha_U^{\mathbf{P}_l}}{\simeq} \mathrm{Hom}_{-T}(Q,U)\}.
$$

Moreover, set

$$
\mathcal{U}_l(\mathbf{m}_T) := \mathrm{Stat}^l({}_TP_S) \cap \mathrm{Adstat}^l({}_SQ_T); \quad \mathcal{U}_r(\mathbf{m}_T) := \mathrm{Stat}^r({}_SQ_T) \cap \mathrm{Adstat}^r({}_TP_S);
$$

$$
\mathcal{V}_l(\mathbf{m}_T) := \mathrm{Stat}^l({}_TP_S) \cap \mathcal{D}_l(\mathbf{m}_T); \quad\quad \mathcal{V}_r(\mathbf{m}_T) := \mathrm{Stat}^r({}_SQ_T) \cap \mathcal{D}_r(\mathbf{m}_T);
$$

$$
\mathbb{V}_l(\mathbf{m}_T) := {}_I\mathcal{C} \cap \mathcal{D}_l(\mathbf{m}_T); \quad\quad\quad\quad \mathbb{V}_r(\mathbf{m}_T) := \mathcal{C}_I \cap \mathcal{D}_r(\mathbf{m}_T);
$$

$$
\widehat{\mathcal{V}}_l(\mathbf{m}_T) := \mathcal{V}_l(\mathbf{m}_T) \cap {}_I\mathcal{L}; \quad\quad\quad\quad \widehat{\mathcal{V}}_r(\mathbf{m}_T) := \mathcal{V}_r(\mathbf{m}_T) \cap \mathcal{L}_I;
$$

$$
\mathcal{W}_l(\mathbf{m}_T) := \mathrm{Adstat}^l({}_SQ_T) \cap \mathcal{D}_l(\mathbf{m}_T); \quad \mathcal{W}_r(\mathbf{m}_T) := \mathrm{Adstat}^r({}_TP_S) \cap \mathcal{D}_r(\mathbf{m}_T);
$$

$$
\mathbb{W}_l(\mathbf{m}_T) := {}_I\mathcal{L} \cap \mathcal{D}_l(\mathbf{m}_T); \quad\quad\quad\quad \mathbb{W}_r(\mathbf{m}_T) := \mathcal{L}_I \cap \mathcal{D}_r(\mathbf{m}_T);
$$

$$
\widehat{\mathcal{W}}_l(\mathbf{m}_T) := \mathcal{W}_l(\mathbf{m}_T) \cap {}_I\mathcal{C}; \quad\quad\quad\quad \widehat{\mathcal{W}}_r(\mathbf{m}_T) := \mathcal{W}_r(\mathbf{m}_T) \cap \mathcal{C}_I;
$$

$$
\mathcal{X}_l(\mathbf{m}_T) := \mathcal{V}_l(\mathbf{m}_T) \cap \mathcal{W}_l(\mathbf{m}_T); \quad\quad \mathcal{X}_r(\mathbf{m}_T) := \mathcal{V}_r(\mathbf{m}_T) \cap \mathcal{W}_r(\mathbf{m}_T);
$$

$$
\mathbb{X}_l(\mathbf{m}_T) := \mathbb{V}_l(\mathbf{m}_T) \cap \mathbb{W}_l(\mathbf{m}_T); \quad\quad \mathbb{X}_r(\mathbf{m}_T) := \mathbb{V}_r(\mathbf{m}_T) \cap \mathbb{W}_r(\mathbf{m}_T).
$$

$$
\mathcal{X}_l^*(\mathbf{m}_T) := \{{}_S(Q \otimes_T U) \mid V \in \mathcal{X}_l(\mathbf{m}_T)\}; \quad \mathcal{X}_r^*(\mathbf{m}_T) := \{(U \otimes_T P)_S \mid V \in \mathcal{X}_r(\mathbf{m}_T)\};
$$

$$
\mathbb{X}_l^*(\mathbf{m}_T) := \{{}_S(Q \otimes_T U) \mid V \in \mathbb{X}_l(\mathbf{m}_T)\}; \quad \mathbb{X}_r^*(\mathbf{m}_T) := \{(U \otimes_T P)_S \mid V \in \mathbb{X}_r(\mathbf{m}_T)\}.
$$

$$
\tag{12}
$$

Given $\mathbf{m}_S = (S,T,Q,P,\langle,\rangle_S,J) \in \mathbb{UMSC}$ one can define analogously, the corresponding intersecting subcategories of ${}_S\mathbb{M}$ and $\mathbb{M}_S$.

As an immediate consequence of Proposition 5.1 we get

Corollary 5.1. *Let $\mathcal{M} = (T, S, P, Q, \langle,\rangle_T, \langle,\rangle_S, I, J) \in \mathbb{IUMC}$ and consider the associated Morita semi-contexts $\mathcal{M}_T$ and $\mathcal{M}_S$ (5).*

1) *If $_I\mathcal{C} \leq \mathcal{D}_l(\mathcal{M}_T)$ and $_J\mathcal{C} \leq \mathcal{D}_l(\mathcal{M}_S)$, then $\mathcal{V}_l(\mathcal{M}_T) \approx \mathcal{V}_l(\mathcal{M}_S)$. Similarly, if $\mathcal{C}_I \leq \mathcal{D}_r(\mathcal{M}_T)$ and $\mathcal{C}_J \leq \mathcal{D}_r(\mathcal{M}_S)$, then $\mathcal{V}_r(\mathcal{M}_T) \approx \mathcal{V}_r(\mathcal{M}_S)$.*
2) *If $_I\mathcal{L} \leq \mathcal{D}_l(\mathcal{M}_T)$ and $_J\mathcal{L} \leq \mathcal{D}_l(\mathcal{M}_S)$, then $\mathcal{W}_l(\mathcal{M}_T) \approx \mathcal{W}_l(\mathcal{M}_S)$. Similarly, if $\mathcal{L}_I \leq \mathcal{D}_r(\mathcal{M}_T)$ and $\mathcal{L}_J \leq \mathcal{D}_r(\mathcal{M}_S)$, then $\mathcal{W}_r(\mathcal{M}_T) \approx \mathcal{W}_r(\mathcal{M}_S)$.*

Starting with a Morita context, the following result was obtained in [Nau1993, Theorem 3.2.]. We restate the result for an arbitrary (not necessarily compatible) Morita datum and *sketch* its proof:

Lemma 5.1. *Let $\mathcal{M} = (T, S, P, Q, \langle,\rangle_T, \langle,\rangle_S, I, J)$ be a unital Morita datum and consider the associated Morita semi-contexts $\mathcal{M}_T$ and $\mathcal{M}_S$ in (5). Then there are equivalences of categories*

$$\mathcal{X}_l(\mathcal{M}_T) \underset{\mathrm{Hom}_{S-}(Q,-)}{\overset{\mathrm{Hom}_{T-}(P,-)}{\approx}} \mathcal{X}_l(\mathcal{M}_S) \quad \text{and} \quad \mathcal{X}_r(\mathcal{M}_T) \underset{\mathrm{Hom}_{-S}(P,-)}{\overset{\mathrm{Hom}_{-T}(Q,-)}{\approx}} \mathcal{X}_r(\mathcal{M}_S).$$

Proof. Let $_TV \in \mathcal{X}_l(\mathcal{M}_T)$. By the equivalence $\mathrm{Stat}^l(_TP_S) \overset{\mathrm{Hom}_T(P,-)}{\approx} \mathrm{Adstat}^l(_TP_S)$ in 5.2 we have $\mathrm{Hom}_{T-}(P, V) \in \mathrm{Adstat}^l(_TP_S)$. Moreover, $V \in \mathcal{D}_l(M)$, hence $\mathrm{Hom}_{T-}(P, V) \simeq Q \otimes_T V$ canonically and it follows then from the equivalence $\mathrm{Adstat}^l(_SQ_T) \overset{Q\otimes_T-}{\approx} \mathrm{Stat}^l(_SQ_T)$ that $\mathrm{Hom}_{T-}(P, V) \in \mathrm{Stat}^l(_SQ_T)$. Moreover, we have the following *natural* isomorphisms

$$P \otimes_S \mathrm{Hom}_{T-}(P, V) \simeq V \simeq \mathrm{Hom}_{S-}(Q, Q \otimes_T V) \simeq \mathrm{Hom}_{S-}(Q, \mathrm{Hom}_{T-}(P, V)),$$
$$(13)$$

i.e., $\mathrm{Hom}_{T-}(P, V) \in \mathcal{D}_l(\mathcal{M}_S)$. Consequently, $\mathrm{Hom}_{T-}(P, V) \in \mathcal{X}_l(\mathcal{M}_S)$. Moreover, (13) yields a natural isomorphism $V \simeq \mathrm{Hom}_{S-}(Q, \mathrm{Hom}_{T-}(P, V))$. Analogously, one can show for every $W \in \mathcal{X}_l(\mathcal{M}_S)$ that $\mathrm{Hom}_{S-}(Q, W) \in \mathcal{X}_l(\mathcal{M}_T)$ and that $W \simeq \mathrm{Hom}_{T-}(P, \mathrm{Hom}_{S-}(Q, W))$ naturally. Consequently, $\mathcal{X}_l(\mathcal{M}_T) \approx \mathcal{X}_l(\mathcal{M}_S)$. The equivalences $\mathcal{X}_r(\mathcal{M}_T) \approx \mathcal{X}_r(\mathcal{M}_S)$ can be proved analogously. $\square$

Proposition 5.2. *Let $\mathcal{M} = (T, S, P, Q, \langle,\rangle_T, \langle,\rangle_S, I, J)$ be a unital injective Morita datum and consider the associated Morita semi-contexts $\mathcal{M}_T$ and $\mathcal{M}_S$ in (5).*

1) *There are equivalences of categories*

$$\mathrm{Stat}^l(_TI_T) \approx \mathrm{Adstat}^l(_TI_T); \quad \mathrm{Stat}^l(_SJ_S) \approx \mathrm{Adstat}^l(_SJ_S);$$
$$\mathrm{Stat}^r(_TI_T) \approx \mathrm{Adstat}^r(_TI_T); \quad \mathrm{Stat}^r(_SJ_S) \approx \mathrm{Adstat}^r(_SJ_S).$$

2) *If $\mathrm{Stat}^l(_TI_T) \leq \mathcal{X}_l^*(\mathcal{M}_S)$ and $\mathrm{Stat}^l(_SJ_S) \leq \mathcal{X}_l^*(\mathcal{M}_T)$, then there are equivalences of categories*

$$\mathrm{Stat}^l(_TI_T) \approx \mathrm{Stat}^l(_SJ_S) \quad \text{and} \quad \mathrm{Adstat}^l(_TI_T) \approx \mathrm{Adstat}^l(_SJ_S).$$

3) *If* $\mathrm{Stat}^r(_T I_T) \leq \mathcal{X}_r^*(\mathcal{M}_S)$ *and* $\mathrm{Stat}^r(_S J_S) \leq \mathcal{X}_r^*(\mathcal{M}_T)$, *then there are equivalences of categories*

$$\mathrm{Stat}^r(_T I_T) \approx \mathrm{Stat}^r(_S J_S) \quad and \quad \mathrm{Adstat}^r(_T I_T) \approx \mathrm{Adstat}^r(_S J_S).$$

Proof. To prove 1), notice that since $\mathcal{M}$ is an injective Morita datum, $P \otimes_S Q \overset{<,>_T}{\simeq} I$ and $Q \otimes_T P \overset{<,>_S}{\simeq} J$ as bimodules and so the four equivalences of categories result from 5.2. To prove 2), one can use an argument similar to that in [Nau1994-b, Theorem 3.9.] to show that the inclusion $\mathrm{Stat}^l(_T I_T) = \mathrm{Stat}^l(_T(P \otimes_S Q)_T) \leq \mathcal{X}_l^*(\mathcal{M}_S)$ implies $\mathrm{Stat}^l(_T I_T) = \mathrm{Stat}^l(_T(P \otimes_S Q)_T) = \mathcal{X}_l(\mathcal{M}_T)$ and that the inclusion $\mathrm{Stat}^l(_S J_S) = \mathrm{Stat}^l(_S(Q \otimes_T P)_S) \leq \mathcal{X}_l^*(\mathcal{M}_T)$ implies $\mathrm{Stat}^l(_S J_S) = \mathrm{Stat}^l(_S(Q \otimes_T P)_S) = \mathcal{X}_l(\mathcal{M}_S)$. The result follows then by Lemma 5.1. The proof of 3) is analogous to that of 2). $\qquad \square$

For injective Morita semi-contexts, several subcategories in (12) are shown to be equal in the following result:

Theorem 5.5. *Let* $\mathbf{m}_T = (T, S, P, Q, <, >_T, I) \in \mathbb{IUMS}$. *Then*

1) $\mathcal{V}_l(\mathbf{m}_T) = \mathbb{V}_l(\mathbf{m}_T)$, $\mathcal{W}_l(\mathbf{m}_T) = \mathbb{W}_l(\mathbf{m}_T)$, *whence*

$$\widehat{\mathcal{V}}_l(\mathbf{m}_T) = \widehat{\mathcal{W}}_l(\mathbf{m}_T) = \mathcal{X}_l(\mathbf{m}_T) = \mathbb{X}_l(\mathbf{m}_T) = {}_I\mathcal{C} \cap \mathcal{D}_l(\mathbf{m}_T) \cap {}_I\mathcal{L}$$

and

$$\mathcal{X}_l^*(\mathbf{m}_T) = \mathbb{X}_l^*(\mathbf{m}_T).$$

2) $\mathcal{V}_r(\mathbf{m}_T) = \mathbb{V}_r(\mathbf{m}_T)$, $\mathcal{W}_r(\mathbf{m}_T) = \mathbb{W}_r(\mathbf{m}_T)$, *whence*

$$\widehat{\mathcal{V}}_r(\mathbf{m}_T) = \widehat{\mathcal{W}}_r(\mathbf{m}_T) = \mathcal{X}_r(\mathbf{m}_T) = \mathbb{X}_r(\mathbf{m}_T) = \mathcal{C}_I \cap \mathcal{D}_r(\mathbf{m}_T) \cap \mathcal{L}_I$$

and

$$\mathcal{X}_r^*(\mathbf{m}_T) = \mathbb{X}_r^*(\mathbf{m}_T).$$

Proof. We prove only 1) as 2) can be proved analogously. Assume the Morita semi-context $\mathbf{m}_T = (T, S, P, Q, \langle, \rangle_T, I)$ is injective. By our assumption we have for every $V \in \mathcal{D}_l(\mathbf{m}_T)$ the commutative diagram

$$
\begin{array}{ccc}
P \otimes_S (Q \otimes_T V) & \xrightarrow[\simeq]{\text{can}} & (P \otimes_S Q) \otimes_T V \\
{\scriptstyle id_P \otimes_S(\alpha_V^{Qr})} \Big\downarrow {\scriptstyle \simeq} & & {\scriptstyle \simeq} \Big\downarrow {\scriptstyle \langle,\rangle_T \otimes_T id_V} \\
P \otimes_S \mathrm{Hom}_{T-}(P, V) \xrightarrow{\omega_{P,V}^l} & V & \xleftarrow{\xi_{I,V}} I \otimes_T V
\end{array}
\qquad (14)
$$

Then it becomes obvious that $\omega_{P,V}^l : P \otimes_S \mathrm{Hom}_T(P, V) \to V$ is an isomorphism if and only if $\xi_{I,V} : I \otimes_T V \to V$ is an isomorphism. Consequently

$$\mathcal{V}(\mathbf{m}_T) = \mathcal{D}_l(\mathbf{m}_T) \cap \mathrm{Stat}^l(_T P_S) = \mathcal{D}_l(\mathbf{m}_T) \cap {}_I\mathcal{C} = \mathbb{V}(\mathbf{m}_T).$$

On the other hand, we have for every $V \in \mathcal{D}_l(\mathbf{m}_T)$ the following commutative diagram

$$
\begin{array}{ccc}
\mathrm{Hom}_{S-}(Q, \mathrm{Hom}_{T-}(P,V)) & \xrightarrow[\simeq]{\mathrm{can}} & \mathrm{Hom}_{T-}(P \otimes_S Q, V) \\
\uparrow{\scriptstyle (Q,\alpha_V^{Q_r})\,\simeq} & & \simeq\,\uparrow{\scriptstyle (\langle,\rangle_T,V)} \\
\mathrm{Hom}_{S-}(Q, Q \otimes_T V) \xleftarrow[\eta_{P,L}^l]{} & V \xrightarrow[\zeta_{I,V}]{} & \mathrm{Hom}_{T-}(I,V)
\end{array}
$$

$$(15)$$

It follows then that $\eta_{P,L}^l : V \to \mathrm{Hom}_S(Q, Q \otimes_T P)$ is an isomorphism if and only if $\zeta_{I,V} : V \to \mathrm{Hom}_T(I,V)$ is an isomorphism. Consequently,

$$\mathcal{W}(\mathbf{m}_T) = \mathcal{D}_l(\mathbf{m}_T) \cap \mathrm{Adstat}^l({}_T P_S) = \mathcal{D}_l(\mathbf{m}_T) \cap_I \mathcal{L} = \mathbb{W}(\mathbf{m}_T).$$

Moreover, we have

$$\widehat{\mathcal{V}_l}(\mathbf{m}_T) := \mathcal{V}_l(\mathbf{m}_T) \cap {}_I\mathcal{L} = \mathbb{V}_l(\mathbf{m}_T) \cap {}_I\mathcal{L} = {}_I\mathcal{C} \cap \mathcal{D}_l(\mathbf{m}_T) \cap {}_I\mathcal{L}$$
$$= {}_I\mathcal{C} \cap \mathbb{W}_l(\mathbf{m}_T) = {}_I\mathcal{C} \cap \mathcal{W}_l(\mathbf{m}_T) = \widehat{\mathcal{W}_l}(\mathbf{m}_T).$$

On the other hand, we have

$$\mathcal{X}_l(\mathbf{m}_T) = \mathcal{V}_l(\mathbf{m}_T) \cap \mathcal{W}_l(\mathbf{m}_T) = \mathbb{V}_l(\mathbf{m}_T) \cap \mathbb{W}_l(\mathbf{m}_T) = \mathbb{X}_l(\mathbf{m}_T)$$

and so the equalities $\widehat{\mathcal{V}_l}(\mathbf{m}_T) = \widehat{\mathcal{W}_l}(\mathbf{m}_T) = \mathcal{X}_l(\mathbf{m}_T) = \mathbb{X}_l(\mathbf{m}_T)$ and $\mathcal{X}_l^*(\mathbf{m}_T) = \mathbb{X}_l^*(\mathbf{m}_T)$ are established. $\qquad\square$

In addition to establishing several other equivalences of intersecting subcategories, the following result reframes the equivalence of categories $\widehat{\mathcal{V}} \approx \widehat{\mathcal{W}}$ in [Nau1994-b, Theorem 4.9.] for an arbitrary (not necessarily compatible) injective Morita datum:

Theorem 5.6. *Let* $\mathcal{M} = (T, S, P, Q, \langle,\rangle_T, \langle,\rangle_S, I, J)$ *be an injective Morita datum and consider the associated Morita semi-contexts $\mathcal{M}_T$ and $\mathcal{M}_S$ (5).*

1) *The following subcategories are mutually equivalent:*

$$\widehat{\mathcal{V}_l}(\mathcal{M}_T) = \widehat{\mathcal{W}_l}(\mathcal{M}_T) = \mathbb{X}_l(\mathcal{M}_T) = \mathcal{X}_l(\mathcal{M}_T)$$
$$\approx \mathcal{X}_l(\mathcal{M}_S) = \mathbb{X}_l(\mathcal{M}_S) = \widehat{\mathcal{W}_l}(\mathcal{M}_S) = \widehat{\mathcal{V}_l}(\mathcal{M}_S).$$

$$(16)$$

2) *If $\mathcal{V}_l(\mathcal{M}_T) \le {}_I\mathcal{L}$ and $\mathcal{W}_l(\mathcal{M}_S) \le {}_J\mathcal{C}$, then $\mathcal{V}_l(\mathcal{M}_T) \approx \mathcal{W}_l(\mathcal{M}_S)$.*
   If $\mathcal{W}_l(\mathcal{M}_T) \le {}_I\mathcal{C}$ and $\mathcal{V}_l(\mathcal{M}_S) \le {}_J\mathcal{L}$, then $\mathcal{W}_l(\mathcal{M}_T) \approx \mathcal{V}_l(\mathcal{M}_S)$.

3) *The following subcategories are mutually equivalent:*

$$\widehat{\mathcal{V}_r}(\mathcal{M}_T) = \widehat{\mathcal{W}_r}(\mathcal{M}_T) = \mathbb{X}_r(\mathcal{M}_T) = \mathcal{X}_r(\mathcal{M}_T)$$
$$\approx \mathcal{X}_r(\mathcal{M}_S) = \mathbb{X}_r(\mathcal{M}_S) = \widehat{\mathcal{W}_r}(\mathcal{M}_S) = \widehat{\mathcal{V}_r}(\mathcal{M}_S).$$

$$(17)$$

4) *If $\mathcal{V}_r(\mathcal{M}_T) \le \mathcal{L}_I$ and $\mathcal{W}_r(\mathcal{M}_T) \le \mathcal{C}_J$, then $\mathcal{V}_r(\mathcal{M}_T) \approx \mathcal{W}_r(\mathcal{M}_S)$.*
   If $\mathcal{W}_r(\mathcal{M}_T) \le \mathcal{C}_J$ and $\mathcal{V}_r(\mathcal{M}_S) \le \mathcal{L}_I$, then $\mathcal{V}_r(\mathcal{M}_S) \approx \mathcal{W}_r(\mathcal{M}_T)$.

Proof. By Lemma 5.1, $\mathcal{X}_l(\mathcal{M}_T) \approx \mathcal{X}_l(\mathcal{M}_S)$ and so 1) follows by Theorem 5.5. If $\mathcal{V}_l(\mathcal{M}_T) \leq {}_I\mathcal{L}$ and $\mathcal{W}_l(\mathcal{M}_S) \leq {}_J\mathcal{C}$, then we have

$$\mathcal{V}_l(\mathcal{M}_T) = \mathcal{V}_l(\mathcal{M}_T) \cap {}_I\mathcal{L} = \widehat{\mathcal{V}_l}(\mathcal{M}_T) \approx \widehat{\mathcal{W}_l}(\mathcal{M}_S) = \mathcal{W}_l(\mathcal{M}_S) \cap {}_J\mathcal{C} = \mathcal{W}_l(\mathcal{M}_S).$$

On the other hand, if $\mathcal{W}_l(\mathcal{M}_T) \leq {}_I\mathcal{L}$ and $\mathcal{V}_l(\mathcal{M}_S) \leq {}_J\mathcal{C}$, then

$$\mathcal{W}_l(\mathcal{M}_T) = \mathcal{W}_l(\mathcal{M}_T) \cap {}_I\mathcal{C} = \widehat{\mathcal{W}_l}(\mathcal{M}_T) \approx \widehat{\mathcal{V}_l}(\mathcal{M}_S) = \mathcal{V}_l(\mathcal{M}_S) \cap {}_J\mathcal{L} = \mathcal{V}_l(\mathcal{M}_S).$$

So we have established 2). The results in 3) and 4) can be obtained analogously. $\square$

6. More applications

In this final section we give more applications of Morita α-(semi-)contexts and injective Morita (semi-)contexts. All rings in this section are *unital*, whence all Morita (semi-)contexts are unital. Moreover, for any ring T we denote with ${}_T\mathbf{E}$ an arbitrary, but fixed, injective cogenerator in ${}_T\mathbb{M}$.

Notation 6.1. *Let T be an A-ring. For any left T-module ${}_TV$, we set $\#V := \mathrm{Hom}_T(V, {}_T\mathbf{E})$. If moreover, ${}_TV_S$ is a (T,S)-bimodule for some B-ring S, then we consider ${}_S^\# V$ with the left S-module structure induced by that of V_S.*

Lemma 6.1. (Compare [Col1990, Lemma 3.2.], [CF2004, Lemmas 2.1.2., 2.1.3.])
Let T be an A-ring, S a B-ring and ${}_TV_S$ a (T,S)-bimodule,

1) *A left T-module ${}_TK$ is V-generated if and only if the canonical T-linear morphism*

$$\omega_{V,K}^l : V \otimes_S \mathrm{Hom}_T(V, K) \to K \tag{18}$$

   is surjective. Moreover, $V \otimes_S W \subseteq \mathrm{Pres}({}_TV) \subseteq \mathrm{Gen}({}_TV)$ for every left S-module ${}_SW$.

2) *A left S-module ${}_SL$ is ${}_S^\# V$-cogenerated if and only if the canonical S-linear morphism*

$$\eta_{V,L}^l : L \to \mathrm{Hom}_T(V, V \otimes_S L) \tag{19}$$

   is injective. Moreover, $\mathrm{Hom}_T(V, M) \subseteq \mathrm{Copres}({}_S^\# V) \subseteq \mathrm{Cogen}({}_S^\# V)$ for every left T-module ${}_TM$.

Remark 6.1. Let T be an A-ring, S a B-ring and ${}_TV_S$ a (T,S)-bimodule. Notice that for any left S-module ${}_SL$ we have

$$\mathrm{ann}_L^\otimes(V_S) := \{l \in L \mid V \otimes_S l = 0\} = \mathrm{Ker}(\eta_{V,L}^l),$$

whence (by Lemma 6.1 2)) V_S is L-faithful if and only if ${}_SL$ is ${}_S^\# V$-cogenerated. It follows then that V_S is completely faithful if and only if ${}_S^\# V$ is a cogenerator.

Localization and colocalization

In what follows we clarify the relations between static (adstatic) modules and subcategories colocalized (localized) by a trace ideal of a Morita context satisfying the α-condition.

Recall that for any (T, S)-bimodule ${}_T P_S$ we have by Lemma 6.1:

$$\operatorname{Stat}^l({}_T P_S) \subseteq \operatorname{Gen}({}_T P) \quad \text{and} \quad \operatorname{Adstat}^l({}_T P_S) \subseteq \operatorname{Cogen}({}^{\#}_S P). \tag{20}$$

Theorem 6.2. *Let* $\mathcal{M} = (T, S, P, Q, \langle , \rangle_T, \langle , \rangle_S, I, J) \in \mathbb{UMC}$. *Then we have*

$$_I \mathcal{C} \subseteq {}_I \mathfrak{D} \subseteq \operatorname{Gen}({}_T P). \tag{21}$$

Assume $\mathbf{P}_r := (Q, P_S) \in \mathcal{P}_r^\alpha(S)$. *Then*

1) $\operatorname{Gen}({}_T P) = \operatorname{Stat}^l({}_T P_S) \subseteq {}_I \mathfrak{F}$.
2) *If* $\operatorname{Gen}({}_T P) \subseteq {}_I \mathcal{C}$, *then* ${}_I \mathcal{C} = {}_I \mathfrak{D} = \operatorname{Gen}({}_T P) = \operatorname{Stat}^l({}_T P_S)$.
3) *If* $\mathbf{Q}_r := (P, Q_T) \in \mathcal{P}_r^\alpha(T)$, *then* ${}_T I \subseteq {}_T T$ *is pure and* ${}_I \mathcal{C} = {}_I \mathfrak{D}$.

Proof. For every left T-module ${}_T K$, consider the following diagram with canonical morphisms and let $\alpha_2 := \zeta_{I,K} \circ \omega_{P,K}^l$. It is easy to see that both rectangles and the two right triangles commutes:

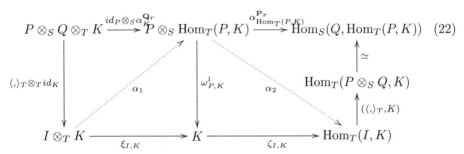

$$P \otimes_S Q \otimes_T K \xrightarrow{id_P \otimes_S \alpha_K^{Q_r}} P \otimes_S \operatorname{Hom}_T(P, K) \xrightarrow{\alpha_{\operatorname{Hom}_T(P,K)}^{P_r}} \operatorname{Hom}_S(Q, \operatorname{Hom}_T(P, K)) \tag{22}$$

It follows directly from the definitions that ${}_I \mathcal{C} \subseteq {}_I \mathfrak{D}$ and $\operatorname{Stat}^l({}_T P_S) \subseteq \operatorname{Gen}({}_T P)$. If ${}_T K$ is I-divisible, then $\xi_{I,K} \circ \langle , \rangle_T \otimes_T id_K = \omega_{P,K}^l \circ id_P \otimes_S \alpha_K^{Q_r}$ is surjective, whence $\omega_{P,K}^l$ is surjective and we conclude that ${}_T K$ is P-generated by Lemma 6.1 1). Consequently, ${}_I \mathfrak{D} \subseteq \operatorname{Gen}({}_T P)$.

Assume now that $\mathbf{P}_r \in \mathcal{P}_r^\alpha(S)$. Considering the canonical map $\rho_Q : T \to \operatorname{End}({}_S Q)^{op}$, the map $\rho_Q \circ \langle , \rangle_T = \alpha_Q^{P_r}$ is injective and so the bilinear map $\langle , \rangle_T$ is injective (i.e., $P \otimes_S Q \overset{\langle , \rangle_T}{\simeq} I$). Define $\alpha_1 := (id_P \otimes_S \alpha_K^{Q_r}) \circ (\langle , \rangle_T \otimes_T id_K)^{-1}$, so that the left triangles commute. Notice that $\alpha_{\operatorname{Hom}_T(P,K)}^{P_r}$ is injective and the commutativity of the upper right triangle in Diagram (22) implies that α_2 is injective (whence $\omega_{P,K}^l$ is injective by the commutativity of the lower right triangle).

1. If $K \in \operatorname{Stat}^l({}_T P_S)$, then the commutativity of the lower right triangle (22) and the injectivity of α_2 show that $\zeta_{I,K}$ is injective; hence, $\operatorname{Stat}^l({}_T P_S) \subseteq {}_I \mathfrak{F}$. On the other hand, if ${}_T K$ is P-generated, then $\omega_{P,K}^l$ is surjective by Lemma 6.1 (1), thence bijective, i.e., $K \in \operatorname{Stat}^l({}_T P_S)$. Consequently, $\operatorname{Gen}({}_T P) = \operatorname{Stat}^l({}_T P_S)$.
2. This follows directly from the inclusions in (21) and 1).
3. Assume $\mathbf{Q}_r := (P, Q_T) \in \mathcal{P}_r^\alpha(T)$. Since $\mathbf{P}_r \in \mathcal{P}_r^\alpha(S)$, it follows by analogy to Proposition 2.1 3) that P_S is flat, hence $id_P \otimes_S \alpha_K^{Q_r}$ is injective. The

commutativity of the upper left triangle in Diagram (22) implies then that α_1 is injective, thence $\xi_{I,K}$ is injective by commutativity of the lower left triangle (i.e., $_TI \subseteq {}_TT$ is K-pure). If $_TK$ is divisible, then $K \otimes_T I \overset{\xi_{I,K}}{\simeq} K$ (i.e., $K \in {}_I\mathcal{C}$). $\qquad\square$

Theorem 6.3. *Let $\mathcal{M} = (T, S, P, Q, \langle,\rangle_T, \langle,\rangle_S, I, J) \in \mathbb{UMC}$. Then we have*

$$_J\mathcal{L} \subseteq {}_J\mathfrak{F} \subseteq \mathrm{Cogen}({}^\#_S P) \quad and \quad \mathrm{Adstat}^l({}_TP_S) \subseteq \mathrm{Cogen}({}^\#_S P).$$

Assume $\mathbf{Q}_r := (P, Q_T) \in \mathcal{P}^\alpha_r(T)$. Then

1) *$J_S \subseteq S_S$ is pure and $_J\mathcal{C} \subseteq \mathrm{Cogen}({}^\#_S P)$.*
2) *If $\mathbf{P}_r := (Q, P_S) \in \mathcal{P}^\alpha_r(S)$, then $_J\mathcal{L} \subseteq \mathrm{Adstat}^l({}_TP_S) \subseteq \mathrm{Cogen}({}^\#_S P) \subseteq {}_J\mathfrak{F}$.*
3) *If $\mathbf{P}_r \in \mathcal{P}^\alpha_r(S)$ and $\mathrm{Cogen}({}^\#_S P) \subseteq {}_J\mathcal{L}$, then $_J\mathcal{L} = \mathrm{Cogen}({}^\#_S P) = \mathrm{Adstat}^l({}_TP_S)$.*

Proof. For every right S-module L consider the commutative diagram with canonical morphisms and let α_3 be so defined, that the left triangles become commutative

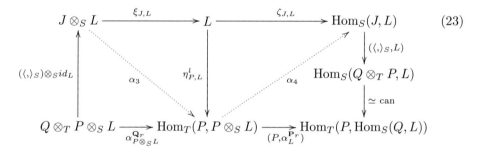

$$(23)$$

By definition $_J\mathcal{L} \subseteq {}_J\mathfrak{F}$ and $\mathrm{Adstat}^l({}_TP_S) \subseteq \mathrm{Cogen}({}^\#_S P)$. If $_SL \in {}_J\mathfrak{F}$, then $\zeta_{J,L}$ is injective and it follows by commutativity of the right rectangle in Diagram (23) that $\eta^l_{P,L}$ is injective, hence $_SL$ is ${}^\#_S P$-cogenerated by Lemma 6.1 2). Consequently, $_J\mathfrak{F} \subseteq \mathrm{Cogen}({}^\#_S P)$.

Assume now that $\mathbf{Q}_r \in \mathcal{P}^\alpha_r(T)$. Then it follows from Lemma 4.1 that $\langle,\rangle_S$ is injective (hence $Q \otimes_T P \overset{\langle,\rangle_S}{\simeq} J$) and so $\alpha_4 := (\mathrm{can} \circ (\langle,\rangle_S, L))^{-1} \circ (P, \alpha^{\mathbf{P}_r}_L)$ is injective.

1. Since α_3 is injective, $\xi_{J,L}$ is also injective for every $_SL$, i.e., $J_S \subseteq S_S$ is pure. If $_SL \in {}_J\mathcal{C}$, then it follows from the commutativity of the left rectangle in Diagram (23) that $\eta^l_{P,L}$ is injective, hence $L \in \mathrm{Cogen}({}^\#_S P)$ by Lemma 6.1 (2).
2. Assume that $\mathbf{P}_r \in \mathcal{P}^\alpha_r(S)$, so that α_4 is injective. If $_SL \in {}_J\mathcal{L}$, then $\zeta_{J,L}$ is an isomorphism, thence $\eta^l_{P,L}$ is surjective (notice that α_4 is injective). Consequently, $_J\mathcal{L} \subseteq \mathrm{Adstat}^l({}_TP_S)$.
3. This follows directly from the assumptions and 2). $\qquad\square$

∗-modules

To the end of this section, we fix a unital ring T, a left T-module $_TP$ and set $S := \mathrm{End}(_TP)^{op}$.

Definition 6.1. ([MO1989]) *We call $_TP$ a ∗-module, iff* $\mathrm{Gen}(_TP) \approx \mathrm{Cogen}(_S^{\#}P)$.

Remark 6.2. It was shown by J. Trlifaj [Trl1994] that all ∗-modules are finitely generated.

By definition, $\mathrm{Stat}^l(_TP_S) \leq {_T}\mathbb{M}$ and $\mathrm{Adstat}^l(_TP_S) \leq {_S}\mathbb{M}$ are the *largest* subcategories between which the adjunction $(P \otimes_S -, \mathrm{Hom}_T(P, -))$ induces an equivalence. On the other hand, Lemma 6.1 shows that $\mathrm{Gen}(_TP) \leq {_T}\mathbb{M}$ and $\mathrm{Cogen}(_S^{\#}P) \leq {_S}\mathbb{M}$ are the *largest* such subcategories (see [Col1990, Section 3] for more details). This suggests the following observation:

Proposition 6.1. ([Xin1999, Lemma 2.3.]) *We have*

$$_TP \text{ is a } \ast\text{-module} \Leftrightarrow \mathrm{Stat}(_TP) = \mathrm{Gen}(_TP) \quad and \quad \mathrm{Adstat}(_TP) = \mathrm{Cogen}(_S^{\#}P).$$

Definition 6.2. *A left T-module $_TU$ is said to be*

$$\textbf{semi-}\sum\textbf{-quasi-projective } (abbr. \text{ s-}\sum\textbf{-quasi-projective}),$$

iff for any left T-module $_TV \in \mathrm{Pres}(_TU)$ and any U-presentation

$$U^{(\Lambda)} \to U^{(\Lambda')} \to V \to 0$$

of $_TV$ (if any), the following induced sequence is exact:

$$\mathrm{Hom}_T(U, U^{(\Lambda)}) \to \mathrm{Hom}_T(U, U^{(\Lambda')}) \to \mathrm{Hom}_T(U, V) \to 0;$$

$$\textbf{weakly-}\sum\textbf{-quasi-projective } (abbr. \text{ w-}\sum\textbf{-quasi-projective}),$$

iff for any left T-module $_TV$ and any short exact sequence

$$0 \to K \to U^{(\Lambda')} \to V \to 0$$

with $K \in \mathrm{Gen}(_TU)$ (if any), the following induced sequence is exact:

$$0 \to \mathrm{Hom}_T(U, K) \to \mathrm{Hom}_T(U, U^{(\Lambda')}) \to \mathrm{Hom}_T(U, V) \to 0;$$

self-tilting, *iff $_TU$ is w-$\sum$-quasi-projective and $\mathrm{Gen}(_TU) = \mathrm{Pres}(_TU)$;*

$\sum$**-self-static**, *iff any direct sum $U^{(\Lambda)}$ is U-static.*

(self)-small, *iff $\mathrm{Hom}_T(U, -)$ commutes with direct sums (of $_TU$);*

Proposition 6.2. *Assume $\mathcal{M} = (T, S, P, Q, \langle, \rangle_T, \langle, \rangle_S)$ is a unital Morita context.*
1) *If $\mathbf{P}_r := (Q, P_S) \in \mathcal{P}_r^{\alpha}(S)$, then:*
   (a) $\mathrm{Gen}(_TP) = \mathrm{Stat}^l(_TP_S)$;
   (b) *there is an equivalence of categories* $\mathrm{Gen}(_TP) \approx \mathrm{Cop}(_S^{\#}P)$;
   (c) $_TP$ *is* $\sum$-*self-static and* $\mathrm{Stat}^l(_TP_S)$ *is closed under factor modules.*
   (d) $\mathrm{Gen}(_TP) = \mathrm{Pres}(_TP)$;
2) *If $\mathcal{M} \in \mathrm{UMC}_r^{\alpha}$ and $\mathrm{Cogen}(_S^{\#}P) \subseteq {_J}\mathcal{L}$, then:*

(a) $\mathrm{Gen}(_T P) = \mathrm{Stat}^l(_T P_S)$ *and* $\mathrm{Cogen}(^\#_S P) = \mathrm{Adstat}^l(_T P_S)$;

(b) *there is an equivalence of categories* $\mathrm{Cogen}(^\#_S P) \approx \mathrm{Gen}(_T P)$;

(c) $_T P$ *is a $*$-module;*

(d) $_T P$ *is self-tilting and self-small.*

Proof. 1) If $\mathbf{P}_r \in \mathcal{P}^\alpha_r(S)$, then it follows by Theorem 6.2 that $\mathrm{Gen}(_T P) = \mathrm{Stat}^l(_T P_S)$, which is equivalent to each of b) and c) by [Wis2000, 4.4.] and to d) by [Wis2000, 4.3.].

2) It follows by the assumptions, Theorems 6.2, 6.3 and 5.2 that $\mathrm{Gen}(_T P) = \mathrm{Stat}^l(_T P_S) \approx \mathrm{Adstat}^l(_T P_S) = \mathrm{Cogen}(^\#_S P)$, whence $\mathrm{Gen}(_T P) \approx \mathrm{Cogen}(^\#_S P)$ (which is the definition of $*$-modules). Hence a) $\Leftrightarrow$ b) $\Leftrightarrow$ c). The equivalence a) $\Leftrightarrow$ d) is evident by [Wis2000, Corollary 4.7.] and we are done. $\qquad\square$

Wide Morita contexts

Wide Morita contexts were introduced by F. Castaño Iglesias and J. Gómez-Torrecillas [C-IG-T1995] and [C-IG-T1996] as an extension of classical *Morita contexts* to Abelian categories.

Definition 6.3. *Let $\mathcal{A}$ and $\mathcal{B}$ be Abelian categories. A* **right** *(***left***)* **wide Morita context** *between $\mathcal{A}$ and $\mathcal{B}$ is a datum $\mathcal{W}_r = (G, \mathcal{A}, \mathcal{B}, F, \eta, \rho)$, where $G : \mathcal{A} \rightleftarrows \mathcal{B} : F$ are right (left) exact covariant functors and $\eta : F \circ G \longrightarrow 1_\mathcal{A}$, $\rho : G \circ F \longrightarrow 1_\mathcal{B}$ $(\eta : 1_\mathcal{A} \longrightarrow F \circ G$, $\rho : 1_\mathcal{B} \longrightarrow G \circ F)$ are natural transformations, such that for every pair of objects $(A, B) \in \mathcal{A} \times \mathcal{B}$ the compatibility conditions $G(\eta_A) = \rho_{G(A)}$ and $F(\rho_B) = \eta_{F(B)}$ hold.*

Definition 6.4. *Let $\mathcal{A}$ and $\mathcal{B}$ be Abelian categories and $\mathcal{W} = (G, \mathcal{A}, \mathcal{B}, F, \eta, \rho)$ be a right (left) wide Morita context. We call $\mathcal{W}$* **injective** *(respectively* **semi-strict**, **strict***), iff η and ρ are monomorphisms (respectively epimorphisms, isomorphisms).*

Remarks 6.3. Let $\mathcal{W} = (G, \mathcal{A}, \mathcal{B}, F, \eta, \rho)$ be a right (left) wide Morita context.

1. It follows by [CDN2005, Propositions 1.1., 1.4.] that if either η or ρ is an epimorphism (monomorphism), then $\mathcal{W}$ is strict, whence $\mathcal{A} \approx \mathcal{B}$.
2. The resemblance of *injective* left wide Morita contexts is with the Morita-Takeuchi contexts for comodules of coalgebras, i.e., the so called *pre-equi-valence data* for categories of comodules introduced in [Tak1977] (see [C-IG-T1998] for more details).

Injective right wide Morita contexts. In a recent work [CDN2005, 5.1.], Chifan, et al. clarified (for module categories) the relation between *classical Morita contexts* and *right wide Morita contexts*. For the convenience of the reader and for later reference, we include in what follows a brief description of this relation.

6.4. Let T, S be rings, $\mathcal{A} := {}_T\mathbb{M}$ and $\mathcal{B} := {}_S\mathbb{M}$. Associated to each Morita context $\mathcal{M} = (T, S, P, Q, \langle, \rangle_T, \langle, \rangle_S)$ is a wide Morita context as follows: Define $G : \mathcal{A} \rightleftarrows \mathcal{B} : F$ by $G(-) = Q \otimes_T -$ and $F(-) = P \otimes_S -$. Then there are natural transformations

$\eta : F \circ G \longrightarrow 1_{T\mathbb{M}}$ and $\rho : G \circ F \longrightarrow 1_{S\mathbb{M}}$ such that for each $_TV$ and W_S :

$$\begin{aligned} \eta_V &: P \otimes_S (Q \otimes_T V) \to V, \quad \sum p_i \otimes_S (q_i \otimes_T v_i) \mapsto \sum \langle p_i, q_i \rangle_T v_i, \\ \rho_W &: Q \otimes_T (P \otimes_S W) \to W, \quad \sum q_i \otimes_T (p_i \otimes_S w_i) \mapsto \sum \langle q_i, p_i \rangle_S w_i. \end{aligned} \tag{24}$$

Then the datum $\mathcal{W}_r(\mathcal{M}) := (G, {}_T\mathbb{M}, {}_S\mathbb{M}, F, \eta, \rho)$ is a right wide Morita context. Conversely, let T', S' be two rings and $\mathcal{W}'_r = (G', {}_{T'}\mathbb{M}, {}_{S'}\mathbb{M}, F', \eta', \rho')$ be a *right wide* Morita context *between* $_{T'}\mathbb{M}$ and $_{S'}\mathbb{M}$ such that *the right exact functors* $G' : {}_{T'}\mathbb{M} \rightleftarrows {}_{S'}\mathbb{M} : F'$ *commute with direct sums*. By Watts' Theorems (e.g., [Gol1979]), there exists a (T, S)-bimodule P' (e.g., $F'(S')$) such that $F' \simeq P' \otimes_{S'} -$, an (S, T)-bimodule Q' such that $G' \simeq Q' \otimes_{T'} -$ and there should exist two bilinear forms

$$\langle,\rangle_{T'} : P' \otimes_{S'} Q' \to T' \quad \text{and} \quad \langle,\rangle_{S'} : Q' \otimes_{T'} P' \to S',$$

such that the natural transformations $\eta' : F' \circ G' \to 1_{T'\mathbb{M}}$, $\rho : G' \circ F' \to 1_{S'\mathbb{M}}$ are given by

$$\eta'_{V'}(p' \otimes_{S'} q' \otimes_{T'} v') = \langle p', q' \rangle_{T'} v' \quad \text{and} \quad \rho'_{W'}(q' \otimes_T p' \otimes_S w') = \langle q', p' \rangle_{S'} w'$$

for all $V' \in {}_{T'}\mathbb{M}$, $W' \in {}_{S'}\mathbb{M}$, $p' \in P'$, $q' \in Q'$, $v' \in V'$ and $w' \in W'$. It can be shown that in this way one obtains a Morita context $\mathcal{M}' = \mathcal{M}'(\mathcal{W}'_r) := (T', S', P', Q', \langle,\rangle_{T'}, \langle,\rangle_{S'})$. Moreover, it turns out that given a wide Morita context $\mathcal{W}_r$, we have $\mathcal{W}_r \simeq \mathcal{W}_r(\mathcal{M}(\mathcal{W}_r))$.

The following result clarifies the relation between *injective Morita contexts* and *injective right wide Morita contexts*.

Theorem 6.5. *Let* $\mathcal{M} = (T, S, P, Q, \langle,\rangle_T, \langle,\rangle_S)$ *be a Morita context,* $\mathcal{A} := {}_T\mathbb{M}$, $\mathcal{B} := {}_S\mathbb{M}$ *and consider the induced right wide Morita context* $\mathcal{W}_r(\mathcal{M}) := (G, \mathcal{A}, \mathcal{B}, F, \eta, \rho)$.

1) *If* $\mathcal{W}_r(\mathcal{M})$ *is an injective right wide Morita context, then* $\mathcal{M}$ *is an injective Morita context.*

2) *If* $\mathcal{M} \in \mathbb{UMC}_r^\alpha$, *then* $\mathcal{W}_r(\mathcal{M})$ *is an injective right wide Morita context.*

Proof. 1) Let $\mathcal{W}_r(\mathcal{M})$ be an injective right wide Morita context. Then in particular, $\langle,\rangle_T = \eta_T$ and $\langle,\rangle_S = \rho_S$ are injective, i.e., $\mathcal{M}$ is an injective Morita context.

2) Assume that $\mathcal{M}$ satisfies the right α-condition. Suppose there exists some $_TV$ and $\sum p_i \otimes_S (q_i \otimes_T v_i) \in \text{Ker}(\eta_V)$. Then for any $q \in Q$ we have

$$\begin{aligned} 0 = q \otimes_T \eta_V\left(\sum (p_i \otimes_S q_i) \otimes_T v_i\right) &= \sum q \otimes_T \langle p_i, q_i \rangle_T v_i \\ &= \sum q \langle p_i, q_i \rangle_T \otimes_T v_i \quad = \sum \langle q, p_i \rangle_S q_i \otimes_T v_i \\ &= \sum \langle q, p_i \rangle_S (q_i \otimes_T v_i) \quad = \alpha_{Q \otimes_T V}^{\mathbf{P}_r}\left(\sum p_i \otimes_S (q_i \otimes_T v_i)\right)(q). \end{aligned}$$

Since $\mathbf{P}_r := (Q, P_S) \in \mathcal{P}_r^\alpha(S)$, the morphism $\alpha_{Q \otimes_T V}^{\mathbf{P}_r}$ is injective and so $\sum p_i \otimes_S (q_i \otimes_T v_i) = 0$, i.e., η_V is injective. Analogously, suppose $\sum q_i \otimes_T (p_i \otimes_S w_i) \in$

$\mathrm{Ker}(\rho_W)$. Then for any $p \in P$ we have

$$
\begin{aligned}
0 = p \otimes_S \rho_W\Big(\sum q_i \otimes_T (p_i \otimes_S w_i)\Big) &= \sum p \otimes_S \langle q_i, p_i \rangle_S w_i \\
= \sum p\langle q_i, p_i \rangle_S \otimes_S w_i \qquad &= \sum \langle p, q_i \rangle_T p_i \otimes_S w_i \\
= \sum \langle p, q_i \rangle_T (p_i \otimes_S w_i) \qquad &= \alpha^{\mathbf{Q}_r}_{P \otimes_S W}\Big(\sum q_i \otimes_T (p_i \otimes_S w_i)\Big)(p).
\end{aligned}
$$

Since $\mathbf{Q}_r := (P, Q_T) \in \mathcal{P}^\alpha_r(T)$, the morphism $\alpha^{\mathbf{Q}_r}_{P \otimes_S W}$ is injective and so $\sum q_i \otimes_T (p_i \otimes_S w_i) = 0$, i.e., ρ_W is injective. Consequently, the induced right wide Morita context $\mathcal{W}_r(\mathcal{M})$ is injective. $\qquad\square$

Acknowledgement

The authors thank the referee for his/her careful reading of the paper and for the fruitful suggestions, comments and corrections, which helped in improving several parts of the paper. Moreover, they acknowledge the excellent research facilities as well as the support of their respective institutions, King Fahd University of Petroleum and Minerals and King AbdulAziz University.

References

[Abr1983] G.D. Abrams, *Morita equivalence for rings with local units*, Comm. Algebra **11** (1983), 801–837.

[Abu2005] J.Y. Abuhlail, *On the linear weak topology and dual pairings over rings*, Topology Appl. **149** (2005), 161–175.

[AF1974] F. Anderson and K. Fuller, *Rings and Categories of Modules*, Springer-Verlag (1974).

[AGH-Z1997] A.V. Arhangeĺskii, K.R. Goodearl and B. Huisgen-Zimmermann, *Kiiti Morita*, (1915–1995), Notices Amer. Math. Soc. **44(6)** (1997), 680–684.

[AG-TL2001] J.Y. Abuhlail, J. Gómez-Torrecillas and F. Lobillo, *Duality and rational modules in Hopf algebras over commutative rings*, J. Algebra **240** (2001), 165–184.

[Alp1990] J.L. Alperin, *Static modules and nonnormal Clifford theory*, J. Austral. Math. Soc. Ser. A **49(3)** (1990), 347–353.

[AM1987] P.N. Ánh and L. Márki, *Morita equivalence for rings without identity*, Tsukuba J. Math **11** (1987), 1–16.

[Ami1971] S.A. Amitsur, *Rings of quotients and Morita contexts*, J. Algebra **17** (1971), 273–298.

[Ber2003] I. Berbee, *The Morita-Takeuchi theory for quotient categories*, Comm. Algebra **31(2)** (2003), 843–858.

[C-IG-T1995] F. Castaño Iglesias and J. Gomez-Torrecillas, *Wide Morita contexts*, Comm. Algebra **23** (1995), 601–622.

[C-IG-T1996] F. Castaño Iglesias and J. Gomez-Torrecillas, *Wide left Morita contexts and equivalences*, Rev. Roum. Math. Pures Appl. **4(1-2)** (1996), 17–26.

[C-IG-T1998] F. Castaño Iglesias and J. Gomez-Torrecillas, *Wide Morita contexts and equivalences of comodule categories*, J. Pure Appl. Algebra **131** (1998), 213–225.

[BW2003] T. Brzeziński and R. Wisbauer, *Corings and Comodules*, Lond. Math. Soc. Lec. Not. Ser. **309,** Cambridge University Press (2003).

[Cae1998] S. Caenepeel, *Brauer Groups, Hopf Algebras and Galois Theory*, Kluwer Academic Publishers (1998).

[C-IG-TW2003] F. Castaño Iglesias, J. Gómez-Torrecillas and R. Wisbauer, *Adjoint functors and equivalence of subcategories*, Bull. Sci. Math. **127** (2003), 279–395.

[CDN2005] N. Chifan, S. Dăscălescu and C. Năstăsescu, *Wide Morita contexts, relative injectivity and equivalence results*, J. Algebra **284** (2005), 705–736.

[Col1990] R. Colpi, *Some remarks on equivalences between categories of modules*, Comm. Algebra **18** (1990), 1935–1951.

[CF2004] R. Colby and K. Fuller, *Equivalence and Duality for Module Categories. With Tilting and Cotilting for Rings*, Cambridge University Press (2004).

[Fai1981] C. Faith, *Algebra I, Rings, Modules and Categories*, Springer-Verlag (1981).

[Gol1979] J. Golan, *An Introduction to Homological Algebra*, Academic Press (1979).

[HS1998] Z. Hao and K.-P. Shum, *The Grothendieck groups of rings of Morita contexts*, Group theory (Beijing, 1996), 88–97, Springer (1998).

[Kat1978] T. Kato, *Duality between colocalization and localization*, J. Algebra **55** (1978), 351–374.

[KO1979] T. Kato and K. Ohtake, *Morita contexts and equivalences*. J. Algebra **61** (1979), 360–366.

[Lam1999] T.Y. Lam, *Lectures on Modules and Rings,* Springer (1999).

[MO1989] C. Mcnini and A. Orsatti, *Representable equivalences between categories of modules and applications*. Rend. Sem. Mat. Univ. Padova **82** (1989), 203–231.

[Mül1974] B.J. Müller, *The quotient category of a Morita context*, J. Algebra **28** (1974), 389–407.

[Nau1990a] S.K. Nauman, *Static modules, Morita contexts, and equivalences*. J. Algebra **135** (1990), 192–202.

[Nau1990b] S.K. Nauman, *Static modules and stable Clifford theory*, J. Algebra **128(2)** (1990), 497–509.

[Nau1993] S.K. Nauman, *Intersecting subcategories of static modules and their equivalences*, J. Algebra **155(1)** (1993), 252–265.

[Nau1994-a] S.K. Nauman, *An alternate criterion of localized modules*, J. Algebra **164** (1994), 256–263.

[Nau1994-b] S.K. Nauman, *Intersecting subcategories of static modules, stable Clifford theory and colocalization-localization*, J. Algebra **170(2)** (1994), 400–421.

[Nau2004] S.K. Nauman, *Morita similar matrix rings and their Grothendieck groups*, Aligarh Bull. Math. **23(1-2)** (2004), 49–60.

[Sat1978] M. Sato, *Fuller's Theorem of equivalences*, J. Algebra **52** (1978), 274–284.

[Tak1977] M. Takeuchi, *Morita theorems for categories of comodules*, J. Fac. Univ. Tokyo **24** (1977), 629–644.

[Trl1994] J. Trlifaj, *Every ∗-module is finitely generated*, J. Algebra **169** (1994), 392–398.

[Ver2006] J. Vercruysse, *Local units versus local dualisations: corings with local structure maps*, Commun. Algebra **34** (2006), 2079–2103.

[Wis1991] R. Wisbauer, *Foundations of Module and Ring Theory, a Handbook for Study and Research*, Gordon and Breach Science Publishers (1991).

[Wis1998] R. Wisbauer, *Tilting in module categories*, in "Abelian groups, module theory and topology" , LNPAM **201** (1998), 421–444.

[Wis2000] R. Wisbauer, *Static modules and equivalences*, in "Interactions between Ring Theory and Representation Theory", Ed. V. Oystaeyen, M. Saorin, Marcel Decker (2000), 423–449.

[Xin1999] Lin Xin, A note on *-modules, Algebra Colloq. **6(2)** (1999), 231–240.

[Z-H1976] B. Zimmermann-Huisgen, *Pure submodules of direct products of free modules*, Math. Ann. **224** (1976), 233–245.

J.Y. Abuhlail
Department of Mathematics & Statistics
King Fahd University of Petroleum & Minerals
Box # 5046
31261 Dhahran (KSA)
e-mail: `abuhlail@kfupm.edu.sa`

S.K. Nauman
Department of Mathematics
King AbdulAziz University
P.O. Box 80203
21589 Jeddah (KSA)
e-mail: `synakhaled@hotmail.com`

Modules and Comodules
Trends in Mathematics, 31–45
© 2008 Birkhäuser Verlag Basel/Switzerland

A Categorical Proof of a Useful Result

A. Ardizzoni and C. Menini

Dedicated to Robert Wisbauer

Abstract. We give a categorical proof of the following equality

$$\bigcap_{i=0}^{n} (V \otimes V_{n-i} + V_i \otimes V) = \sum_{i=1}^{n} V_i \otimes V_{n+1-i}$$

which holds for any chain $\{0\} = V_0 \subseteq V_1 \subseteq V_2 \subseteq \cdots$ of subspaces of a vector space V.

Keywords. Monoidal categories, abelian categories, colimits.

1. Introduction

Let (C, Δ, ε) be a coalgebra over a field K, let $C_0 = \operatorname{Corad}(C)$ and let

$$C_i = \wedge^{i+1} C_0 = \Delta^{-1} (C \otimes C_{i-1} + C_0 \otimes C).$$

A key step in the proof that $(C_i)_{i \in \mathbb{N}}$ is a coalgebra filtration, namely that

$$\Delta (C_n) \subseteq \sum_{i=0}^{n} C_i \otimes C_{n-i}$$

is based (see [Sw, Theorem 9.1.6 page 191] and [Mo, Theorem 5.2.2 page 60]) on the following result (see [Sw, Lemma 9.1.5 page 190]): for any ascending chain of subspaces $\{0\} = V_0 \subseteq V_1 \subseteq V_2 \subseteq \cdots$ of a vector space V, the following equality holds for every $n \in \mathbb{N}$:

$$\bigcap_{i=0}^{n} (V \otimes V_{n-i} + V_i \otimes V) = \sum_{i=1}^{n} V_i \otimes V_{n+1-i}.$$

The aim of this paper is to give a categorical proof of this result in the framework of monoidal categories.

Some applications of this result can be found in [AM].

This paper was written while both the authors were members of G.N.S.A.G.A. with partial financial support from M.i.U.R..

2. Preliminaries and notations

Notations. Let $[(X, i_X)]$ be a subobject of an object E in an abelian category $\mathcal{C}$, where $i_X = i_X^E : X \hookrightarrow E$ is a monomorphism and $[(X, i_X)]$ is the associated equivalence class. By abuse of language, we will say that (X, i_X) is a subobject of E and we will write $(X, i_X) = (Y, i_Y)$ to mean that $(Y, i_Y) \in [(X, i_X)]$. The same convention applies to cokernels. If (X, i_X) is a subobject of E then we will write $(E/X, p_X) = \mathrm{Coker}(i_X)$, where $p_X = p_X^E : E \to E/X$.
Let $(X_1, i_{X_1}^{Y_1})$ be a subobject of Y_1 and let $(X_2, i_{X_2}^{Y_2})$ be a subobject of Y_2. Let $x : X_1 \to X_2$ and $y : Y_1 \to Y_2$ be morphisms such that $y \circ i_{X_1}^{Y_1} = i_{X_2}^{Y_2} \circ x$. Then there exists a unique morphism, which we denote by $y/x = \frac{y}{x} : Y_1/X_1 \to Y_2/X_2$, such that $\frac{y}{x} \circ p_{X_1}^{Y_1} = p_{X_2}^{Y_2} \circ y$:

$\delta_{u,v}$ will denote the Kronecker symbol for every $u, v \in \mathbb{N}$.

Given a family of morphisms $(f_a)_{a \in I}$, where $f_a : A \to A_a$, $\Delta\left[(f_a)_{a \in I}\right]$ will denote the associated diagonal morphism. Analogously $\nabla\left[(g_a)_{a \in I}\right]$ will denote the codiagonal morphism associated to a family $(g_a)_{a \in I}$, where $g_a : A_a \to A$.

2.1. Monoidal categories. Recall that (see [Ka, Chap. XI]) a *monoidal category* is a category $\mathcal{M}$ endowed with an object $\mathbf{1} \in \mathcal{M}$ (called *unit*), a functor $\otimes : \mathcal{M} \times \mathcal{M} \to \mathcal{M}$ (called *tensor product*), and functorial isomorphisms $a_{X,Y,Z} : (X \otimes Y) \otimes Z \to X \otimes (Y \otimes Z)$, $l_X : \mathbf{1} \otimes X \to X$, $r_X : X \otimes \mathbf{1} \to X$, for every X, Y, Z in $\mathcal{M}$. The functorial morphism a is called the *associativity constraint* and satisfies the *Pentagon Axiom,* that is the following relation

$$(U \otimes a_{V,W,X}) \circ a_{U,V \otimes W,X} \circ (a_{U,V,W} \otimes X) = a_{U,V,W \otimes X} \circ a_{U \otimes V,W,X}$$

holds true, for every U, V, W, X in $\mathcal{M}$. The morphisms l and r are called the *unit constraints* and they obey the *Triangle Axiom,* that is $(V \otimes l_W) \circ a_{V,\mathbf{1},W} = r_V \otimes W$, for every V, W in $\mathcal{M}$.

It is well known that the Pentagon Axiom completely solves the consistency problem arising out of the possibility of going from $((U \otimes V) \otimes W) \otimes X$ to $U \otimes (V \otimes (W \otimes X))$ in two different ways (see [Maj, page 420]). This allows the notation $X_1 \otimes \cdots \otimes X_n$ forgetting the brackets for any object obtained from $X_1, \ldots, X_n$ using $\otimes$. Also, as a consequence of the coherence theorem, the constraints take care of themselves and can then be omitted in any computation involving morphisms in $\mathcal{M}$.

Thus, for sake of simplicity, from now on, we will omit the associativity constraints.

2.2. We fix the following assumptions.

- $\mathcal{M}$ is a monoidal category which is abelian with additive tensor functors.
- $((X_i)_{i\in\mathbb{N}}, (\xi_i^j)_{i,j\in\mathbb{N}})$ is a direct system in $\mathcal{M}$ where, for $i \leq j$, $\xi_i^j : X_i \to X_j$.
- $(\xi_i : X_i \to X)_{i\in\mathbb{N}}$ is a compatible family of morphisms with respect to the given direct system.

Fix $n \in \mathbb{N}$ and assume that

- $\xi_i^{i+1} : X_i \to X_{i+1}$ is a split monomorphism for every $0 \leq i \leq n-1$, i.e., there exists a morphism $\lambda_{i+1}^i : X_{i+1} \to X_i$ such that $\lambda_{i+1}^i \circ \xi_i^{i+1} = \mathrm{Id}_{X_i}$.
- $X_0 = 0$,
- $\xi_n : X_n \to X$ is a monomorphism.
- $\xi_n \otimes X, X \otimes \xi_n, \xi_n \otimes \frac{X}{X_n}, \frac{X}{X_n} \otimes \xi_n, \frac{\xi_n}{X_a} \otimes \frac{\xi_n}{X_b}$ are monomorphisms for every $0 \leq a,b \leq n$.

Denote by $\tau_i : X \to \frac{X}{X_i}$ the canonical projection for every $i \in \mathbb{N}$.

For $i = j$ let $\lambda_j^i := \mathrm{Id}_{X_j}$ and for $i < j \leq n$, let $\lambda_j^i : X_j \to X_i$ be the composition

$$\lambda_{i+1}^i \circ \cdots \circ \lambda_j^{j-1} : X_j \to X_i.$$

Clearly,

$$\lambda_b^a \circ \xi_a^b = \mathrm{Id}_{X_a}, \text{ for every } a \leq b \leq n. \tag{1}$$

Set

$$\tau_a^b := \left(X_n \xrightarrow{\lambda_n^b} X_b \xrightarrow{\omega_a^b} \frac{X_b}{X_a} \right), \text{ for every } a \leq b \leq n \tag{2}$$

where ω_a^b denotes the canonical projection. Note that $\tau_a^n = \omega_a^n$. Since the following sequence

$$0 \to X_a \xrightarrow{\xi_a^b} X_b \xrightarrow{\omega_a^b} \frac{X_b}{X_a} \to 0 \tag{3}$$

is exact and in view of (1), there exists a unique morphism $\sigma_b^a : \frac{X_b}{X_a} \to X_b$ such that

$$\sigma_b^a \circ \omega_a^b + \xi_a^b \circ \lambda_b^a = \mathrm{Id}_{X_b}. \tag{4}$$

Moreover

$$\omega_a^b \circ \sigma_b^a = \mathrm{Id}_{\frac{X_b}{X_a}}, \text{ for every } a \leq b \leq n. \tag{5}$$

Clearly also the following sequence

$$0 \to \frac{X_b}{X_a} \xrightarrow{\sigma_b^a} X_b \xrightarrow{\lambda_b^a} X_a \to 0 \tag{6}$$

is exact.

Lemma 2.1. *The following relations hold.*

$$\tau_a^b \circ \xi_t^n = 0, \text{ for every } t \leq a \leq b \leq n. \tag{7}$$

$$\tau_a^b \circ \xi_b^n \circ \sigma_b^a = \mathrm{Id}_{X_b/X_a}, \text{ for every } a \leq b \leq n. \tag{8}$$

$$\lambda_b^a \circ \xi_u^b = \lambda_u^a, \text{ for every } a \leq u \leq b. \tag{9}$$

$$\tau_{a-1}^a \circ \xi_u^n \circ \sigma_u^{u-1} = \delta_{a,u}\mathrm{Id}_{X_a/X_{a-1}} \text{ for every } 0 \leq a, u \leq n. \tag{10}$$

$$\sum_{1 \leq a \leq t} \xi_a^t \circ \sigma_a^{a-1} \circ \omega_{a-1}^a \circ \lambda_t^a = \mathrm{Id}_{X_t}. \tag{11}$$

$$\omega_t^n = \sum_{u+1 \leq b \leq n} \omega_t^n \xi_b^n \sigma_b^{b-1} \tau_{b-1}^b, \text{ for every } 0 \leq u \leq t \leq n. \tag{12}$$

$$\omega_j^n \xi_i^n = 0, \text{ for every } i \leq j. \tag{13}$$

$$\sigma_n^j \omega_j^n \xi_i^n \sigma_i^{i-1} = \xi_i^n \sigma_i^{i-1}, \text{ for every } j+1 \leq i. \tag{14}$$

Proof. (7) follows by the following argument:

$$\tau_a^b \circ \xi_t^n \overset{(2)}{=} \omega_a^b \circ \lambda_n^b \circ \xi_t^n = \omega_a^b \circ \lambda_n^b \circ \xi_b^n \circ \xi_t^b$$

$$\overset{(1)}{=} \omega_a^b \circ \xi_t^b = \omega_a^b \circ \xi_a^b \circ \xi_t^a$$

$$\overset{(3)}{=} 0.$$

(8) follows by the following argument:

$$\tau_a^b \circ \xi_b^n \circ \sigma_b^a \overset{(2)}{=} \omega_a^b \circ \lambda_n^b \circ \xi_b^n \circ \sigma_b^a \overset{(1),(5)}{=} \mathrm{Id}_{X_b/X_a}.$$

(9) follows by the following argument:

$$\lambda_b^a \circ \xi_u^b = \lambda_u^a \circ \lambda_b^u \circ \xi_u^b \overset{(1)}{=} \lambda_u^a.$$

Let us prove that (10) holds. If $a = u$ the formula follows by (8). If $u \leq a - 1$ the formula follows by (7). Assume $a < u$. Then

$$\tau_{a-1}^a \circ \xi_u^n \circ \sigma_u^{u-1} \overset{(2)}{=} \omega_{a-1}^a \circ \lambda_n^a \circ \xi_u^n \circ \sigma_u^{u-1}$$

$$\overset{(9)}{=} \omega_{a-1}^a \circ \lambda_u^a \circ \sigma_u^{u-1}$$

$$= \omega_{a-1}^a \circ \lambda_{u-1}^a \circ \lambda_u^{u-1} \circ \sigma_u^{u-1}$$

$$\overset{(6)}{=} 0.$$

Hence (10) is true.

Let us prove (11) by induction on $t \geq 1$.

$t = 1$: Then

$$\sum_{1 \le a \le t} \xi_a^t \circ \sigma_a^{a-1} \circ w_{a-1}^a \circ \lambda_t^a = \xi_1^1 \circ \sigma_1^0 \circ w_0^1 \circ \lambda_1^1 = \mathrm{Id}_{X_1}.$$

$t \Rightarrow t+1$: We have

$$\sum_{1 \le a \le t+1} \xi_a^{t+1} \circ \sigma_a^{a-1} \circ w_{a-1}^a \circ \lambda_{t+1}^a$$

$$= \xi_{t+1}^{t+1} \circ \sigma_{t+1}^t \circ w_t^{t+1} \circ \lambda_{t+1}^{t+1} + \sum_{1 \le a \le t} \xi_a^{t+1} \circ \sigma_a^{a-1} \circ w_{a-1}^a \circ \lambda_{t+1}^a$$

$$= \sigma_{t+1}^t \circ w_t^{t+1} + \xi_t^{t+1} \circ \left(\sum_{1 \le a \le t} \xi_a^t \circ \sigma_a^{a-1} \circ w_{a-1}^a \circ \lambda_t^a \right) \circ \lambda_{t+1}^t$$

$$\overset{(11)}{=} \sigma_{t+1}^t \circ w_t^{t+1} + \xi_t^{t+1} \circ \lambda_{t+1}^t \overset{(4)}{=} \mathrm{Id}_{X_{t+1}}.$$

Let us prove (12). For every $1 \le u \le t \le n$ we have

$$w_t^n \overset{(11)}{=} w_t^n \sum_{1 \le b \le n} \xi_b^n \sigma_b^{b-1} \tau_{b-1}^b = \sum_{1 \le b \le n} w_t^n \xi_b^n \sigma_b^{b-1} \tau_{b-1}^b$$

$$= \sum_{\substack{1 \le b \le n \\ b \ge u+1}} w_t^n \xi_b^n \sigma_b^{b-1} \tau_{b-1}^b + \sum_{\substack{1 \le b \le n \\ b \le u}} w_t^n \xi_b^n \sigma_b^{b-1} \tau_{b-1}^b$$

$$= \sum_{u+1 \le b \le n} w_t^n \xi_b^n \sigma_b^{b-1} \tau_{b-1}^b + \sum_{1 \le b \le n, b \le u} w_t^n \xi_t^n \xi_b^t \sigma_b^{b-1} \tau_{b-1}^b$$

$$\overset{(3)}{=} \sum_{u+1 \le b \le n} w_t^n \xi_b^n \sigma_b^{b-1} \tau_{b-1}^b.$$

Note that the first equality in the above computation tells us that (12) holds also in the case $u = 0$.

(13) follows by the following argument:

$$w_j^n \xi_i^n = w_j^n \xi_j^n \xi_i^j \overset{(3)}{=} 0.$$

(14) follows by the following argument:

$$_n^j w_j^n \xi_i^n \sigma_i^{i-1} \overset{(4)}{=} \left(\mathrm{Id}_{X_n} - \xi_j^n \lambda_n^j \right) \xi_i^n \sigma_i^{i-1} = \xi_i^n \sigma_i^{i-1} - \xi_j^n \lambda_n^j \xi_i^n \sigma_i^{i-1}$$

$$\overset{(9)}{=} \xi_i^n \sigma_i^{i-1} - \xi_j^n \lambda_i^j \sigma_i^{i-1} = \xi_i^n \sigma_i^{i-1} - \xi_j^n \lambda_{i-1}^j \lambda_i^{i-1} \sigma_i^{i-1}$$

$$\overset{(6)}{=} \xi_i^n \sigma_i^{i-1}. \qquad \square$$

3. The main result

Proposition 3.1. *The following sequence is exact.*

$$\bigoplus_{a+b=n+1} X_a \otimes X_b \xrightarrow{\nabla\left[(\xi_a^n \otimes \xi_b^n)_{a+b=n+1}\right]} X_n \otimes X_n \xrightarrow{\Delta\left[\left(\tau_{a-1}^a \otimes \tau_{b-1}^b\right)_{\substack{1 \le a,b \le n, \\ a+b \ge n+2}}\right]}$$

$$\bigoplus_{\substack{1 \le a,b \le n \\ a+b \ge n+2}} \frac{X_a}{X_{a-1}} \otimes \frac{X_b}{X_{b-1}} \to 0.$$

Proof. Let us prove that $\Delta\left[\left(\tau_{a-1}^a \otimes \tau_{b-1}^b\right)_{\substack{1 \le a,b \le n, \\ a+b \ge n+2}}\right]$ is a split epimorphism:

$$\Delta\left[\left(\tau_{a-1}^a \otimes \tau_{b-1}^b\right)_{\substack{1 \le a,b \le n, \\ a+b \ge n+2}}\right] \circ \nabla\left[\left(\xi_u^n \circ \sigma_u^{u-1} \otimes \xi_v^n \circ \sigma_v^{v-1}\right)_{\substack{1 \le u,v \le n, \\ u+v \ge n+2}}\right]$$

$$= \Delta\left[\nabla\left[\left[\left(\tau_{a-1}^a \circ \xi_u^n \circ \sigma_u^{u-1} \otimes \tau_{b-1}^b \circ \xi_v^n \circ \sigma_v^{v-1}\right)\right]_{\substack{1 \le u,v \le n, \\ u+v \ge n+2}}\right]_{\substack{1 \le a,b \le n, \\ a+b \ge n+2}}\right]$$

$$\overset{(10)}{=} \Delta\left[\nabla\left[\left[\left(\delta_{a,u}\mathrm{Id}_{X_a/X_{a-1}} \otimes \delta_{b,v}\mathrm{Id}_{X_b/X_{b-1}}\right)\right]_{\substack{1 \le u,v \le n, \\ u+v \ge n+2}}\right]_{\substack{1 \le a,b \le n, \\ a+b \ge n+2}}\right]$$

$$= \mathrm{Id}_{\bigoplus_{\substack{1 \le a,b \le n \\ a+b \ge n+2}} \frac{X_a}{X_{a-1}} \otimes \frac{X_b}{X_{b-1}}}.$$

Now we have

$$\Delta\left[\left(\tau_{a-1}^a \otimes \tau_{b-1}^b\right)_{\substack{1 \le a,b \le n, \\ a+b \ge n+2}}\right] \circ \nabla\left[\left(\xi_u^n \otimes \xi_v^n\right)_{u+v=n+1}\right]$$

$$= \Delta\left[\nabla\left[\left(\tau_{a-1}^a \xi_u^n \otimes \tau_{b-1}^b \xi_v^n\right)_{u+v=n+1}\right]_{\substack{1 \le a,b \le n, \\ a+b \ge n+2}}\right].$$

Since $u+v = n+1$ and $a+b \ge n+2$ we deduce that either $u \le a-1 \le a \le n$ or $v \le b-1 \le b \le n$. In view of (7), we have that $\tau_{a-1}^a \xi_u^n \otimes \tau_{b-1}^b \xi_v^n = 0$ so that the composition above is zero.

Let now $f : X_n \otimes X_n \to Z$ be a morphism such that

$$f \circ \nabla \left[(\xi_a^n \otimes \xi_b^n)_{a+b=n+1} \right] = 0.$$

Thus

$$f \circ (\xi_a^n \otimes \xi_b^n) = 0 \quad \text{for every} \quad 0 \leq a, b \leq n+1, a+b = n+1.$$

Moreover, if $0 \leq a, b \leq n+1$ and $a+b \leq n+1$, then

$$f \circ (\xi_a^n \otimes \xi_b^n) = f \circ (\xi_a^n \otimes \xi_{n+1-a}^n) \circ (X_a \otimes \xi_b^{n+1-a}) = 0$$

so that

$$f \circ (\xi_a^n \otimes \xi_b^n) = 0 \quad \text{for every} \quad 0 \leq a, b \leq n+1$$
$$\text{such that} \quad a+b \leq n+1. \tag{15}$$

We have

$$f \circ \nabla \left[\left(\xi_u^n \circ \sigma_u^{u-1} \otimes \xi_v^n \circ \sigma_v^{v-1} \right)_{\substack{1 \leq u, v \leq n, \\ u+v \geq n+2}} \right] \circ \Delta \left[\left(\tau_{a-1}^a \otimes \tau_{b-1}^b \right)_{\substack{1 \leq a, b \leq n, \\ a+b \geq n+2}} \right]$$

$$= \sum_{\substack{1 \leq a, b \leq n, \\ a+b \geq n+2}} f \circ \left(\xi_a^n \circ \sigma_a^{a-1} \tau_{a-1}^a \otimes \xi_b^n \circ \sigma_b^{b-1} \tau_{b-1}^b \right).$$

By applying (11) to the case $t = n$, we get

$$\sum_{1 \leq a \leq n} \xi_a^n \circ \sigma_a^{a-1} \tau_{a-1}^a = \sum_{1 \leq a \leq n} \xi_a^n \circ \sigma_a^{a-1} \circ \omega_{a-1}^a \circ \lambda_n^a \overset{(11)}{=} \mathrm{Id}_{X_n}$$

so that

$$\sum_{1 \leq a, b \leq n} \left(\xi_a^n \circ \sigma_a^{a-1} \tau_{a-1}^a \otimes \xi_b^n \circ \sigma_b^{b-1} \tau_{b-1}^b \right)$$

$$= \left(\sum_{1 \leq a \leq n} \xi_a^n \circ \sigma_a^{a-1} \tau_{a-1}^a \right) \otimes \left(\sum_{1 \leq b \leq n} \xi_b^n \circ \sigma_b^{b-1} \tau_{b-1}^b \right) = \mathrm{Id}_{X_n \otimes X_n}.$$

Therefore

$$f = f \circ \mathrm{Id}_{X_n \otimes X_n} = f \circ \sum_{1 \leq a, b \leq n} \left(\xi_a^n \circ \sigma_a^{a-1} \tau_{a-1}^a \otimes \xi_b^n \circ \sigma_b^{b-1} \tau_{b-1}^b \right)$$

$$= \sum_{\substack{1 \leq a, b \leq n \\ a+b \leq n+1}} f \circ \left(\xi_a^n \circ \sigma_a^{a-1} \tau_{a-1}^a \otimes \xi_b^n \circ \sigma_b^{b-1} \tau_{b-1}^b \right)$$

$$+ \sum_{\substack{1 \leq a, b \leq n \\ a+b \geq n+2}} f \circ \left(\xi_a^n \circ \sigma_a^{a-1} \tau_{a-1}^a \otimes \xi_b^n \circ \sigma_b^{b-1} \tau_{b-1}^b \right)$$

$$\overset{(15)}{=} \sum_{\substack{1 \leq a, b \leq n \\ a+b \geq n+2}} f \circ \left(\xi_a^n \circ \sigma_a^{a-1} \tau_{a-1}^a \otimes \xi_b^n \circ \sigma_b^{b-1} \tau_{b-1}^b \right)$$

$$= f \circ \nabla \left[\left(\xi_u^n \circ \sigma_u^{u-1} \otimes \xi_v^n \circ \sigma_v^{v-1} \right)_{\substack{1 \le u,v \le n, \\ u+v \ge n+2}} \right] \circ \Delta \left[\left(\tau_{a-1}^a \otimes \tau_{b-1}^b \right)_{\substack{1 \le a,b \le n, \\ a+b \ge n+2}} \right].$$

$\square$

Lemma 3.1. *Consider the following commutative diagram*

$$
\begin{array}{ccccc}
\bigoplus\limits_{a+b=n+1} X_a \otimes X_b & \xrightarrow{\nabla\left[(\xi_a^n \otimes \xi_b^n)_{a+b=n+1}\right]} & X_n \otimes X_n & \xrightarrow{\Delta\left[(\tau_a^n \otimes \tau_b^n)_{a+b=n}\right]} & \bigoplus\limits_{a+b=n} \frac{X_n}{X_a} \otimes \frac{X_n}{X_b} \\
\text{Id} \downarrow & & \downarrow \xi_n \otimes \xi_n & & \downarrow \bigoplus\limits_{a+b=n} \frac{\xi_n}{X_a} \otimes \frac{\xi_n}{X_b} \\
\bigoplus\limits_{a+b=n+1} X_a \otimes X_b & \xrightarrow{\nabla\left[(\xi_a \otimes \xi_b)_{a+b=n+1}\right]} & X \otimes X & \xrightarrow{\Delta\left[(\tau_a \otimes \tau_b)_{a+b=n}\right]} & \bigoplus\limits_{a+b=n} \frac{X}{X_a} \otimes \frac{X}{X_b}.
\end{array}
$$

Then the lower sequence is exact whenever the upper one is exact.

Proof. Let $f : Z \to X \otimes X$ be a morphism such that $\Delta\left[(\tau_a \otimes \tau_b)_{a+b=n}\right] \circ f = 0$. Then

$$(\tau_a \otimes \tau_b) \circ f = 0, \qquad \text{for every } a+b=n. \tag{16}$$

Since $\xi_n \otimes X$ is a monomorphism, we get the exact sequence

$$0 \to X_n \otimes X \xrightarrow{\xi_n \otimes X} X \otimes X \xrightarrow{\tau_n \otimes X} \frac{X}{X_n} \otimes X.$$

Since $X_0 = 0$, we have

$$(\tau_n \otimes X) \circ f = (\tau_n \otimes \tau_0) \circ f \overset{(16)}{=} 0.$$

By the universal property of the kernel, there exists a unique morphism $f' : Z \to X_n \otimes X$ such that

$$(\xi_n \otimes X) \circ f' = f. \tag{17}$$

Since $X_n \otimes \xi_n$ is a monomorphism, we have an exact sequence

$$0 \to X_n \otimes X_n \xrightarrow{X_n \otimes \xi_n} X_n \otimes X \xrightarrow{X_n \otimes \tau_n} X_n \otimes \frac{X}{X_n}.$$

Since $X_0 = 0$, we obtain

$$\left(\xi_n \otimes \frac{X}{X_n}\right) \circ (X_n \otimes \tau_n) \circ f' = (X \otimes \tau_n) \circ (\xi_n \otimes X) \circ f'$$

$$\overset{(17)}{=} (X \otimes \tau_n) \circ f = (\tau_0 \otimes \tau_n) \circ f \overset{(16)}{=} 0.$$

Since $\xi_n \otimes \frac{X}{X_n}$ is a monomorphism, we get $(X_n \otimes \tau_n) \circ f' = 0$. Thus by the universal property of the kernel, there exists a unique morphism $\overline{f} : Z \to X_n \otimes X_n$ such that $(X_n \otimes \xi_n) \circ \overline{f} = f'$. Hence

$$(\xi_n \otimes \xi_n) \circ \overline{f} = (\xi_n \otimes X) \circ (X_n \otimes \xi_n) \circ \overline{f} = (\xi_n \otimes X) \circ f' \overset{(17)}{=} f.$$

Therefore

$$\left(\bigoplus_{a+b=n} \frac{\xi_n}{X_a} \otimes \frac{\xi_n}{X_b}\right) \circ \Delta\left[(\tau_a^n \otimes \tau_b^n)_{a+b=n}\right] \circ \overline{f} = \Delta\left[(\tau_a \otimes \tau_b)_{a+b=n}\right] \circ (\xi_n \otimes \xi_n) \circ \overline{f}$$

$$= \Delta\left[(\tau_a \otimes \tau_b)_{a+b=n}\right] \circ f = 0.$$

Since $\frac{\xi_n}{X_a} \otimes \frac{\xi_n}{X_b}$ is a monomorphism, the finite coproduct $\left(\bigoplus_{a+b=n} \frac{\xi_n}{X_a} \otimes \frac{\xi_n}{X_b}\right)$ is a monomorphism too so that we obtain $\Delta\left[(\tau_a^n \otimes \tau_b^n)_{a+b=n}\right] \circ \overline{f} = 0$. Write

$$\nabla\left[(\xi_a^n \otimes \xi_b^n)_{a+b=n+1}\right] = u_n \circ u_n' \qquad \text{and} \qquad \nabla\left[(\xi_a \otimes \xi_b)_{a+b=n+1}\right] = v_n \circ v_n'$$

where

$$u_n : \operatorname{Im}\left\{\nabla\left[(\xi_a^n \otimes \xi_b^n)_{a+b=n+1}\right]\right\} \to X_n \otimes X_n,$$
$$v_n : \operatorname{Im}\left\{\nabla\left[(\xi_a \otimes \xi_b)_{a+b=n+1}\right]\right\} \to X_n \otimes X_n$$

are monomorphisms and

$$u_n' : \bigoplus_{a+b=n+1} X_a \otimes X_b \to \operatorname{Im}\left\{\nabla\left[(\xi_a^n \otimes \xi_b^n)_{a+b=n+1}\right]\right\},$$
$$v_n' : \bigoplus_{a+b=n+1} X_a \otimes X_b \to \operatorname{Im}\left\{\nabla\left[(\xi_a \otimes \xi_b)_{a+b=n+1}\right]\right\}$$

are epimorphisms. From the commutativity of

$$\begin{array}{ccc}
\displaystyle\bigoplus_{a+b=n+1} X_a \otimes X_b & \xrightarrow{\nabla\left[(\xi_a^n \otimes \xi_b^n)_{a+b=n+1}\right]} & X_n \otimes X_n \\
\text{Id} \downarrow & & \downarrow \xi_n \otimes \xi_n \\
\displaystyle\bigoplus_{a+b=n+1} X_a \otimes X_b & \xrightarrow{\nabla\left[(\xi_a \otimes \xi_b)_{a+b=n+1}\right]} & X \otimes X
\end{array}$$

we get that there exists a unique morphism

$$w_n : \operatorname{Im}\left\{\nabla\left[(\xi_a^n \otimes \xi_b^n)_{a+b=n+1}\right]\right\} \to \operatorname{Im}\left\{\nabla\left[(\xi_a \otimes \xi_b)_{a+b=n+1}\right]\right\}$$

that makes the left square of the diagram below commutative

$$\begin{array}{ccccccc}
0 \to \operatorname{Im}\left\{\nabla\left[(\xi_a^n \otimes \xi_b^n)_{a+b=n+1}\right]\right\} & \xrightarrow{u_n} & X_n \otimes X_n & \xrightarrow{\Delta\left[(\tau_a^n \otimes \tau_b^n)_{a+b=n}\right]} & \displaystyle\bigoplus_{a+b=n} \frac{X_n}{X_a} \otimes \frac{X_n}{X_b} \\
w_n \downarrow & & \downarrow \xi_n \otimes \xi_n & & \downarrow \displaystyle\bigoplus_{a+b=n} \frac{\xi_n}{X_a} \otimes \frac{\xi_n}{X_b} \\
0 \to \operatorname{Im}\left\{\nabla\left[(\xi_a \otimes \xi_b)_{a+b=n+1}\right]\right\} & \xrightarrow{v_n} & X \otimes X & \xrightarrow{\Delta\left[(\tau_a \otimes \tau_b)_{a+b=n}\right]} & \displaystyle\bigoplus_{a+b=n} \frac{X}{X_a} \otimes \frac{X}{X_b}.
\end{array}$$

Note that by assumption the upper sequence in the diagram above is exact.

Thus, since $\Delta\left[(\tau_a^n \otimes \tau_b^n)_{a+b=n}\right] \circ \overline{f} = 0$, by the universal property of the kernel, there exists a unique morphism

$$\widehat{f} : Z \to \operatorname{Im}\left\{\nabla\left[(\xi_a^n \otimes \xi_b^n)_{a+b=n+1}\right]\right\}$$

such that $u_n \circ \widehat{f} = \overline{f}$. We have

$$v_n \circ \left(w_n \circ \widehat{f} \right) = (\xi_n \otimes \xi_n) \circ u_n \circ \widehat{f} = (\xi_n \otimes \xi_n) \circ \overline{f} = f.$$

Since v_n is a monomorphism this is enough to conclude that the lower sequence in the diagram above is exact too. $\square$

Theorem 3.1. *Let $\mathcal{M}$ be a monoidal category which is abelian and with additive tensor functors. Let $((X_i)_{i \in \mathbb{N}}, (\xi_i^j)_{i,j \in \mathbb{N}})$ be a direct system in $\mathcal{M}$ where, for $i \leq j$, $\xi_i^j : X_i \to X_j$.*

Let $(\xi_i : X_i \to X)_{i \in \mathbb{N}}$ be a compatible family of morphisms with respect to the given direct system.

Fix $n \in \mathbb{N}$ and assume that

- *$\xi_i^{i+1} : X_i \to X_{i+1}$ is a split monomorphism for every $0 \leq i \leq n-1$,*
- *$X_0 = 0$,*
- *$\xi_n : X_n \to X$ is a monomorphism,*
- *$\xi_n \otimes X, X \otimes \xi_n, \xi_n \otimes \frac{X}{X_n}, \frac{X}{X_n} \otimes \xi_n, \frac{\xi_n}{X_a} \otimes \frac{\xi_n}{X_b}$ are monomorphisms for every $0 \leq a, b \leq n$ (e.g., the tensor product functors are left exact).*

Denote by $\tau_i : X \to \frac{X}{X_i}$ the canonical projection for every $i \in \mathbb{N}$.
Then the following sequence is exact.

$$\bigoplus_{a+b=n+1} X_a \otimes X_b \xrightarrow{\nabla\left[(\xi_a \otimes \xi_b)_{a+b=n+1}\right]} X \otimes X \xrightarrow{\Delta\left[(\tau_a \otimes \tau_b)_{a+b=n}\right]} \bigoplus_{a+b=n} \frac{X}{X_a} \otimes \frac{X}{X_b}.$$

Proof. In view of Lemma 3.1, it remains to prove that the following sequence is exact

$$\bigoplus_{a+b=n+1} X_a \otimes X_b \xrightarrow{\nabla\left[(\xi_a^n \otimes \xi_b^n)_{a+b=n+1}\right]} X_n \otimes X_n \xrightarrow{\Delta\left[(\tau_a^n \otimes \tau_b^n)_{a+b=n}\right]} \bigoplus_{a+b=n} \frac{X_n}{X_a} \otimes \frac{X_n}{X_b}$$

Denote by

$$\gamma_u : \frac{X_n}{X_u} \otimes \frac{X_n}{X_{n-u}} \to \bigoplus_{a+b=n} \frac{X_n}{X_a} \otimes \frac{X_n}{X_b}$$

the canonical inclusion for every $0 \leq u \leq n$.

We compute

$$\nabla\left[\left[\sum_{0 \leq t \leq n} \gamma_t \left(\omega_t^n \xi_u^n \sigma_u^{u-1} \otimes \omega_{n-t}^n \xi_v^n \sigma_v^{v-1}\right)\right]_{\substack{1 \leq u,v \leq n, \\ u+v \geq n+2}}\right] \circ \Delta\left[\left(\tau_{a-1}^a \otimes \tau_{b-1}^b\right)_{\substack{1 \leq a,b \leq n, \\ a+b \geq n+2}}\right]$$

$$= \sum_{\substack{1 \leq a,b \leq n, \\ a+b \geq n+2}} \sum_{0 \leq t \leq n} \gamma_t \left(\omega_t^n \xi_a^n \sigma_a^{a-1} \otimes \omega_{n-t}^n \xi_b^n \sigma_b^{b-1}\right) \circ \left(\tau_{a-1}^a \otimes \tau_{b-1}^b\right)$$

$$= \begin{bmatrix} \sum_{\substack{1\le a,b\le n,\\ a+b\ge n+2}} \sum_{0\le t\le a-1} \gamma_t \left(\omega_t^n \xi_a^n \sigma_a^{a-1} \otimes \omega_{n-t}^n \xi_b^n \sigma_b^{b-1} \right) \circ \left(\tau_{a-1}^a \otimes \tau_{b-1}^b \right) + \\ + \sum_{\substack{1\le a,b\le n,\\ a+b\ge n+2}} \sum_{a\le t\le n} \gamma_t \left(\omega_t^n \xi_a^n \sigma_a^{a-1} \otimes \omega_{n-t}^n \xi_b^n \sigma_b^{b-1} \right) \circ \left(\tau_{a-1}^a \otimes \tau_{b-1}^b \right) \end{bmatrix}$$

$$= \begin{bmatrix} \sum_{\substack{1\le a,b\le n,\\ a+b\ge n+2}} \sum_{0\le t\le a-1} \gamma_t \left(\omega_t^n \xi_a^n \sigma_a^{a-1} \otimes \omega_{n-t}^n \xi_b^n \sigma_b^{b-1} \right) \circ \left(\tau_{a-1}^a \otimes \tau_{b-1}^b \right) + \\ + \sum_{\substack{1\le a,b\le n,\\ a+b\ge n+2}} \sum_{a\le t\le n} \gamma_t \left(\omega_t^n \xi_t^n \xi_a^n \sigma_a^{a-1} \otimes \omega_{n-t}^n \xi_b^n \sigma_b^{b-1} \right) \circ \left(\tau_{a-1}^a \otimes \tau_{b-1}^b \right) \end{bmatrix}$$

$$\overset{(3)}{=} \sum_{\substack{1\le a,b\le n,\\ a+b\ge n+2}} \sum_{0\le t\le a-1} \gamma_t \left(\omega_t^n \xi_a^n \sigma_a^{a-1} \tau_{a-1}^a \otimes \omega_{n-t}^n \xi_b^n \sigma_b^{b-1} \tau_{b-1}^b \right)$$

$$= \sum_{1\le a\le n} \sum_{0\le t\le a-1} \gamma_t \left(\omega_t^n \xi_a^n \sigma_a^{a-1} \tau_{a-1}^a \otimes \omega_{n-t}^n \left(\sum_{\substack{1\le b\le n,\\ b\ge n+2-a}} \xi_b^n \sigma_b^{b-1} \tau_{b-1}^b \right) \right)$$

$$\overset{(12)}{=} \sum_{1\le a\le n} \sum_{0\le t\le a-1} \gamma_t \left(\omega_t^n \xi_a^n \sigma_a^{a-1} \tau_{a-1}^a \otimes \omega_{n-t}^n \right)$$

$$= \sum_{0\le t\le n-1} \sum_{t+1\le a\le n} \gamma_t \left(\omega_t^n \xi_a^n \sigma_a^{a-1} \tau_{a-1}^a \otimes \omega_{n-t}^n \right)$$

$$= \sum_{0\le t\le n-1} \gamma_t \left(\omega_t^n \left(\sum_{t+1\le a\le n} \xi_a^n \sigma_a^{a-1} \tau_{a-1}^a \right) \otimes \omega_{n-t}^n \right) \overset{(12)}{=} \sum_{0\le t\le n-1} \gamma_t \left(\omega_t^n \otimes \omega_{n-t}^n \right)$$

$$= \Delta \left[(\omega_a^n \otimes \omega_b^n)_{a+b=n} \right] = \Delta \left[(\tau_a^n \otimes \tau_b^n)_{a+b=n} \right]$$

In conclusion we have proved that

$$\nabla \left[\left[\sum_{0\le t\le n} \gamma_t \left(\omega_t^n \xi_u^n \sigma_u^{u-1} \otimes \omega_{n-t}^n \xi_v^n \sigma_v^{v-1} \right) \right]_{\substack{1\le u,v\le n,\\ u+v\ge n+2}} \right] \circ \Delta \left[(\tau_{a-1}^a \otimes \tau_{b-1}^b)_{\substack{1\le a,b\le n,\\ a+b\ge n+2}} \right]$$

$$= \Delta \left[(\tau_a^n \otimes \tau_b^n)_{a+b=n} \right]$$

In view of Proposition 3.1, in order to conclude we will prove that

$$\nabla \left[\left[\sum_{0\le t\le n} \gamma_t \left(\omega_t^n \xi_u^n \sigma_u^{u-1} \otimes \omega_{n-t}^n \xi_v^n \sigma_v^{v-1} \right) \right]_{\substack{1\le u,v\le n,\\ u+v\ge n+2}} \right]$$

$$= \nabla \left[\left[\Delta \left[\left(\omega_t^n \xi_u^n \sigma_u^{u-1} \otimes \omega_{n-t}^n \xi_v^n \sigma_v^{v-1} \right)_{0\le t\le n} \right] \right]_{\substack{1\le u,v\le n,\\ u+v\ge n+2}} \right]$$

is a monomorphism.

We have

$$
\left[\Delta\left[\nabla\left[\left(\omega_{u-1}^u\lambda_n^u\xi_{a+1}^n\lambda_n^{a+1}\sigma_n^a\otimes\omega_{v-1}^v\lambda_n^v\sigma_n^b\right)_{a+b=n}\right]\right]_{\substack{1\leq u,v\leq n,\\u+v\geq n+2}}\circ\nabla\left[\Delta\left[\left(\omega_t^n\xi_c^n\sigma_c^{c-1}\otimes\omega_{n-t}^n\xi_d^n\sigma_d^{d-1}\right)_{0\leq t\leq n}\right]\right]_{\substack{1\leq c,d\leq n,\\c+d\geq n+2}}\right]
$$

$$
=\Delta\left[\left[\nabla\left[\begin{array}{c}\left(\omega_{u-1}^u\lambda_n^u\xi_{a+1}^n\lambda_n^{a+1}\sigma_n^a\right.\\\otimes\omega_{v-1}^v\lambda_n^v\sigma_n^b\big)_{a+b=n}\\\circ\Delta\left[\left(\omega_t^n\xi_c^n\sigma_c^{c-1}\otimes\omega_{n-t}^n\xi_d^n\sigma_d^{d-1}\right)_{0\leq t\leq n}\right]\end{array}\right]_{\substack{1\leq c,d\leq n,\\c+d\geq n+2}}\right]_{\substack{1\leq u,v\leq n,\\u+v\geq n+2}}\right]
$$

$$
=\Delta\left[\left[\nabla\left[\sum_{a+b=n}\begin{array}{c}\left(\omega_{u-1}^u\lambda_n^u\xi_{a+1}^n\lambda_n^{a+1}\sigma_n^a\omega_a^n\xi_c^n\sigma_c^{c-1}\right.\\\otimes\omega_{v-1}^v\lambda_n^v\sigma_n^b\omega_b^n\xi_d^n\sigma_d^{d-1}\big)\end{array}\right]_{\substack{1\leq c,d\leq n,\\c+d\geq n+2}}\right]_{\substack{1\leq u,v\leq n,\\u+v\geq n+2}}\right]
$$

Consider $\omega_{u-1}^u\lambda_n^u\xi_{a+1}^n\lambda_n^{a+1}\sigma_n^a\omega_a^n\xi_c^n\sigma_c^{c-1}$ and $\omega_{v-1}^v\lambda_n^v\sigma_n^b\omega_b^n\xi_d^n\sigma_d^{d-1}$.

If $d\leq b$, then $\omega_{v-1}^v\lambda_n^v\sigma_n^b\omega_b^n\xi_d^n\sigma_d^{d-1}\overset{(13)}{=}0$.

If $b+1\leq d$, then $\omega_{v-1}^v\lambda_n^v\sigma_n^b\omega_b^n\xi_d^n\sigma_d^{d-1}\overset{(14)}{=}\omega_{v-1}^v\lambda_n^v\xi_d^n\sigma_d^{d-1}$.

If $c\leq a$, then $\omega_{u-1}^u\lambda_n^u\xi_{a+1}^n\lambda_n^{a+1}\sigma_n^a\omega_a^n\xi_c^n\sigma_c^{c-1}\overset{(13)}{=}0$.

If $a+1\leq c$, then $\omega_{u-1}^u\lambda_n^u\xi_{a+1}^n\lambda_n^{a+1}\sigma_n^a\omega_a^n\xi_c^n\sigma_c^{c-1}\overset{(14)}{=}\omega_{u-1}^u\lambda_n^u\xi_{a+1}^n\lambda_n^{a+1}\xi_c^n\sigma_c^{c-1}$
$\overset{(9)}{=}\omega_{u-1}^u\lambda_n^u\xi_{a+1}^n\lambda_c^{a+1}\sigma_c^{c-1}$.

In order to write the last term we distinguish between two cases.

If $a+1\leq c-1$, then $\omega_{u-1}^u\lambda_n^u\xi_{a+1}^n\lambda_c^{a+1}\sigma_c^{c-1}=\omega_{u-1}^u\lambda_n^u\xi_{a+1}^n\lambda_{c-1}^{a+1}\lambda_c^{c-1}\sigma_c^{c-1}\overset{(6)}{=}0$.

If $a+1=c$, then $\omega_{u-1}^u\lambda_n^u\xi_{a+1}^n\lambda_c^{a+1}\sigma_c^{c-1}=\omega_{u-1}^u\lambda_n^u\xi_c^n\lambda_c^c\sigma_c^{c-1}=\omega_{u-1}^u\lambda_n^u\xi_c^n\sigma_c^{c-1}$
$$=\tau_{u-1}^u\xi_c^n\sigma_c^{c-1}\overset{(10)}{=}\delta_{u,c}\mathrm{Id}_{X_u/X_{u-1}}$$

We have so proved that

$$\omega_{u-1}^u\lambda_n^u\xi_{a+1}^n\lambda_n^{a+1}\sigma_n^a\omega_a^n\xi_c^n\sigma_c^{c-1}=\delta_{u,c}\delta_{a+1,c}\mathrm{Id}_{X_u/X_{u-1}},$$

$$\omega_{v-1}^v\lambda_n^v\sigma_n^b\omega_b^n\xi_d^n\sigma_d^{d-1}=\begin{cases}0&d\leq b\\\omega_{v-1}^v\lambda_n^v\xi_d^n\sigma_d^{d-1}&d\geq b+1\end{cases}$$

and hence

$$
\Delta\left[\left[\nabla\left[\left[\sum_{a+b=n}\begin{pmatrix}\omega_{u-1}^u\lambda_n^u\xi_{u-2}^n\lambda_n^{u-2}\sigma_n^a\omega_a^n\xi_c^n\sigma_c^{c-1}\\ \otimes\,\omega_{v-1}^v\lambda_n^v\sigma_n^b\omega_b^n\xi_d^n\sigma_d^{d-1}\end{pmatrix}\right]_{\substack{1\le c,d\le n,\\ c+d\ge n+2}}\right]\right]_{\substack{1\le u,v\le n,\\ u+v\ge n+2}}\right]
$$

$$
=\Delta\left[\left[\nabla\left[\left[\sum_{a+b=n}\begin{pmatrix}\delta_{u,c}\delta_{a+1,c}\mathrm{Id}_{X_u/X_{u-1}}\\ \otimes\,\omega_{v-1}^v\lambda_n^v\sigma_n^b\omega_b^n\xi_d^n\sigma_d^{d-1}\end{pmatrix}\right]_{\substack{1\le c,d\le n,\\ c+d\ge n+2}}\right]\right]_{\substack{1\le u,v\le n,\\ u+v\ge n+2}}\right]
$$

$$
=\Delta\left[\left[\nabla\left[\left[\begin{pmatrix}\delta_{u,c}\mathrm{Id}_{X_u/X_{u-1}}\\ \otimes\,\omega_{v-1}^v\lambda_n^v\sigma_n^{n-c+1}\omega_{n-c+1}^n\xi_d^n\sigma_d^{d-1}\end{pmatrix}\right]_{\substack{1\le c,d\le n,\\ c+d\ge n+2}}\right]\right]_{\substack{1\le u,v\le n,\\ u+v\ge n+2}}\right]
$$

$$
\overset{d\not\le n-c+1}{=}\Delta\left[\left[\nabla\left[\left[\left(\delta_{u,c}\mathrm{Id}_{X_u/X_{u-1}}\otimes\omega_{v-1}^v\lambda_n^v\xi_d^n\sigma_d^{d-1}\right)\right]_{\substack{1\le c,d\le n,\\ c+d\ge n+2}}\right]\right]_{\substack{1\le u,v\le n,\\ u+v\ge n+2}}\right]
$$

$$
=\Delta\left[\left[\nabla\left[\left[\left(\delta_{u,c}\mathrm{Id}_{X_u/X_{u-1}}\otimes\tau_{v-1}^v\xi_d^n\sigma_d^{d-1}\right)\right]_{\substack{1\le c,d\le n,\\ c+d\ge n+2}}\right]\right]_{\substack{1\le u,v\le n,\\ u+v\ge n+2}}\right]
$$

$$
\overset{(10)}{=}\Delta\left[\left[\nabla\left[\left[\left(\delta_{u,c}\mathrm{Id}_{X_u/X_{u-1}}\otimes\delta_{v,d}\mathrm{Id}_{X_d/X_{d-1}}\right)\right]_{\substack{1\le c,d\le n,\\ c+d\ge n+2}}\right]\right]_{\substack{1\le u,v\le n,\\ u+v\ge n+2}}\right]
$$

$$
=\mathrm{Id}\underset{\substack{1\le c,d\le n,\\ c+d\ge n+2}}{\bigoplus}\frac{X_c}{X_{c-1}}\otimes\frac{X_d}{X_{d-1}}.
$$

$\square$

Theorem 3.2. *Let $\mathcal{M}$ be a monoidal category which is abelian and with additive tensor functors. Let $((X_i)_{i\in\mathbb{N}}, (\xi_i^j)_{i,j\in\mathbb{N}})$ be a direct system in $\mathcal{M}$ where, for $i \leq j$, $\xi_i^j : X_i \to X_j$.*

Let $(\xi_i : X_i \to X)_{i\in\mathbb{N}}$ be a compatible family of morphisms with respect to the given direct system.

Fix $n \in \mathbb{N}$ and assume that

- *$\xi_i^{i+1} : X_i \to X_{i+1}$ is a split monomorphism for every $0 \leq i \leq n-1$,*
- *$X_0 = 0$,*
- *there exists $\lambda_n : X \to X_n$ such that*

$$\lambda_n \circ \xi_j = \xi_j^n, \text{ for every } j \leq n \tag{18}$$

and denote by $\tau_i : X \to \frac{X}{X_i}$ the canonical projection for every $i \in \mathbb{N}$.

Then the following sequence is exact.

$$\bigoplus_{a+b=n+1} X_a \otimes X_b \xrightarrow{\nabla\left[(\xi_a\otimes\xi_b)_{a+b=n+1}\right]} X \otimes X \xrightarrow{\Delta\left[(\tau_a\otimes\tau_b)_{a+b=n}\right]} \bigoplus_{a+b=n} \frac{X}{X_a} \otimes \frac{X}{X_b}.$$

Proof. In view of (18), $\lambda_n \circ \xi_n = \xi_n^n = \mathrm{Id}_{X_n}$ so that $\xi_n, \xi_n \otimes X, X \otimes \xi_n, \xi_n \otimes \frac{X}{X_n}, \frac{X}{X_n} \otimes \xi_n$ are split monomorphisms. We can apply Theorem 3.1 once proved that $\frac{\xi_n}{X_a} \otimes \frac{\xi_n}{X_b}$ is a monomorphism for every $0 \leq a, b \leq n$. Note that in view of (18) we can consider the morphism $\frac{\lambda_n}{X_a}$ and we have

$$\frac{\lambda_n}{X_a} \circ \frac{\xi_n}{X_a} = \frac{\lambda_n \circ \xi_n}{X_a} \overset{(18)}{=} \frac{\mathrm{Id}_{X_n}}{X_a} = \mathrm{Id}_{\frac{X_n}{X_a}}. \qquad \square$$

Corollary 3.1. *Let $\mathcal{M}$ be a monoidal category which is abelian and with additive tensor functors. Let $((X_i)_{i\in\mathbb{N}}, (\xi_i^j)_{i,j\in\mathbb{N}})$ be a direct system in $\mathcal{M}$ where, for $i \leq j$, $\xi_i^j : X_i \to X_j$.*

Let $(X, \xi_i) := \varinjlim X_i$, with $\xi_i : X_i \to X$. Assume that

- *$\xi_i^j : X_i \to X_j$ is a split monomorphism for every $i \leq j$,*
- *$X_0 = 0$,*

and denote by $\tau_i : X \to \frac{X}{X_i}$ the canonical projection for every $i \in \mathbb{N}$.
Then, for every $n \in \mathbb{N}$, the following sequence is exact.

$$\bigoplus_{a+b=n+1} X_a \otimes X_b \xrightarrow{\nabla\left[(\xi_a\otimes\xi_b)_{a+b=n+1}\right]} X \otimes X \xrightarrow{\Delta\left[(\tau_a\otimes\tau_b)_{a+b=n}\right]} \bigoplus_{a+b=n} \frac{X}{X_a} \otimes \frac{X}{X_b}.$$

Proof. Mimicking the proof of [Ar, Lemma 3.12], for every $i \in \mathbb{N}$, one proves there exists $\lambda_i : X \to X_i$ such that $\lambda_i \circ \xi_j = \xi_j^i$, for every $j \leq i$. Now to conclude, apply Theorem 3.2. $\qquad \square$

References

[AM] A. Ardizzoni and C. Menini, *Braided Bialgebras of Type One: Applications*, submitted. (arXiv:0704.2106v1)

[Ar] A. Ardizzoni, *Wedge Products and Cotensor Coalgebras in Monoidal Categories*, Algebr. Represent. Theory, to appear. (arXiv:math.CT/0602016)

[Ka] C. Kassel, *Quantum Groups,* Graduate Text in Mathematics **155**, Springer, 1995.

[Maj] S. Majid, *Foundations of quantum group theory*, Cambridge University Press, 1995.

[Mo] S. Montgomery, *Hopf Algebras and their actions on rings,* CMBS Regional Conference Series in Mathematics **82**, 1993.

[Sw] M. Sweedler, *Hopf Algebras*, Benjamin, New York, 1969.

A. Ardizzoni and C. Menini
University of Ferrara
Department of Mathematics
Via Machiavelli 35
I-44100 Ferrara, Italy
e-mail: rdzlsn@unife.it
e-mail: men@unife.it

Modules and Comodules
Trends in Mathematics, 47–64
© 2008 Birkhäuser Verlag Basel/Switzerland

On Nichols Algebras with Generic Braiding

Nicolás Andruskiewitsch and Iván Ezequiel Angiono

Abstract. We extend the main result of [AS3] to braided vector spaces of generic diagonal type using results of Heckenberger.

Mathematics Subject Classification (2000). 17B37; 16W20, 16W30 .

Keywords. Quantized enveloping algebras, Nichols algebras, automorphisms of non-commutative algebras.

1. Introduction

We fix an algebraically closed field $\Bbbk$ of characteristic 0; all vector spaces, Hopf algebras and tensor products are considered over $\Bbbk$. If H is a Hopf algebra, then $G(H) := \{x \in H - 0 : \Delta(x) = x \otimes x\}$ is a subgroup of the group of units of H; this is a basic invariant of Hopf algebras. Recall that H is pointed if the coradical of H equals $\Bbbk G(H)$, or equivalently if any irreducible H-comodule is one-dimensional.

The purpose of this paper is to show the validity of the following classification theorem.

Theorem 1.1. *Let H be a pointed Hopf algebra with finitely generated abelian group $G(H)$, and generic infinitesimal braiding (see page 50). Then the following are equivalent:*

(a) *H is a domain with finite Gelfand-Kirillov dimension.*
(b) *The group $\Gamma := G(H)$ is free abelian of finite rank, and there exists a generic datum $\mathcal{D}$ for Γ such that $H \simeq U(\mathcal{D})$ as Hopf algebras.*

We refer to the Appendix for the definitions of generic datum and $U(\mathcal{D})$; see [AS3] for a detailed exposition. The general scheme of the proof is exactly the same as for the proof of [AS3, Th. 5.2], an analogous theorem but assuming "positive" instead of "generic" infinitesimal braiding. The main new feature is the following result.

Lemma 1.2. *Let (V, c) be a finite-dimensional braided vector space with generic braiding. Then the following are equivalent:*

(a) *$\mathcal{B}(V)$ has finite Gelfand-Kirillov dimension.*
(b) *(V, c) is twist-equivalent to a braiding of DJ-type with finite Cartan matrix.*

Rosso has proved this assuming "positive" instead of "generic" infinitesimal braiding [R1, Th. 21]. Once we establish Lemma 1.2, the proofs of Lemma 5.1 and Th. 5.2 in [AS3] extend immediately to the generic case. Why can Lemma 1.2 be proved now? Because of the fundamental result of Heckenberger [H1] on Nichols algebras of Cartan type, see Th. 3.10 below. Heckenberger's theorem has as starting point Kharchenko's theory of PBW-bases in a class of pointed Hopf algebras. Besides, Heckenberger introduced the important notion of Weyl groupoid, crucial in the proof of [H1, Th. 1].

Here is the plan of this note. In Section 2 we overview Kharchenko's theory of PBW-bases. It has to be mentioned that related results were announced by Rosso [R2]. See also [U] for a generalization. We sketch a proof of [R1, L. 19] using PBW-bases. Section 3 is devoted to the Weyl groupoid. We discuss its definition and the proof of [H1, Th. 1] in our own terms. Then we prove Lemma 1.2.

2. PBW-basis of Nichols algebras of diagonal type

The goal is to describe an appropriate PBW-basis of the Nichols algebra $\mathcal{B}(V)$ of a braided vector space of diagonal type. The argument is as follows. First, there is a basis of the tensor algebra of a vector space V (with a fixed basis) by Lyndon words, appropriately chosen monomials on the elements of the basis. Any Lyndon word has a canonical decomposition as a product of a pair of smaller words, called the Shirshov decomposition. If V has a braiding c, then for any Lyndon word l there is a polynomial $[l]_c$ called an hyperletter[1], defined by induction on the length as a braided commutator of the hyperletters corresponding to the words in the Shirshov decomposition. The hyperletters form a PBW-basis of $T(V)$ and their classes form a PBW-basis of $\mathcal{B}(V)$.

2.1. Lyndon words

Let X be a finite set with a fixed total order: $x_1 < \cdots < x_\theta$. Let $\mathbb{X}$ be the corresponding vocabulary – the set of words with letters in X – with the lexicographical order. This order is stable by left, but not by right, multiplication: $x_1 < x_1 x_2$ but $x_1 x_3 > x_1 x_2 x_3$. However, if $u < v$ and u does not "begin" by v, then $uw < vt$, for all $w, t \in \mathbb{X}$.

Definition 2.1. *A Lyndon word is $u \in \mathbb{X}$, $u \neq 1$, such that u is smaller than any of its proper ends: if $u = vw$, $v, w \in \mathbb{X} - \{1\}$, then $u < w$. We denote by L the set of Lyndon words.*

Here are some relevant properties of Lyndon words.

(a) If $u \in L$ and $s \geq 2$, then $u^s \notin L$.

(b) Let $u \in \mathbb{X} - X$. Then, u is Lyndon if and only if for any representation $u = u_1 u_2$, with $u_1, u_2 \in \mathbb{X}$ not empty, one has $u_1 u_2 = u < u_2 u_1$.

---

[1]Kharchenko baptised this elements as superletters, but we suggest to call them hyperletters to avoid confusions with the theory of supermathematics.

(c) Any Lyndon word begins by its smallest letter.
(d) If $u_1, u_2 \in L, u_1 < u_2$, then $u_1 u_2 \in L$.
(e) If $u, v \in L, u < v$ then $u^k < v$, for all $k \in \mathbb{N}$.
(f) Let $u, u_1 \in L$ such that $u = u_3 u_2$ and $u_2 < u_1$. Then $uu_1 < u_3 u_1$, $uu_1 < u_2 u_1$.

Theorem 2.2. (*Lyndon*). *Any word* $u \in \mathbb{X}$ *can be written in a unique way as a product of non increasing Lyndon words:* $u = l_1 l_2 \dots l_r$, $l_i \in L$, $l_r \leq \dots \leq l_1$. □

The *Lyndon decomposition* of $u \in \mathbb{X}$ is the unique decomposition given by the Theorem; the $l_i \in L$ appearing in the decomposition are called the *Lyndon letters* of u.

The lexicographical order of $\mathbb{X}$ turns out to be the same as the lexicographical order on the Lyndon letters. Namely, if $v = l_1 \dots l_r$ is the Lyndon decomposition of v, then $u < v$ if and only if:

(i) the Lyndon decomposition of u is $u = l_1 \dots l_i$, for some $1 \leq i < r$, or
(ii) the Lyndon decomposition of u is $u = l_1 \dots l_{i-1} l l'_{i+1} \dots l'_s$, for some $1 \leq i < r$, $s \in \mathbb{N}$ and $l, l'_{i+1}, \dots, l'_s$ in L, with $l < l_i$.

Here is another characterization of Lyndon words. See [Sh, Kh] for more details.

Theorem 2.3. *Let* $u \in \mathbb{X} - X$. *Then,* $u \in L$ *if and only if there exist* $u_1, u_2 \in L$ *with* $u_1 < u_2$ *such that* $u = u_1 u_2$. □

Let $u \in L - X$. A decomposition $u = u_1 u_2$, with $u_1, u_2 \in L$, is not unique. A very useful decomposition is singled out in the following way.

Definition 2.4. [Sh]. *Let* $u \in L - X$. *A decomposition* $u = u_1 u_2$, *with* $u_1, u_2 \in L$ *such that* u_2 *is the smallest end among those proper non-empty ends of* u *is called the* Shirshov decomposition *of* u.

Let $u, v, w \in L$ be such that $u = vw$. Then, $u = vw$ is the Shirshov decomposition of u if and only if either $v \in X$, or else if $v = v_1 v_2$ is the Shirshov decomposition of v, then $w \leq v_2$.

2.2. Braided vector spaces of diagonal type

We briefly recall some notions we shall work with; we refer to [AS2] for more details. A braided vector space is a pair (V, c), where V is a vector space and $c \in \text{Aut}(V \otimes V)$ is a solution of the braid equation: $(c \otimes \text{id})(\text{id} \otimes c)(c \otimes \text{id}) = (\text{id} \otimes c)(c \otimes \text{id})(\text{id} \otimes c)$. We extend the braiding to $c : T(V) \otimes T(V) \to T(V) \otimes T(V)$ in the usual way. If $x, y \in T(V)$, then the braided bracket is

$$[x, y]_c := \text{multiplication} \circ (\text{id} - c)(x \otimes y). \tag{1}$$

Assume that $\dim V < \infty$ and pick a basis $x_1, \dots, x_\theta$ of V, so that we may identify $\mathbb{k}\mathbb{X}$ with $T(V)$. The algebra $T(V)$ has different gradings:

(i) As usual, $T(V) = \oplus_{n \geq 0} T^n(V)$ is $\mathbb{N}_0$-graded. If ℓ denotes the length of a word in $\mathbb{X}$, then $T^n(V) = \oplus_{x \in \mathbb{X}, \ell(x) = n} \mathbb{k} x$.

(ii) Let $\mathbf{e}_1, \ldots, \mathbf{e}_\theta$ be the canonical basis of $\mathbb{Z}^\theta$. Then $T(V)$ is also $\mathbb{Z}^\theta$-graded, where the degree is determined by $\deg x_i = \mathbf{e}_i$, $1 \leq i \leq \theta$.

We say that a braided vector space (V, c) is of *diagonal type* with respect to the basis $x_1, \ldots x_\theta$ if there exist $q_{ij} \in \Bbbk^\times$ such that $c(x_i \otimes x_j) = q_{ij} x_j \otimes x_i$, $1 \leq i, j \leq \theta$. Let $\chi : \mathbb{Z}^\theta \times \mathbb{Z}^\theta \to \Bbbk^\times$ be the bilinear form determined by $\chi(\mathbf{e}_i, \mathbf{e}_j) = q_{ij}$, $1 \leq i, j \leq \theta$. Then

$$c(u \otimes v) = q_{u,v} v \otimes u \qquad (2)$$

for any $u, v \in X$, where $q_{u,v} = \chi(\deg u, \deg v) \in \Bbbk^\times$. Here and elsewhere the degree is with respect to the $\mathbb{Z}^\theta$ grading, see above. In this case, the braided bracket satisfies a "braided" Jacobi identity as well as braided derivation properties, namely

$$[[u, v]_c, w]_c = [u, [v, w]_c]_c - q_{u,v}[u, w]_c v + q_{vw}[u, w]_c v, \qquad (3)$$

$$[u, v\, w]_c = [u, v]_c w + q_{u,v} v [u, w]_c, \qquad (4)$$

$$[u\, v, w]_c = q_{v,w}[u, w]_c v + u [v, w]_c, \qquad (5)$$

for any homogeneous $u, v, w \in T(V)$.

Let (V, c) be a braided vector space. Let $\widetilde{I}(V)$ be the largest homogeneous Hopf ideal of the tensor algebra $T(V)$ that has no intersection with $V \oplus \Bbbk$. The Nichols algebra $\mathcal{B}(V) = T(V)/\widetilde{I}(V)$ is a braided Hopf algebra with very rigid properties; it appears naturally in the structure of pointed Hopf algebras. If (V, c) is of diagonal type, then the ideal $\widetilde{I}(V)$ is $\mathbb{Z}^\theta$-homogeneous hence $\mathcal{B}(V)$ is $\mathbb{Z}^\theta$-graded. See [AS2] for details. The following statement, that we include for later reference, is well known.

Lemma 2.5.

(a) *If q_{ii} is a root of unit of order $N > 1$, then $x_i^N = 0$. In particular, if $\mathcal{B}(V)$ is an integral domain, then $q_{ii} = 1$ or it is not a root of unit, $i = 1, \ldots, \theta$.*

(b) *If $i \neq j$, then $(ad_c x_i)^r (x_j) = 0$ if and only if $(r)!_{q_{ii}} \prod_{0 \leq k \leq r-1} (1 - q_{ii}^k q_{ij} q_{ji}) = 0$.*

(c) *If $i \neq j$ and $q_{ij} q_{ji} = q_{ii}^r$, where $0 \leq -r < \mathrm{ord}(q_{ii})$ (which could be infinite), then $(ad_c x_i)^{1-r}(x_j) = 0$.* □

We shall say that a braiding c is *generic* if it is diagonal with matrix (q_{ij}) where q_{ii} is not a root of 1, for any i.

Finally, we recall that the infinitesimal braiding of a pointed Hopf algebra H is the braided vector space arising as the space of coinvariants of the Hopf bimodule H_1/H_0, where $H_0 \subset H_1$ are the first terms of the coradical filtration of H. We refer to [AS2] for a detailed explanation.

2.3. PBW-basis on the tensor algebra of a braided vector space of diagonal type

We begin by the formal definition of PBW-basis.

Definition 2.6. *Let A be an algebra, $P, S \subset A$ and $h : S \to \mathbb{N} \cup \{\infty\}$. Let also $<$ be a linear order on S. Let us denote by $B(P, S, <, h)$ the set*

$$\{p\, s_1^{e_1} \ldots s_t^{e_t} : t \in \mathbb{N}_0, \quad s_1 > \cdots > s_t, \quad s_i \in S, \quad 0 < e_i < h(s_i), \quad p \in P\}.$$

If $B(P, S, <, h)$ is a basis of A, then we say that $(P, S, <, h)$ is a set of PBW generators with height h, and that $B(P, S, <, h)$ is a PBW-basis of A. If P, $<$, h are clear from the context, then we shall simply say that S is a PBW-basis of A.

Let us start with a finite-dimensional braided vector space (V, c); we fix a basis $x_1, \dots, x_\theta$ of V, so that we may identify $\Bbbk \mathbb{X}$ with $T(V)$. Recall the braided bracket (1). Let us consider the graded map $[-]_c$ of $\Bbbk \mathbb{X}$ given by

$$[u]_c := \begin{cases} u, & \text{if } u = 1 \text{ or } u \in X; \\ [[v]_c, [w]_c]c, & \text{if } u \in L, \ \ell(u) > 1 \text{ and } u = vw \\ & \quad \text{is the Shirshov decomposition;} \\ [u_1]_c \dots [u_t]_c, & \text{if } u \in \mathbb{X} - L \\ & \quad \text{with Lyndon decomposition } u = u_1 \dots u_t; \end{cases}$$

Let us now assume that (V, c) is of diagonal type with respect to the basis $x_1, \dots, x_\theta$, with matrix (q_{ij}).

Definition 2.7. *Given $l \in L$, the polynomial $[l]_c$ is called a hyperletter. Consequently, a hyperword is a word in hyperletters, and a monotone hyperword is a hyperword of the form $W = [u_1]_c^{k_1} \dots [u_m]_c^{k_m}$, where $u_1 > \dots > u_m$.*

Let us collect some facts about hyperletters and hyperwords.

(a) Let $u \in L$. Then $[u]_c$ is a homogeneous polynomial with coefficients in $\mathbb{Z}[q_{ij}]$ and $[u]_c \in u + \Bbbk \mathbb{X}_{>u}^{\ell(u)}$.

(b) Given monotone hyperwords W, V, one has
$$W = [w_1]_c \dots [w_m]_c > V = [v_1]_c \dots [v_t]_c,$$
where $w_1 \geq \dots \geq w_r, v_1 \geq \dots \geq v_s$, if and only if
$$w = w_1 \dots w_m > v = v_i \dots v_t.$$

Furthermore, the principal word of the polynomial W, when decomposed as sum of monomials, is w with coefficient 1.

The following statement is due to Rosso.

Theorem 2.8. *[R2]. Let $m, n \in L$, with $m < n$. Then $[[m]_c, [n]_c]_c$ is a $\mathbb{Z}[q_{ij}]$-linear combination of monotone hyperwords $[l_1]_c \dots [l_r]_c, l_i \in L$, whose hyperletters satisfy $n > l_i \geq mn$, and such that $[mn]_c$ appears in the expansion with non-zero coefficient. Moreover, for any hyperword*
$$\deg(l_1 \dots l_r) = \deg(mn). \quad \square$$

The next technical Lemma is crucial in the proof of Theorem 2.10 below and also in the next subsection. Part (a) appears in [Kh], part (b) in [R2].

Lemma 2.9.

(a) *Any hyperword W is a linear combination of monotone hyperwords bigger than W, $[l_1] \dots [l_r], l_i \in L$, such that $\deg(W) = \deg(l_1 \dots l_r)$, and whose* *hyperletters are between the biggest and the lowest hyperletter of the given word.*

(b) *For any Lyndon word l, let W_l be the vector subspace of $T(V)$ generated by the monotone hyperwords in hyperletters $[l_i]_c$, $l_i \in L$ such that $l_i \geq l$. Then W_l is a subalgebra.* $\square$

From this, it can be deduced that the set of monotone hyperwords is a basis of $T(V)$, or in other words that our first goal is achieved.

Theorem 2.10. [Kh]. *The set of hyperletters is a PBW-basis of $T(V)$.* $\square$

2.4. PBW Basis on quotients of the tensor algebra of a braided vector space of diagonal type

We are next interested in Hopf algebra quotients of $T(V)$. We begin by describing the comultiplication of hyperwords.

Lemma 2.11. [R2]. *Let $u \in \mathbb{X}$, and $u = u_1 \ldots u_r v^m$, $v, u_i \in L, v < u_r \leq \cdots \leq u_1$ the Lyndon decomposition of u. Then,*

$$\Delta([u]_c) = 1 \otimes [u]_c + \sum_{i=0}^{m} \binom{n}{i}_{q_{v,v}} [u_1]_c \ldots [u_r]_c [v]_c^i \otimes [v]_c^{n-i}$$

$$+ \sum_{\substack{l_1 \geq \cdots \geq l_p > v, l_i \in L \\ 0 \leq j \leq m}} x^{(j)}_{l_1,\ldots,l_p} \otimes [l_1]_c \ldots [l_p]_c [v]_c^j;$$

each $x^{(j)}_{l_1,\ldots,l_p}$ is $\mathbb{Z}^\theta$-homogeneous, and $\deg(x^{(j)}_{l_1,\ldots,l_p}) + \deg(l_1 \ldots l_p v^j) = \deg(u)$. $\square$

The following definition appears in [Ha] and is used implicitly in [Kh].

Definition 2.12. *Let $u, v \in \mathbb{X}$. We say that $u \succ v$ if and only if either $\ell(u) < \ell(v)$, or else $\ell(u) = \ell(v)$ and $u > v$ (lexicographical order). This $\succ$ is a total order, called the deg-lex order.*

Note that the empty word 1 is the maximal element for $\succ$. Also, this order is invariant by right and left multiplication.

Let now I be a proper Hopf ideal of $T(V)$, and set $H = T(V)/I$. Let $\pi : T(V) \to H$ be the canonical projection. Let us consider the subset of $\mathbb{X}$:

$$G_I := \{u \in \mathbb{X} : u \notin \mathbb{X}_{\succ u} + I\}.$$

Proposition 2.13. [Kh], *see also* [R2]. *The set $\pi(G_I)$ is a basis of H.* $\square$

Notice that

(a) If $u \in G_I$ and $u = vw$, then $v, w \in G_I$.
(b) Any word $u \in G_I$ factorizes uniquely as a non-increasing product of Lyndon words in G_I.

Towards finding a PBW-basis the quotient H of $T(V)$, we look at the set

$$S_I := G_I \cap L. \tag{6}$$

We then define the function $h_I : S_I \rightarrow \{2, 3, \dots\} \cup \{\infty\}$ by

$$h_I(u) := \min \left\{ t \in \mathbb{N} : u^t \in \Bbbk \mathbb{X}_{\succ u^t} + I \right\}. \tag{7}$$

With these conventions, we are now able to state the main result of this subsection.

Theorem 2.14. [Kh]. *Keep the notation above. Then*

$$B_I' := B\left(\{1 + I\}, [S_I]_c + I, <, h_I\right)$$

is a PBW-basis of $H = T(V)/I$. □

The next consequences of the Theorem 2.14 are used later. See [Kh] for proofs.

Corollary 2.15. *A word u belongs to G_I if and only if the corresponding hyperletter $[u]_c$ is not a linear combination, modulo I, of greater hyperwords of the same degree as u and of hyperwords of lower degree, where all the hyperwords belong to B_I.* □

Proposition 2.16. *In the conditions of the Theorem 2.14, if $v \in S_I$ is such that $h_I(v) < \infty$, then $q_{v,v}$ is a root of unit. In this case, if t is the order of $q_{v,v}$, then $h_I(v) = t$.* □

Corollary 2.17. *If $h_I(v) := h < \infty$, then $[v]^h$ is a linear combination of monotone hyperwords, in greater hyperletters of length $h\ell(v)$, and of monotone hyperwords of lower length.* □

2.5. PBW Basis on the Nichols algebra of a braided vector space of diagonal type

Keep the notation of the preceding subsection. By Theorem 2.14, the Nichols algebra $\mathcal{B}(V)$ has a PBW-basis consisting of homogeneous elements (with respect to the $\mathbb{Z}^\theta$-grading). As in [H1], we can even assume that

⊛ The height of a PBW-generator $[u], \deg(u) = d$, is finite if and only if $2 \leq \operatorname{ord}(q_{u,u}) < \infty$, and in such case, $h_{\widetilde{I}(V)}(u) = \operatorname{ord}(q_{u,u})$.

This is possible because if the height of $[u], \deg(u) = d$, is finite, then $2 \leq \operatorname{ord}(q_{u,u}) = m < \infty$, by Proposition 2.16. And if $2 \leq \operatorname{ord}(q_{u,u}) = m < \infty$, but $h_{\widetilde{I}(V)}(u)$ is infinite, we can add $[u]^m$ to the PBW basis: in this case, $h_{\widetilde{I}(V)}(u) = \operatorname{ord}(q_{u,u})$, and $q_{u^m, u^m} = q_{u,u}^{m^2} = 1$.

Let $\Delta^+(\mathcal{B}(V)) \subseteq \mathbb{N}^n$ be the set of degrees of the generators of the PBW-basis, counted with their multiplicities and let also $\Delta(\mathcal{B}(V)) = \Delta^+(\mathcal{B}(V)) \cup (-\Delta^+(\mathcal{B}(V)))$. We now show that $\Delta^+(\mathcal{B}(V))$ is independent of the choice of the PBW-basis with the property ⊛, a fact repeatedly used in [H1].

Let $R := \Bbbk[x_1^{\pm 1}, \dots, x_\theta^{\pm 1}]$, resp. $\widehat{R} := \Bbbk[[x_1^{\pm 1}, \dots, x_\theta^{\pm 1}]]$, the algebra of Laurent polynomials in θ variables, resp. formal Laurent series in θ variables. If $n = (n_1, \dots, n_\theta) \in \mathbb{Z}^\theta$, then we set $X^n = X_1^{n_1} \cdots X_\theta^{n_\theta}$. If $T \in \operatorname{Aut}(\mathbb{Z}^\theta)$, then we denote by the same letter T the algebra automorphisms $T : R \to R$, $T : \widehat{R} \to \widehat{R}$, $T(X^n) = X^{T(n)}$, for all $n \in \mathbb{Z}^\theta$. We also set

$$\mathfrak{q}_h(t) := \frac{t^h - 1}{t - 1} \in \Bbbk[t], \quad h \in \mathbb{N}; \quad \mathfrak{q}_\infty(t) := \frac{1}{1 - t} = \sum_{s=0}^\infty t^s \in \Bbbk[[t]].$$

We say that a $\mathbb{Z}^\theta$-graded vector space $V = \oplus_{n \in \mathbb{Z}^\theta} V^n$ is *locally finite* if $\dim V^n < \infty$, for all $n \in \mathbb{Z}^\theta$. In this case, the Hilbert or Poincaré series of V is $\mathcal{H}_V = \sum_{n \in \mathbb{Z}^\theta} \dim V^n X^n$. If V, W are $\mathbb{Z}^\theta$ graded, then $V \otimes W = \oplus_{n \in \mathbb{Z}^\theta} \left(\oplus_{p \in \mathbb{Z}^\theta} V^p \otimes W^{n-p} \right)$ is $\mathbb{Z}^\theta$-graded. If V, W are locally finite and additionally $V^n = W^n = 0$, for all $n < M$, for some $M \in \mathbb{Z}^\theta$, then $V \otimes W$ is locally finite, and $\mathcal{H}_{V \otimes W} = \mathcal{H}_V \mathcal{H}_W$.

Lemma 2.18. *Let $\chi : \mathbb{Z}^\theta \times \mathbb{Z}^\theta \to \Bbbk^\times$ be a bilinear form and set $q_\alpha := \chi(\alpha, \alpha)$, $h_\alpha := \operatorname{ord} q_\alpha$, $\alpha \in \mathbb{Z}^\theta$. Let $N, M \in \mathbb{N}$ and $\alpha_1, \ldots, \alpha_N, \beta_1, \ldots, \beta_M \in \mathbb{N}_0^\theta \setminus \{0\}$ such that*

$$\prod_{1 \leq i \leq N} \mathfrak{q}_{h_{\alpha_i}}(X^{\alpha_i}) = \prod_{1 \leq j \leq M} \mathfrak{q}_{h_{\beta_j}}(X^{\beta_j}). \tag{8}$$

Then $N = M$ and exists $\sigma \in \mathbb{S}_N$ such that $\alpha_i = \beta_{\sigma(i)}$.

Proof. If $\gamma \in \mathbb{N}_0^\theta \setminus \{0\}$, then set

$$C_\gamma := \left\{ (s_1, \ldots, s_N) : \sum_{i=1}^N s_i \alpha_i = \gamma, 0 \leq s_i < h_{\alpha_i}, \quad 1 \leq i \leq N \right\},$$

$c_\gamma := \#C_\gamma \in \mathbb{N}_0$. Then the series in (8) equals $1 + \sum_{\mathbb{N}_0^\theta \setminus \{0\}} c_\gamma X^\gamma$. Let $m_1 := \min\{|\gamma| : c_\gamma \neq 0\}$. Then $m_1 = c_{\gamma_1}$, for some $\gamma_1 \in \mathbb{N}_0^\theta \setminus \{0\}$, and $s = (s_1, \ldots, s_N) \in \mathbb{C}_{\gamma_1}$ should belong to the canonical basis. Let $I := \{i : \alpha_i = \gamma_1\} \subseteq \{1, \ldots N\}$, $J := \{j : \beta_j = \gamma_1\} \subseteq \{1, \ldots M\}$. Since $c_{\gamma_1} = \#I = \#J$, there exists a bijection from I to J, and moreover, $\prod_{1 \leq i \leq N, i \notin I} \mathfrak{q}_{h_{\alpha_i}}(X^{\alpha_i}) = \prod_{1 \leq j \leq M, j \notin J} \mathfrak{q}_{h_{\beta_j}}(X^{\beta_j})$. The Lemma then follows by induction on $k = \min\{N, M\}$. □

Hence, if $V = \oplus_{n \in \mathbb{Z}^\theta} V^n$ is locally finite, and $\mathcal{H}_V = \prod_{1 \leq i \leq N} \mathfrak{q}_{h_{\alpha_i}}(X^{\alpha_i})$, then the family $\alpha_1, \ldots, \alpha_N$ is unique up to a permutation.

We now sketch a proof of [R1, Lemma 19] using the PBW-basis; see [A] for a complete exposition.

Lemma 2.19. *Let V be a braided vector space of diagonal type with matrix $q_{ij} \in \Bbbk^\times$. If $\mathcal{B}(V)$ is a domain and its Gelfand-Kirillov dimension is finite, then for any pair $i, j \in \{1, \ldots, \theta\}, i \neq j$, there exists $m_{ij} \geq 0$ such that*

$$(ad_c x_i)^{m_{ij}+1}(x_j) = 0.$$

Proof. Let $i \in \{1, \ldots, \theta\}$. By Lemma 2.5, either q_{ii} is not a root of the unit or else it is 1. Suppose that there are $i \neq j \in \{1, \ldots, \theta\}$ such that $(ad_c x_i)^m(x_j) \neq 0$, for all $m > 0$; say, $i = 1$, $j = 2$. Hence $q_{11}^m q_{12} q_{21} \neq 1$, $m \in \mathbb{N}$. Then one can show by induction that $x_1^m x_2 \in S_I$, using Corollary 2.15. Next, assume that $[x_1^m x_2]_c$ has finite height. Then, using Corollary 2.17, necessarily $[x_1^m x_2]_c^k = 0$ for some k. Let $c_k = \prod_{p=0}^{k-1}(1 - q_{11}^p q_{12} q_{21})$. Using skew-derivations, we obtain that

$$q_{21}^{-k\frac{(m+1)m}{2}} (m)_{q_{22}^{-1}}! c_k^m (km)_{q_{11}^{-1}}! = 0,$$

a contradiction. Thus each $x_1^k x_2$ has infinite height. Thus, $[x_1^k x_2]_c$ is a PBW-generator of infinite height. If $r \leq n$ and $0 \leq n_1 \leq \cdots \leq n_r, n_i \in \mathbb{N}$ such that $\sum_{j=1}^r n_j = n - r$, then $[x_1^{n_1} x_2] \ldots [x_1^{n_r} x_2] \in \mathcal{B}^n(V)$ is an element of B'_I. This collection is linearly independent, hence

$$\dim \mathcal{B}^n(V) \geq \sum_{r=1}^n \binom{(n-r)+r-1}{n-r} = \sum_{r=0}^{n-1} \binom{n-1}{r} = 2^{n-1}.$$

Therefore, $\mathrm{GKdim}(\mathcal{B}(V)) = \infty$. $\qquad\square$

3. The Weyl groupoid

3.1. Groupoids

There are several alternative definitions of a groupoid; let us simply say that a groupoid is a category (whose collection of all arrows is a set) where all the arrows are invertible. Let $\mathcal{G}$ be a groupoid; it induces an equivalence relation $\approx$ on the set of objects (or points) $\mathcal{P}$ by $x \approx y$ iff there exists an arrow $g \in \mathcal{G}$ going from x to y. If $x \in \mathcal{P}$, then $\mathcal{G}(x) = $ all arrows going from x to itself, is a group. A groupoid is essentially determined by

- the equivalence relation $\approx$, and
- the family of groups $\mathcal{G}(x)$, where x runs in a set of representatives of the partition associated to $\approx$.

A relevant example of a groupoid (for the purposes of this paper) is the *transformation groupoid*: if G is a group acting on a set X, then $\mathcal{G} = G \times X$ is a groupoid with operation $(g, x)(h, y) = (gh, y)$ if $x = h(y)$, but undefined otherwise. Thus the set of points in $\mathcal{G}$ is X and an arrow (g, x) goes from x to $g(x)$:

$$x \xrightarrow{(g,x)} g(x)$$

In this example, $\mathcal{G}(x)$ is just the isotropy group of x. Thus, if G acts freely on X (that is, all the isotropy groups are trivially) then $\mathcal{G}$ is just the equivalence relation whose classes are the orbits of the action. This is the case if

$$G = GL(\theta, \mathbb{Z}), \quad X = \text{ set of all ordered bases of } \mathbb{Z}^\theta, \tag{9}$$

with the natural action.

3.2. The ith reflection

For $i, j \in \{1, \ldots, \theta\}$, $i \neq j$, we consider

$$M_{i,j} := \{(ad_c x_i)^m (x_j) : m \in \mathbb{N}\},$$
$$m_{ij} := \min\{m \in \mathbb{N} : (m+1)_{q_{ii}} (1 - q_{ii}^m q_{ij} q_{ji}) = 0\}.$$

By Lemma 2.5, $m_{i,j} < \infty$ if and only if $M_{i,j}$ is finite. In this case,

$$(ad_c x_i)^{m_{ij}} (x_j) \neq 0, \quad (ad_c x_i)^{m_{ij}+1} (x_j) = 0.$$

Let $i \in \{1, \ldots, \theta\}$. Set $m_{ii} = -2$. If any set $M_{i,j}$ is finite, for all $j \in \{1, \ldots, \theta\}$, $i \neq j$, then we define a linear map $s_i : \mathbb{Z}^\theta \to \mathbb{Z}^\theta$, by $s_i(e_j) = e_j + m_{ij} e_i$, $j \in \{1, \ldots, \theta\}$. Note that $s_i^2 = \mathrm{id}$.

We recall that there are operators $y_i^L, y_i^R : \mathcal{B}(V) \to \mathcal{B}(V)$, $i = 1, \ldots, n$ that play the role of left and right invariant derivations. There is next a Hopf algebra $H_i := \Bbbk\left[y_i^R\right] \# \Bbbk\left[e_i, e_i^{-1}\right]$; $\mathcal{B}(V)$ is an H_i-module algebra. We explicitly record the following equality in $\mathcal{A}_i := (\mathcal{B}(V)^{op} \# H_i^{cop})^{op}$:

$$(\rho \# 1) \cdot_{op} \left(1 \# y_i^R\right) = \left(1 \# y_i^R\right) \cdot_{op} (e_i^{-1} \triangleright \rho) \# 1 + y_i^R(\rho) \# 1, \quad \rho \in \mathcal{B}(V). \tag{10}$$

In the setting above, the following Lemma is crucial for the proof of Theorem 3.2. See [A] for a complete proof, slightly different from the argument sketched in [H1].

Lemma 3.1. $\mathcal{B}(V) \cong \ker(y_i^L) \otimes \Bbbk[x_i]$ *as graded vector spaces. Moreover,* $\ker(y_i^L)$ *is generated as algebra by* $\cup_{j \neq i} M_{i,j}$. $\qquad\square$

The next result is the basic ingredient of the Weyl groupoid. We discuss some details of the proof that are implicit in [H1].

Theorem 3.2. [H1, Prop. 1]. *Let* $i \in \{1, \ldots, \theta\}$ *such that* $M_{i,j}$ *is finite, for all* $j \in \{1, \ldots, \theta\}$, $i \neq j$. *Let* V_i *be the vector subspace of* $\mathcal{A}_i$ *generated by*

$$\{(ad_c x_i)^{m_{ij}} (x_j) : j \neq i\} \cup \{y_i^R\}.$$

The subalgebra $\mathcal{B}_i$ *of* $\mathcal{A}_i$ *spanned by* V_i *is isomorphic to* $\mathcal{B}(V_i)$, *and*

$$\Delta^+(\mathcal{B}_i) = \left\{s_i\left(\Delta^+(\mathcal{B}(V))\right) \setminus \{-e_i\}\right\} \cup \{e_i\}.$$

Proof. We just comment the last statement. The algebra H_i is $\mathbb{Z}^\theta$-graded, with $\deg y_i^R = -\mathbf{e}_i, \deg e_i^{\pm 1} = 0$. Hence, the algebra $\mathcal{A}_i$ is $\mathbb{Z}^\theta$-graded, because $\mathcal{B}(V)$ and H_i are graded, and (10) holds.

Hence, consider the abstract basis $\{u_j\}_{j \in \{1, \ldots, \theta\}}$ of V_i, with the grading $\deg u_j = \mathbf{e}_j$, $\mathcal{B}(V_i)$ is $\mathbb{Z}^\theta$-graded. Consider also the algebra homomorphism $\Omega : \mathcal{B}(V_i) \to \mathcal{B}_i$ given by

$$\Omega(u_j) := \begin{cases} (ad_c x_i)^{m_{ij}} (x_j) & j \neq i, \\ y_i^R & j = i. \end{cases}$$

By the first part of the Theorem, proved in [H1], Ω is an isomorphism.

Note:

- $\deg \Omega(u_j) = \deg \left((ad_c x_i)^{m_{ij}} (x_j) \right) = \mathbf{e}_j + m_{ij} \mathbf{e}_i = s_i(\deg \mathbf{u}_j)$, if $j \neq i$;
- $\deg \Omega(u_i) = \deg \left(y_i^R \right) = -\mathbf{e}_i = s_i(\deg \mathbf{u}_i)$.

As Ω is an algebra homomorphism, $\deg(\Omega(\mathbf{u})) = s_i(\deg(\mathbf{u}))$, for all $\mathbf{u} \in \mathcal{B}(V_i)$. As $s_i^2 = \mathrm{id}$, $s_i(\deg(\Omega(\mathbf{u}))) = \deg(\mathbf{u})$, for all $\mathbf{u} \in \mathcal{B}(V_i)$, and $\mathcal{H}_{\mathcal{B}(V_i)} = s_i(\mathcal{H}_{\mathcal{B}_i})$.

Suppose first that $\operatorname{ord} x_i = h_i < \infty$. Then

$$
\mathcal{H}_{\mathcal{B}_i} = \mathcal{H}_{\ker y_i^L} \mathcal{H}_{\Bbbk[y_i^R]} = \mathcal{H}_{\ker y_i^L} \mathfrak{q}_{h_i}(X_i^{-1}) = \frac{\mathcal{H}_{\mathcal{B}(V)}}{\mathfrak{q}_{h_i}(X_i)} \mathfrak{q}_{h_i}(X_i^{-1}),
$$

the first equality because of $\Delta(\mathcal{B}(V)) \subseteq \mathbb{N}_0^\theta$, the second since $\operatorname{ord} x_i = \operatorname{ord} y_i^R$, and the last by Proposition 3.1. As s_i is an algebra homomorphism, we have

$$
\mathcal{H}_{\mathcal{B}(V_i)} = s_i(\mathcal{H}_{\mathcal{B}_i}) = s_i \left(\mathcal{H}_{\mathcal{B}(V)} \right) \frac{\mathfrak{q}_{h_i}(X_i)}{\mathfrak{q}_{h_i}(X_i^{-1})}.
$$

But

$$
s_i \left(\mathcal{H}_{\mathcal{B}(V)} \right) = \prod_{\alpha \in \Delta^+(\mathcal{B}(V))} s_i \left(\mathfrak{q}_{h_\alpha}(X^\alpha) \right)
$$

$$
= \left(\prod_{\alpha \in \Delta^+(\mathcal{B}(V)) \setminus \{\mathbf{e}_i\}} \mathfrak{q}_{h_\alpha}(X^{s_i(\alpha)}) \right) \mathfrak{q}_{h_i}(X_i^{-1});
$$

thus

$$
\mathcal{H}_{\mathcal{B}(V_i)} = \left(\prod_{\alpha \in \Delta^+(\mathcal{B}(V)) \setminus \{\mathbf{e}_i\}} \mathfrak{q}_{h_\alpha}(X^{s_i(\alpha)}) \right) \mathfrak{q}_{h_i}(X_i).
$$

By Lemma 2.18, $\Delta^+(\mathcal{B}_i) = \{s_i \left(\Delta^+(\mathcal{B}(V)) \right) \setminus \{-\mathbf{e}_i\}\} \cup \{\mathbf{e}_i\}$.

Suppose now that $\operatorname{ord} x_i = h_i = \infty$. We have to manipulate somehow the Hilbert series, because $\mathcal{A}_i$ is not locally finite. For this, we introduce an extra variable X_0, corresponding to an extra generator $\widetilde{e}_0$ of $\mathbb{Z}^{\theta+1}$ (whose canonical basis is denoted $\widetilde{e}_0, \widetilde{e}_1, \ldots, \widetilde{e}_\theta$), and consider $\Lambda = \frac{1}{2}\mathbb{Z}^{\theta+1}$. We then define a Λ-grading in $\mathcal{B}(V)$, by

$$
\widetilde{\deg}(x_j) = \begin{cases} \widetilde{e}_j & j \neq i, \\ \frac{1}{2}(\widetilde{e}_i - \widetilde{e}_0), & j = i. \end{cases}
$$

Let $\widetilde{s}_i : \Lambda \to \Lambda$ given by

$$
\widetilde{s}_i(\widetilde{e}_j) = \begin{cases} \widetilde{e}_j + \frac{m_{ij}}{2}(\widetilde{e}_i - \widetilde{e}_0) & j \neq i, 0, \\ \widetilde{e}_0, & j = i, \\ \widetilde{e}_i, & j = 0. \end{cases}
$$

Consider the homomorphism $\Xi : \Lambda \to \mathbb{Z}^\theta$, given by

$$
\Xi(\widetilde{e}_j) = \begin{cases} e_j & j \neq 0, \\ -e_i, & j = 0. \end{cases}
$$

Hence $\widetilde{\deg}(x_j) = \Xi(\deg x_j)$, for each $j \in \{1,\dots,\theta\}$. Note that

- $\Xi(\widetilde{s}_i(\widetilde{e}_j)) = \Xi(\widetilde{e}_j + \frac{m_{ij}}{2}(\widetilde{e}_i - \widetilde{e}_0)) = e_j + m_{ij}e_i = s_i(e_j) = s_i(\Xi(\widetilde{e}_j))$, if $j \neq i$,
- $\Xi(\widetilde{s}_i(\widetilde{e}_i)) = \Xi(\widetilde{e}_0) = -e_i = s_i(e_i) = s_i(\Xi(\widetilde{e}_i))$,
- $\Xi(\widetilde{s}_i(\widetilde{e}_0)) = \Xi(\widetilde{e}_i) = e_i = s_i(-e_i) = s_i(\Xi(\widetilde{e}_0))$;

thus $\Xi(\widetilde{s}_i(\alpha)) = s_i(\Xi(\alpha))$, for all $\alpha \in \Lambda$. With respect to grading, we can repeat the previous argument, defining $\widetilde{\Delta}^+(\mathcal{B}(V)) \subseteq \Lambda$ in analogous way to $\Delta^+(\mathcal{B}(V))$. We get

$$\widetilde{\Delta}^+(\mathcal{B}(V_i)) = \left(\widetilde{s}_i(\widetilde{\Delta}^+(\mathcal{B}(V)) \setminus \{-\widetilde{e}_i\}\right) \cup \{\widetilde{e}_i\};$$

as $\Xi(\widetilde{\Delta}^+(\mathcal{B}(V))) = \Delta^+(\mathcal{B}(V))$, $\Xi(\widetilde{\Delta}^+(\mathcal{B}(V_i))) = \Delta^+(\mathcal{B}(V_i))$, we have

$$\begin{aligned}
\Delta^+(\mathcal{B}(V_i)) &= \Xi\left(\left(\widetilde{s}_i(\widetilde{\Delta}^+(\mathcal{B}(V)) \setminus \{-\widetilde{e}_i\}\right) \cup \{\widetilde{e}_i\}\right) \\
&= s_i\left(\Xi\left(\widetilde{\Delta}^+(\mathcal{B}(V)) \setminus \{\widetilde{e}_i\}\right) \cup \{e_i\}\right) \\
&= \left(s_i\left(\Delta^+(\mathcal{B}(V))\right) \setminus \{-e_i\}\right) \cup \{e_i\}.
\end{aligned}$$

The proof now follows from Lemma 2.18. $\qquad\square$

By Theorem 3.2, the initial braided vector space with matrix $(q_{kj})_{1 \leq k,j \leq \theta}$, is transformed into another braided vector space of diagonal type V_i, with matrix $(\bar{q}_{kj})_{1 \leq k,j \leq \theta}$, where $\bar{q}_{jk} = q_{ii}^{m_{ij}m_{ik}} q_{ik}^{m_{ij}} q_{ji}^{m_{ik}} q_{jk}$, $j,k \in \{1,\dots,\theta\}$.

If $j \neq i$, then $\overline{m_{ij}} = \min\left\{m \in \mathbb{N} : (m+1)_{\bar{q}_{ii}}\left(\bar{q}_{ii}^m \bar{q}_{ij} \bar{q}_{ji} = 0\right)\right\} = m_{ij}$. Thus, the previous transformation is invertible.

3.3. Definition of the Weyl groupoid

Let $E = (\mathbf{e}_1,\dots,\mathbf{e}_\theta)$ be the canonical basis of $\mathbb{Z}^\theta$. Let $(q_{ij})_{1 \leq i,j \leq \theta} \in (\mathbb{C}^\times)^{\theta \times \theta}$. We fix once and for all the bilinear form $\chi : \mathbb{Z}^\theta \times \mathbb{Z}^\theta \to \mathbb{C}^\times$ given by

$$\chi(\mathbf{e}_i, \mathbf{e}_j) = q_{ij}, \qquad 1 \leq i,j \leq \theta.$$

Let $F = (\mathbf{f}_1,\dots,\mathbf{f}_\theta)$ be an arbitrary ordered basis of $\mathbb{Z}^\theta$ and let $\widetilde{q}_{ij} = \chi(\mathbf{f}_i, \mathbf{f}_j)$, $1 \leq i,j \leq \theta$, the *braiding matrix with respect to the basis* F. Fix $i \in \{1,\dots,\theta\}$. If $1 \leq i \neq j \leq \theta$, then we consider the set

$$\{m \in \mathbb{N}_0 : (m+1)_{\widetilde{q}_{ii}}(\widetilde{q}_{ii}^m \widetilde{q}_{ij} \widetilde{q}_{ji} - 1) = 0\}.$$

This set might well be empty, for instance if $\widetilde{q}_{ii} = 1 \neq \widetilde{q}_{ij}\widetilde{q}_{ji}$ for all $j \neq i$. If this set is nonempty, then its minimal element is denoted m_{ij} (which of course depends on the basis F). Set also $m_{ii} = -2$. Let $s_{i,F} \in GL(\mathbb{Z}^\theta)$ be the pseudo-reflection given by $s_{i,F}(\mathbf{f}_j) := \mathbf{f}_j + m_{ij}\mathbf{f}_i$, $j \in \{1,\dots,\theta\}$.

Let us compute the braiding matrix with respect to the matrix $s_{i,F}(F)$. Let $\mathbf{u}_j := s_{i,F}(\mathbf{f}_j)$ and $\widehat{q}_{rj} = \chi(\mathbf{u}_r, \mathbf{u}_j)$. If we also set $m_{ii} := -2$, $1 \leq i \leq \theta$ by convenience, then

$$\widehat{q}_{rs} = \widetilde{q}_{ii}^{m_{ir}m_{is}} \widetilde{q}_{ri}^{m_{is}} \widetilde{q}_{is}^{m_{ir}} \widetilde{q}_{rs}, \qquad 1 \leq r,s \leq \theta. \tag{11}$$

In particular $\widehat{q}_{ii} = \widetilde{q}_{ii}$ and $\widehat{q}_{jj} = \left(\widetilde{q}_{ii}^{m_{ij}}\widetilde{q}_{ji}\widetilde{q}_{ij}\right)^{m_{ij}}\widetilde{q}_{jj}$, $1 \leq j \leq \theta$. Thus, even if m_{ij} are defined for the basis F and for all $i \neq j$, they need not be defined for the basis $s_{i,F}(F)$. For example, if

$$\begin{pmatrix} \widetilde{q}_{ii} & \widetilde{q}_{ij} \\ \widetilde{q}_{ji} & \widetilde{q}_{jj} \end{pmatrix} = \begin{pmatrix} -1 & -\xi \\ \xi & \xi^{-2} \end{pmatrix},$$

where ξ is a root of 1 of order > 4, then $\begin{pmatrix} \widehat{q}_{ii} & \widehat{q}_{ij} \\ \widehat{q}_{ji} & \widehat{q}_{jj} \end{pmatrix} = \begin{pmatrix} -1 & \xi^{-1} \\ -\xi^{-1} & 1 \end{pmatrix}$ and m_{ji} is not defined with respect to the new basis $s_{i,F}(F)$. However, for an arbitrary F and i such that m_{ij} for F is defined, then m_{ij} is defined for the new basis $s_{i,F}(F)$ and coincides with the old one, so that

$$s_{i,s_{i,F}(F)} = s_{i,F}. \tag{12}$$

Notice that formula (11) and a variation thereof appear in [H2].

Definition 3.3. *The Weyl groupoid $W(\chi)$ of the bilinear form χ is the smallest subgroupoid of the transformation groupoid (9) with respect to the following properties:*

- *$(\mathrm{id}, E) \in W(\chi)$,*
- *if $(\mathrm{id}, F) \in W(\chi)$ and $s_{i,F}$ is defined, then $(s_{i,F}, F) \in W(\chi)$.*

In other words, $W(\chi)$ is just a set of bases of $\mathbb{Z}^\theta$: the canonical basis E, then all bases $s_{i,E}(E)$ provided that $s_{i,E}$ is defined, then all bases $s_{j,s_i(E)}s_{i,E}(E)$ provided that $s_{i,E}$ and $s_{j,s_i(E)}$ are defined, and so on.

Here is an alternative description of the Weyl groupoid. Consider the set of all pairs $(F, (\widetilde{q}_{ij})_{1 \leq i,j \leq \theta})$ where F is an ordered basis of $\mathbb{Z}^\theta$ and the $\widetilde{q}_{ij}$'s are non-zero scalars. Let us say that

$$(F, (\widetilde{q}_{ij})_{1 \leq i,j \leq \theta}) \sim (U, (\widehat{q}_{ij})_{1 \leq i,j \leq \theta})$$

if there exists and index i such that m_{ij} exists for all $1 \leq i \neq j \leq \theta$, $U = s_{i,F}(F)$ and $\widehat{q}_{ij}$ is obtained from $\widetilde{q}_{ij}$ by (11). By (12) this is reflexive; consider the equivalence relation $\approx$ generated by $\sim$. Then $W(\chi)$ is the equivalence class of $(E, (q_{ij})_{1 \leq i,j \leq \theta})$ with respect to $\approx$. Actually, if

$$(F, (\widetilde{q}_{ij})_{1 \leq i,j \leq \theta}) \quad \text{and} \quad (U, (\widehat{q}_{ij})_{1 \leq i,j \leq \theta})$$

belong to the equivalence class, there will be a unique $s \in GL(\mathbb{Z}^\theta)$, which is a product of suitable s_i's, such that $(s, F) \in W(\chi)$ and $s(F) = U$.

The equivalence class of $(F, (\widetilde{q}_{ij})_{1 \leq i,j \leq \theta})$ is denoted $W(F, (\widetilde{q}_{ij})_{1 \leq i,j \leq \theta})$. Furthermore, if χ is a fixed bilinear form as above, $F = (f_i)_{1 \leq i \leq \theta}$ and $\widetilde{q}_{ij} = \chi(f_i, f_j)$, $1 \leq i, j \leq \theta$, then we denote $W(F, \chi) := W(F, (\widetilde{q}_{ij})_{1 \leq i,j \leq \theta})$; and $W(\chi) := W(E, \chi)$ where E is the canonical basis.

From this viewpoint, it is natural to introduce the following concept.

Definition 3.4. *We say that* $(\widetilde{q}_{ij})_{1\leq i,j\leq \theta}$ *and* $(\widehat{q}_{ij})_{1\leq i,j\leq \theta} \in \Bbbk^{\times}$ *are* Weyl equivalent *if there exist ordered bases* F *and* U *such that*

$$(F, (\widetilde{q}_{ij})_{1\leq i,j\leq \theta}) \approx (U, (\widehat{q}_{ij})_{1\leq i,j\leq \theta}).$$

Now recall that $(\widetilde{q}_{ij})_{1\leq i,j\leq \theta}$ *and* $(\widehat{q}_{ij})_{1\leq i,j\leq \theta} \in \Bbbk^{\times}$ *are* twist equivalent *if* $\widetilde{q}_{ii} = \widehat{q}_{ii}$ *and* $\widetilde{q}_{ij}\widetilde{q}_{ji} = \widehat{q}_{ij}\widehat{q}_{ji}$ *for all* $1 \leq i, j \leq \theta$.

    It turns out that it is natural to consider the equivalence relation $\approx_{WH}$ *generated by twist- and Weyl-equivalence, see* [H2, Def. 3], *see also* [H3, Def. 2]. *We propose to call* $\approx_{WH}$ *the* Weyl-Heckenberger *equivalence; note that this is the "Weyl equivalence" in* [H2]. *We suggest this new terminology because the Weyl groupoid is really an equivalence relation.*

    The Weyl groupoid is meant to generalize the set of basis of a root system. For convenience we set $\mathfrak{P}(\chi) = \{F : (\mathrm{id}, F) \in W(\chi)\}$, the set of points of the groupoid $W(\chi)$. Then the *generalized root system*[2] associated to χ is

$$\Delta(\chi) = \bigcup_{F\in\mathfrak{P}(\chi)} F. \tag{13}$$

We record for later use the following evident fact.

Remark 3.5. *The following are equivalent:*

1. *The groupoid* $W(\chi)$ *is finite.*
2. *The set* $\mathfrak{P}(\chi)$ *is finite.*
3. *The generalized root system* $\Delta(\chi)$ *is finite.*   □

    Let also $\mathrm{\mu} : W(\chi) \to GL(\theta, \mathbb{Z})$, $\mathrm{\mu}(s, F) = s$ if $(s, F) \in W(\chi)$. We denote by $W_0(\chi)$ the subgroup generated by the image of $\mathrm{\mu}$. Compare with [Se].

    Let us say that χ is *standard* if for any $F \in \mathfrak{P}(\chi)$, the integers m_{rj} are defined, for all $1 \leq r, j \leq \theta$, and the integers m_{rj} for the bases $s_{i,F}(F)$ coincide with those for F for all i, r, j (clearly it is enough to assume this for the canonical basis E).

Proposition 3.6. *Assume that* χ *is standard. Then*

$$W_0(\chi) = \langle s_{i,E} : 1 \leq i \leq \theta \rangle.$$

Furthermore $W_0(\chi)$ *acts freely and transitively on* $\mathfrak{P}(\chi)$.   □

    The first claim says that $W_0(\chi)$ is a Coxeter group. The second implies that $W_0(\chi)$ and $\mathfrak{P}(\chi)$ have the same cardinal.

---

[2]Actually this is a little misleading, since in the case of braidings of symmetrizable Cartan type, one would get just the real roots.

Proof. Let $F \in \mathfrak{P}(\chi)$. Since χ is standard, for any $1 \leq i, j \leq \theta$

$$s_{j,s_{i,F}(F)} = s_{i,F} s_{j,F} s_{i,F}. \tag{14}$$

Hence $W_0(\chi) \subseteq \langle s_{i,E} : 1 \leq i \leq \theta \rangle$; the other inclusion being clear, the first claim is established. Now, by the very definition of the Weyl groupoid, there exists a unique $w \in W_0(\chi)$ such that $w(E) = F$. Thus, to prove the second claim we only need to check that the action is well defined; and for this, it is enough to prove: if $w \in W_0(\chi)$, then $w(E) \in \mathfrak{P}(\chi)$. We proceed by induction on the length of w, the case of length one being obvious. Let $w' = w s_{i,E}$, with length of $w' =$ length of $w + 1$. Then $F = w(E) \in \mathfrak{P}(\chi)$. The matrix of $s_{i,F}$ in the basis E is $\| s_{i,F} \|_E = \| \operatorname{id} \|_{F,E} \| s_{i,F} \|_F \| \operatorname{id} \|_{E,F}$ and since χ is standard, we conclude that $s_{i,F} = w s_{i,E} w^{-1}$. [3] That is, $w' = s_{i,F} w$ and $w'(E) = s_{i,F}(F) \in \mathfrak{P}(\chi)$. $\qquad \square$

Remark 3.7. *Assume that χ is standard. Then the following are equivalent:*

1. *The groupoid $W(\chi)$ is finite.*
2. *The set $\mathfrak{P}(\chi)$ is finite.*
3. *The generalized root system $\Delta(\chi)$ is finite.*
4. *The group $W_0(\chi)$ is finite.*

If this holds, then $W_0(\chi)$ is a finite Coxeter group; and thus belongs to the well-known classification list in [B].

3.4. Nichols algebras of Cartan type

Definition 3.8. *A braided vector space (V, c) is of* Cartan type *if it is of diagonal type with matrix $(q_{ij})_{1 \leq i,j \leq \theta}$ and for any $i, j \in \{1, \ldots, \theta\}$, $q_{ii} \neq 1$, and there exists $a_{ij} \in \mathbb{Z}$ such that*

$$q_{ij} q_{ji} = q_{ii}^{a_{ij}}.$$

The integers a_{ij} are uniquely determined by requiring that $a_{ii} = 2$, $0 \leq -a_{ij} < \operatorname{ord}(q_{ii})$, $1 \leq i \neq j \leq \theta$. Thus $(a_{ij})_{1 \leq i,j \leq \theta}$ is a generalized Cartan matrix [K].

If (V, c) is a braided vector space of Cartan type with generalized Cartan matrix $(a_{ij})_{1 \leq i,j \leq \theta}$, then for any $i, j \in \{1, \ldots, \theta\}$, $j \neq i$, $m_{ij} = -a_{ij}$. It is easy to see that a braiding of Cartan type is standard, see the first part of the proof of [H1, Th. 1]. Hence we have from Remark 3.7:

Lemma 3.9. *Assume that χ is of Cartan type with symmetrizable Cartan matrix C. Then the following are equivalent:*

1. *The generalized root system $\Delta(\chi)$ is finite.*
2. *The Cartan matrix C is of finite type.* $\qquad \square$

We are now ready to sketch a proof of the main theorem in [H1].

---

[3]Here, one uses that the matrix $\| \operatorname{id} \|_{F,E}$ when seen as transformation of $\mathbb{Z}^\theta$, sends $\mathbf{e}_i$ to $\mathbf{f}_i$ for all i.

Theorem 3.10. *Let V be a braided vector space of Cartan type with Cartan matrix C. Then, the following are equivalent.*

1. *The set $\Delta(\mathcal{B}(V))$ is finite.*
2. *The Cartan matrix C is of finite type.*

Proof. (1) $\Rightarrow$ (2). As $\Delta(\chi) \subseteq \Delta(\mathcal{B}(V))$, $\Delta(\chi)$ is finite. If C is symmetrizable, we apply Lemma 3.9. If C is not symmetrizable, one reduces as in [AS1] to the smallest possible cases, see [H1]. See [H1] for the proof of (2) $\Rightarrow$ (1). □

We can now prove Lemma 1.2: (b) $\implies$ (a) was already discussed in [AS3]. (a) $\implies$ (b): it follows from Lemma 2.19 that (V,c) is of Cartan type; hence it is of finite Cartan type, by Theorem 3.10. To prove that (V,c) is of DJ-type – see [AS3, p. 84] – it is enough to assume that C is irreducible; then the result follows by inspection.

We readily get the following Corollary, as in [AS3, Th.2.9] – that really follows from results of Lusztig and Rosso. Let V be a braided vector space of diagonal type with matrix $q_{ij} \in \mathbb{k}^\times$. Let $m_{ij} \geq 0$ be as in Lemma 2.19.

Corollary 3.11. *If $\mathcal{B}(V)$ is a domain and its Gelfand-Kirillov dimension is finite, then $\mathcal{B}(V) \simeq \mathbb{k} < x_1, \ldots, x_\theta : (ad_c x_i)^{m_{ij}+1}(x_j) = 0, \quad i \neq j >$.* □

Notice that the hypothesis "$\mathcal{B}(V)$ is a domain" is equivalent to "$q_{ii} = 1$ or it is not a root of 1, for all i", cf. [Kh].

Appendix A. Generic data and the definition of $U(\mathcal{D})$

In this Appendix, we briefly recall the main definitions and results from [AS3] needed for Theorem 1.1. Everything below belongs to [AS3]; see *loc. cit.* for more details. Below, we shall refer to the following terminology.

- Γ is a free abelian group of finite rank s.
- $(a_{ij}) \in \mathbb{Z}^{\theta \times \theta}$ is a Cartan matrix of finite type [K]; we denote by $(d_1, \ldots, d_\theta)$ a diagonal matrix of positive integers such that $d_i a_{ij} = d_j a_{ji}$, which is minimal with this property.
- $\mathcal{X}$ is the set of connected components of the Dynkin diagram corresponding to the Cartan matrix (a_{ij}); if $i, j \in \{1, \ldots, \theta\}$, then $i \sim j$ means that they belong to the same connected component.
- $(q_I)_{I \in \mathcal{X}}$ is a family of elements in $\mathbb{k}$ which are not roots of 1.
- $g_1, \ldots, g_\theta$ are elements in Γ, $\chi_1, \ldots, \chi_\theta$ are characters in $\widehat{\Gamma}$, and all these satisfy

$$\langle \chi_i, g_i \rangle = q_I^{d_i}, \quad \langle \chi_j, g_i \rangle \langle \chi_i, g_j \rangle = q_I^{d_i a_{ij}}, \quad \text{for all } 1 \leq i, j \leq \theta, \quad i \in I.$$

We say that two vertices i and j are *linkable* (or that i is *linkable to* j) if $i \not\sim j$, $g_i g_j \neq 1$ and $\chi_i \chi_j = \varepsilon$.

Definition A.1. *A* linking datum *for*

$$\Gamma, \quad (a_{ij}), \quad (q_I)_{I \in \mathcal{X}}, \quad g_1, \ldots, g_\theta \quad and \ \chi_1, \ldots, \chi_\theta$$

is a collection $(\lambda_{ij})_{1 \leq i < j \leq \theta, i \nsim j}$ *of elements in* $\{0,1\}$ *such that* λ_{ij} *is arbitrary if* i *and* j *are linkable but 0 otherwise. Given a linking datum, we say that two vertices* i *and* j *are linked if* $\lambda_{ij} \neq 0$. *The collection*

$$\mathcal{D} = \mathcal{D}((a_{ij}), (q_I), (g_i), (\chi_i), (\lambda_{ij})),$$

where (λ_{ij}) *is a linking datum, will be called a generic datum of finite Cartan type for* Γ.

In the next Definition, ad_c is the "braided" adjoint representation, see [AS3].

Definition A.2. *Let* $\mathcal{D} = \mathcal{D}((a_{ij}), (q_I), (g_i), (\chi_i), (\lambda_{ij}))$ *be a generic datum of finite Cartan type for* Γ. *Let* $U(\mathcal{D})$ *be the algebra presented by generators* $a_1, \ldots, a_\theta,$ $y_1^{\pm 1}, \ldots, y_s^{\pm 1}$ *and relations*

$$y_m^{\pm 1} y_h^{\pm 1} = y_h^{\pm 1} y_m^{\pm 1}, \quad y_m^{\pm 1} y_m^{\mp 1} = 1, \qquad 1 \leq m, h \leq s, \tag{15}$$

$$y_h a_j = \chi_j(y_h) a_j y_h, \qquad 1 \leq h \leq s, 1 \leq j \leq \theta, \tag{16}$$

$$(\mathrm{ad}_c a_i)^{1-a_{ij}} a_j = 0, \qquad 1 \leq i \neq j \leq \theta, \quad i \sim j, \tag{17}$$

$$a_i a_j - \chi_j(g_i) a_j a_i = \lambda_{ij}(1 - g_i g_j), \qquad 1 \leq i < j \leq \theta, \quad i \nsim j. \tag{18}$$

The relevant properties of $U(\mathcal{D})$ are stated in the following result.

Theorem A.3. [AS3, Th. 4.3]. *The algebra* $U(\mathcal{D})$ *is a pointed Hopf algebra with structure determined by*

$$\Delta y_h = y_h \otimes y_h, \qquad \Delta a_i = a_i \otimes 1 + g_i \otimes a_i, \qquad 1 \leq h \leq s, 1 \leq i \leq \theta. \tag{19}$$

Furthermore, $U(\mathcal{D})$ *has a PBW-basis given by monomials in the root vectors, that are defined by an iterative procedure. The coradical filtration of* $U(\mathcal{D})$ *is given by the ascending filtration in powers of those root vectors. The associated graded Hopf algebra* $\mathrm{gr}\, U(\mathcal{D})$ *is isomorphic to* $\mathcal{B}(V)\#\Bbbk\Gamma$; $U(\mathcal{D})$ *is a domain with finite Gelfand-Kirillov dimension.*

Acknowledgements

We thank István Heckenberger for some useful remarks on a first version of this paper. We also thank the referee for his/her interesting remarks.

References

[AS1] N. ANDRUSKIEWITSCH and H.-J. SCHNEIDER, *Finite quantum groups and Cartan matrices.* Adv. Math 54, 1–45 (2000).

[AS2] N. ANDRUSKIEWITSCH and H.-J. SCHNEIDER, *Pointed Hopf Algebras*, in "New directions in Hopf algebras", 1–68, Math. Sci. Res. Inst. Publ. **43**, Cambridge Univ. Press, Cambridge, 2002.

[AS3] N. ANDRUSKIEWITSCH and H.-J. SCHNEIDER, *A characterization of quantum groups.* J. Reine Angew. Math. **577**, 81–104 (2004).

[A]   I. ANGIONO, *Álgebras de Nichols sobre grupos abelianos*, Tesis de licenciatura, Univ. Nac. de La Plata (2007). Available at `http://www.mate.uncor.edu/angiono`.

[B]   N. BOURBAKI, *Groupes et algèbres de Lie*, Chapitres 4, 5 et 6, Hermann, Paris, 1968.

[Ha]   M. HALL, *A basis for free Lie rings and higher commutators in free groups*, Proc. Am. Math. Soc., **1**, 575–581 (1950).

[H1]   I. HECKENBERGER, *The Weyl groupoid of a Nichols algebra of diagonal type*, Inventiones Math. **164**, 175–188 (2006).

[H2]   I. HECKENBERGER, *Weyl equivalence for rank 2 Nichols algebras of diagonal type*, Ann. Univ. Ferrara – Sez. VII – Sc. Mat., vol. **LI**, pp. 281–289 (2005).

[H3]   I. HECKENBERGER, *Rank 2 Nichols algebras with finite arithmetic root system*. Algebr. Represent. Theory, to appear.

[Kh]   V. KHARCHENKO, *A quantum analog of the Poincaré-Birkhoff-Witt theorem*, Algebra and Logic, 38, (1999), 259–276.

[K]   V. KAC, *Infinite-dimensional Lie algebras*, 3rd Edition, Cambridge University Press, Cambridge, 1990.

[R1]   M. ROSSO, *Quantum groups and quantum shuffles*, Invent. Math. **133** (1998), 399–416.

[R2]   M. ROSSO, *Lyndon words and Universal R-matrices*, talk at MSRI, October 26, 1999, available at `http://www.msri.org`; *Lyndon basis and the multiplicative formula for R-matrices*, preprint (2003).

[Se]   V. SERGANOVA, *On generalizations of root systems*, Commun. Algebra **24** (1996), 4281–4299.

[Sh]   A.I. SHIRSHOV, *On free Lie rings*, Mat. Sb., **45**(87), No. 2, 113–122 (1958).

[U]   S. UFER, *PBW bases for a class of braided Hopf algebras*, J. Alg. 280 (2004) 84–119.

Nicolás Andruskiewitsch and Iván Ezequiel Angiono
Facultad of Matemática, Astronomía y Física
Universidad Nacional of Córdoba
CIEM – CONICET
(5000) Ciudad Universitaria
Córdoba, Argentina
e-mail: `andrus@mate.uncor.edu`
e-mail: `angiono@mate.uncor.edu`

Modules and Comodules
Trends in Mathematics, 65–85
© 2008 Birkhäuser Verlag Basel/Switzerland

Dual Algebras of Some Semisimple Finite-dimensional Hopf Algebras

V.A. Artamonov and I.A. Chubarov

To Prof. Robert Wisbauer on the occasion of his 65th anniversary

Abstract. In this paper we establish properties of dual Hopf algebras for two series of finite-dimensional semisimple Hopf algebras. It is shown none of dual algebra belong to this class.

1. Introduction

This article as the previous one [A] was motivated by a result by G.M. Seitz who characterized finite groups G having only one irreducible complex representation of degree $n > 1$. Such a group G is either an extraspecial 2-group of order 2^{2m+1}, $n = 2^m$, or $|G| = n(n+1)$, where $n+1 = p^f$, p a prime [H][Theorem 7.10]. So in [A] semisimple finite-dimensional Hopf algebras H were studied which had only one irreducible representation Φ of degree n greater than 1. One could expect that a restriction of Φ to the subgroup of group-like elements $G(H)$ would be irreducible, so G could be one of the groups characterized by Seitz. But it turns out to be not the case.

Let H be a finite-dimensional semisimple Hopf algebra over an algebraically closed field k, such that $\mathrm{char}\,k$ and $\dim H$ are coprime.

Throughout the paper we shall keep to notations from [M]. For example there is a left and right action $f \rightharpoonup x$, $x \leftharpoonup f$ of $f \in H^*$ on $x \in H$, defined as follows: if $\Delta(x) = \sum_x x_{(1)} \otimes x_{(2)}$ then

$$f \rightharpoonup x = \sum_x x_{(1)} \langle f, x_{(2)} \rangle, \quad x \leftharpoonup f = \sum_x \langle f, x_{(1)} \rangle x_{(2)}$$

Suppose that a semisimple finite-dimensional Hopf H as an algebra has up to an isomorphism only one irreducible representation. Denote by $G = G(H^*)$ the group

Research partially supported by grants RFBR 06-01-00037, NSh-5666.2006.1.

of group-like elements in the dual Hopf algebra H^*. Then H as a k-algebra has a semisimple direct decomposition

$$H = \oplus_{g \in G} k e_g \oplus \mathrm{Mat}(n, k) E, \tag{1}$$

where $\{e_g,\ g \in G,\ E\}$ is a system of central orthogonal idempotents in H. Here E is the identity matrix in $\mathrm{Mat}(n, k)$. Moreover $1 = \sum_{g \in G} e_g + E$ in H. It follows from Nichols-Zoeller Theorem [NZ] that the order $|G|$ of G divides n^2.

Theorem 1.1 ([A]). *Let $G = G(H^*)$ be the group of group-like elements in the dual Hopf algebra H^*, and H have direct decomposition* (1) *where e_g, E are corresponding central idempotents. Then comultiplication Δ, counit ε and antipode S in H have the form*

$$\Delta(x) = \sum_{g \in G} [(g \rightharpoonup x) \otimes e_g + e_g \otimes (x \leftharpoonup g)] + \Delta'(x),$$

$$\Delta(e_t) = \sum_{g,h \in G,\ gh=t} e_g \otimes e_h + \Delta_t.$$

Here $x \in \mathrm{Mat}(n, k)$, $t \in G$ and $\Delta'(x)$, $\Delta_t \in \mathrm{Mat}(n, k) \otimes \mathrm{Mat}(n, k)$. The elements $\Delta'(E)$ and Δ_g, $g \in G$, is a set of orthogonal idempotents, satisfying the following relations:

$$[1 \otimes (g \rightharpoonup)] \Delta_t = \Delta_{tg^{-1}}; \quad [(\leftharpoonup g) \otimes 1] \Delta_t = \Delta_{g^{-1}t}.$$

Moreover if $g \in G$, then $S(e_g) = e_{g^{-1}}$. If $\mu : H \otimes H \to H$ is the multiplication map then

$$\mu(1 \otimes S)\Delta_g = \mu(S \otimes 1)\Delta_g = \delta_{g,1} e_g,$$
$$\mu(1 \otimes S)\Delta'(x) = \mu(S \otimes 1)\Delta'(x) = 0$$

for all $g \in G$ and for all $x \in \mathrm{Mat}(n, k)$.

Under additional assumption that $|G| = n^2$, equivalently $\Delta' = 0$, it is proved in [A] that up to a Hopf algebra isomorphism H belongs to one of the following two series. In the modified form the main result of [A, Theorems 4.1 and 5.1] can be formulated in the following form.

Theorem 1.2 ([A]). *Suppose that H is from* (1), *$G = G(H^*)$. Then $|G| = n^2$ if and only if $\Delta' = 0$ in Theorem 1.1. Moreover by [TY, Corollary 3.3] the group G is Abelian.*

Let $U = (u_{ij})$, $V = (v_{ij})$ be matrices from $\mathrm{GL}(n, k)$ such that $V = \frac{1}{n}U^{-1}$ and either $U = E, V = \frac{1}{n}E$ or $U = S, V = -\frac{1}{n}S$, where

$$S = \begin{pmatrix} T & 0 & \cdots & 0 \\ 0 & T & \cdots & 0 \\ \multicolumn{4}{c}{\dotfill} \\ 0 & 0 & \cdots & T \end{pmatrix}, \quad T = \begin{pmatrix} 0 & -1 \\ 1 & 0 \end{pmatrix}.$$

In both cases comultiplication Δ, counit ε and antipode S are as follows:

$$\Delta(e_g) = \sum_{h \in G} e_h \otimes e_{h^{-1}g} + \Delta_g, \quad \Delta_g \in \mathrm{Mat}(n,k) \otimes \mathrm{Mat}(n,k),$$

$$\Delta(x) = \sum_{g \in G} \left[(g \rightharpoonup x) \otimes e_g + e_g \otimes (x \leftharpoonup g) \right], \quad x \in \mathrm{Mat}(n,k)$$

$$\varepsilon(e_g) = \delta_{g,1}, \quad \varepsilon(x) = 0, \quad x \in \mathrm{Mat}(n,k)$$

$$S(y) = \begin{cases} e_{g^{-1}}, & y = g \in G, \\ nU^t y V = U\,{}^t y U^{-1} & y \in \mathrm{Mat}(n,k). \end{cases}$$

where

$$\Delta_g = \sum_{i,j,p,q} \left(E_{ij} \leftharpoonup g^{-1} \right) \otimes u_{ip} v_{qj} E_{pq} = \sum_{i,j,p,q} E_{ij} \otimes u_{ip} v_{qj} \left(g^{-1} \rightharpoonup E_{pq} \right)$$

Moreover there exists a projective representation $g \mapsto A_g \in \mathrm{GL}(n,k)$ of degree n such that

$$g \rightharpoonup x = A_g x A_{g^{-1}},$$

$$x \leftharpoonup g = n^2 U\,{}^t A_g V x\, U^t A_{g^{-1}} V = U\,{}^t A_g U^{-1} x\, U^t A_{g^{-1}} U^{-1},$$

$$A_g U\,{}^t A_h U^{-1} A_g^{-1} U\,{}^t A_h^{-1} U^{-1} = [A_g,\, U\,{}^t A_h^{-1} U^{-1}] = \mu_{g,h} E, \quad \mu_{g,h} \in k^*,$$

$$\Delta_g = \sum_{i,j,p,q,r,s=1}^{n} E_{ij} \otimes u_{ip} a_{rp}(g^{-1}) a_{qs}(g) v_{qj} E_{rs}$$

for any $g \in G$. Moreover $\mathrm{tr} A_g = n \delta_{g,1}$ and

$$\mathcal{R} = \sum_{i,j} E_{ij} \otimes E_{ji} = \sum_{g \in G} U^t A_{g^{-1}} \otimes {}^t A_g V \tag{2}$$

Conversely, if H has semisimple decomposition (1) with comultiplication as above then H is a Hopf algebra.

Here E_{ij} are matrix unit elements and ${}^t A_g$ is the transpose of a matrix A_g.

Following [OM] we shall call the matrix S from Theorem 1.2 *hyperbolic*.

In [TY, Theorem 3.2] there is given a classification of Hopf algebras H in terms of category of representations and in terms of bicharacters of G. As an application in [TY, Theorem 4.1] the authors proved that in the case $n = 2$ there exist up to equivalence four classes of Hopf algebras H. Theorem 1.2 gives an explicit form of comultiplication in H for arbitrary n in terms of projective representations up to an orthogonal (symplectic) equivalence. This form allows us to carry our calculations in dual Hopf algebra H^* and prove Theorem 6.1.

Observe that (2) can be presented in another form, namely

$$n(\operatorname{tr}Z)E = n^2 \left[\operatorname{tr}(VZU)\right]E = n^2 \sum_{ij} E_{ij}(VZU)E_{ji}$$

$$= n^2 \sum_{g \in G} U\,^t A_{g^{-1}} VZU\,^t A_g V = \sum_{g \in G}(Z \leftharpoonup g^{-1}) = Z \leftharpoonup \left(\sum_{g \in G} g\right) \tag{3}$$

for any matrix $Z \in \operatorname{Mat}(n,k)$.

Theorem 1.3 ([A]). *Let H be from Theorem 1.2. Then*

$$w = \sum_{g \in G} \chi_{g,w} e_g + Z_w \in H, \ Z_w \in \operatorname{Mat}(n,k), \ \chi_{g,w} \in k$$

is a group-like element from H if and only if the following conditions are satisfied

1) $\chi_{gh,w} = \chi_{g,w}\chi_{h,w}$ *for all $g,h \in G$ which means that $\chi_{*,w}$ is a one-dimensional character of G;*
2) $g \rightharpoonup Z_w = \chi_{g,w} Z_w = Z_w \leftharpoonup g$ *for every $g \in G$.*
3) $Z_w U\,^t Z_w = U$.

Moreover if w, Y are from Theorem 1.3 then $Z_w \otimes Z_w = \sum_{g \in G} \chi_{g,w} \Delta_g$.
Since $U^2 = \pm E$ it follows from the property 1.3) in Theorem 1.3 that

$$Z_w^{-1} = U\,^t Z_w U^{-1} = U^{-2}\left(U\,^t Z_w U^{-1}\right)U^2 = U^{-1}\,^t Z_w U.$$

It was shown in [A] that H is a group algebra if and only if the following equivalent conditions are satisfied:

1. H is cocommutative,
2. $g \rightharpoonup x = x \leftharpoonup g$ for all $g \in G$, $x \in \operatorname{Mat}(n,k)$.
3. $A_g = \pm\,^t A_g$ for each $g \in G$.

In Theorem 4.7 we specify the form of all matrices Z_w keeping invariant all other data of H. We show that taking an isomorphic copy of H we can assume that matrices Z_w, $w \in G(H)$, have block-diagonal form where sizes of blocks are 1, 2, 4 of a prescribed form.

If H is a Hopf algebra from Theorem 1.2 then the dual Hopf algebra H^* is also semisimple [M1]. The aim of this paper is to give an answer to the following question raised by N. Andruskiewitsch: does there exist a Hopf algebra H such that both H and H^* belong to series from Theorem 1.1 or Theorem 1.2, provided $n > 2$. If $n = 2$ then there exists an example of H isomorphic to H^* due to G. Kac and V. Paljutkin [KP] The first main result of the paper Theorem 6.1 shows that if H is from Theorem 1.2 then a semisimple Hopf algebra H^* could not isomorphic to any algebra from Theorem 1.1.

In the second part of the paper which starts from §4 we specify in Theorem 4.7 the structure of a group $G(H)$ of group-like elements and in Proposition 5.2 one-dimensional direct summands in the dual Hopf algebra H^* where H is from Theorem 1.2. Moreover in Proposition 5.3 we find an explicit form of direct sum of non-commutative simple direct summands in semisimple decomposition of H^*.

The authors thank Prof. S. Natale for useful discussions and comments. She indicated papers [N1], [N2], [T], [TY] where the similar class of Hopf algebras was considered from another point of view, namely in [N2, §3.4]. In the present paper we develop group-theoretical approach and direct calculations which could be useful for other purposes. We are also grateful to the referee for very useful remarks.

2. Antipode

In this section we are going to consider properties of an antipode in a semisimple Hopf algebra from Theorem 1.1. An antipode S is an involutive anti-automorphism of the semisimple finite-dimensional Hopf algebra H from (1). Then $\mathrm{Mat}(n,k)$ is S-invariant. By [P, §12.7] there exists an invertible matrix $U \in \mathrm{GL}(n,k)$ such that $S(x) = U\,{}^t x U^{-1}$, where ${}^t x$ is the transpose of x. Since $S^2 = 1$ for any matrix x we have

$$x = S\left(S(x)\right) = U^{\,t}\left(U\,{}^t x U^{-1}\right)U^{-1} = U\,{}^t U^{-1} x\,{}^t U\,U^{-1}$$

and therefore $U\,{}^t U^{-1} = \lambda E$ for some $\lambda \in k^*$. It follows that $U = \lambda\,{}^t U$ and $\lambda = \pm 1$. Hence U is (skew)symmetric. We have proved

Proposition 2.1. *There is a (skew)symmetric matrix $U \in \mathrm{GL}(n,k)$ such that $S(x) = U\,{}^t x U^{-1}$ for all $x \in \mathrm{Mat}(n,k)$. In particular*

$$\mathrm{tr}\left(S(x)\right) = \mathrm{tr}x$$

for any matrix x.

If $\Delta(x) = \sum_x x_{(1)} \otimes x_{(2)}$ then $\Delta(S(x)) = \sum_x S(x_{(2)}) \otimes S(x_{(1)})$ and therefore

$$f \rightharpoonup S(x) = \sum_x S(x_{(2)})\langle f, S(x_{(1)})\rangle = \sum_x S(x_{(2)})\langle S(f), x_{(1)}\rangle$$

$$= S\left(\sum_x x_{(2)}\langle f^{-1}, x_{(1)}\rangle\right) = S\left(x \leftharpoonup f^{-1}\right) \tag{4}$$

for all $f \in G$ and all $x \in \mathrm{Mat}(n,k)$. Similarly

$$S(x) \leftharpoonup f = S\left(f^{-1} \rightharpoonup x\right). \tag{5}$$

Proposition 2.2. *For any matrix x and any $g \in G$ we have*

$$\mathrm{tr}(g \rightharpoonup x) = \mathrm{tr}(x \leftharpoonup g) = \mathrm{tr}(x).$$

Proof. If $g \in G$ then the map $x \mapsto g \rightharpoonup x$ is an automorphism of the matrix algebra $\mathrm{Mat}(n,k)$. Since each automorphism of $\mathrm{Mat}(n,k)$ is inner, it preserves traces. □

3. Dual algebra H^*

The Hopf algebra H from Theorem 1.1 has a base e_g, $g \in G$, and E_{ij}, $1 \leqslant i, j \leqslant n$, where $\{E_{ij} \mid 1 \leqslant i, j \leqslant n\}$ is a system of matrix units in $\mathrm{Mat}(n, k)$. The space $\mathrm{Mat}(n, k)$ is equipped with a bilinear form

$$\langle A, B \rangle = \mathrm{tr}\,(A \cdot S(B)) = \mathrm{tr}(A \cdot U\,{}^t BU^{-1}), \tag{6}$$

which is non-degenerate because n is not divisible by $\mathrm{char}k$. Here S is the antipode from Proposition 2.1. Note that

$$\langle U E_{pq} U^{-1}, E_{ij} \rangle = \mathrm{tr}(E_{pq} E_{ji}) = \delta_{qj} \delta_{pi}.$$

So the base

$$\left\{ E^{ij} = U E_{ij} U^{-1} \mid i, j = 1, \ldots, n \right\} \tag{7}$$

of $\mathrm{Mat}(n, k)$ is dual to the base $\{E_{ij} \mid i, j = 1, \ldots, n\}$ with respect to the form (6). Denote by $\mathrm{Mat}(n, k)^*$ the linear span of the base (7) which is viewed as the dual space to $\mathrm{Mat}(n, k)$. So the identification of $\mathrm{Mat}(n, k)^* \to \mathrm{Mat}(n, k)$ is given by the map $\langle X, - \rangle \rightsquigarrow X$.

Expand each E^{ji} to the element of H^* such that $\langle E^{ji}, e_g \rangle = 0$ for $g \in G$. In particular $\langle X, E_{ij} \rangle$ is equal to the (i, j)th entry of the matrix $U^{-1} XU$. Hence as a linear space H^* is a direct sum $H^* = kG \oplus \mathrm{Mat}(n, k)^*$, where kG is the group algebra of the group G.

Proposition 3.1. *Let $g \in G$ and $X, Y \in \mathrm{Mat}(n, k)$. Then*

$$\langle X, Y \leftharpoonup g \rangle = \langle g \rightharpoonup X, Y \rangle, \quad \langle X \leftharpoonup g, Y \rangle = \langle X, g \rightharpoonup Y \rangle,$$

$$\langle X, Y \rangle = \langle Y, X \rangle.$$

It means that the operators $g \rightharpoonup$, $\leftharpoonup g$ are adjoint with respect to the symmetric bilinear form (6).

Proof. Applying (4), (5), Proposition 2.1 and Proposition 2.2 we obtain

$$\langle X, Y \rangle = \mathrm{tr}\,(X \cdot S(Y)) = \mathrm{tr}\,[S\,(X \cdot S(Y))] = \mathrm{tr}\,(Y \cdot S(X)) = \langle Y, X \rangle;$$

$$\langle X, Y \leftharpoonup g \rangle = \mathrm{tr}\,(X \cdot S(Y \leftharpoonup g)) = \mathrm{tr}\,\left[X \left(g^{-1} \rightharpoonup S(Y) \right) \right]$$

$$= \mathrm{tr}\,\left[g \rightharpoonup \left(X \left(g^{-1} \rightharpoonup S(Y) \right) \right) \right] = \mathrm{tr}\,[(g \rightharpoonup X) S(Y)] = \langle g \rightharpoonup X, Y \rangle.$$

The second equality can be proved in a similar way. $\square$

Corollary 3.2. *If $w, w' \in G(H)$ in Theorem 1.3 and $w \notin w'K$ then $\langle Z_w, Z'_w \rangle = 0$. Moreover $\langle Z_w, Z_w \rangle = n$. Here K is the subgroup of all $v \in G(H)$ such that $\chi_{g,v} = 1$ for all $g \in G$, see Section 4, Propositions 4.2 and 4.3.*

Proof. Applying Theorem 1.2, Theorem 1.3, property 1.3) and Proposition 4.3 we obtain

$$\langle Z_w, Z'_w \rangle = \mathrm{tr}\,\left(Z_w U\,{}^t Z_{w'} U^{-1} \right) = \mathrm{tr}\,\left(Z_{w'} Z_w^{-1} \right)$$

$$= \mathrm{tr}\,\left(Z_{w'w^{-1}} \right) = \begin{cases} 0, & w \notin w'K, \\ n, & w = w'. \end{cases}$$

$\square$

Proposition 3.3. *Let Δ^* be the comultiplication in H^*. Then*

$$\Delta^*(E^{pq}) = \sum_{j=1}^{n} E^{pj} \otimes E^{jq}$$

and if $g \in G$ then $\Delta^*(g) = g \otimes g$.

Proof. Recall that $(H \otimes H)^* = H^* \otimes H^*$ and therefore $\langle \Delta^* f, x \otimes y \rangle = \langle f, xy \rangle$ for all $f \in H^*$ and all $x, y \in H$ [M]. Let

$$a = \sum_{h \in G} \alpha_h e_h + X \in H, \quad \alpha_h \in k, \ X \in \mathrm{Mat}(n, k),$$

$$b = \sum_{h \in G} \beta_h e_h + Y \in H, \quad \beta_h \in k, \ Y \in \mathrm{Mat}(n, k).$$

Then for any $E_{pq} \in \mathrm{Mat}(n, k)$, $g \in G$ we have

$$\langle \Delta^*(E^{pq}), a \otimes b \rangle = \langle E^{pq}, XY \rangle = \mathrm{tr}\left(E_{pq} U\, {}^t(XY) U^{-1} \right)$$

$$= \left(U\,{}^tY\,{}^tX U^{-1} \right)_{qp} = \sum_{j=1}^{n} \left(U\,{}^tY U^{-1} \right)_{qj} \left(U\,{}^tX U^{-1} \right)_{jp}$$

$$= \sum_{j=1}^{n} \mathrm{tr}\left(E_{jq} U\,{}^tY U^{-1} \right) \mathrm{tr}\left(E_{pj} U\,{}^tX U^{-1} \right) = \sum_{j=1}^{n} \langle E^{pj}, X \rangle \langle E^{jq}, Y \rangle,$$

$$\langle \Delta^*(g), a \otimes b \rangle = \langle g, ab \rangle = \alpha_g \beta_g = \langle g, a \rangle \langle g, b \rangle. \qquad \square$$

Corollary 3.4. *Let $w \in G(H)$ in Theorem 1.3 and $g \in G$. Then $w \rightharpoonup g = \chi_{g,w} g = g \leftharpoonup w$.*

Proof. We know that $\Delta^*(g) = g \otimes g$. Hence $w \rightharpoonup g = g \langle w, g \rangle = \chi_{g,w} g$ by Theorem 1.3. $\qquad \square$

Let $a * b$ denote the convolution multiplication in H^*. Note that ε is the unit element in H^* and it is equal to $1 \in G$.

Proposition 3.5. *Suppose that H is from Theorem 1.2. If $g, h \in G$ and $X, Y \in \mathrm{Mat}(n, k)^*$ then*

$$g * h = gh, \quad g * X = g \rightharpoonup X, \quad X * g = X \leftharpoonup g,$$

$$X * Y = \frac{1}{n} \sum_{g \in G} \langle Y \leftharpoonup g^{-1}, X \rangle g = \frac{1}{n} \sum_{g \in G} \langle Y * g^{-1}, X \rangle g.$$

Proof. Let $\Delta_g = \sum_i a_i \otimes b_i$, where $a_i, b_i \in \mathrm{Mat}(n, k)$ and $Z \in \mathrm{Mat}(n, k)$. Applying Theorem 1.2 and Proposition 3.1 we obtain for $f \in G$

$$\langle g * h, e_f \rangle = \sum_{r \in G} \langle g, e_r \rangle \langle h, e_{r^{-1} f} \rangle + \sum_i \langle g, a_i \rangle \langle g, b_i \rangle = \langle gh, e_f \rangle;$$

$$\langle g * h, X \rangle = \sum_{f \in G} \left[\langle g, f \rightharpoonup X \rangle \langle h, e_f \rangle + \langle g, e_f \rangle \langle h, X \leftharpoonup f \rangle \right] = 0;$$

$$\langle g * X, e_h \rangle = \sum_{r \in G} \langle g, e_r \rangle \langle X, e_{r^{-1}h} \rangle + \sum_{i,j=1}^{n} \langle g, a_i \rangle \langle X, b_i \rangle$$

$$= 0 = \langle X * g, e_f \rangle;$$

$$\langle g * X, Y \rangle = \sum_{f \in G} [\langle g, f \rightharpoonup Y \rangle \langle X, e_f \rangle + \langle g, e_f \rangle \langle X, Y \leftharpoonup f \rangle]$$

$$= \langle X, Y \leftharpoonup g \rangle = \langle g \rightharpoonup X, Y \rangle;$$

$$\langle X * g, Y \rangle = \sum_{f \in G} [\langle X, f \rightharpoonup Y \rangle \langle g, e_f \rangle + \langle X, e_f \rangle \langle g, Y \leftharpoonup f \rangle]$$

$$= \langle X, g \rightharpoonup Y \rangle = \langle X \leftharpoonup g, Y \rangle;$$

$$\langle X * Y, Z \rangle = \sum_{f \in G} [\langle X, f \rightharpoonup Z \rangle \langle Y, e_f \rangle + \langle X, e_f \rangle \langle Y, Z \leftharpoonup f \rangle] = 0.$$

Set $M_{ij} = \sum_{p,q} u_{ip} E_{pq} v_{qj} \in \mathrm{Mat}(n, k)$ in notations of Theorem 1.2. Then

$$\langle X * Y, e_g \rangle = \sum_{r \in G} \langle X, e_r \rangle \langle Y, e_{r^{-1}g} \rangle + \sum_{i,j=1}^{n} \langle X, E_{i,j} \rangle \langle Y, g^{-1} \rightharpoonup M_{i,j} \rangle$$

$$= \sum_{i,j=1}^{n} \langle X, E_{i,j} \rangle \langle Y \leftharpoonup g^{-1}, M_{i,j} \rangle$$

$$= \sum_{i,j,p,q=1}^{n} \langle X, E_{i,j} \rangle u_{ip} v_{qj} \langle Y \leftharpoonup g^{-1}, E_{p,q} \rangle$$

$$= \sum_{i,j,p,q=1}^{n} (U^{-1}XU)_{ij} u_{ip} v_{qj} \left(U^{-1}(Y \leftharpoonup g^{-1})U\right)_{pq}$$

$$= \frac{1}{n} \mathrm{tr}\left(UU^{-1}(Y \leftharpoonup g^{-1})UU^{-1\,t}\left(U^{-1}XU\right)\right)$$

$$= \frac{1}{n} \mathrm{tr}\left((Y \leftharpoonup g^{-1})U^{\,t}XU^{-1}\right) = \frac{1}{n} \langle Y \leftharpoonup g^{-1}, X \rangle. \qquad \square$$

Corollary 3.6. *Let H be from Theorem 1.2. Then H^* is a $\mathbb{Z}_2$-graded algebra with the grading*

$$H^* = H_0^* \oplus H_1^*, \qquad H_0^* = kG, \ H_1^* = \mathrm{Mat}(n, k)^*.$$

In particular $\mathrm{Mat}(n, k)^*$ is a free cyclic left and right kG-module.

The last statement follows from [M, §3.1].

4. Group-like elements in H

Starting from this section we shall assume that H is from Theorem 1.2. In this section we shall specify the form of element from the group $G(H)$.

Each element of the group of group-like elements $G(H)$ is characterized in Theorem 1.3.

Proposition 4.1. *The maps*

$$G \to k^*, \qquad g \mapsto \chi_{g,w}, \quad w \in G(H) \text{ is fixed;}$$
$$G(H) \to k^*, \qquad w \mapsto \chi_{g,w}, \quad g \in G \text{ is fixed;}$$
$$G(H) \to \mathrm{GL}(n,k), \quad w \mapsto Z_w.$$

are group homomorphisms. The last homomorphism is injective.

Proof. The first map is a group homomorphism by Theorem 1.3. Consider the second and the third maps. If

$$w = \sum_g \chi_{g,w} e_g + Z_w, \ u = \sum_g \chi_{g,u} e_g + Z_u \in G(H)$$

then $wu = \sum_g \chi_{g,w} \chi_{g,u} e_g + Z_w Z_u \in G(H)$. Hence all maps are homomorphisms.

If $Z_w = E$, then $g \rightharpoonup E = E$ for all $g \in G$ and therefore $\chi_{g,w} = 1$ for all $g \in G$. It means that w is the identity element of $G(H)$.   □

Let F be the set all $g \in G$ such that $\chi_{g,w} = 1$ for all $w \in G(H)$. It is easy to check that F is a normal subgroup in G. Similarly let K be the set of all $w \in G(H)$ such that $\chi_{g,w} = 1$ for all $g \in G$. It is easy again to check that K is a normal subgroup in $G(H)$ containing $G(H)'$. If $w \in K$ then $w = \sum_g e_g + Z_w$. An element $w = 1$ with $Z_w = E$ exists so the other one is $-E$, [A]. Hence K is a central subgroup in $G(H)$ of order 2 consisting of elements $\sum_g e_g \pm E$. The group $G(H)/K$ is isomorphic via the map $w \mapsto \chi_{*,w}$ to a subgroup of the group of one-dimensional characters of G. Since the group G is Abelian by [TY] the group of one-dimensional characters of G is isomorphic to G. We have proved

Proposition 4.2. *The group $G(H)$ contains a central subgroup K of order 2. The group $G(H)/K$ is isomorphic to a subgroup of the group of one-dimensional characters of G. In particular, $G(H)$ is nilpotent of class 2 and $G(H)/K$ is Abelian and isomorphic to a subgroup of G. The group $G(H)$ has an Abelian normal subgroup N, the centralizer of K, of index 1 or 2.*

Proposition 4.1 and Theorem 1.3 mean that there is a faithful n-dimensional representation $w \mapsto Z_w$ of groups

$$G(H) \to \begin{cases} \mathrm{O}(n,k), & U = E \\ \mathrm{Sp}(n,k) & U = \mathcal{S}, \end{cases}$$

where $\mathrm{O}(n,k)$ stands for the group of orthogonal matrices of size n and $\mathrm{Sp}(n,k)$ denotes the group of symplectic matrices of even size n.

Proposition 4.3. *If $w \in G(H) \setminus K$ then $\mathrm{tr} Z_w = 0$. In particular the order of w in $G(H)/K$ is less than n.*

Proof. Take $Z = Z_w$ in (3). Then by Theorem 1.3, property 1.3)

$$n\mathrm{tr}(Z_w)E = \sum_{g \in G} (Z \leftharpoonup g) = \left(\sum_{g \in G} \chi_{g,w} \right) Z_w.$$

If $w \notin K$ then $\chi_{g,w}$ is nontrivial and therefore by orthogonality relation $\sum_{g \in G} \chi_{g,w} = 0$.

Suppose that the order of w in $G(H)/K$ is greater than n. The matrix Z_w has a finite order and therefore it is conjugate to a diagonal matrix

$$\begin{pmatrix} \lambda_1 & 0 & 0 \\ 0 & \lambda_2 & 0 \\ \cdots\cdots\cdots\cdots \\ 0 & 0 & \lambda_n \end{pmatrix}$$

where $\lambda_1, \ldots, \lambda_n$ are roots of 1 in k. It follows that each matrix Z_w^j for $j = 1, \ldots, n$ is conjugate to

$$\begin{pmatrix} \lambda_1^j & 0 & 0 \\ 0 & \lambda_2^j & 0 \\ \cdots\cdots\cdots\cdots \\ 0 & 0 & \lambda_n^j \end{pmatrix}$$

and by preceding considerations

$$\begin{cases} \operatorname{tr} Z_w & = \lambda_1 + \cdots + \lambda_n & = 0, \\ \cdots\cdots\cdots\cdots\cdots\cdots\cdots \\ \operatorname{tr} Z_w^n & = \lambda_1^n + \cdots + \lambda_n^n & = 0. \end{cases}$$

From Newton's formulae in the theory of symmetric functions it follows that $\lambda_1 = \cdots = \lambda_n = 0$ which is impossible. $\qquad\square$

Let L be the space of columns of height n equipped with a (skew)symmetric bilinear form

$$(x, y) = {}^t x U y, \quad x, y \in L. \tag{8}$$

Then each matrix Z_w, $w \in G(H)$, is an isometry of L with respect to the form (8).

Proposition 4.4. *If $\lambda \in k$ is an eigenvalue of a matrix Z_w, $w \in G(H)$, then λ^{-1} is an eigenvalue of Z_w with the same multiplicity.*

Proof. By the property 1.3) from Theorem 1.3 ${}^t Z_w^{-1} = U^{-1} Z_w U$. Hence the characteristic polynomials of the matrices Z_w and Z_w^{-1} coincide. $\qquad\square$

Proposition 4.5. *Let $Z_w x = \lambda x$, $Z_w y = \mu y$ where $\lambda, \mu \in k$, $\lambda\mu \neq 1$ and $x, y \in L$. Then $(x, y) = 0$. In particular if $\lambda = \mu \neq \pm 1$ then $(x, y) = 0$.*

Proof. We have $(x, y) = (Z_w x, Z_w y) = \lambda\mu(x, y)$ and the proof follows. $\qquad\square$

Fix a maximal Abelian subgroup N in $G(H)$. By Proposition 4.2 either $N = G(H)$ or N is a normal subgroup of index 2 in $G(H)$. If $N \neq G(H)$ fix a matrix $Z = Z_{w'}$, $w \in G \backslash N$. Then there exists a matrix Z_w such that $Z Z_w Z^{-1} = [Z, Z_w]Z_w = -Z_w$. Since the group N is Abelian the space L has a direct decomposition $L = \oplus_\psi L_\psi$ where each ψ is a character of N and

$$L_\psi = \{ x \in L \mid (\forall w \in N) \ Z_w x = \psi(w)x \}.$$

By Proposition 4.5 we know that L_ψ and $L_{\psi'}$ are orthogonal with respect to (8) provided $\psi(w)\psi'(w) \neq 1$ for some $w \in N$. Hence L has an orthogonal direct decomposition

$$L = \oplus_\psi^\perp \left[L_\psi \oplus L_{\psi^{-1}} \right]. \tag{9}$$

Suppose first that $N = G(H)$. Then L has the orthogonal decomposition (9) and therefore a restriction of the form (8) to each summand $L_\psi \oplus L_{\psi^{-1}}$ is non-degenerate.

Suppose that $U = E$ and there exists an element $e \in L_\psi$ such that $(e,e) = 1$. Then the restriction of the form (8) to ke is non-degenerate and therefore ke has an orthogonal complement in $L_\psi \oplus L_{\psi^{-1}}$ which is $G(H)$-invariant.

Suppose that the restriction of the form (8) to each of the spaces $L_\psi, L_{\psi^{-1}}$ vanishes. Since the form (8) in $L_\psi \oplus L_{\psi^{-1}}$ is non-degenerate there exists an element $e = a + b$, $a \in L_\psi$, $b \in L_{\psi^{-1}}$ such that $(e,e) = 2$. Then $(a,b) = 1$. Hence the form (8) is non-degenerate on $ka + kb$ and $ka + kb$ is $G(H)$-invariant, namely $Z_w a = \psi(w)a$, $Z_w b = \psi^{-1}(w)b$. The orthogonal complement to $ka + kb$ in $L_\psi \oplus L_{\psi^{-1}}$ is also $G(H)$-invariant. Finally take some orthonormal base in $ka + kb$.

Suppose now that $U = S$ is hyperbolic. Let there exist elements $a, b \in L_\psi$ such that $(a,b) = 1$. Proposition 4.5 implies $\psi = \psi^{-1}$. The restriction of the form (8) to $G(H)$-invariant space $ka + kb$ is non-degenerate. So does its orthogonal complement in $L_\psi = L_{\psi^{-1}}$.

Suppose that the restriction of the form (8) to $L_\psi, L_{\psi^{-1}}$ vanishes. Then as above there exist elements $a \in L_\psi$, $b \in L_{\psi^{-1}}$ such that $(a,b) = 1$. Then $ka + kb$ is $G(H)$-invariant and the restriction of form (8) to $ka + kb$ is non-degenerate.

Suppose now that $G(H)$ is non-Abelian and therefore $N \neq G(H)$.

Proposition 4.6. *Let $w \in G(H)$ and $w' \in G(H) \setminus N$. If $Z = Z_{w'}$ then for any ψ we have $Z(L_\psi) = L_{\psi'}$ where*

$$\psi'(w) = \begin{cases} \psi(w), & [Z, Z_w] = E, \\ -\psi(w), & [Z, Z_w] = -E. \end{cases}$$

In particular $L_\psi \perp L_{\psi'}$.

Proof. Let $w \in N$. If $[Z, Z_w] = -E$ and therefore for any $x \in L_\psi$ we have $Z_w(Z(x)) = [Z_w, Z]Z(Z_w(x)) = -\psi(w)Z(x)$. Otherwise Z and Z_w commute and $Z_w(Z(x)) = Z(Z_w(x)) = \psi(w)Z(x)$. In all cases

$$\psi(w)\psi'(w) = \begin{cases} \psi^2(w), & [Z, Z_w] = 1, \\ -\psi^2(w), & [Z, Z_w] = -E. \end{cases}$$

Hence $\psi' \neq \psi^{-1}$ and hence Proposition 4.5 implies $L_\psi \perp L_{\psi'}$. □

If ψ' is from Proposition 4.6 then $\psi' \neq \psi$ provided $N \neq G(H)$. Moreover $\left(\psi^{-1}\right)' = (\psi')^{-1}$. So the space

$$\widehat{L} = L_\psi + L_{\psi'} + L_{\psi^{-1}} + L_{(\psi^{-1})'}$$

is $G(H)$-invariant and by (9) it has an orthogonal $G(H)$-invariant direct complement in L. So the restriction of the form (8) to the space $\widehat{L}$ is nondegenerate [B,
§4.2.]. If $N \neq G(H)$ by previous considerations we have $\psi \neq \psi' \neq \psi^{-1} \neq (\psi')^{-1}$.
If $(\psi')^{-1} = \psi$, then $\psi^{-1} = \psi'$ which is impossible. So we can conclude that there
are only the following situations:

(i) $\psi = \psi^{-1} \neq \psi' = (\psi')^{-1}$;

(ii) ψ, ψ', ψ^{-1}, $(\psi')^{-1}$ are different.

In the case i) L_ψ and $L_{\psi'}$ are orthogonal and $\widehat{L} = L_\psi \oplus^\perp L_{\psi'}$ and L_ψ has an
orthogonal complement in L. It means that the restriction of the form (8) to L_ψ
is non-degenerate. Then there exists a base $\{e_j\}$ in L_ψ such that the Gram matrix
of the form (8) in this base is either equal to the identity matrix in the symmetric
case $(U = E)$ or is hyperbolic in the symplectic case $(U = \mathcal{S})$.

The elements $\{Z(e_i)\}$ form a base of $L_{\psi'}$ and in this base the Gram matrix
of the form (8) is equal to the Gram matrix in the base $\{e_i\}$. Moreover each
$e_j \perp Z(e_i)$ and $Z^2(e_j) = \psi(Z^2)e_j$ since $Z^2 \in N$.

In the symmetric case the linear span $\langle e_i, Z(e_i) \rangle$ is $G(H)$-invariant and it is
a direct orthogonal summand of L. In the base $e_i, Z(e_i)$ the Gram matrix is the
identity matrix of size 2, the matrix of Z and of any Z_w, $w \in G(H)$, is equal to

$$\begin{pmatrix} 0 & \psi(Z^2) \\ 1 & 0 \end{pmatrix} \text{ and to } \begin{pmatrix} \psi(w) & 0 \\ 0 & \psi'(w) \end{pmatrix}, \tag{10}$$

respectively.

In the symplectic case take a linear span

$$\langle e_{2i+1}, e_{2i+2}, Z(e_{2i+1}), Z(e_{2i+2}) \rangle$$

of dimension 4. The Gram matrix in the span of the chosen base has the form

$$\begin{pmatrix} 0 & -1 & 0 & 0 \\ 1 & 0 & 0 & 0 \\ 0 & 0 & 0 & -1 \\ 0 & 0 & 1 & 0 \end{pmatrix}.$$

In this base the matrix of Z and of any Z_w, $w \in N$, has the form

$$\begin{pmatrix} 0 & 0 & \psi(Z^2) & 0 \\ 0 & 0 & 0 & \psi(Z^2) \\ 1 & 0 & 0 & 0 \\ 0 & 1 & 0 & 0 \end{pmatrix} \text{ and } \begin{pmatrix} \psi(w) & 0 & 0 & 0 \\ 0 & \psi(w) & 0 & 0 \\ 0 & 0 & \psi'(w) & 0 \\ 0 & 0 & 0 & \psi'(w) \end{pmatrix}, \tag{11}$$

respectively.

In the case ii) we have a direct decomposition

$$\widehat{L} = \left(L_\psi \oplus L_{\psi^{-1}} \right) \oplus^\perp \left(L_{\psi'} \oplus L_{(\psi')^{-1}} \right).$$

A restriction of the form (8) to each of the spaces $L_\psi, L_{\psi^{-1}}, L_{\psi'}, L_{(\psi')^{-1}}$ is a zero
form by Proposition 4.5 but its restriction to $L_\psi \oplus L_{\psi^{-1}}$ is non-degenerate [B,

§4.2.]. Hence we can choose a base $\{e_i\}$ in L_ψ and a base $\{e_i'\}$ in $L_{\psi'}$ such that the Gram matrix in the base $\{e_i, e_i'\}$ is equal to

$$\begin{pmatrix} 0 & \pm 1 \\ 1 & 0 \end{pmatrix}.$$

The sign $+$ is taken in the symmetric case and the sign $-$ is taken in the symplectic case. Then elements $\{Z(e_i), Z(e_i')\}$ form a base of $L_{\psi'} \oplus L_{(\psi')^{-1}}$ and

$$Z^2(e_i) = \psi(Z^2)e_i, \quad Z^2(e_i') = \psi'(Z^2)e_i' = \psi(Z^2)e_i'$$

by Proposition 4.6. So the Gram matrix of the system of vectors $e_i, e_i', Z(e_i), Z(e_i')$ is equal to

$$\begin{pmatrix} 0 & \pm 1 & 0 & 0 \\ 1 & 0 & 0 & 0 \\ 0 & 0 & 0 & \pm 1 \\ 0 & 0 & 1 & 0 \end{pmatrix}$$

with the same rule for signs as above. The matrix of Z in this base is equal to the first matrix in (11). On the other side each Z_w, $w \in N$, is a diagonal matrix with diagonal entries $\psi(w)$, $\psi(w)^{-1}$, $\psi'(w)$, $(\psi; (w))^{-1}$, respectively.

We have proved

Theorem 4.7. *Let N be a maximal Abelian subgroup in $G(H)$. There exists an isomorphic copy of H such that the following properties are satisfied.*

1) *If $N = G(H)$ then each matrix Z_w is diagonal.*
2) *If $N \neq G(H)$ then N has index 2 in $G(H)$. Let $Z = Z_{w'}$ where $w' \in G(H) \backslash N$. Then each matrix Z_w, $w \in N$, has diagonal form. The matrix Z has block-diagonal form with the followings blocks.*
   a) *In the symmetric case each block of Z has size 2 and is equal to the first matrix from (10) for some ψ.*
   b) *In the symplectic case each block of Z has size 4 and equal to the first matrix from (11) for some ψ.*

Theorem 4.7 can be considered as a refinement of Theorem 1.3.

5. Central elements

In this section we shall characterize central elements in H^* where H is from Theorem 1.2 and find one-dimensional direct summands in H^*. Consider an element

$$c = a + X \in H^*, \quad \text{where } a \in kG, \ X \in \mathrm{Mat}(n, k)^* \tag{12}$$

and find criterion of its centrality. By Proposition 3.5 for each $h \in G$ we have $h * c = ha + (h * X) = c * h = ah + (X * h)$. Hence a is a central element in kG and $h * X = X * h$. Similarly applying Proposition 3.1 we obtain

$$Y * c = Y * a + Y * X = (Y * a) + \frac{1}{n} \sum_{g \in G} \langle X \leftharpoonup g^{-1}, Y \rangle g$$

$$c * Y = a * Y + X * Y = (a * Y) + \frac{1}{n} \sum_{g \in G} \langle Y \leftharpoonup g^{-1}, X \rangle g$$

$$= (a * Y) + \frac{1}{n} \sum_{g \in G} \langle g^{-1} \rightharpoonup X, Y \rangle g$$

for any matrix $Y \in \mathrm{Mat}(n, k)^*$. Hence for all $Y \in \mathrm{Mat}(n, k)^*$ and for all $g \in G$ we have

$$Y * a = a * Y, \quad X * g = g * X. \tag{13}$$

Hence we have proved

Proposition 5.1. *An element* (12) *is central if and only if* a *is a central element in* kG *and* (13) *holds for all* $Y \in \mathrm{Mat}(n, k)^*$ *and for all* $g \in G$. *In particular the center*

$$Z(H^*) = [Z(H^*) \cap kG] \oplus [Z(H^*) \cap \mathrm{Mat}(n, k)^*].$$

If $a \in Z(H^*) \cap kG$ *then* (13) *holds for any* $Y \in \mathrm{Mat}(n, k)^*$. $Z(H^*) \cap \mathrm{Mat}(n, k)^*$ *consists of all matrices* $X \in \mathrm{Mat}(n, k)^*$ *satisfying* (13) *for all* $g \in G$. *In particular if we identify* $\mathrm{Mat}(n, k)^*$ *with* $\mathrm{Mat}(n, k)$ *then the image of* $Z(H^*) \cap \mathrm{Mat}(n, k)^*$ *in* $\mathrm{Mat}(n, k)$ *is a subalgebra in* $\mathrm{Mat}(n, k)$, *containing the identity matrix.*

Proof. It suffices to prove the last affirmation. Let $X_1, X_2 \in Z(H^*) \cap \mathrm{Mat}(n, k)^*$. Then for any $h \in G$ we have

$$(X_1 X_2) * h = X_1 X_2 \leftharpoonup h = (X_1 \leftharpoonup h)(X_2 \leftharpoonup h)$$
$$= (h \rightharpoonup X_1)(h \rightharpoonup X_1) = h \rightharpoonup X_1 X_2 = h * (X_1 X_2). \qquad \square$$

Consider now central idempotents in H^* corresponding to one-dimensional representations of H^*. There exists one-to-one correspondence between these central idempotents and elements from $G(H)$. Namely if $w = \sum_{g \in G} \chi_{g,w} e_g + Z_w$ from Theorem 1.3 is a group-like element in H then there exists a unique central idempotent $e_w = \sum_{g \in G} \tau_{g,w} g + X_w \in H^*$ such that $z * e_w = e_w * z = \langle z, w \rangle e_w$ for all $z \in H^*$ and $\langle e_w, w \rangle = 1$.

Taking $c = e_w$ in Proposition 5.1 we obtain in addition to the conditions from Proposition 5.1 the following relations:

$$\langle h, w \rangle e_w = \chi_{h,w} e_w = \sum_{f \in G} \chi_{h,w} \tau_{f,w} f + \chi_{h,w} X_w$$

$$= h * e_w = \sum_g \tau_{g,w} h g + (h \rightharpoonup X_w);$$

$$\langle e_w, w \rangle = \sum_g \tau_{g,w} \chi_{g,w} + \langle X_w, Z_w \rangle = 1;$$

$$\langle Y, w \rangle e_w = \langle Y, Z_w \rangle e_w = \sum_g \langle Y, Z_w \rangle \tau_{g,w} g + \langle Y, Z_w \rangle X_w$$

$$= Y * e_w = \sum_g \tau_{g,w} (Y \leftharpoonup g) + Y * X_w$$

$$= \sum_g \left[\tau_{g,w}(Y \leftharpoonup g) + \frac{1}{n}\langle X_w \leftharpoonup g^{-1}, Y \rangle g \right]$$

for all $h \in G$ and any matrix Y. It follows that for all $g, h \in G$ we have

$$\tau_{g,w} = \chi_{h,w}\tau_{hg,w},$$

$$\chi_{h,w}X_w = h \rightharpoonup X_w = X_w \leftharpoonup h,$$

$$1 = \sum_g \tau_{g,w}\chi_{,w}g + \langle X_w, Z_w \rangle,$$

$$\tau_{g,w}\langle Y, Z_w \rangle = \frac{1}{n}\langle X_w \leftharpoonup g^{-1}, Y \rangle = \frac{1}{n}\chi_{g,w}\langle X_w, Y \rangle,$$

$$\langle Y, Z_w \rangle X_w = \sum_g \tau_{g,w}(Y \leftharpoonup g) = Y \leftharpoonup \left(\sum_{g \in G} \tau_{g,w}g \right).$$

Equivalently by Proposition 5.1 and Theorem 1.2 we can write for any $Y \in \mathrm{Mat}(n,k)^*$

$$\tau_{g,w} = \chi_{g^{-1},w}\tau_{1,w},$$

$$\chi_{h,w}X_w = h \rightharpoonup X_w = X_w \leftharpoonup h, \tag{14}$$

$$1 = \tau_{1,w}\sum_g \chi^{-1}_{g,w}\chi_{g,w} + \langle X_w, Z_w \rangle = \tau_1 n^2 + \langle X_w, Z_w \rangle, \tag{15}$$

$$\tau_{1,w}\chi^{-1}_{g,w}\langle Y, Z_w \rangle = \frac{\chi^{-1}_{g,w}}{n}\langle X_w, Y \rangle = \frac{\chi^{-1}_{g,w}}{n}\langle Y, X_w \rangle, \tag{16}$$

$$\langle Y, Z_w \rangle X_w = \tau_{1,w}Y \leftharpoonup \left(\sum_g \chi^{-1}_{g,w}g \right). \tag{17}$$

Since Y is arbitrary the equation (16) has the form

$$n\tau_{1,w}Z_w = X_w. \tag{18}$$

If we put $Y = X_w$ in (17) then we obtain $\langle X_w, Z_w \rangle X_w = \tau_{1,w}n^2 X_w$ by (14). Note that $X_w \neq 0$ by (18) and Theorem 1.3. Hence $\langle X_w, Z_w \rangle = \tau_{1,w}n^2$ and (15) implies $\tau_{1,w} = \frac{1}{2n^2}$ and $X_w = \frac{1}{2n}Z_w$.

Now (17) has the form

$$n\langle Y, Z_w \rangle Z_w = Y \leftharpoonup \left(\sum_g \chi^{-1}_{g,w}g \right) \tag{19}$$

for all $Y \in \mathrm{Mat}(n,k)$.

Proposition 5.2. *The central idempotent e_w corresponding to a group-like element $w = \sum_g \chi_{g,w}e_g + Z_w \in H$ has the form*

$$e_w = \frac{1}{2n^2}\sum_g \chi^{-1}_{g,w}g + \frac{1}{2n}Z_w.$$

Moreover (19) *holds and in particular*

$$\frac{1}{2n^2} \sum_g \chi_{g,w'}^{-1} g - \frac{1}{2n} Z_w = \begin{cases} 0, & w' \notin wK, \\ \frac{1}{2n} Z_w, & w' = w. \end{cases} \tag{20}$$

The algebra H^* has a direct decomposition

$$H^* = \left(\oplus_{w \in G(H)} ke_w \right) \oplus P \tag{21}$$

where the ideal P is a direct sum of full matrix algebras of sizes greater than 1.

Proof. In fact applying (19) and Corollary 3.2 we prove the second half of (20). The first equality in (20) follows from the fact that corresponding to w and to w' idempotents in kG are orthogonal. $\qquad \square$

As was noticed in [A] for any w there exists another group-like element $\overline{w} = \sum_{g \in G} \chi_{g,w}^{-1} e_g - Z_w$. Then $\overline{e_w} = \frac{1}{2n^2} \sum_g \chi_{g,w}^{-1} g - \frac{1}{2n} Z_w$, and therefore

$$e_w + \overline{e_w} = \frac{1}{n^2} \sum_g \chi_{g,w}^{-1} g \in kG \subset H^*.$$

Let $K = \left\{ \sum_{g \in G} e_g \pm E \right\}$ be a central subgroup of order 2 in $G(H)$. Then

$$1 - \sum_{w \in G(H)} e_w = 1 - \frac{1}{n^2} \sum_{w \in G(H)/K} \sum_{g \in G} \chi_{g,w}^{-1} g$$

$$= 1 - \frac{1}{n^2} \sum_{g \in G} \sum_{w \in G(H)/K} \chi_{g,w}^{-1} g = 1 - \frac{1}{|G|} \sum_{g \in G} \sum_{w \in G(H)/K} \chi_{g,w}^{-1} g.$$

Note that by Proposition 4.1

$$\sum_{w \in G(H)/K} \chi_{g,w}^{-1} = \begin{cases} |G(H)/K| = \frac{|G(H)|}{2}, & \chi_{g,w}^{-1} = 1 \text{ for all } w \in G(H)/K, \\ 0, & \text{otherwise.} \end{cases}$$

Let as above F be the set of all $g \in G$ such that $\chi_{g,w} = 1$ for all $w \in G(H)$. Then F is a normal subgroup in G, containing the derived subgroup G'. It follows that

$$e = 1 - \sum_{w \in G(H)} e_w = 1 - \frac{|G(H)/K|}{|G|} \sum_{g \in F} g = 1 - \frac{|G(H)|}{2|G|} \sum_{g \in F} g \tag{22}$$

is a sum of central indecomposable idempotents of H^* corresponding to simple noncommutative components of H^*. Then this element has the property that it belongs to $\ker \varepsilon$. Hence $1 = \frac{|G(H)||F|}{2|G|}$ that is $2|G| = |G(H)| \cdot |F|$.

Proposition 5.3. *The following assertions take place.*

1) $\mathrm{Mat}(n, k)^*$ *as a vector space has a decomposition*

$$\mathrm{Mat}(n, k)^* = M' \oplus M,$$

$$M' = \oplus_{w \in G(H)/K} kZ_w, \quad M = e * \mathrm{Mat}(n, k)^* \tag{23}$$

where e is from (22). Moreover M' is a subalgebra in $\mathrm{Mat}(n,k)$ of dimension $\frac{|G(H)|}{2}$. If $f \in F$ then $f * X = X$ for any $X \in M'$.

2) If P is the ideal from Proposition 5.2 in the algebra H^*, then $P = e\,kG \oplus M$, where

$$\dim e\,kG = \dim M = n^2 - \frac{|G(H)|}{2}.$$

Proof. The equality (21) and Proposition 5.2 imply the decomposition (23) of $\mathrm{Mat}(n,k)^*$ as a vector space. By Proposition 4.1 the set M' is a subalgebra in $\mathrm{Mat}(n,k)^*$. The dimension of M' is equal to the maximal number of linearly independent elements among Z_w. If $Z_w \neq -Z_{w'}$ then by Theorem 1.3 they are eigenvectors for the action of G with different characters. Hence they are independent and therefore

$$\dim M' = \frac{|G(H)|}{2}.$$

If $f \in F$ then $\chi_{g,w} = 1$ for all $w \in G(H)$. Hence by the property 1.3) from Theorem 1.2 and by Proposition 3.5 $f * Z_w = f \rightharpoonup Z_w = \chi_{g,w} Z_w = Z_w$.

(22) shows that $\dim e\,kG = |G| - \frac{|G(H)|}{2}$ and $\dim P = \dim H^* - |G(H)| = 2|G| - |G(H)|$. Thus

$$\dim M = \dim P - \dim e\,kG = |G| - \frac{|G(H)|}{2}. \qquad \square$$

Note that M' by Corollary 3.2 has orthogonal base Z_w, $w \in G(H)/K$ with respect to the from (6).

Proposition 5.4. *M consists of all matrices X such that $\mathrm{tr}\,(XZ_w) = 0$ for all $w \in G(H)$. In particular M is the orthogonal complement to M' with respect to the bilinear form (6).*

Proof. By (21) M consists of all matrices X which are annihilated by each element e_w. Applying Proposition 5.2, Proposition 3.5 and Theorem 1.3 we obtain

$$0 = X * e_w = X * \left(\frac{1}{2n^2} \sum_g \chi_{g,w}^{-1} g + \frac{1}{2n} Z_w \right)$$

$$= \frac{1}{2n^2} \sum_g \chi_{g,w}^{-1} X * g + \frac{1}{2n} \sum_{h \in G} \langle Z_w * h^{-1}, X \rangle h$$

$$= \frac{1}{2n^2} \sum_g \chi_{g,w}^{-1} X * g + \frac{1}{2n} \sum_{h \in G} \chi_{h,w}^{-1} \langle Z_w, X \rangle h.$$

In other words $0 = \sum_g \chi_{g,w}^{-1} X * g = \langle X, Z_w \rangle$. It follows by Theorem 1.3 that

$$0 = \langle X, Z_w \rangle = \mathrm{tr}\,\left(U^{-1} X U\,^t Z_w \right) = \mathrm{tr}\,\left(X U\,^t Z_w U^{-1} \right) = \mathrm{tr}\,\left(X Z_w^{-1} \right).$$

By (19) the equality $\mathrm{tr}(X Z_w) = 0$ for all w implies $0 = \sum_g \chi_{g,w}^{-1} X * g$. $\qquad \square$

Proposition 5.5. *If M is from Proposition 5.3 then $\dim M \geqslant n^2 - n - t$ for some integer $0 \leqslant t \leqslant \frac{n}{2}$. If $G(H)$ is Abelian then $t = 0$.*

Proof. The set of matrices $Z_w, w \in G(H)$, by Proposition 4.1 define a faithful representation of the group $G(H)$. By Proposition 4.2 the group $G(H)$ has an Abelian subgroup of index 2. Hence by [AS, Chapter 3.5, Theorem 3.7] dimensions of irreducible representation of $G(H)$ are 1 or 2 and therefore there exists an invertible matrix $S \in \mathrm{GL}(n, k)$ such that each matrix $S^{-1} Z_w S$ has block diagonal form

$$S^{-1} Z_w S = \begin{pmatrix} Y_{w,1} & 0 & 0 \\ 0 & Y_{w,2} & 0 \\ \cdots\cdots\cdots\cdots\cdots \\ 0 & 0 & Y_{w,s} \end{pmatrix}$$

where blocks $Y_{w,j}$ has size 1 or 2 which corresponds to an irreducible representation of $G(H)$.

If $X \in M$ and $S^{-1} X S$ has corresponding partition into blocks

$$S^{-1} X S = \begin{pmatrix} X_{11} & X_{12} & X_{1s} \\ X_{21} & X_{22} & X_{2s} \\ \cdots\cdots\cdots\cdots\cdots \\ X_{s1} & X_{s2} & X_{ss} \end{pmatrix}, \tag{24}$$

then Proposition 5.4 means that $\sum_{j=1}^{s} \mathrm{tr}\, (X_{jj} Y_{w,j}) = 0$ for any $w \in G(H)$. Hence M contains all matrices X such that $X_{jj} = 0$ in (24) for all j. Suppose that t diagonal blocks in (24) have size 2 and the other $s - t$ have size 1. Then $n = 2t + s$ and the dimension of the space of matrices from (24) with zero diagonal blocks is equal to $n^2 - 4t - (s - t) = n^2 - 3t - s$. Since $s = n - 2t$ we have $n^2 - 2t - s = n^2 - 3t - n + 2t = n^2 - n - t$. $\qquad\square$

Combining with Proposition 5.3 we can prove

Corollary 5.6. $|G(H)| \leqslant 3n$. *If $G(H)$ is Abelian then $|G(H)| \leqslant 2n$.*

Proof. By Proposition 5.3 $\dim M = n^2 - \frac{|G(H)|}{2}$. Hence

$$n^2 - \frac{|G(H)|}{2} \geqslant n^2 - n - t$$

and therefore $\frac{|G(H)|}{2} \leqslant n + t$ or $|G(H)| \leqslant 2n + 2t \leqslant 3n$. If $G(H)$ is Abelian then $t = 0$. $\qquad\square$

6. Non-isomorphism of H and of $\tilde{H}^*$

In this section we shall prove that if H and $\tilde{H}$ are Hopf algebras from Theorem 1.2 then $\tilde{H}^*$ and H are non-isomorphic Hopf algebras.

Suppose the opposite that there exists a Hopf algebra isomorphism $\xi : H^* \to \tilde{H}$. Then H and $\tilde{H}$ have the same dimension $2n^2$ and ξ induces group isomorphism $\xi : G = G(H^*) \to G(\tilde{H})$. So $|G(\tilde{H})| = n^2$. Applying Corollary 5.6 we can conclude that $n^2 \leqslant 3n$ or $n \leqslant 3$. In this case the group $G(\tilde{H})$ has order 4, 9 because $n > 1$ and therefore $G(\tilde{H})$ is Abelian. Hence $G(\tilde{H})$ has order 4. So we have proved

Theorem 6.1. *Let $n > 2$. Then H^* is not isomorphic to any Hopf algebra $\tilde{H}$ belonging to the class of Hopf algebras from Theorem 1.2.*

It means that the Hopf algebra H^* of dimension $2n^2 > 8$ where H is from Theorem 1.2 is a series of semisimple Hopf algebras which is not included in Theorem 1.2.

Notice that Theorem 6.1 could be proved in another way. If H^* is an algebra from Theorem 1.1 then the ideal $P = eH^*$ from Proposition 5.2 and Proposition 5.3 is a 2-graded ideal in H^* and it is simple. Following [BZ] we can classify 2-gradings in P.

7. Triangularity

A Hopf algebra H is *triangular* if there exists an invertible element $R \in H \otimes H$ such that $R\Delta(x)R^{-1} = \tau(\Delta(x))$ for all $x \in H$ where $\tau : H \otimes H \to H \otimes H$ is a twist $\tau(a \otimes b) = b \otimes a$ for all $a, b \in H$.

Theorem 7.1. *Let H be from Theorem 1.1. Suppose that there exists an invertible matrix T such that $T(g \rightharpoonup x)T^{-1} = x \leftharpoonup g$ for all $x \in \mathrm{Mat}(n,k)$ and all $g \in G$. Then H is triangular.*

Proof. Let

$$\mathcal{R} = \sum_{i,j} E_{ij} \otimes E_{ji} \in \mathrm{Mat}(n,k) \otimes \mathrm{Mat}(n,k).$$

Is was shown in [A][§5] that $\mathcal{R}^2 = E \otimes E$ and $(A \otimes B)\mathcal{R} = \mathcal{R}(B \otimes A)$ for all $A, B \in \mathrm{Mat}(n,k)$. Let $e = \sum_{g \in G} e_g \in kG$ and

$$R = e \otimes e + T \otimes e + e \otimes T^{-1} + \mathcal{R}$$

We claim that $R^{-1} = e \otimes e + T^{-1} \otimes e + e \otimes T + \mathcal{R}$. In fact

$$\begin{aligned}
R\,R^{-1} &= (e \otimes e)^2 + (T \otimes e)\left(T^{-1} \otimes e\right) + \left(e \otimes T^{-1}\right)(e \otimes T) + \mathcal{R}^2 \\
&= e \otimes e + E \otimes e + e \otimes E + E \otimes E = (e + E) \otimes (e + E) \\
&= 1 \otimes 1.
\end{aligned}$$

Suppose that $u = \sum_{g \in G} \alpha_g e_g + x \in H$ where $\alpha_g \in k$ and $x \in \mathrm{Mat}(n,k)$. Then by Theorem 1.1

$$R\Delta(u)R^{-1} = \left(e \otimes e + T \otimes e + e \otimes T^{-1} + \mathcal{R}\right)$$
$$\times \left[\left(\sum_{g \in G} \alpha_g \left(\sum_{h \in H} e_h \otimes e_{h^{-1}g} + \Delta_g \right) \right) \right.$$
$$\left. + \sum_{g \in G} ((g \rightharpoonup x) \otimes e_g + e_g \otimes (x \leftharpoonup g)) + \Delta'(x) \right]$$
$$\times \left(e \otimes e + T^{-1} \otimes e + e \otimes T + \mathcal{R}\right)$$

$$= \sum_{g \in G} \alpha_g \left(\sum_{h \in H} e_h \otimes e_{h^{-1}g} \right)$$

$$+ \sum_{g \in G} \left(T(g \rightharpoonup x)T^{-1} \otimes e_g + e_g \otimes T(x \leftharpoonup g)T^{-1} \right)$$

$$+ \sum_g \alpha_g \mathcal{R} \Delta_g \mathcal{R} + \mathcal{R} \Delta'(x) \mathcal{R}$$

$$= \tau \left(\Delta \left(\sum_g \alpha_g e_g \right) + \Delta'(x) \right)$$

$$+ \sum_{g \in G} \left(T(g \rightharpoonup x)T^{-1} \otimes e_g + e_g \otimes T^{-1}(x \leftharpoonup g)T \right).$$

By the assumption $T(g \rightharpoonup x)T^{-1} = x \leftharpoonup g$ and therefore $T^{-1}(x \leftharpoonup g)T = g \rightharpoonup x$ for all $x \in \mathrm{Mat}(n,k)$ and all $g \in G$. $\qquad\square$

References

[A] Artamonov V.A. On semisimple finite dimensional Hopf algebras, Math. Sbornik (to appear).

[AS] Artamonov V.A., Slovokhotov Yu.L. Group theory and their applications in physics, chemistry and crystallography, Publishing center "Academia", Moscow, 2005.

[B] Bourbaki N., Algebra, Ch. IX. Paris, Hermann.

[BZ] Bahturin Y.A., Zaicev M.V., Group gradings on associative algebras, J. Algebra, 241(2001), 677–698.

[CR] Curtis Ch.W., Reiner I. Representation theory of finite groups and associative algebras. Interscience Publ., Witney & Sons, New York, London, 1962.

[JHLMYYZ] Jiang-Hua Lu, Min Yanm, Youngchang Zhu, Quasi-triangular structure on Hopf algebras with positive bases. Contemp. Math., 267(2000), 339–356.

[H] Huppert B., Character theory of finite groups, de Gruyter Expositions in Mathematics 25, 1998.

[KP] Kac, G., Paljutkin, V., Finite ring groups, Trudy Moscow Math. Obschestva, 15 (1966), 224–261.

[LR] Larson, R., Radford, D., Finite-dimensional cosemisimple Hopf algebras in characteristic zero are semisimple. J. Algebra, 117(1988), 267–289.

[M] Montgomery, S., Hopf Algebras and Their Actions on Rings, Regional Conf. Ser. Math. Amer. Math. Soc., Providence RI, 1993.

[M1] Montgomery S., Classifying finite-dimensional semisimple Hopf algebras, Contemp. Math., 229(1998), 265–279.

[N1] Natale S., On group theoretical algebras and exact factorizations of finite groups, J. Algebra, 270 (2003), 190–211.

[N2] Natale S., Semisolvability of semisimple Hopf algebras of low dimension, Memoirs of AMS vol. 186, 2007.

[NZ] Nichols W., Zoeller M., A Hopf algebra freeness theorem, Amer. J. Math., 111(1989), 381–385.

[OM] O'Meara O.T., Symplectic groups, Providence RI, Amer. Math. Soc., 1978.

[P]   Pierce, R.S., Associative algebras, Springer-Verlag, New York, Heidelberg, Berlin, 1982.

[S]   Schneider H.-J., Lectures on Hopf algebras, Universidad de Cordoba, Trabajos de Matematica, N 31/95, Cordoba (Argentina), 1995.

[Se] Seitz G.M., Finite groups having only one irreducible representation of degree greater than one. *Proc. Amer. Math. Soc.* 19(1968), 459–461.

[T]   Tambara D., Representations of tensor categories with fusion rules of self-duality for finite abelian groups, Israel J. Math., 118(2000), 29–60.

[TY] Tambara D., Yamagami S., Tensor categories with fusion rules of self-duality for finite abelian groups, J.Algebra 209(1998), 692–707.

V.A. Artamonov and I.A. Chubarov
Department of Algebra
Faculty of Mechanics and Mathematics
Moscow State University
e-mail: artamon@bk.ru
e-mail: igrek@dubki.ru

Modules and Comodules

Trends in Mathematics, 87–99

Balanced Bilinear Forms for Corings

M. Beattie, D. Bulacu and Ş. Raianu

Dedicated to Robert Wisbauer on the occasion of his 65th birthday

Abstract. We review the role that balanced bilinear forms play in the definitions of properties of corings and suggest a definition for a coring to be symmetric.

1. Introduction

Let k be a commutative ring and A a finite dimensional k-algebra. An element e in $A \otimes_k A$ is called A-central, or a Casimir element [3, Section 1.3], if $ae = ea$ for all $a \in A$. Various properties of A are defined in terms of A-central elements. For example, A is k-separable if and only if A has an A-central element e such that $\pi(e) = 1$ where $\pi : A \otimes_k A \to A$ is the usual map $\pi(a \otimes_k b) = ab$. A is a Frobenius k-algebra if and only if there exists an A-central element e and a map $\epsilon \in A^*$ such that $(\epsilon \otimes_k A)(e) = (A \otimes_k \epsilon)(e) = 1$. Equivalently, A is Frobenius if and only if there is a nondegenerate bilinear map $B : A \times A \to k$ such that $B(xy, z) = B(x, yz)$. The algebra A is symmetric if A is Frobenius and $B(x, y) = B(y, x)$.

For C a coalgebra over k, various analogous properties of C may be defined in terms of balanced bilinear forms from $C \otimes_k C$ to k, generalizing the idea of A-central element. A C^*-balanced form is a k-bilinear form B from $C \otimes_k C$ to k such that $B(c \leftharpoonup c^*, d) = B(c, c^* \rightharpoonup d)$ for all $c, d \in C$, $c^* \in C^*$ with the usual actions of C^* on C. For k a field, the idea of a symmetric coalgebra was recently defined in [4], namely that a k-coalgebra C is symmetric if and only if there is a nondegenerate symmetric balanced bilinear form B from $C \otimes_k C$ to k.

For C an A-coring, where A is not necessarily commutative, the situation is complicated by the presence of left and right duals. However, the idea of balanced bilinear forms from $C \otimes_A C$ to A still makes sense and is used to define various

---

Support for the first author's research, and partial support for the second author, came from an NSERC Discovery Grant. The second author held a postdoctoral fellowship at Mount Allison University during this project. He would like to thank Mount Allison University for their warm hospitality.

properties of corings analogous to those for coalgebras. We recall some of these properties, and, in the last section, suggest a working definition for the notion of a symmetric coring along with some examples.

We will work over a commutative ring k and all maps are assumed to be k-linear. Throughout this paper, A will denote a not necessarily commutative k-algebra. We will use the Sweedler summation notation for comultiplication, but omitting the summation sign. For background on coalgebras over a field we refer the reader to [5]. The first chapter of [2] contains basics on coalgebras over a commutative ring. The identity map on a k-module X is denoted simply as X.

2. Corings

Recall that for A a not necessarily commutative ring, an A-coring $\mathcal{C}$ is defined to be a coalgebra in the monoidal category of (A,A)-bimodules, $({}_A\mathcal{M}_A, \otimes_A, A)$. More precisely, $\mathcal{C}$ is an (A,A)-bimodule, together with (A,A)-bimodule maps $\Delta_{\mathcal{C}} : \mathcal{C} \to \mathcal{C} \otimes_A \mathcal{C}$ and $\epsilon_{\mathcal{C}} : \mathcal{C} \to A$ such that $\Delta_{\mathcal{C}}$ is co-associative and the compatibility conditions $(\epsilon_{\mathcal{C}} \otimes_A \mathcal{C}) \circ \Delta_{\mathcal{C}}(c) = c$ and $(\mathcal{C} \otimes_A \epsilon_{\mathcal{C}}) \circ \Delta_{\mathcal{C}}(c) = c$, for all $c \in \mathcal{C}$, hold.

For definitions and details about corings, we refer the reader either to [2] or, for the original definition, to [9]. We will normally write (A,A)-bimodule actions on a module $M \in {}_A\mathcal{M}_A$ by juxtaposition, i.e., we write amb for the left action of a and the right action of b on m.

If $A = k$, then we recover the definition of a coalgebra over the commutative ring k. Simple examples of corings include the following.

Examples 2.1.

(i) **Trivial coring.** For A a ring, let $\mathcal{C} = A$ itself and define $\Delta_{\mathcal{C}}(a) = a \otimes_A 1 = 1 \otimes_A a$ and $\epsilon_{\mathcal{C}}(a) = a$.

(ii) **Matrix coring.** (See [2, 17.7].) For A a ring, let $\mathcal{C} = M_n^c(A)$, n by n matrices over A with A-basis e_{ij}, $1 \le i,j \le n$ and $ae_{ij} = e_{ij}a$. Then $(\mathcal{C}, \Delta_{\mathcal{C}}, \epsilon_{\mathcal{C}})$ is an A-coring where $\Delta_{\mathcal{C}}(e_{ij}) = \sum_{k=1}^n e_{ik} \otimes_A e_{kj}$ and $\epsilon_{\mathcal{C}}(e_{ij}) = \delta_{i,j}$, and the maps are extended by A-linearity.

(iii) **An entwining structure example.** Let k be a field, let H be a k-Hopf algebra, and A a right H-comodule algebra via $a \mapsto a_{[0]} \otimes_k a_{[1]}$. Then $A \otimes_k H$ becomes an A-coring as follows: the left A-module structure is given by multiplication on the first component, and the right A-module structure is given by $(a \otimes_k h)b = ab_{[0]} \otimes_k hb_{[1]}$. The comultiplication is $\Delta : A \otimes_k H \longrightarrow (A \otimes_k H) \otimes_A (A \otimes_k H) \simeq A \otimes_k H \otimes_k H$, $\Delta(a \otimes_k h) = (a \otimes_k h_{(1)}) \otimes_A (1_A \otimes_k h_{(2)})$, and the counit is $\epsilon : A \otimes_k H \longrightarrow A$, $\epsilon(a \otimes_k h) = \epsilon(h)a$.

(iv) **Opposite coring.** Let A^o be the opposite algebra of A. If M is an (A,A)-bimodule, then M is also an (A^o, A^o)-bimodule, denoted M^o, as usual, via $a^o m^o b^o = (bma)^o$. The twist map $\tau : M \otimes_A M \to M^o \otimes_{A^o} M^o$ defined by $\tau(m \otimes_A n) = n^o \otimes_{A^o} m^o$ is a k-module isomorphism. The opposite coring, denoted $\mathcal{C}^o$, is defined [6, 1.7] to be the A^o-coring $(\mathcal{C}, \Delta_{\mathcal{C}}^o, \epsilon_{\mathcal{C}})$, where comultiplication $\Delta_{\mathcal{C}}^o(c^o) = \tau \circ \Delta_{\mathcal{C}}(c) = (c_2)^o \otimes_{A^o} (c_1)^o$.

If $A = k$, and C is a k-coalgebra, then the dual C^* is an algebra via the convolution map. However, for C a coring, there are right and left dual rings associated with C, denoted C^* and $^*C$.

Following [2, 17.8], we write $C^* := \mathrm{Hom}_A(C, A)$, the right A-module homomorphisms from C to A. C^* has a ring structure with associative multiplication $*^r$ given by $f *^r g(c) = g(f(c_1)c_2)$ and unit ϵ_C. There is a ring morphism from A^o to C^* by $a^o \mapsto \epsilon_C(a-)$. Thus C^* has left A^o-action via $(\epsilon_C(b-) *^r c^*)(c) = c^*(bc)$ and, similarly, a right A^o-action. Then C^* is an (A, A)-bimodule via $(ac^*b)(c) = ac^*(bc)$.

Similarly, $^*C := {}_A\mathrm{Hom}(C, A)$, the left A-module homomorphisms from C to A. $^*C$ has a ring structure with associative multiplication $*^l$ given by $f *^l g(c) = f(c_1 g(c_2))$ and unit ϵ_C. There is a ring morphism from A^o to $^*C$ given by $a \mapsto \epsilon_C(-a)$. Then $^*C$ is an (A, A)-bimodule via $(ac^*b)(c) = c^*(ca)b$.

Note that convolution is not well defined on either C^* or $^*C$. The problem is that Δ_C maps to $C \otimes_A C$ and A-linearity may fail.

Again, using notation from [2], we denote $^*C^* := {}_A\mathrm{Hom}_A(C, A) = {}^*C \cap C^*$ to be the set of (A, A)-bimodule maps from C to A. On $^*C^*$ the associative multiplications $*^l$ and $*^r$ both equal the convolution multiplication, so that $^*C^*$ with the convolution multiplication $f * g(c) = f(c_1)g(c_2)$ is an associative ring with unit ϵ_C.

Remark 2.2. $^*C^*$ is not a left or a right A-module under either the A-module structures of $^*C$ or of C^*. For suppose f lies in $^*C^*$ and we attempt to define af by $(af)(c) = f(ca)$ (the left A-module structure on $^*C$). Now af may not lie in $^*C^* \subset C^*$ since $(af)(cb) = f(cba) = f(c)ba$ which is not, in general, equal to $(af)(c)b = f(c)ab$. Similarly, if we let $(af)(c) = af(c)$, then af may not lie in $^*C^*$. However, if $a \in Z(A)$, the centre of A, then the definitions above of left module structure agree and $af \in {}^*C^*$, so that $^*C^*$ is a left $Z(A)$-module. Similarly, the right module structures on $^*C$ and C^* also may not induce right A-module structures on $^*C^*$ but do induce a right $Z(A)$-module structure.

For C a coalgebra over a commutative ring k, it is well known that C is a (C^*, C^*)-bimodule. For corings the situation is somewhat different.

The coring C is a right C^*-module via $c \leftharpoonup c^* = c^*(c_1)c_2$ and the right A-action on C commutes with the right C^*-action. However, C is not an (A, C^*)-bimodule in general since, for $c^* \in C^*$, we need not have equality of $c^*(ac_1)c_2$ and $ac^*(c_1)c_2$. Similarly, there is a left $^*C$-action on C which commutes with the left A-action given by $c^* \rightharpoonup c = c_1 c^*(c_2)$. It follows from the coassociativity of Δ_C and the fact that Δ_C is an (A, A)-bimodule map, that C is a $(^*C, C^*)$-bimodule with the above actions.

3. Balanced bilinear forms for corings

Recall that $M \in {}_A\mathcal{M}$ is locally projective as a left A-module if and only if for any finite set S of elements of M, there exist $x_1, \ldots, x_n \in M, f_1, \ldots, f_n \in {}_A\mathrm{Hom}(M, A) = {}^*M$ such that $m = \sum_{i=1}^n f_i(m)x_i$, for any $m \in S$. Any object of $\mathcal{M}^{\mathcal{C}}$, the category of right $\mathcal{C}$-comodules, can be viewed as an object of $_{*\mathcal{C}}\mathcal{M}$, the category of left $^*\mathcal{C}$-modules, and if $\mathcal{C}$ is locally projective, then $\mathcal{M}^{\mathcal{C}}$ is a full subcategory of $_{*\mathcal{C}}\mathcal{M}$. Moreover, the rational functor $\mathrm{Rat}^{\mathcal{C}} : {_{*\mathcal{C}}\mathcal{M}} \to \mathcal{M}^{\mathcal{C}}$ is defined. Similar statements hold for right local projectivity.

Suppose that $\mathcal{C}$ is locally projective as a left A-module. Recall from [2, Section 20], that if $M \in \mathcal{M}_A$ is a left $^*\mathcal{C}$-module, $\mathrm{Rat}^{\mathcal{C}}(M)$, the rational submodule of M, may be defined as the set of rational elements of M, where $m \in M$ is called rational if there exists $\sum_{i=1}^{n} m_i \otimes_A c_i \in M \otimes_A \mathcal{C}$ such that $\phi \cdot m = \sum_i m_i \phi(c_i)$, for all ϕ in $^*\mathcal{C}$. By the locally projective condition on $\mathcal{C}$, $\sum_i m_i \otimes_A c_i$ is uniquely determined and so these elements define a right $\mathcal{C}$-comodule structure on $\mathrm{Rat}^{\mathcal{C}}(M)$. For $M = {^*\mathcal{C}}$ we have that $\mathrm{Rat}^{\mathcal{C}}(^*\mathcal{C})$ is an ideal of $^*\mathcal{C}$ (and thus also an A-subbimodule of $^*\mathcal{C}$).

For $\mathcal{C}$ locally projective as a right A-module, similar statements hold. Here the rational functor is denoted $^{\mathcal{C}}\mathrm{Rat} : \mathcal{M}_{\mathcal{C}^*} \to {^{\mathcal{C}}\mathcal{M}}$, and $^{\mathcal{C}}\mathrm{Rat}(\mathcal{C}^*)$ is an ideal of $\mathcal{C}^*$.

With the (A, A)-bimodule structure on $\mathcal{C} \otimes_A \mathcal{C}$ given by $a(c \otimes_A c')b = ac \otimes_A c'b$, the set of balanced bilinear forms on $\mathcal{C}$ is defined as follows.

Definition 3.1. *The set of balanced bilinear forms on $\mathcal{C}$, denoted* $\mathrm{bbf}(\mathcal{C})$, *is defined to be the set of* $\sigma \in {_A\mathrm{Hom}_A}(\mathcal{C} \otimes_A \mathcal{C}, A)$ *such that*

$$(\sigma \otimes_A \mathcal{C}) \circ (\mathcal{C} \otimes_A \Delta_{\mathcal{C}}) = (\mathcal{C} \otimes_A \sigma) \circ (\Delta_{\mathcal{C}} \otimes_A \mathcal{C})$$

or, equivalently,

$$\sigma(c \otimes_A d_1)d_2 = c_1 \sigma(c_2 \otimes_A d), \quad \text{for all } c, d \in \mathcal{C}. \tag{3.1}$$

By [2, 6.4], if $\mathcal{C} = C$ is a locally projective k-coalgebra, then (3.1) is equivalent to the defining relation for C^*-balanced forms B given in the introduction.

As usual, $\sigma \in \mathrm{bbf}(\mathcal{C})$ is called right nondegenerate if $\sigma(c \otimes_A \mathcal{C}) = 0$ implies $c = 0$, and left nondegenerate if $\sigma(\mathcal{C} \otimes_A c) = 0$ implies $c = 0$.

Examples 3.2.

(i) For $\mathcal{C} = A$, the trivial coring of Examples 2.1, let $\sigma = \epsilon_{\mathcal{C}} \otimes_A \epsilon_{\mathcal{C}}$. It is easy to see that $\sigma \in \mathrm{bbf}(\mathcal{C})$ and is nondegenerate.

(ii) For $\mathcal{C}$ the n by n matrix coring of Examples 2.1, define σ to be the A-linear map from $\mathcal{C} \otimes_A \mathcal{C}$ to k defined on generators by $\sigma(e_{ij} \otimes_A e_{kl}) = \delta_{i,l}\delta_{j,k}$. Again, it is easily checked that $\sigma \in \mathrm{bbf}(\mathcal{C})$ and is nondegenerate.

(iii) Let $\mathcal{C}$ be the coring $A \otimes_k H$ from Examples 2.1 and suppose that H has a left integral t in H^*. It is well known that $B(h, g) = t(hS(g))$ where S is the antipode of H, is a balanced form on H since

$$h_1 t(h_2 S(g)) = h_1 S(g_2) g_3 t(h_2 S(g_1)) = t(hS(g_1))g_2 \quad \text{for all} \quad g, h \in H.$$

Now define $\sigma : \mathcal{C} \otimes_A \mathcal{C} \to A$ by

$$\sigma((a \otimes_k h) \otimes_A (b \otimes_k g)) = ab_{[0]} t(hb_{[1]} S(g)).$$

It is easy to check that σ is A-bilinear and the balanced property comes from that of B.

(iv) Let $\mathrm{bbf}(\mathcal{C}^o)$ denote the set of balanced bilinear forms for the A^o-coring $\mathcal{C}^o$. For $\sigma \in \mathrm{bbf}(\mathcal{C})$, define $\sigma^o : \mathcal{C}^o \otimes_{A^o} \mathcal{C}^o \to A^o$ by $\sigma^o(c^o \otimes_{A^o} d^o) = (\sigma \circ \tau(c^o \otimes_{A^o} d^o))^o = (\sigma(d \otimes_A c))^o$. It is easily checked that $\sigma^o \in \mathrm{bbf}(\mathcal{C}^o)$.

Now, let $\sigma \in \mathrm{bbf}(\mathcal{C})$. Since σ is a left and right A-module map, σ induces a well-defined left A-module map $\sigma^r : \mathcal{C} \to \mathcal{C}^*$, given by $\sigma^r(c)(d) = \sigma(c \otimes_A d)$. It is easily checked that σ^r is also a right A-module map and a right $\mathcal{C}^*$-module map. Similarly, $\sigma^l : \mathcal{C} \to {}^*\mathcal{C}$ defined by $\sigma^l(d)(c) = \sigma(c \otimes_A d)$ is well defined, a left ${}^*\mathcal{C}$-module and an (A, A)-bimodule map.

The next lemma is a straightforward generalization to corings of well-known facts for coalgebras. Statements (i) and (ii) can be found in [8, Proposition 1] for coalgebras over a field, and in [2, Section 6.6] for locally projective coalgebras over a commutative ring.

Lemma 3.3. *We have the following bijective correspondences.*

(i) *Let $\mathcal{C}$ be an A-coring which is locally projective as a right and as a left A-module. There is a bijective correspondence between $\mathrm{bbf}(\mathcal{C})$ and the set of (A, A)-bimodule, right $\mathcal{C}^*$-module maps from $\mathcal{C}$ to ${}^{\mathcal{C}}\mathrm{Rat}(\mathcal{C}^*)$. Under this correspondence, right nondegenerate forms correspond to monomorphisms from $\mathcal{C}$ to ${}^{\mathcal{C}}\mathrm{Rat}(\mathcal{C}^*)$.*

(ii) *For $\mathcal{C}$ an A-coring which is locally projective as a right and as a left A-module, there is a bijective correspondence between $\mathrm{bbf}(\mathcal{C})$ and the set of (A, A)-bimodule, left ${}^*\mathcal{C}$-module maps from $\mathcal{C}$ to $\mathrm{Rat}^{\mathcal{C}}({}^*\mathcal{C})$. Under this correspondence, left nondegenerate forms correspond to monomorphisms from $\mathcal{C}$ to $\mathrm{Rat}^{\mathcal{C}}({}^*\mathcal{C})$.*

(iii) *For any A-coring $\mathcal{C}$, (not necessarily locally projective), there is a bijective correspondence between $\mathrm{bbf}(\mathcal{C})$ and the set of $(\mathcal{C}, \mathcal{C})$-bicomodule maps from $\mathcal{C} \otimes_A \mathcal{C}$ to $\mathcal{C}$, where $\mathcal{C} \otimes_A \mathcal{C}$ is a left (right) $\mathcal{C}$-comodule via $\triangle_{\mathcal{C}} \otimes_A \mathcal{C}$ ($\mathcal{C} \otimes_A \triangle_{\mathcal{C}}$, respectively).*

Proof. (i) First we show that there is a bijective correspondence between $\mathrm{bbf}(\mathcal{C})$ and ${}_A\mathrm{Hom}_A(\mathcal{C}, \mathcal{C}^*) \cap \mathrm{Hom}_{\mathcal{C}^*}(\mathcal{C}, \mathcal{C}^*)$. Suppose that $\varphi : \mathcal{C} \to \mathcal{C}^*$ is an (A, A)-bimodule, right $\mathcal{C}^*$-module map. Define $\sigma := \sigma_\varphi : \mathcal{C} \otimes_A \mathcal{C} \to A$ by $\sigma_\varphi(c \otimes_A d) = \varphi(c)(d)$, for all $c, d \in \mathcal{C}$. We must show that σ_φ is a well-defined (A, A)-bimodule map and is balanced.

Since φ is a right A-module map, we have

$$\sigma(ca \otimes_A d) = \varphi(ca)(d) = (\varphi(c)a)(d) = \varphi(c)(ad) = \sigma(c \otimes_A ad),$$

and so σ is well defined. Next, note that for $a, a' \in A$, since $\varphi(ac) = a\varphi(c)$ and $\varphi(ac) \in \mathcal{C}^*$, then

$$\sigma(ac \otimes_A da') = \varphi(ac)(da') = a(\varphi(c)(d))a',$$

so that σ is an (A, A)-bimodule map. Finally, to see that $\sigma \in \mathrm{bbf}(\mathcal{C})$, note that for $c \in \mathcal{C}$, $c^* \in \mathcal{C}^*$, since $\varphi(c \leftharpoonup c^*) = \varphi(c) *^r c^*$, then

$$c^*(c_1 \sigma(c_2 \otimes_A d)) = c^*(\sigma(c \otimes_A d_1)d_2),$$

for all $d \in \mathcal{C}$. Hence $c_1\sigma(c_2 \otimes_A d)$ and $\sigma(c \otimes_A d_1)d_2$ have the same image under every $c^* \in \mathcal{C}^*$ and thus, since $\mathcal{C}$ is locally projective as an A-module, they are equal.

Conversely, given $\sigma \in \mathrm{bbf}(\mathcal{C})$, define the right $\mathcal{C}^*$-module map σ^r as above. It is easy to check that $\sigma \mapsto \sigma^r$ and $\varphi \mapsto \sigma_\varphi$ define inverse bijections between $\mathrm{bbf}(\mathcal{C})$ and ${}_A\mathrm{Hom}_A(\mathcal{C},\mathcal{C}^*) \cap \mathrm{Hom}_{\mathcal{C}^*}(\mathcal{C},\mathcal{C}^*)$. Clearly, right nondegenerate forms correspond to monomorphisms.

Now we show that any map in ${}_A\mathrm{Hom}_A(\mathcal{C},\mathcal{C}^*) \cap \mathrm{Hom}_{\mathcal{C}^*}(\mathcal{C},\mathcal{C}^*)$ has its image in ${}^\mathcal{C}\mathrm{Rat}(\mathcal{C}^*)$ so that ${}_A\mathrm{Hom}_A(\mathcal{C},\mathcal{C}^*) \cap \mathrm{Hom}_{\mathcal{C}^*}(\mathcal{C},\mathcal{C}^*)$ is the set of (A,A) bimodule, right $\mathcal{C}^*$-module maps from $\mathcal{C}$ to ${}^\mathcal{C}\mathrm{Rat}(\mathcal{C}^*)$. Let $\sigma \in \mathrm{bbf}(\mathcal{C})$ and $\varphi = \sigma^r$; we show that every element $\varphi(c)$ is rational, i.e., for each $c \in \mathcal{C}$, there exists $\sum_i c_i \otimes_k x_i \in \mathcal{C} \otimes_k \mathcal{C}^*$ such that $\varphi(c) *^r c^* = \sum_i c^*(c_i)x_i$, for all $c^* \in \mathcal{C}^*$. Let $\sum_i c_i \otimes_k x_i = c_1 \otimes_k \varphi(c_2)$. We have that for any $c^* \in \mathcal{C}^*$ and $c, d \in \mathcal{C}$,

$$
\begin{aligned}
(\varphi(c) *^r c^*)(d) &= c^*(\varphi(c)(d_1)d_2) = c^*(\sigma(c \otimes_A d_1)d_2) \\
&= c^*(c_1\sigma(c_2 \otimes_A d)) = c^*(c_1)\varphi(c_2)(d) \text{ (since } c^* \in \mathcal{C}^*),
\end{aligned}
$$

so that $\varphi(c) *^r c^* = c^*(c_1)\varphi(c_2)$, showing that $\varphi(c)$ is rational.

(ii) For $\gamma : \mathcal{C} \to {}^*\mathcal{C}$ an (A,A)-bimodule, left ${}^*\mathcal{C}$-module map, define $\sigma_\gamma : \mathcal{C} \otimes_A \mathcal{C} \to A$ by $\sigma_\gamma(c \otimes_A d) = \gamma(d)(c)$. Conversely, for $\sigma \in \mathrm{bbf}(\mathcal{C})$, define σ^l as above. The proof that these provide inverse bijections and that for $\gamma \in {}_A\mathrm{Hom}_A(\mathcal{C}, {}^*\mathcal{C}) \cap {}_{*\mathcal{C}}\mathrm{Hom}(\mathcal{C}, {}^*\mathcal{C})$ then the image of γ lies in $\mathrm{Rat}^\mathcal{C}({}^*\mathcal{C})$ is analogous to the proof of (i).

(iii) Take $\sigma \in \mathrm{bbf}(\mathcal{C})$ and define $m_\sigma : \mathcal{C} \otimes_A \mathcal{C} \to \mathcal{C}$ by $m_\sigma(c \otimes_A d) = c_1\sigma(c_2 \otimes_A d) = \sigma(c \otimes_A d_1)d_2$. Conversely, given a $(\mathcal{C},\mathcal{C})$-bicomodule map $m : \mathcal{C} \otimes_A \mathcal{C} \to \mathcal{C}$, let $\sigma = \epsilon_\mathcal{C} \circ m$. The verification that m_σ is a well-defined $(\mathcal{C},\mathcal{C})$-bicomodule map (note that by convention a $(\mathcal{C},\mathcal{C})$-bicomodule map must be (A,A)-linear), that $\sigma = \epsilon_\mathcal{C} \circ m \in \mathrm{bbf}(\mathcal{C})$ and that the correspondence is bijective is straightforward. $\qquad\square$

Remark 3.4. Let $\mathcal{C}$ be an A-coring.

(i) For $\sigma \in \mathrm{bbf}(\mathcal{C})$, the map $m = m_\sigma$ in Lemma 3.3 (iii) is associative since for $c, d, e \in \mathcal{C}$,

$$
\begin{aligned}
m(c \otimes_A m(d \otimes_A e)) &= m(c \otimes_A d_1\sigma(d_2 \otimes_A e)) &= m(c \otimes_A d_1)\sigma(d_2 \otimes_A e) \\
&= \sigma(c \otimes_A d_1)d_2\sigma(d_3 \otimes_A e) &= \sigma(c \otimes_A d_1)m(d_2 \otimes_A e) \\
&= m(\sigma(c \otimes_A d_1)d_2 \otimes_A e) &= m(m(c \otimes_A d) \otimes_A e).
\end{aligned}
$$

Then $(\mathcal{C}, m_\sigma)$ is an associative ring, in general without a unit.

(ii) For $\sigma \in \mathrm{bbf}(\mathcal{C})$ and σ^r as above, we have that σ^r is a ring homomorphism from $(\mathcal{C}, m_\sigma)$ to $\mathcal{C}^{*op} = {}^*(\mathcal{C}^o)$. To see this, we compute for all $c, d, e \in \mathcal{C}$,

$$
\begin{aligned}
\sigma^r(m_\sigma(c \otimes_A d))(e) &= \sigma^r(\sigma(c \otimes_A d_1)d_2)(e) &= \sigma(c \otimes_A d_1)\sigma(d_2 \otimes_A e) \\
&= \sigma(c \otimes_A d_1\sigma(d_2 \otimes_A e)) &= \sigma(c \otimes_A \sigma(d \otimes_A e_1)e_2) \\
&= \sigma^r(c)(\sigma(d \otimes_A e_1)e_2) &= \sigma^r(c)(\sigma^r(d)(e_1)e_2) \\
&= \sigma^r(d) *^r \sigma^r(c)(e).
\end{aligned}
$$

Similarly, σ^l is a ring homomorphism from $(\mathcal{C}, m_\sigma)$ to $({}^*\mathcal{C})^o$.

(iii) If M is a right $\mathcal{C}$-comodule (i.e., $M \in \mathcal{M}_A$ with a coaction $\rho^M : M \to M \otimes_A \mathcal{C}$ which is a coassociative right A-module map), then M is a right $(\mathcal{C}, m_\sigma)$ module via $m \cdot c = m_0 \sigma(m_1 \otimes_A c)$. The computation to show associativity is in [2, 26.7] or is a straightforward exercise. Unless $\sigma \circ \Delta_{\mathcal{C}} = \epsilon_{\mathcal{C}}$, we need not have that $M \otimes_{(\mathcal{C}, m_\sigma)} \mathcal{C} \cong M$.

Finally, we note that the notion of balanced bilinear forms for corings is integral to the definition of coseparable corings and co-Frobenius corings.

Coseparable corings form an important class of corings for which forgetful functors are separable functors. Recall [2, Section 26] that an A-coring $\mathcal{C}$ is called coseparable if there exists a $(\mathcal{C}, \mathcal{C})$-bicomodule map $\pi : \mathcal{C} \otimes_A \mathcal{C} \to \mathcal{C}$ such that $\pi \circ \Delta_{\mathcal{C}} = \mathcal{C}$. By Lemma 3.3 (iii) or as noted in [2, 26.1(b)], such a map π exists if and only if there exists $\sigma \in \mathrm{bbf}(\mathcal{C})$ (called a cointegral in [2]) such that $\sigma \circ \Delta_{\mathcal{C}} = \epsilon_{\mathcal{C}}$, i.e., $(\mathcal{C}, m_\sigma)$ is a nonunital ring whose product has a section. In other words, $(\mathcal{C}, m_\sigma)$ is a separable A-ring in the sense of [1] or [2, Section 26].

In [2, 27.15], a left (right) co-Frobenius coring is an A-coring $\mathcal{C}$ such that there is an injective morphism from $\mathcal{C}$ to $^*\mathcal{C}$ ($\mathcal{C}^*$ respectively). Thus, by Lemma 3.3 (iii), $\mathcal{C}$ is left (right) co-Frobenius via an injective morphism which is an (A, A)-bimodule map if and only if there is a left (right) nondegenerate $\sigma \in \mathrm{bbf}(\mathcal{C})$. Note that this latter is precisely the definition of co-Frobenius in [7, Definition 2.4] (i.e., the A-bilinearity is specified) and that there the opposite multiplication to that of [2] is used on the left and right duals so that A (not A^o) embeds in $^*\mathcal{C}$ and $\mathcal{C}^*$.

Now, a consistent definition of symmetric coring would require that the coring is co-Frobenius on the left and on the right with some compatibility conditions between the two structures.

4. The notion of a symmetric coring

In this section, we explore whether the notion of a symmetric coring is a sensible one, and suggest one possible definition which takes into account the fact that A is not necessarily commutative.

The idea of a symmetric coalgebra C over a field k was given in [4] where the authors proved the following.

Theorem 4.1. *Let C be a k-coalgebra, k a field. Then the following are equivalent:*

(i) *There exists an injective morphism $\alpha : C \to C^*$ of (C^*, C^*)-bimodules.*

(ii) *There exists a bilinear form $B : C \times C \to k$ which is symmetric, nondegenerate and C^*-balanced.*

A symmetric coalgebra over a field is then defined to be one satisfying the equivalent conditions of Theorem 4.1.

One might expect that a suitable definition for a symmetric coring would be that an A-coring $\mathcal{C}$ is symmetric if there exists a right and left nondegenerate symmetric $\sigma \in \mathrm{bbf}(\mathcal{C})$. We next show the equivalence of four conditions which mimic those of Theorem 4.1 in case $\mathcal{C}$ is right and left locally projective over A.

Lemma 4.2. *Let C be an A-coring. Let $\sigma \in \mathrm{bbf}(C)$ and consider the k-module map σ^{op} from $C \otimes_k C$ to A defined to be the composite of the twist map $\tau : C \otimes_k C \to C \otimes_k C$ and the surjection from $C \otimes_k C$ to $C \otimes_A C$ followed by σ. Then $\sigma^{op} \in \mathrm{bbf}(C)$ if and only if the following three conditions hold.*

(i) $\mathrm{Im}(\sigma) \subseteq Z(A)$.

(ii) $\sigma(ca \otimes_A d) = a\sigma(c \otimes_A d)$, *for all* $a \in A, c, d \in C$.

(iii) $\sigma(c_1 \otimes_A d)c_2 = d_1\sigma(c \otimes_A d_2)$, *for all* $c, d \in C$.

Proof. If $\sigma^{op} \in \mathrm{bbf}(C)$, then σ^{op} must be well defined on $C \otimes_A C$. Thus we must have $\sigma^{op}(ca \otimes_A d) = \sigma^{op}(c \otimes_A ad)$, i.e., $\sigma(d \otimes_A ca) = \sigma(ad \otimes_A c)$. This holds if and only if $\sigma(d \otimes_A c)a = a\sigma(d \otimes_A c)$, so that $\mathrm{Im}(\sigma) \subseteq Z(A)$. Also, σ^{op} must be an (A, A)-bimodule map, so that $\sigma(c \otimes_A ad) = \sigma(c \otimes_A d)a$ or, equivalently, using the facts that $\sigma \in \mathrm{bbf}(C)$ and $\mathrm{Im}(\sigma) \subseteq Z(A)$, $\sigma(ca \otimes_A d) = a\sigma(c \otimes_A d)$. The last condition is equivalent to the fact that σ^{op} must satisfy the balanced property (3.1) of Definition 3.1. $\qquad\square$

If $A = k$, then (i) and (ii) in the lemma above are automatic and (iii) holds if and only if $\sigma \in \mathrm{bbf}(C^o)$.

Next we fix some notation.

Let Γ denote the subring of $^*C^*$ of maps from C to A with image in $Z(A)$, the centre of A. On Γ the left (right) A-actions induced from those on $^*C$ and C^* coincide. Thus we may define an (A, A)-bimodule structure on Γ by $(ac^*b)(c) = c^*(bca) = bc^*(c)a$. Note that if $f \in \Gamma$, then $f(c)(ab - ba) = 0$, i.e., the image of f annihilates $[A, A]$, the additive commutator of A.

Proposition 4.3. *Let C be an A-coring. If C is left and right locally projective over A, then the following conditions are equivalent:*

(i) *There exists an injective morphism α of (A, A)-bimodules from C to the subring Γ of $^*C^*$ defined above such that $\mathrm{Im}(\alpha)$ is a $(^*C, C^*)$-bimodule with left and right actions given by $*^l$ and $*^r$, and α is a $(^*C, C^*)$-bimodule map.*

(ii) *There exists an injective map $\alpha : C \to {}^*C^*$ such that $\alpha_1 := i_1 \circ \alpha$ is an (A, A)-bimodule, right C^*-module morphism from C to C^*, and $\alpha_2 := i_2 \circ \alpha$ is an (A, A)-bimodule, left $^*C$-module morphism from C to $^*C$ where $i_1 : {}^*C^* \hookrightarrow C^*$ and $i_2 : {}^*C^* \hookrightarrow {}^*C$ are the inclusion maps.*

(iii) *There exists a right nondegenerate $\sigma \in \mathrm{bbf}(C)$ such that $\sigma^{op} \in \mathrm{bbf}(C)$ also.*

(iv) *There exists a right and left nondegenerate $\sigma \in \mathrm{bbf}(C)$ such that $\sigma = \sigma^{op}$.*

Proof. (i) implies (ii). It is straightforward to check that if a map α satisfies (i), then it also satisfies (ii).

(ii) implies (iii). Suppose that (ii) holds. Then by Lemma 3.3, $\alpha_1 = \sigma^r$ for some right nondegenerate $\sigma \in \mathrm{bbf}(C)$ and $\alpha_2 = \omega^l$ for some left nondegenerate $\omega \in \mathrm{bbf}(C)$, i.e., $\alpha(c)(d) = \sigma(c \otimes_A d) = \omega(d \otimes_A c)$. Then $\omega = \sigma^{op}$ and $\sigma^{op} \in \mathrm{bbf}(C)$.

(iii) implies (i). Suppose $\sigma \in \mathrm{bbf}(C)$ is right nondegenerate and $\sigma^{op} \in \mathrm{bbf}(C)$. Then σ^r is an (A, A)-bimodule map from C to C^*, and $(\sigma^{op})^l$ is an (A, A)-bimodule map from C to $^*C$. Since $\sigma^r(c)(d) = \sigma(c \otimes_A d) = \sigma^{op}(d \otimes_A c) = (\sigma^{op})^l(c)(d)$, then

$\sigma^r = (\sigma^{op})^l$. Define $\alpha = \sigma^r = (\sigma^{op})^l$. Then α maps $\mathcal{C}$ to $*\mathcal{C}^*$ and since σ is right nondegenerate, α is injective. Also $Im(\alpha)$ is a right $\mathcal{C}^*$-module and α is a right $\mathcal{C}^*$-module map by Lemma 3.3 since $\sigma^r(c) *^r c^* = \sigma^r(c \leftharpoonup c^*) \in Im(\alpha)$. Similarly, $Im(\alpha)$ is a left $*\mathcal{C}$-module and α is a left $*\mathcal{C}$-module map. By Lemma 4.2, $Im(\sigma) \subseteq Z(A)$ so that $Im(\alpha) \subseteq \Gamma$. Finally, since for $f \in *\mathcal{C}, g \in \mathcal{C}^*, c \in \mathcal{C}$, we have $(f *^l \alpha(c)) *^r g = \alpha(f \rightharpoonup c) *^r g = \alpha((f \rightharpoonup c) \leftharpoonup g) = \alpha((f \rightharpoonup (c \leftharpoonup g)) = f *^l (\alpha(c) *^r g)$ so that α is a $(*\mathcal{C}, \mathcal{C}^*)$-bimodule map.

Thus we have shown the equivalence of (i),(ii) and (iii). Clearly (iv) implies (iii) and it remains to show that the equivalent conditions (i), (ii) and (iii) imply (iv).

Let α satisfy (i), (ii). By Remark 3.4, we have that $\alpha = \sigma^r$ is a multiplication preserving isomorphism from $(\mathcal{C}, m_\sigma)$ to $Im(\alpha)$ and $\alpha = (\sigma^{op})^l$ is also a multiplication preserving isomorphism from $(\mathcal{C}, m_{\sigma^{op}})$ to $Im(\alpha)$. Thus $(\sigma^r)^{-1} \circ (\sigma^{op})^l = \alpha^{-1} \circ \alpha = Id_\mathcal{C}$ is a multiplication preserving bijection from $(\mathcal{C}, m_{\sigma^{op}})$ to $(\mathcal{C}, m_\sigma)$. Then $m_\sigma(c \otimes_A d) = m_{\sigma^{op}}(c \otimes_A d)$, i.e., $c_1 \sigma(c_2 \otimes_A d) = c_1 \sigma(d \otimes_A c_2)$ so that, applying $\epsilon_\mathcal{C}$, we see that $\sigma = \sigma^{op}$. $\qquad\square$

Note that Proposition 4.3 provides a new proof of the equivalence of the statements in Theorem 4.1 which does not depend on the fact that the rational dual of a co-Frobenius coalgebra over a field has local units.

A definition of symmetric coring parallel to the definition of symmetric coalgebra over a field would be to require that the coring satisfy condition (iv) in Proposition 4.3. However, this seems very restrictive, depending on commutativity of elements in A. Instead, we suggest the following.

Definition 4.4. *Let A be a ring, not necessarily commutative, and let A' be the ideal of A generated by the additive commutator $[A, A] = \{ab - ba \mid a, b \in A\}$. Let $\mathcal{C}$ be an A-coring. We say that $\mathcal{C}$ is symmetric if there exists $\sigma \in bbf(\mathcal{C})$ such that σ is left and right nondegenerate and for all c, d in $\mathcal{C}$, we have that $\sigma(c \otimes_A d) - \sigma(d \otimes_A c) \in A'$.*

Examples 4.5.

(i) The trivial coring and the matrix coring from Examples 2.1, with the non-degenerate balanced forms defined in Examples 3.2 are both symmetric in the sense of Definition 4.4 but do not satisfy the equivalent conditions of Proposition 4.3, unless A is commutative.

(ii) The coring $\mathcal{C} = A \otimes_k H$ from Examples 2.1 with nontrivial H-coaction on A and with σ as in Examples 3.2 is not symmetric in either sense even if $t(hS(g)) = t(gS(h))$ for all $g, h \in H$. (Of course, if the coaction is trivial, then the coring is clearly symmetric in the sense of Definition 4.4 if $B(h, g) = t(hS(g))$ is a symmetric form for H.)

Corings satisfying the equivalent conditions of Proposition 4.3 clearly are symmetric in the sense of Definition 4.4.

The next example builds a symmetric coring from a symmetric coseparable k-coalgebra, k a field.

Example 4.6. Let k be a field and let C be a k-coalgebra with a C^*-balanced nondegenerate symmetric bilinear form $B : C \otimes_k C \to k$ such that $B \circ \Delta_C = \epsilon_C$; in other words, C is a symmetric coalgebra via B and also a coseparable coalgebra via B. For example, let C be a cosemisimple involutory k-Hopf algebra H with antipode S. Then if λ is a left and right integral for H in H^* such that $\lambda(1) = 1$, then $B : H \otimes_k H \to k$ defined by $B(h, g) = \lambda(hS(g))$ satisfies the conditions above.

Let $A = C^*$, the dual algebra of C with convolution multiplication $*$ and let $\mathcal{C} = C$; we show first that $\mathcal{C}$ has the structure of an A-coring.

$\mathcal{C}$ is an (A, A)-bimodule with the standard C^*-bimodule structure on C, namely $c^* \rightharpoonup c \leftharpoonup d^* = d^*(c_1)c_2c^*(c_3)$, for all $c \in C$, $c^*, d^* \in C^*$. Define the coproduct on $\mathcal{C}$ to be the composite of the co-opposite comultiplication of C and the surjection from $C \otimes_k C$ to $C \otimes_A C$, namely $\Delta_{\mathcal{C}}(c) = c_2 \otimes_A c_1$. Since $\Delta_{\mathcal{C}}(d^*(c_1)c_2c^*(c_3)) = c^*(c_4)c_3 \otimes_A c_2d^*(c_1) = c^* \rightharpoonup c_2 \otimes_A c_1 \leftharpoonup d^*$, then $\Delta_{\mathcal{C}}$ is an (A, A)-bimodule map and the coassociativity follows from that of Δ_C.

Now, we define the counit map for $\mathcal{C}$. For B the balanced bilinear form for the k-coalgebra C as above, define $\epsilon_{\mathcal{C}} : C \to C^*$ by $\epsilon_{\mathcal{C}}(c) = B(c, -) = B(-, c)$. Note that $\epsilon_{\mathcal{C}}$ is injective since B is left and right nondegenerate. Then

$$\epsilon_{\mathcal{C}}(d^* \rightharpoonup c)(x) = B(x, c_1)d^*(c_2) = d^*(x_1)B(x_2, c) = (d^* * \epsilon_{\mathcal{C}})(x),$$

so $\epsilon_{\mathcal{C}}$ is a left A-module map and, similarly, $\epsilon_{\mathcal{C}}$ is a right A-module map.

The compatibility of $\Delta_{\mathcal{C}}$ and $\epsilon_{\mathcal{C}}$ follows from the coseparability of C by

$$(\epsilon_{\mathcal{C}} \otimes_A \mathcal{C})(c_2 \otimes_A c_1) = \epsilon_{\mathcal{C}}(c_2) \rightharpoonup c_1 = c_1B(c_2, c_3) = c$$

and, similarly, $(\mathcal{C} \otimes_A \epsilon_{\mathcal{C}}) \circ \Delta_{\mathcal{C}} = \mathcal{C}$. Thus we have shown that $\mathcal{C}$ is an A-coring.

Now, define $\mathcal{B} : \mathcal{C} \otimes_A \mathcal{C} \to A$ by $\mathcal{B}(c \otimes_A d) = \epsilon_{\mathcal{C}}(c) * \epsilon_{\mathcal{C}}(d)$. Since $\epsilon_{\mathcal{C}}$ is an (A, A)-bimodule map, then $\mathcal{B}$ is also. We show that $\mathcal{B}$ is a well-defined balanced form. To see that $\mathcal{B}$ is well defined, we compute for $c, d, x \in \mathcal{C}$, $c^* \in A$,

$$\begin{aligned}
\mathcal{B}(c \leftharpoonup c^* \otimes_A d)(x) &= c^*(c_1)\mathcal{B}(c_2 \otimes_A d)(x) \\
&= c^*(c_1)B(c_2, x_1)B(x_2, d) \\
&= B(c, x_1)c^*(x_2)B(x_3, d) \\
&= B(c, x_1)B(x_2, d_1)c^*(d_2) \\
&= \mathcal{B}(c \otimes_A c^* \rightharpoonup d)(x).
\end{aligned}$$

To see that $\mathcal{B}$ is balanced, we must show that $c_2 \leftharpoonup \mathcal{B}(c_1 \otimes_A d) = \mathcal{B}(c \otimes_A d_2) \rightharpoonup d_1$, for all $c, d \in \mathcal{C}$. We already showed that $c_2 \leftharpoonup \epsilon_{\mathcal{C}}(c_1) = c$ and thus we have that

$$c_2 \leftharpoonup \mathcal{B}(c_1 \otimes_A d) = c_2 \leftharpoonup \epsilon_{\mathcal{C}}(c_1) * \epsilon_{\mathcal{C}}(d) = c \leftharpoonup \epsilon_{\mathcal{C}}(d) = B(d, c_1)c_2.$$

Similarly, $\mathcal{B}(c \otimes_A d_2) \rightharpoonup d_1 = d_1B(c, d_2)$ and these are equal since B is balanced and symmetric.

We now show that $\mathcal{B}$ is right nondegenerate, i.e., that $\mathcal{B}(d \otimes_A -) = 0$ implies $d = 0$. Since $\mathcal{B}(d \otimes_A c) = B(d, -) * B(-, c)$, if $\mathcal{B}(d \otimes_A -) = 0$, then $B(d, x_1)B(c, x_2) = 0$ for all $c, x \in C$, and, in particular, $0 = B(d, c_1)B(c_3, c_2) = $

$B(d, c)$, for all $c \in C$, contradicting the fact that B is right nondegenerate. Similarly, $\mathcal{B}$ is left nondegenerate.

Clearly, $\mathcal{B}$ is symmetric in the sense of Definition 4.4 but unless $B(c, -) * B(d, -) = B(d, -) * B(c, -)$ in $A = C^*$, for all $c, d \in C$, the equivalent conditions of Proposition 4.3 do not hold.

Finally, we show that if $\mathcal{C}$ is an A-coring which is symmetric in the sense of Definition 4.4 via a map $\sigma \in \mathrm{bbf}(\mathcal{C})$ satisfying a further nondegeneracy condition, then the coring induced by the surjection from A to A/A' satisfies Proposition 4.3 (iv).

Theorem 4.7. *Let $\mathcal{C}$ be an A-coring and let $\sigma \in \mathrm{bbf}(\mathcal{C})$ such that*
  (i) *σ is left and right nondegenerate.*
  (ii) *For all $c, d \in C$, we have that $\sigma(c \otimes_A d) - \sigma(d \otimes_A c) \in A'$.*
  (iii) *$\sigma(c \otimes_A d) \in A'$, for all $d \in C$, implies that $c \in A'C + CA'$.*
    Then the surjection from A to $B = A/A'$ induces a B-coring structure on $\mathcal{D} := B \otimes_A \mathcal{C} \otimes_A B$ such that $\mathcal{D}$ satisfies condition (iv) *of Proposition 4.3.*

Proof. Let $\bar{a}$ denote the image of $a \in A$ in $B = A/A'$; for $\bar{1}$, we may write 1_B. We note first that $\mathcal{D} = B \otimes_A \mathcal{C} \otimes_A B = 1_B \otimes_A \mathcal{C} \otimes_A 1_B$ since for $a \in A$, $\bar{a} = (1_B)a = a(1_B)$. In particular, if $c \in A'C + CA'$, then $1_B \otimes_A c \otimes_A 1_B = 0$ in $\mathcal{D}$. Recall [2, 17.2] that $\mathcal{D}$ is a B-coring with counit $\epsilon_{\mathcal{D}}$ and comultiplication $\Delta_{\mathcal{D}}$ defined by

$$\epsilon_{\mathcal{D}}(1_B \otimes_A c \otimes_A 1_B) = \overline{\epsilon_{\mathcal{C}}(c)}, \qquad \text{and}$$

$$\Delta_{\mathcal{D}}(1_B \otimes_A c \otimes_A 1_B) = (1_B \otimes_A c_1 \otimes_A 1_B) \otimes_B (1_B \otimes_A c_2 \otimes_A 1_B).$$

Note that if $a' \in A'$ then for $c, d \in C$ we have $\sigma(c \otimes_A a'd) = \sigma(a'd \otimes_A c) + a''$, for some $a'' \in A'$, and so $\sigma(c \otimes_A a'd) = a'\sigma(d \otimes_A c) + a'' \in A'$ and

$$\overline{\sigma(c \otimes_A a'd)} = \overline{\sigma(ca' \otimes_A d)} = \bar{0}.$$

Thus the (B, B)-bimodule map $\mathcal{B} : \mathcal{D} \otimes_B \mathcal{D} \to B$ given by

$$\mathcal{B}((1_B \otimes_A c \otimes_A 1_B) \otimes_B (1_B \otimes_A d \otimes_A 1_B)) = \overline{\sigma(c \otimes_A d)},$$

is well defined. For if $x' - x \in A'$, then

$$\sigma(cx' \otimes_A d) - \sigma(cx \otimes_A d) = \sigma(c(x' - x) \otimes_A d) \in A'.$$

Furthermore, $\mathcal{B}$ is balanced. To see this, we compute for $c, d \in C$,

$$\mathcal{B}((1_B \otimes_A c \otimes_A 1_B) \otimes_B (1_B \otimes_A d_1 \otimes_A 1_B))(1_B \otimes_A d_2 \otimes_A 1_B)$$
$$= \overline{\sigma(c \otimes_A d_1)} \otimes_A d_2 \otimes_A 1_B$$
$$= 1_B \otimes_A \sigma(c \otimes_A d_1)d_2 \otimes_A 1_B$$
$$= 1_B \otimes_A c_1\sigma(c_2 \otimes_A d) \otimes_A 1_B$$
$$= 1_B \otimes_A c_1 \otimes_A \overline{\sigma(c_2 \otimes_A d)}$$
$$= (1_B \otimes_A c_1 \otimes_A 1_B)\mathcal{B}((1 \otimes_A c_2 \otimes_A 1) \otimes_B (1_B \otimes_A d \otimes_A 1_B)).$$

Thus, we have that $\mathcal{B} \in \mathrm{bbf}(\mathcal{D})$. We show that $\mathcal{B}$ is left and right nondegenerate and that $\mathcal{B} = \mathcal{B}^{op}$.

Suppose that $\mathcal{B}((1_B \otimes_A c \otimes_A 1_B) \otimes_B -)$ maps all elements of $\mathcal{D}$ to $\bar{0} \in B$. Then $\sigma(c \otimes_A d) \in A'$, for all $d \in \mathcal{C}$. By (iii) in the statement of the theorem, we must have that $c \in A'\mathcal{C} + \mathcal{C}A'$. But then $1_B \otimes_A c \otimes_A 1_B = 0 \in \mathcal{D}$, and so $\mathcal{B}$ is right nondegenerate. Similarly, $\mathcal{B}$ is left nondegenerate.

Finally, we note that $\mathcal{B} = \mathcal{B}^{op}$ is straightforward, so the proof is complete.

$\square$

Example 4.8. We noted in Example 4.5 that the trivial coring and matrix coring do not necessarily satisfy the conditions in Proposition 4.3. However, in both cases, Theorem 4.7 applies.

Further questions

(i) Another equivalent condition for a coalgebra C over a field k to be symmetric involves the multiplication on C induced by the multiplication on the ring with local units $C^{*\mathrm{rat}}$ [4, Theorem 3.3 (3)]. This parallels an equivalent condition for a finite-dimensional algebra over a field to be symmetric [4, Theorem 3.1 (3)], namely that a k-algebra A is symmetric if there exists a k-linear map $f : A \to k$ such that $f(xy) = f(yx)$ for $x, y \in A$, and $\mathrm{Ker}(f)$ does not contain a non-zero left ideal. An analogous condition should hold in the coring case.

(ii) In [4, Section 5], a functorial characterization of symmetric coalgebras over a field is given. It is unclear what the corresponding results (if any) for symmetric corings should be.

Acknowledgment

Many thanks to the referee for careful reading of this note and for very helpful comments.

References

[1] T. Brzeziński, L. Kadison and R. Wisbauer, On coseparable and biseparable corings, In: Hopf algebras in non-commutative geometry and physics, S. Caenepeel, F. Van Oystaeyen (eds.), Lecture Notes in Pure and Applied Mathematics 239, Marcel Dekker, New York, 2004, p. 71–88 (arXiv:math.RA/0208122).

[2] T. Brzeziński and R. Wisbauer, Corings and Comodules, London Math. Soc. Lecture Note Series 309, Cambridge University Press, 2003.

[3] S. Caenepeel, G. Militaru and S. Zhu, Frobenius and Separable Functors for Generalized Module Categories and Nonlinear Equations, LNM 1787, Springer Berlin-Heidelberg, 2002.

[4] F. Castaño Iglesias, S. Dăscălescu and C. Năstăsescu, Symmetric coalgebras, J. Algebra 279 (2004), 326-344.

[5] S. Dăscălescu, C. Năstăsescu and Ş. Raianu, Hopf algebras: an introduction, Marcel Dekker, Pure and Applied Mathematics 235, 2001.

[6] J. Gómez-Torrecillas and A. Louly, Coseparable corings, Communications in Algebra 31 (2003), 4455–4471.

[7] M. Iovanov and J. Vercruysse, CoFrobenius corings and adjoint functors, J. Pure Appl. Alg., to appear. (arXiv:math.RA/0610853)

[8] B.I. Lin, Semiperfect coalgebras, J. Algebra 49 (1977), 357–373.

[9] M.E. Sweedler, The predual theorem to the Jacobson-Bourbaki theorem, Transactions of the Amer. Math. Soc. 213 (1975), 391–406.

M. Beattie
Department of Mathematics and Computer Science
Mount Allison University
Sackville, NB E4L 1E6, Canada
e-mail: mbeattie@mta.ca

D. Bulacu
Faculty of Mathematics and Informatics
University of Bucharest
Str. Academiei 14
RO-010014 Bucharest 1, Romania
e-mail: dbulacu@al.math.unibuc.ro

Ş. Raianu
Mathematics Department
California State University
Dominguez Hills
1000 E Victoria St
Carson CA 90747, USA
e-mail: sraianu@csudh.edu

Modules and Comodules

Trends in Mathematics, 101–111

© 2008 Birkhäuser Verlag Basel/Switzerland

Ring Hulls of Semiprime Homomorphic Images

Gary F. Birkenmeier, Jae Keol Park and S. Tariq Rizvi

Dedicated to Professor Robert Wisbauer on his 65th birthday

Abstract. We investigate connections between the right FI-extending right ring hulls of semiprime homomorphic images of a ring R and the right FI-extending right rings of quotients of R by considering ideals of R which are essentially closed and contain the prime radical $\mathbf{P}(R)$. As an application of our results, we show that the bounded central closure of a unital C^*-algebra A contains a nonzero homomorphic image of A/K for every nonessential ideal K of A.

Throughout this paper all rings are associative with unity and R denotes such a ring. Subrings and overrings preserve the unity of the base ring. Ideals without the adjective "right" or "left" mean two-sided ideals. All modules are assumed to be unital.

From [6], a ring R is called *right FI-extending* if for any ideal I of R there exists an idempotent $e \in R$ such that I is right essential in eR. Recall from [14] that a ring R is called *quasi-Baer* if the right annihilator of every ideal is generated by an idempotent of R as a right ideal. In [9, Theorem 4.7] it is shown that a semiprime ring is right FI-extending if and only if it is quasi-Baer. We use $\mathfrak{FJ}$ and $\mathfrak{qB}$ to denote the class of right FI-extending rings and the class of quasi-Baer rings, respectively.

For a ring R, we use $Q(R)$, $\mathbf{I}(R)$, and $\mathbf{B}(R)$ to denote the maximal right ring of quotients of R, the set of all idempotents of R, and the set of all central idempotents of R, respectively.

For an arbitrary ring R, the right FI-extending right ring hull or the quasi-Baer right ring hull may not exist or may not be unique in case they do exist, even when R is right nonsingular. On the other hand, in [12] it is shown that, for a semiprime ring R, there exists the smallest right FI-extending right ring of quotients (i.e., right FI-extending ring hull), $\widehat{Q}_{\mathfrak{FJ}}(R)$, of R which coincides with its quasi-Baer right ring hull, $\widehat{Q}_{\mathfrak{qB}}(R)$ (see Theorem 4). Also it is proved in [2] and [12] that lying over, going up, and incomparability of prime ideals, various regularity conditions, and classical Krull dimension transfer between R and $\widehat{Q}_{\mathfrak{FJ}}(R)$.

Moreover, an example of a semiprime ring R is provided in [12] which shows that this transference, in general, fails between a semiprime ring R and its right rings of quotients which properly contains $\widehat{Q}_{\mathfrak{FJ}}(R)$. Indeed, let $\mathbb{Z}[G]$ be the group ring of the group $G = \{1, g\}$ over the ring $\mathbb{Z}$. Then $\mathbb{Z}[G]$ is semiprime and $Q(\mathbb{Z}[G]) = \mathbb{Q}[G]$, where $\mathbb{Q}$ is the field of rational numbers. Now

$$\mathbb{Z}[G] \subsetneq \widehat{Q}_{\mathfrak{FJ}}(\mathbb{Z}[G]) = \{(a + c/2 + d/2) + (b + c/2 - d/2)g \mid a, b, c, d \in \mathbb{Z}\}$$
$$\subsetneq \mathbb{Z}[1/2][G] \subseteq \mathbb{Q}[G],$$

where $\mathbb{Z}[1/2]$ is the subring of $\mathbb{Q}$ generated by $\mathbb{Z}$ and $1/2$. In this case, for example, LO (lying over) does not hold between $\mathbb{Z}[G]$ and $\mathbb{Z}[1/2][G]$. Assume to the contrary that LO holds. From [22, Theorem 4.1], LO holds between $\mathbb{Z}$ and $\mathbb{Z}[G]$. Hence there exists a prime ideal P of $\mathbb{Z}[G]$ such that $P \cap \mathbb{Z} = 2\mathbb{Z}$. By LO, there is a prime ideal K of $\mathbb{Z}[1/2][G]$ such that $K \cap \mathbb{Z}[G] = P$. Now $K \cap \mathbb{Z}[1/2] = K_0$ is a prime ideal of $\mathbb{Z}[1/2]$. So $K_0 \cap \mathbb{Z} = K \cap \mathbb{Z}[1/2] \cap \mathbb{Z} = K \cap \mathbb{Z} = 2\mathbb{Z}$. Thus $2 \in K_0$. But since K_0 is an ideal of $\mathbb{Z}[1/2]$, $1 = 2 \cdot (1/2) \in K_0$, a contradiction. Next, $\mathbb{Q}[G]$ is (von Neumann) regular but $\mathbb{Z}[G]$ is not, so the transference of (von Neumann) regularity does not hold for right rings of quotients properly containing $\widehat{Q}_{\mathfrak{FJ}}(R)$. For further details, see [12, Example 3.7].

Thereby, semiprime rings exhibit optimal behavior with respect to the existence (and uniqueness) of right FI-extending right ring hulls and the transference of various interesting properties between the ring and its right FI-extending ring hull. Thus, for an arbitrary ring R, it seems natural to investigate connections between the right FI-extending right ring hulls of semiprime homomorphic images of R and the right FI-extending right rings of quotients of R.

In this paper, we make such a connection by considering ideals of R which are essentially closed and contain $\mathbf{P}(R)$, the prime radical of R. For example, the ideal $\bar{\rho}(R)$ discussed in [5] is an essentially closed ideal and an essential extension of a supernilpotent radical of R. Also the $\mathcal{S}$-closure, $\mathbf{P}^*(R)$, of $\mathbf{P}(R)$ (see [18, pp. 42–48] and [23]) is an essentially closed ideal containing $\mathbf{P}(R)$.

Let $Q(R)$ denote the maximal right ring of quotients of R. For an essentially closed proper ideal I of R such that $\mathbf{P}(R) \subseteq I$ and $e \in \mathbf{I}(Q(R))$ with I_R is essential in $(1 - e)Q(R)_R$, we show that: (1) if $Q(R) = E(R_R)$, then the right FI-extending ring hull of R/I, $\widehat{Q}_{\mathfrak{FJ}}(R/I)$, is isomorphic to the subring of $eQ(R)e$ generated by eRe and the set of central idempotents $\mathbf{B}(eQ(R)e)$ of $eQ(R)e$, $\langle eRe \cup \mathbf{B}(eQ(R)e)\rangle_{eQ(R)e}$; (2) if $Z(R_R) = 0$ and T is a right FI-extending right ring of quotients of R, then $\langle eTe\rangle_{eQ(R)e}$ is a semiprime right FI-extending right ring of quotients of an isomorphic copy of $\widehat{Q}_{\mathfrak{FJ}}(R/I)$. We apply our results to the study of homomorphic images of a C^*-algebra. In fact, we show that the bounded central closure of a unital C^*-algebra A contains a nonzero homomorphic image of A/K for every nonessential ideal K of A. Some of the results in this paper are related to, but not corollaries of, those in [17].

For a right R-module M_R and a submodule N_R, $N_R \leq^{\text{ess}} M_R$ and $N_R \trianglelefteq M_R$ (or simply, $N \trianglelefteq M$ when the context is clear) mean that N_R is essential in M_R and N_R is a fully invariant submodule of M_R, respectively. Let I be a subset of a ring R. We use $I \trianglelefteq R$ to denote that I is an ideal of R. For a nonempty subset X of a ring R, we let $\ell_R(X)$, $r_R(X)$, and $\langle X \rangle_R$ denote the left annihilator of X in R, the right annihilator of X in R, and the subring of R generated by X, respectively. We use $Z(R_R)$ and $Z_2(R_R)$ to denote the right singular ideal and the second right singular ideal of R, respectively.

According to [3] an idempotent e of a ring R is called *left* (resp., *right*) *semicentral* if $ae = eae$ (resp., $ea = eae$) for all $a \in R$. Equivalently, an idempotent e is left (resp., right) semicentral if and only if eR (resp., Re) is an ideal of R. For a ring R, we use $\mathbf{S}_\ell(R)$ (resp., $\mathbf{S}_r(R)$) to denote the set of all left (resp., right) semicentral idempotents. Recall from [7] that a ring R is called *semicentral reduced* if $\mathbf{S}_\ell(R) = \{0, 1\}$ (i.e., 1 is a semicentral reduced idempotent of R). It can be seen that $\mathbf{S}_\ell(eRe) = \{0, e\}$ if and only if $\mathbf{S}_r(eRe) = \{0, e\}$. Note that $\mathbf{B}(R) = \mathbf{S}_\ell(R) \cap \mathbf{S}_r(R)$.

From [18], recall that by definition, the $\mathcal{S}$-closure $\mathbf{P}^*(R)$ of $\mathbf{P}(R)$, is the smallest submodule of R_R containing $\mathbf{P}(R)$ such that $R/\mathbf{P}^*(R)$ is nonsingular as a right R-module. Then we see that $\mathbf{P}^*(R)$ is an ideal of R [18, p. 47, Exercise 1]. From [18, Proposition 2.4] and [23, Proposition 4.11], it follows that $\mathbf{P}^*(R)$ is an essentially closed submodule of R_R and $[\mathbf{P}(R) + Z_2(R_R)]_R \leq^{\text{ess}} (\mathbf{P}^*(R))_R$.

Henceforth, we assume that all right essential overrings of R are R-submodules of a fixed injective hull $E(R_R)$ of R_R and all right rings of quotients of R are subrings of a fixed maximal right ring of quotients $Q(R)$ of R.

Definition 1. ([11, Definition 2.1]) Let $\mathfrak{K}$ denote a class of rings. For a ring R, let S be a right essential overring of R and T an overring of R. Consider the following conditions.

(i) $S \in \mathfrak{K}$.

(ii) If $T \in \mathfrak{K}$ and T is a subring of S, then $T = S$.

(iii) If S and T are subrings of a ring V and $T \in \mathfrak{K}$, then S is a subring of T.

(iv) If $T \in \mathfrak{K}$ and T is a right essential overring of R, then S is a subring of T.

If S satisfies (i) and (ii), then we say that S is a $\mathfrak{K}$ *right ring hull* of R, denoted by $\widetilde{Q}_{\mathfrak{K}}(R)$ (i.e., $\widetilde{Q}_{\mathfrak{K}}(R)$ is minimal among right essential overrings of R). If S satisfies (i) and (iii), then we say that S is the $\mathfrak{K}$ *absolute to V right ring hull* of R, denoted by $Q_{\mathfrak{K}}^V(R)$ (i.e., $Q_{\mathfrak{K}}^V(R)$ is the smallest right essential overring of R in $\mathfrak{K}$ that is a subring of V); for the $\mathfrak{K}$ absolute to $Q(R)$ right ring hull, we use the notation $\widehat{Q}_{\mathfrak{K}}(R)$. If S satisfies (i) and (iv), then we say that S is the $\mathfrak{K}$ *absolute right ring hull* of R, denoted by $Q_{\mathfrak{K}}(R)$ (i.e., $Q_{\mathfrak{K}}(R)$ is the smallest right essential overring of R in $\mathfrak{K}$). Thus when $Q_{\mathfrak{K}}(R)$ exists, it is the intersection of all right essential overrings of R in $\mathfrak{K}$. Observe that if $Q(R) = E(R_R)$, then $\widehat{Q}_{\mathfrak{K}}(R) = Q_{\mathfrak{K}}(R)$. Left sided versions can be defined similarly.

When the class $\mathfrak{K} = \mathfrak{F}\mathfrak{I}$, $Q_{\mathfrak{F}\mathfrak{I}}(R)$ and $\widehat{Q}_{\mathfrak{F}\mathfrak{I}}(R)$ denote the absolute right FI-extending right ring hull and the absolute to $Q(R)$ right FI-extending right ring hull of R, respectively. Similarly, $Q_{\mathfrak{q}\mathfrak{B}}(R)$ and $\widehat{Q}_{\mathfrak{q}\mathfrak{B}}(R)$ denote the absolute quasi-Baer right ring hull and the the absolute to $Q(R)$ quasi-Baer right ring hull of R, respectively.

Lemma 2. Let C be a proper ideal of a ring R such that C is a complement of a right ideal of R.

(i) If $\mathbf{P}(R) \subseteq C$, then R/C is a semiprime ring.

(ii) If R is semiprime, then $R/Z_2(R_R)$ is semiprime and right nonsingular.

(iii) $R^* = R/\mathbf{P}^*(R)$ is semiprime and right nonsingular.

Proof. (i) Let X be a right ideal of R such that C is a complement of X. First we show that $(C \oplus X)/C$ is essential in R/C as a right R/C-module. To see this, assume to the contrary that there exists a nonzero right R/C-submodule Y/C of R/C such that $[(C \oplus X)/C] \cap (Y/C) = 0$. There exists $y \in Y$ such that $y \notin C$. Then $(C + yR) \cap X \neq 0$. So there exist $c \in C$, $r \in R$, and $0 \neq x \in X$ such that $c + yr = x$. Then $yr = -c + x \in (C \oplus X) \cap Y \subseteq C$. Hence $x \in C \cap X = 0$, a contradiction. Therefore $(C \oplus X)/C$ is essential in R/C as a right R/C-module.

Next, let $0 \neq H/C \trianglelefteq R/C$ such that $(H/C)^2 = 0$. Then $H^2 \subseteq C$. Since $(C \oplus X)/C$ is essential in R/C as a right R/C-module, $(H/C) \cap [(C \oplus X)/C] \neq 0$. So $H \cap (C \oplus X) = C \oplus (H \cap X) \not\subseteq C$. Thus $0 \neq H \cap X$. But $(H \cap X)^2 \subseteq C \cap X = 0$. So $H \cap X \subseteq X \cap \mathbf{P}(R) \subseteq X \cap C = 0$, a contradiction. Therefore R/C is a semiprime ring.

Parts (ii) and (iii) follow as direct consequences of part (i). $\qquad\square$

Corollary 3. Let $R^\natural = R/\mathbf{P}(R)$. Then we have the following.

(i) $R^\natural/Z_2(R^\natural{}_{R^\natural})$ is a semiprime right nonsingular ring.

(ii) $Z_2(R^\natural{}_{R^\natural}) = \mathbf{P}^*(R^\natural) \subseteq \mathbf{P}^*(R)/\mathbf{P}(R)$. Hence $R/\mathbf{P}^*(R)$ is a homomorphic image of $R^\natural/Z_2(R^\natural_{R^\natural})$.

Proof. (i) It is a direct consequence of Lemma 2.

(ii) Since $Z_2(R^\natural_{R^\natural}) + \mathbf{P}(R^\natural) \subseteq \mathbf{P}^*(R)$, then $Z_2(R^\natural_{R^\natural}) \subseteq \mathbf{P}^*(R)$. Note that $R^\natural/Z_2(R^\natural_{R^\natural})$ is right nonsingular and so $Z_2(R^\natural_{R^\natural})$ is $\mathcal{S}$-closed in $R^\natural_{R^\natural}$. Since $\mathbf{P}(R^\natural) = 0 \subseteq Z_2(R^\natural_{R^\natural})$, then $\mathbf{P}^*(R) \subseteq Z_2(R^\natural_{R^\natural})$. Thus $Z_2(R^\natural_{R^\natural}) = \mathbf{P}^*(R^\natural)$. Now note that $(R/\mathbf{P}(R))/(\mathbf{P}^*(R)/\mathbf{P}(R)) \cong R/\mathbf{P}^*(R)$ is right nonsingular. So $\mathbf{P}^*(R)/\mathbf{P}(R)$ is $\mathcal{S}$-closed in $R^\natural_{R^\natural}$, hence $\mathbf{P}^*(R^\natural) = Z_2(R^\natural_{R^\natural}) \subseteq \mathbf{P}^*(R)/\mathbf{P}(R)$. $\qquad\square$

When R is a ring, the ring $\langle R \cup \mathbf{B}(Q(R)) \rangle_{Q(R)}$ has been called the *idempotent closure* of R in [2] by Beidar and Wisbauer. We use $R\mathbf{B}(Q(R))$ to denote the idempotent closure of R.

The next result from [12] shows the existence and uniqueness of the quasi-Baer and right FI-extending right ring hulls of a semiprime ring. In 1971, Mewborn [20] proved the existence of a Baer (absolute) ring hull for a semiprime commutative ring. Our result also generalizes Memborn's result since every commutative quasi-Baer ring is a Baer ring.

Theorem 4. (cf. [12, Theorem 3.3]) Let R be a semiprime ring. Then $\widehat{Q}_{\mathfrak{FJ}}(R) = RB(Q(R))$ and $\widehat{Q}_{\mathfrak{qB}}(R) = RB(Q(R))$.

In the following theorem, we provide a connection between the right FI-extending right ring hulls of semiprime homomorphic images of R and the right FI-extending right rings of quotients of R.

Theorem 5. Let a ring R be either semiprime or $Q(R) = E(R_R)$. Assume that I is a proper ideal of R such that I_R is essentially closed in R_R. Then we have the following.

(i) There exists $e \in \mathbf{I}(Q(R))$ such that $I_R \leq^{\mathrm{ess}} (1-e)Q(R)_R$ and $I = R \cap (1-e)Q(R)$.

(ii) $eR = eRe$ and $R(1-e) = (1-e)R(1-e)$.

(iii) R/I is ring isomorphic to eRe.

(iv) If R is semiprime, then $eQ(R)e \subseteq Q(eRe)$.

(v) If $E(R_R) = Q(R)$, then $E(eRe_{eRe}) = eQ(R)e$ and $eQ(R)e = Q(eRe)$.

(vi) If $\mathbf{P}(R) \subseteq I$, then R/I is semiprime and $\widehat{Q}_{\mathfrak{FJ}}(R/I) \cong \widehat{Q}_{\mathfrak{FJ}}(eRe) = \langle eRe \cup \mathbf{B}(eQ(R)e)\rangle_{eQ(R)e}$.

(vii) If R is semiprime (resp., right nonsingular and semiprime), then $\widehat{Q}_{\mathfrak{FJ}}(R/I) \cong e\widehat{Q}_{\mathfrak{FJ}}(R)e$ (resp., $Q_{\mathfrak{FJ}}(R/I) \cong eQ_{\mathfrak{FJ}}(R)e$).

Proof. (i) If R is semiprime, use [13, Remark 2.2(i) and Theorem 2.10]. If $Q(R) = E(R_R)$, then the proof is routine.

(ii) If R is semiprime, the proof of this part is clear since $e \in \mathbf{B}(Q(R))$. For $Q(R) = E(R_R)$, let $r \in R$ with $er(1-e) \neq 0$. Since R_R is dense in $Q(R)_R$, there exists $s \in R$ such that $er(1-e)s \neq 0$ and $(1-e)s \in R$. Then $(1-e)s \in R \cap (1-e)Q(R) = I$. Hence $0 \neq er(1-e)s \in eI = 0$, a contradiction. So $eR(1-e) = 0$. Therefore $eR = eRe$ and $R(1-e) = (1-e)R(1-e)$.

(iii) Define $f : R/I \to eRe$ by $f(r+I) = er$. Then f is well defined because $eI = 0$. Clearly f is a ring epimorphism. If $x + I \in \mathrm{Ker}(f)$, then $x \in (1-e)Q(R) \cap R$. By part (i), $x \in I$. Hence $\mathrm{Ker}(f) = 0$. Thus f is a ring isomorphism.

(iv) Since $e \in \mathbf{B}(Q(R))$, eRe_{eRe} is dense in $eQ(R)e_{eRe}$. Thus $eQ(R)e \subseteq Q(eRe)$.

(v) Assume that $E(R_R) = Q(R)$. Let X be a right ideal of eRe and $g : X \to eQ(R)e$ an eRe-module homomorphism. From part (ii) X, eRe, and $eQ(R)e$ are right R-modules, and g is an R-module homomorphism. Since $eQ(R)e \subseteq eQ(R)$ and $eQ(R)$ is the injective hull of eR_R, then g can be extended to an R-module homomorphism $\bar{g} : eR \to eQ(R)$. Now $\bar{g}$ can be extended to an R-module homomorphism $\tilde{g} : eQ(R) \to eQ(R)$. Therefore $\tilde{g}$ is a $Q(R)$-module homomorphism from [19, p.95]. Thus $\bar{g}$ is an eRe-module homomorphism. Since $eR = eRe$ (part (ii)),

$$\bar{g}(eR) = \bar{g}(eRe) = \bar{g}(eRe)e = \bar{g}(eR)e \subseteq eQ(R)e.$$

By the Baer criterion, $eQ(R)e$ is an injective right eRe-module. Since eRe is dense as a right eRe-submodule of $eQ(R)e$, $eQ(R)e$ is the injective hull of eRe as a right eRe-module and $eQ(R)e = Q(eRe)$.

(vi) This is a consequence of parts (iii) and (iv), Lemma 2(i), and Theorem 4.

(vii) Assume that R is semiprime. By [13, Remark 2.2(i) and Theorem 2.10], $e \in \mathbf{B}(Q(R))$. Hence $\mathbf{B}(eQ(R)e) = e\mathbf{B}(Q(R))e$. So

$$\widehat{Q}_{\mathfrak{FJ}}(R/I) \cong \langle eRe \cup \mathbf{B}(eQ(R)e)_{eQ(R)e} = eR\mathbf{B}(Q(R))e = e\widehat{Q}_{\mathfrak{FJ}}(R)e$$

using Theorem 4 and part (v). If additionally $Z(R_R) = 0$, then R/I is right nonsingular by [18, Propositions 1.28 and 2.4] and eRe is right nonsingular. The result now follows from the fact that for any right nonsingular ring T, $\widehat{Q}_{\mathfrak{FJ}}(T) = Q_{\mathfrak{FJ}}(T)$ since $Q(T) = E(T_T)$. □

Corollary 6. Let $\overline{R} = R/Z_2(R_R)$ and $\overline{R}^* = \overline{R}/\mathbf{P}^*(\overline{R})$. Then we have the following.

(i) There exists $\overline{e} = \overline{e}^2 \in Q(\overline{R})$ such that $\mathbf{P}(\overline{R})_{\overline{R}} \leq^{\mathrm{ess}} (1-\overline{e})Q(\overline{R})_{\overline{R}}$ and $\mathbf{P}^*(\overline{R}) = \overline{R} \cap (1-\overline{e})Q(\overline{R})$.

(ii) $R/\mathbf{P}^*(R) \cong \overline{R}/\mathbf{P}^*(\overline{R})$.

(iii) $\widehat{Q}_{\mathfrak{FJ}}(R/\mathbf{P}^*(R)) \cong \langle \overline{e}\overline{R}\overline{e} \cup \mathbf{B}(\overline{e}Q(\overline{R})\overline{e})\rangle_{\overline{e}Q(\overline{R})\overline{e}}$.

Proof. (i) Since $\overline{R}$ is right nonsingular, all parts of Theorem 5 hold when R is replaced by $\overline{R}$.

(ii) First observe that $R/\mathbf{P}^*(R) \cong (R/Z_2(R_R))/(\mathbf{P}^*(R)/Z_2(R_R))$. By Lemma 2(iii), we have that $\mathbf{P}(\overline{R}) = \mathbf{P}(R/Z_2(R_R)) \subseteq \mathbf{P}^*(R)/Z_2(R_R)$ and $\mathbf{P}^*(R)/Z_2(R_R)$ is an $\mathcal{S}$-closed submodule of $\overline{R}$. Let $0 \neq x + Z_2(R_R) \in \mathbf{P}^*(R)/Z_2(R_R)$. Since $(\mathbf{P}(R) + Z_2(R_R))_R \leq^{\mathrm{ess}} \mathbf{P}^*(R)_R$, there exists $L_R \leq^{\mathrm{ess}} R_R$ such that $Z_2(R_R) \subseteq L$ and $xL \subseteq \mathbf{P}(R) + Z_2(R_R)$. Note that $\overline{R}_R$ is nonsingular. Thus [18, Proposition 1.28] yields that $(L/Z_2(R_R))_R \leq^{\mathrm{ess}} \overline{R}_R$. Since $\overline{R}_R$ is right nonsingular, $xL \not\subseteq Z_2(R_R)$. So $0 \neq (x + Z_2(R_R))(L/Z_2(R_R)) \subseteq [\mathbf{P}(R) + Z_2(R_R)]/Z_2(R_R)$. Hence $[(\mathbf{P}(R) + Z_2(R_R))/Z_2(R_R)]_R \leq^{\mathrm{ess}} (\mathbf{P}^*(R)/Z_2(R_R))_R$. By [18, Propositions 2.3 and 2.4], $\mathbf{P}^*(\overline{R}) = \mathbf{P}^*(R)/Z_2(R_R)$. Therefore $R/\mathbf{P}^*(R) \cong \overline{R}/\mathbf{P}^*(\overline{R})$.

(iii) It follows from an application of Theorem 5 to $\overline{R}$. □

Corollaries 3 and 6 motivate one to ask: *Is $R^\natural/\mathbf{P}^*(R^\natural)$ isomorphic to $\overline{R}/\mathbf{P}^*(\overline{R})$?*

Our next example shows that this question, in general, has a negative answer.

Example 7. Let F be a field and

$$R = \begin{pmatrix} F & F[x]/x^2F[x] \\ 0 & F[x]/x^2F[x] \end{pmatrix}.$$

Then $\mathbf{P}^*(R) = Z_2(R_R) = R$. So $R^\natural/\mathbf{P}^*(R^\natural) \cong F \oplus F$, but $\overline{R}/\mathbf{P}^*(\overline{R}) = 0$. Also $\mathbf{P}^*(R^\natural) = 0 \neq \mathbf{P}^*(R)/\mathbf{P}(R)$.

Our next corollary shows that every nonzero homomorphic image of a semiprime ring R has a nonzero homomorphic image in $\widehat{Q}_{\mathfrak{FJ}}(R)$.

Corollary 8. Let R be a semiprime ring and K a proper ideal of R. Then we have the following.

   (i) K has a unique essential closure $I = R \cap eQ(R)$ where $e \in \mathbf{B}(Q(R))$ and $K_R \leq^{\mathrm{ess}} eQ(R)_R$.

   (ii) $R/K \xrightarrow{h} R/I \xrightarrow{\iota} \widehat{Q}_{\mathfrak{F}\mathfrak{J}}(R/I) \cong e\widehat{Q}_{\mathfrak{F}\mathfrak{J}}(R)e$, where h is the canonical ring homomorphism and ι is the inclusion homomorphism.

Proof. (i) This part follows from [13, Remark 2.2(i) and Theorem 2.10].

(ii) This part is a consequence of part (i) and Theorem 5(vii). □

Lemma 9. Let A be a unital semiprime Banach algebra. Then every essentially closed ideal of A is norm closed.

Proof. Let I be an essentially closed ideal of A. Let $\overline{I}$ denote the norm closure of I. Clearly $\ell_A(\overline{I}) \subseteq \ell_A(I)$. Take $a \in \ell_A(I)$ and $x \in \overline{I}$. Then $x = \lim x_n$, where each $x_n \in I$. Hence $ax = a(\lim x_n) = \lim ax_n = 0$. Thus $\ell_A(\overline{I}) = \ell_A(I)$. Now let $0 \neq y \in \overline{I}$. Then, since A is semiprime, $yI \neq 0$. Hence $I_A \leq^{\mathrm{ess}} \overline{I}_A$. Therefore $I = \overline{I}$, so I is norm closed. □

Let A be a (not necessarily unital) C^*-algebra. Then the set $\mathcal{I}_{ce}$ of all norm closed essential ideals of A forms a filter directed downwards by inclusion. The ring $Q^b(A)$ denotes the algebraic direct limit of $\{M(I)\}_{I \in \mathcal{I}_{ce}}$, where $M(I)$ denotes the C^*-algebra multipliers of I; and $Q^b(A)$ is called the *bounded symmetric algebra of quotients* of A in [1, p.57, Definition 2.23]. The norm closure $M_{\mathrm{loc}}(A)$ of $Q^b(A)$ (i.e., the C^*-algebra direct limit $M_{\mathrm{loc}}(A)$ of $\{M(I)\}_{I \in \mathcal{I}_{ce}}$) is called the *local multiplier algebra* of A [1, p.65, Definition 2.3.1]. The local multiplier algebra $M_{\mathrm{loc}}(A)$ was first used in [16] and [21] to show the innerness of certain $*$-automorphisms and derivations. Its structure has been extensively studied in [1]. For more details on $M_{\mathrm{loc}}(A)$ and $Q^b(A)$, see [1], [16], and [21]. Note that C^*-algebras are always semiprime and right nonsingular.

Proposition 10. Let A be a unital C^*-algebra and I a proper ideal of A such that I_A is essentially closed in A_A. Then we have the following.

   (i) A/I is a C^*-algebra.

   (ii) $Q_{\mathfrak{q}\mathfrak{B}}(A/I)$ is $*$-isomorphic to $eQ_{\mathfrak{q}\mathfrak{B}}(A)e$ for some $e \in \mathbf{B}(Q(A))$.

Proof. (i) By Lemma 9, I is norm closed. From [15, p.20, Proposition 1.8.2], A/I is a C^*-algebra.

(ii) By Corollary 8(i), $I = A \cap eQ(A)$ for some $e \in \mathbf{B}(Q(A))$. By [1, p.59, Remark 2.2.9; p.72, Lemma 3.1.2], $\mathbf{B}(M_{\mathrm{loc}}(A)) = \mathbf{B}(Q(A))$. Hence $e \in \mathbf{B}(M_{\mathrm{loc}}(A))$. From Theorem 5(iii), $f : A/I \to eAe$ defined by $f(a + I) = eae$ for $a \in A$ is an isomorphism. Since $e \in \mathbf{B}(M_{\mathrm{loc}}(A))$, e is a projection by [1, p.59, Remark 2.2.9]. Thus $f((a + I)^*) = f(a^* + I) = ea^*e = (eae)^* = f(a + I)^*$. Hence f is a $*$-isomorphism. Note that $Q_{\mathfrak{q}\mathfrak{B}}(A) = Q_{\mathfrak{F}\mathfrak{J}}(A)$ by Theorem 4 because A is semiprime. Now f induces a $*$-isomorphism from $Q_{\mathfrak{q}\mathfrak{B}}(A/I)$ to $eQ_{\mathfrak{q}\mathfrak{B}}(A)e$. □

For a unital C^*-algebra A, Corollary 8 and Proposition 10 yield that $Q_{q\mathfrak{B}}(A)$ (hence the bounded central closure of A [12, Theorem 4.15]) contains a nonzero homomorphic image of A/K for every nonessential ideal K of A.

A ring R is called *right strongly FI-extending* [9] if every ideal of R is right essential in a fully invariant direct summand of R_R. Thus R is right strongly FI-extending if and only if for any $I \trianglelefteq R$ there is $e \in \mathbf{S}_\ell(R)$ such that $I_R \leq^{\mathrm{ess}} eR_R$. Right strongly FI-extending rings are right FI-extending, but the converse does not hold. In fact, let $\mathbb{Z}_3[S_3]$ be the group ring of the symmetric group S_3 on $\{1, 2, 3\}$ over the field $\mathbb{Z}_3$ of three elements. Then $\mathbb{Z}_3[S_3]$ is self-injective (hence right FI-extending). But it is not right strongly FI-extending by [10].

Proposition 11.

  (i) A quasi-Baer right FI-extending ring is right strongly FI-extending.

  (ii) ([11, Proposition 1.2(ii)]) A right nonsingular right FI-extending ring is quasi-Baer (hence right strongly FI-extending).

Proof. We give the proof of part (i). Let R be a quasi-Baer right FI-extending ring. Take $I \trianglelefteq R$. Since R is quasi-Baer, there exists $e \in \mathbf{S}_\ell(R)$ such that $\ell_R(I) = R(1 - e)$. Thus $I \subseteq r_R(\ell_R(I)) = eR$. Because R is right FI-extending, there exists $c = c^2 \in R$ such that $I_R \leq^{\mathrm{ess}} cR_R$. Then $I_R \leq^{\mathrm{ess}} eR \cap cR = ceR$. But $ce = (ce)^2$. So $ceR = cR \subseteq eR$. Since $I \subseteq cR$, $R(1 - c) = \ell_R(cR) \subseteq \ell_R(I) = R(1 - e)$. Hence $eR \subseteq cR$, thus $cR = eR$. Therefore $c \in \mathbf{S}_\ell(R)$, so R is right strongly FI-extending. $\square$

Proposition 12. ([11, Lemma 1.4]) Let T be a right ring of quotients of a ring R. Then we have the following.

  (i) For right ideals X and Y of T, if $X_T \leq^{\mathrm{ess}} Y_T$, then $X_R \leq^{\mathrm{ess}} Y_R$.

  (ii) If X is a fully invariant R-submodule of T, $X_R \leq^{\mathrm{ess}} TXT_R$.

Theorem 13. Assume that R is a right nonsingular ring and I is a proper ideal of R with I_R essentially closed in R_R and $\mathbf{P}(R) \subseteq I$. Let $e \in \mathbf{I}(Q(R))$ such that $I_R \leq^{\mathrm{ess}} (1 - e)Q(R)_R$. Then for any right FI-extending right ring of quotients T of R, $\langle eTe \rangle_{eQ(R)e}$ is a semiprime right FI-extending right ring of quotients of an isomorphic copy of $Q_{\mathfrak{FJ}}(R/I)$.

Proof. Let $S = \langle eTe \rangle_{eQ(R)e}$. Take $0 \neq X \trianglelefteq S$. Then S and X are both eRe-modules. From Theorem 5(ii), S and X are both right R-modules as well. Also $(X \cap R)_R \leq^{\mathrm{ess}} X_R$. Since T is right FI-extending and right nonsingular, T is right strongly FI-extending by Proposition 11. Thus there exists $b \in \mathbf{S}_\ell(T)$ such that $T(X \cap T)T_T \leq^{\mathrm{ess}} bT_T$. Therefore $T(X \cap T)T_R \leq^{\mathrm{ess}} bT_R$ by Proposition 12(i).

Claim 1. $ebe \in \mathbf{I}(S)$.

Proof of Claim 1. Suppose that there exists $y \in eR$ with $by(1 - b) \neq 0$. Since R_R is dense in $Q(R)_R$, there exists $r_1 \in R$ such that $by(1 - b)r_1 \neq 0$, and $y(1 - b)r_1 \in R$. Let $y_1 = y(1 - b)r_1$. Then $y_1 \in eR \cap R$ and $0 \neq by_1 \in bT$. There exists $r_2 \in R$ such that $0 \neq by_1 r_2 \in T(X \cap T)T$. Now also there exist t_i, $v_i \in T$ and $x_i \in X \cap T$ such that $by_1 r_2 = \sum t_i x_i v_i$. Since $1 - b \in \mathbf{S}_r(T)$, it follows that

$$by_1 r_2 = by(1-b)r_1 r_2 = by(1-b)r_1 r_2(1-b) = by_1 r_2(1-b) = \sum t_i x_i v_i(1-b).$$

We show that $XT(1 - b) = 0$. To see this, assume to the contrary that $0 \neq xv(1 - b)$ for some $x \in X$ and $v \in T$. Then there exists $r_3 \in R$ such that $0 \neq xv(1 - b)r_3$ and $v(1 - b)r_3 \in R$ since R_R is dense in T_R. Furthermore, there exists $r_4 \in R$ such that $0 \neq z = xv(1 - b)r_3 r_4 \in X \cap R \subseteq T(X \cap T)T \subseteq bT$. Thus $z = bz(1 - b) \in bT(1 - b)$. Note that $b \in \mathbf{S}_\ell(T)$. Thus $bT(1 - b) \trianglelefteq T$. Since $(bT(1-b))^2 = 0$, $(bT(1-b) \cap R)^2 = 0$ and $bT(1-b) \cap R \trianglelefteq R$. So $bT(1-b) \cap R \subseteq \mathbf{P}(R)$. Now $z \in bT(1 - b) \cap R \subseteq \mathbf{P}(R) \subseteq (1 - e)Q(R)$. Hence $0 \neq z \in eR \cap (1 - e)Q(R)$, a contradiction. Thus $XT(1 - b) = 0$. Therefore $by_1 r_2 = 0$, a contradiction. So $by(1-b) = 0$. Consequently, $beR(1-b) = 0$, so $be = beb$. Thus $ebe = ebebe = (ebe)^2$.

Claim 2. $X_R \leq^{\mathrm{ess}} ebeS_R$.

Proof of Claim 2. First we prove that $X \subseteq ebeS$. For this, we need to see that $(1 - b)X = 0$. Now suppose that $(1 - b)x \neq 0$ for some $x \in X$. Since R_R is dense in $Q(R)_R$, there is $r \in R$ such that $0 \neq (1 - b)xr$, and $xr \in R$. Thus $0 \neq (1 - b)xr = (1 - b)xr(1 - b) \in (1 - b)XT(1 - b) = 0$ because $1 - b \in \mathbf{S}_r(T)$ and $XT(1-b) = 0$. So we get a contradiction. Therefore $(1-b)X = 0$. Hence $X = bX$. Since $X = eX$, $X = bX = beX$. Thus $X = eX = e(beX) = ebeX$. Therefore $X \subseteq ebeS$.

Let $0 \neq s \in ebeS$. Then there exists $r_5 \in R$ with $0 \neq sr_5 \in R \cap eR$. Note that $s = ebes = ebs$. Therefore $sr_5 = ebsr_5$. Since R_R is dense in $Q(R)_R$, there exists $a_1 \in R$ such that $ebsr_5 a_1 \neq 0$, and $bsr_5 a_1 \in R$. Now note that $bsr_5 a_1 \in R \cap bR \subseteq bT$. Since $T(X \cap T)T_R \leq^{\mathrm{ess}} bT_R$, there is $a_2 \in R$ such that $0 \neq bsr_5 a_1 a_2 \in T(X \cap T)T \cap R$. Take $r_6 = a_1 a_2 \in R$. Then $0 \neq bsr_5 r_6 = bebsr_5 r_6 \in R$ because $s = ebs$. So $ebsr_5 r_6 = ebesr_5 r_6 \neq 0$.

By Theorem 5(ii), $0 \neq sr_5 r_6 = ebsr_5 r_6 = e(bsr_5 r_6)e \in e(T(X \cap T)T \cap R)e$ because $bsr_5 r_6 \in R$. Since $X \subseteq S$,

$$eT(X \cap T)Te = eT(e(X \cap T)e)Te = (eTe)(X \cap T)(eTe) \subseteq X.$$

So $0 \neq sr_5 r_6 \in X \cap eRe$. Thus there exists $r_7 \in R$ such that $0 \neq sr_5 r_6 r_7 \in R$. In this case, $sr_5 r_6 r_7 = ebe(sr_5 r_6)r_7 = ebe(sr_5 r_6)er_7 e = sr_5 r_6(er_7 e) \in X$ since $eR = eRe$ by Theorem 5(ii) and $sr_5 r_6 \in X$. Thus $0 \neq sr_5 r_6 r_7 \in (X \cap R)_R$. Hence $(X \cap R)_R \leq^{\mathrm{ess}} ebeS_R$. Now since $(X \cap R)_R \leq^{\mathrm{ess}} X_R \leq ebeS_R$, it follows that $X_R \leq^{\mathrm{ess}} ebeS_R$.

Now, using Theorem 5(ii), $X_R \leq^{\mathrm{ess}} ebeS_R$ implies $X_{eRe} \leq^{\mathrm{ess}} ebeS_{eRe}$. Hence $X_S \leq^{\mathrm{ess}} ebeS_S$. Therefore S is right FI-extending. The fact that S is a right ring of quotients of an isomorphic copy of $\widehat{Q}_{\mathfrak{FJ}}(R/I)$ follows from Theorem 5(iii). Since R/I is semiprime from Theorem 5(vi), R/I and $\widehat{Q}_{\mathfrak{FJ}}(R/I)$ are semiprime. So S is semiprime. $\qquad\square$

Our previous results suggest that when I is an essentially closed proper ideal of R containing $\mathbf{P}(R)$ and R is either quasi-Baer or right FI-extending then R/I should enjoy the same property. This can be shown by the use of Theorem 5 when $Q(R) = E(R_R)$. The following proposition includes this result, however we prove it without appealing to either Theorem 5 or Theorem 13.

Proposition 14. Let R be either a quasi-Baer or a right FI-extending ring. If I is a proper ideal of R such that I_R is essentially closed in R_R and $\mathbf{P}(R) \subseteq I$, then R/I is a semiprime quasi-Baer ring.

Proof. From [8, Corollary 1.3] and by the FI-extending property, there is $e \in \mathbf{S}_\ell(R)$ such that $I = eR$. Then note that $1 - e \in \mathbf{S}_r(R)$. By Lemma 2(i), R/I is semiprime; and by [4, Lemma 2.1], R/I is quasi-Baer. □

Open question. For a quasi-Baer ring or a right FI-extending ring R, when is $R/\mathbf{P}(R)$ also quasi-Baer?

Acknowledgments

The authors thank the referee for his/her comments for the improvement of the paper. Also the authors appreciate the gracious hospitality they received at each others' institutions. The partial support received from the Mathematics Research Institute, Columbus, an OSU-Lima grant, and Busan National University, Busan, South Korea is appreciated.

References

[1] P. Ara and M. Mathieu, *Local Multipliers of C^*-Algebras*, Springer Monographs in Math., Springer-Verlag, London, 2003

[2] K. Beidar and R. Wisbauer, *Strongly and properly semiprime modules and rings*, Ring Theory, Proc. Ohio State-Denison Conf. (S.K. Jain and S.T. Rizvi (eds.)), World Scientific, Singapore (1993), 58–94

[3] G.F. Birkenmeier, *Idempotents and completely semiprime ideals*, Comm. Algebra 11 (1983), 567–580

[4] G.F. Birkenmeier, *Decompositions of Baer-like rings*, Acta Math. Hungar. 59 (1992), 319–326

[5] G.F. Birkenmeier, *When does a supernilpotent radical split off?*, J. Algebra 172 (1995), 49–60

[6] G.F. Birkenmeier, G. Călugăreanu, L. Fuchs and H.P. Goeters, *The fully-invariant-extending property for Abelian groups*, Comm. Algebra 29 (2001), 673–685

[7] G.F. Birkenmeier, H.E. Heatherly, J.Y. Kim and J.K. Park, *Triangular matrix representations*, J. Algebra 230 (2000), 558–595

[8] G.F. Birkenmeier, J.Y. Kim and J.K. Park, *Quasi-Baer ring extensions and biregular rings*, Austral. Math. Soc. 61 (2000), 39–52

[9] G.F. Birkenmeier, B.J. Müller and S.T. Rizvi, *Modules in which every fully invariant submodule is essential in a direct summand*, Comm. Algebra 30 (2002), 1395–1415

[10] G.F. Birkenmeier, J.K. Park and S.T. Rizvi, *Modules with fully invariant submodules essential in fully invariant summands*, Comm. Algebra 30 (2002), 1833–1852

[11] G.F. Birkenmeier, J.K. Park and S.T. Rizvi, *Ring hulls and applications*, J. Algebra 304 (2006), 633–665

[12] G.F. Birkenmeier, J.K. Park and S.T. Rizvi, *Hulls of semiprime rings with applications to C*-algebras*, Preprint

[13] G.F. Birkenmeier, J.K. Park and S.T. Rizvi, *The structure of rings of quotients*, Preprint

[14] W.E. Clark, *Twisted matrix units semigroup algebras*, Duke Math. J. 34 (1967), 417–424

[15] J. Dixmier, *C*-Algebras*, North-Holland (1977)

[16] G.A. Elliott, *Automorphisms determined by multipliers on ideals of a C*-algebra*, J. Funct. Anal. 23 (1976), 1–10

[17] C. Faith and Y. Utumi, *Maximal quotient rings*, Proc. Amer. Math. Soc. 16 (1965), 1084–1089

[18] K.R. Goodearl, *Ring Theory: Nonsingular Rings and Modules*, Marcel Dekker, New York (1976)

[19] J. Lambek, *Lectures on Rings and Modules*, Chelsea, New York (1986)

[20] A.C. Mewborn, *Regular rings and Baer rings*, Math. Z. 121 (1971), 211–219

[21] G.K. Pedersen, *Approximating derivations on ideals of C*-algebras*, Invent. Math. 45 (1978), 299–305

[22] J.C. Robson and L.W. Small, *Liberal extensions*, Proc. London Math. Soc. (3)42 (1981), 83–103

[23] C.L. Walker and E.A. Walker, *Quotient categories and rings of quotients*, Rocky Mountain J. Math. 2 (1972), 513–555

Gary F. Birkenmeier
Department of Mathematics
University of Louisiana at Lafayette
Lafayette, LA 70504-1010, USA
e-mail: gfb1127@louisiana.edu

Jae Keol Park
Department of Mathematics
Busan National University
Busan 609-735
South Korea
e-mail: jkpark@pusan.ac.kr

S. Tariq Rizvi
Department of Mathematics
Ohio State University
Lima, OH 45804-3576, USA
e-mail: rizvi.1@osu.edu

Modules and Comodules
Trends in Mathematics, 113–124
© 2008 Birkhäuser Verlag Basel/Switzerland

Notes on Formal Smoothness

Tomasz Brzeziński

Dedicated to Robert Wisbauer on the occasion of his 65th birthday

Abstract. The definition of an S-category is proposed by weakening the axioms of a Q-category introduced by Kontsevich and Rosenberg. Examples of Q- and S-categories and (co)smooth objects in such categories are given.

Mathematics Subject Classification (2000). 16W30; 18A40; 16D90.

1. Introduction

In [12] Kontsevich and Rosenberg introduced the notion of a Q-category as a framework for developing non-commutative algebraic geometry. Relative to such a Q-category they introduced and studied the notion of a *formally smooth object*. Depending on the choice of Q-category this notion captures, e.g., that of a *smooth algebra* of [16], which arose a considerable interest since its role in non-commutative geometry was revealed in [8].

The aim of these notes is to give a number of examples of Q-categories, and their weaker version which we term S-categories, of interest in module, coring and comodule theories, and to give examples of smooth objects in these Q-categories. Crucial to the definition of an S-category is the notion of a *separable functor* introduced in [14]. In these notes we consider only the separability of functors with adjoints. This case is fully described by the Rafael Theorem [15]: A functor which has a right (resp. left) adjoint is separable if and only if the unit (resp. counit) of adjunction is a natural section (resp. retraction). For a detailed discussion of separable functors we refer to [7].

Throughout these notes, by a category we mean a set-category (i.e., in which morphisms form sets), by functors we mean covariant functors. All rings are unital and associative. For an A-coring $\mathcal{C}$, $\Delta_{\mathcal{C}}$ denotes the coproduct and $\varepsilon_{\mathcal{C}}$ denotes the counit. Whenever needed, we use the standard Sweedler notation for a coproduct $\Delta_{\mathcal{C}}(c) = \sum c_{(1)} \otimes_A c_{(2)}$ and for a coaction $\varrho^M(m) = \sum m_{(0)} \otimes_A m_{(1)}$.

2. Smoothness and cosmoothness in Q- and S-categories

Here we gather definitions of categories and objects we study in these notes.

Definition 2.1. *An* S-category *is a pair of functors* $\mathbb{X} = (\, \bar{\mathfrak{X}} \underset{u^*}{\overset{u_*}{\rightleftarrows}} \mathfrak{X} \,)$ *such that* u^* *is separable and left adjoint of* u_*.

This means that in an S-category $\mathbb{X} = (\, \bar{\mathfrak{X}} \underset{u^*}{\overset{u_*}{\rightleftarrows}} \mathfrak{X} \,)$ the unit of adjunction $\eta : \mathfrak{X} \to u_* u^*$ has a natural retraction $\nu : u_* u^* \to \mathfrak{X}$. Therefore, for all objects x of $\mathfrak{X}$ and y of $\bar{\mathfrak{X}}$, there exist morphisms

$$\bar{\mathfrak{X}}(y, u^*(x)) \to \mathfrak{X}(u_*(y), x), \qquad g \mapsto \nu_x \circ u_*(g).$$

The notion of an S-category is a straightforward generalisation of that of a Q-category, introduced in [12]. The latter is defined as a pair of functors $\mathbb{X} = (\, \bar{\mathfrak{X}} \underset{u^*}{\overset{u_*}{\rightleftarrows}} \mathfrak{X} \,)$ such that u^* is full and faithful and left adjoint of u_*. In a Q-category the unit of adjunction η is a natural isomorphism, hence, in particular, a section. Thus any Q-category is also an S-category. Following the Kontsevich-Rosenberg terminology (prompted by algebraic geometry) the functors u_* and u^* constituting an S-category are termed the *direct image* and *inverse image* functors, respectively.

Definition 2.2. *We say that an S-category* $\mathbb{X} = (\, \bar{\mathfrak{X}} \underset{u^*}{\overset{u_*}{\rightleftarrows}} \mathfrak{X} \,)$ *is* supplemented *if there exists a functor* $u_! : \bar{\mathfrak{X}} \to \mathfrak{X}$ *and a natural transformation* $\bar{\eta} : \bar{\mathfrak{X}} \to u^* u_!$.

In particular, an S-category $\mathbb{X} = (\, \bar{\mathfrak{X}} \underset{u^*}{\overset{u_*}{\rightleftarrows}} \mathfrak{X} \,)$ is supplemented if u^* has a left adjoint. Furthermore, $\mathbb{X}$ is supplemented if the functor u_* is separable, since, in this case, the counit of adjunction has a section which we can take for $\bar{\eta}$ (and $u_! = u_*$). This supplemented S-category is termed a *self-dual supplemented S-category*.

In a supplemented S-category, for any $y \in \bar{\mathfrak{X}}$, there is a canonical morphism in $\mathfrak{X}$, natural in y,

$$r_y : u_*(y) \to u_!(y),$$

defined as a composition

$$r_y : \; u_*(y) \xrightarrow{\; u_*(\bar{\eta}_y) \;} u_* u^* u_!(y) \xrightarrow{\; \nu_{u_!(y)} \;} u_!(y) \,.$$

The existence of canonical morphisms r_y allows us to make the following

Definition 2.3. *Given a supplemented S-category* $\mathbb{X} = (\, \bar{\mathfrak{X}} \underset{u^*}{\overset{u_*}{\rightleftarrows}} \mathfrak{X} \,)$, *with the natural map* $r : u_* \to u_!$, *an object* x *of* $\mathfrak{X}$ *is said to be:*

(a) *formally* $\mathbb{X}$-smooth *if, for any* $y \in \bar{\mathfrak{X}}$, *the mapping* $\mathfrak{X}(x, r_y)$ *is surjective;*
(b) *formally* $\mathbb{X}$-cosmooth *if, for any* $y \in \bar{\mathfrak{X}}$, *the mapping* $\mathfrak{X}(r_y, x)$ *is surjective.*

Remark 2.1. We would like to stress that the notion of formal $\mathbb{X}$-(co)smoothness is relative to the choice of the retraction of the unit of adjunction, and the choice of $u_!$ and $\bar{\eta}$, since the definition of r depends on all these data.

Dually to S- and Q-categories one defines S°-categories and Q°-categories.

Definition 2.4. *An* S°-*category* (*respectively* Q°-*category*) *is a pair of functors* $\mathbb{X} = (\bar{\mathfrak{X}} \xrightarrow[u^*]{\overset{u_*}{\rightleftarrows}} \mathfrak{X})$ *such that* u^* *is separable* (*resp. fully faithful*) *and right adjoint of* u_*.

Thus an adjoint pair of separable functors gives rise to a supplemented S- and S°-category. In these notes (with a minor exception) we concentrate on S-categories.

3. Examples of Q- and S-categories

The following generic example of a Q-category was constructed by Kontsevich and Rosenberg in [12].

Example 3.1 (The Q-category of morphisms). Let $\mathfrak{X}$ be any category, and let $\mathfrak{X}^2$ be the category of morphisms in $\mathfrak{X}$ defined as follows. The objects of $\mathfrak{X}^2$ are morphisms f, g in $\mathfrak{X}$. Morphisms in $\mathfrak{X}^2$ are commutative squares

$$
\begin{array}{ccc}
x & \xrightarrow{f} & y \\
\downarrow & & \downarrow \\
x' & \xrightarrow{g} & y'
\end{array}
$$

where the vertical arrows are in $\mathfrak{X}$. Now, set $\bar{\mathfrak{X}} = \mathfrak{X}^2$. The inverse image functor u^* is

$$
u^* : x \mapsto \left(x \xrightarrow{x} x \right), \qquad \left(x \xrightarrow{f} y \right) \mapsto \left(\begin{array}{ccc} x & \xrightarrow{x} & x \\ f\downarrow & & \downarrow f \\ y & \xrightarrow{y} & y \end{array} \right).
$$

The direct image functor u_* is defined by

$$
u_* : \left(x \xrightarrow{f} y \right) \mapsto x, \qquad \begin{array}{ccc} x & \xrightarrow{f} & y \\ \downarrow & & \downarrow \\ x' & \xrightarrow{g} & y' \end{array} \mapsto \left(\begin{array}{c} x \\ \downarrow \\ x' \end{array} \right).
$$

Note that, for all objects x and morphisms f in $\mathfrak{X}$,

$$
u_* u^*(x) = u_*(x \xrightarrow{x} x) = x, \qquad u_* u^*(f) = f.
$$

Hence, for all objects x in $\mathfrak{X}$, there is an isomorphism (natural in x), $\eta_x : x \to u_* u^*(x)$, $\eta_x = x$.

Note further that for all objects $x \xrightarrow{f} y$ in $\mathfrak{X}^2$, $u^* u_*(f) = x$, and we can define a morphism $\varepsilon_f : u^* u_*(f) \to f$ by

$$\varepsilon_f = \begin{pmatrix} x \xrightarrow{\ x\ } x \\ x\downarrow \qquad \downarrow f \\ x \xrightarrow{\ f\ } y \end{pmatrix}.$$

In this way, u_* is the right adjoint of u^* with counit ε and unit η. The unit is obviously a natural isomorphism, hence u^* is full and faithful and, thus, a Q-category $\mathbb{X} = (\ \bar{\mathfrak{X}} \underset{u^*}{\overset{u_*}{\rightleftarrows}} \mathfrak{X}\ )$ is constructed. $\mathbb{X}$ is supplemented, since u^* has a left adjoint

$$u_! : \left(x \xrightarrow{\ f\ } y \right) \mapsto y, \qquad \begin{pmatrix} x \xrightarrow{\ f\ } y \\ \downarrow \qquad \downarrow \\ x' \xrightarrow{\ g\ } y' \end{pmatrix} \mapsto \begin{pmatrix} y \\ \downarrow \\ y' \end{pmatrix}.$$

The unit of the adjunction $u_! \dashv u^*$ is, for all $f : x \to y$,

$$\bar{\eta}_f = \begin{pmatrix} x \xrightarrow{\ f\ } y \\ f\downarrow \qquad \downarrow y \\ y \xrightarrow{\ y\ } y \end{pmatrix},$$

and thus the corresponding maps r come out as

$$r_f = f.$$

Consequently, an object $x \in \mathfrak{X}$ is formally $\mathbb{X}$-smooth (when $\mathbb{X}$ is supplemented by $u_!$ and $\bar{\eta}$) provided, for all $y \xrightarrow{f} z \in \bar{\mathfrak{X}}$, the mapping

$$\mathfrak{X}(x, y) \to \mathfrak{X}(x, z), \qquad g \mapsto f \circ g,$$

is surjective. Similarly, x is formally $\mathbb{X}$-cosmooth if and only if the mappings

$$\mathfrak{X}(z, x) \to \mathfrak{X}(y, x), \qquad g \mapsto g \circ f,$$

are surjective.

This generic example has a useful modification whereby one takes for $\bar{\mathfrak{X}}$ any full subcategory of $\mathfrak{X}^2$ which contains all the identity morphisms in $\mathfrak{X}$.

Example 3.2 (The Wisbauer Q-category). Let R by a ring and M be a left R-module. Following [17, Section 15] $\sigma[M]$ denotes a full subcategory of the category $_R\mathfrak{M}$ of left R-modules, consisting of objects subgenerated by M. Since $\sigma[M]$ is a full subcategory of $_R\mathfrak{M}$, the inclusion functor

$$u^* : \sigma[M] \to {_R\mathfrak{M}},$$

is full and faithful. It also has the right adjoint, the trace functor (see [17, 45.11] or [5, 41.1]),

$$u_* = T^M : {}_R\mathfrak{M} \to \sigma[M], \quad T^M(L) = \sum \{f(N) \mid N \in \sigma[M], \ f \in \operatorname{Hom}_R(M, L)\}.$$

Hence there is a Q-category $\mathbb{X} = (\ \bar{\mathfrak{X}} \underset{u^*}{\overset{u_*}{\rightleftarrows}} \mathfrak{X}\ )$ with $\mathfrak{X} = \sigma[M]$ and $\bar{\mathfrak{X}} = {}_R\mathfrak{M}$.

All the remaining examples come from the theory of corings.

Example 3.3 (Comodules of a locally projective coring). This is a special case of Example 3.2. Let $(\mathcal{C}, \Delta_\mathcal{C}, \varepsilon_\mathcal{C})$ be an A-coring which is locally projective as a left A-module. Let $R = {}^*\mathcal{C} = \operatorname{Hom}_{A-}(\mathcal{C}, A)$ be a left dual ring of $\mathcal{C}$ with the unit $\varepsilon_\mathcal{C}$ and product, for all $r, s \in R$,

$$rs: \ \mathcal{C} \xrightarrow{\Delta_\mathcal{C}} \mathcal{C} \otimes_A \mathcal{C} \xrightarrow{\mathcal{C} \otimes_A s} \mathcal{C} \xrightarrow{r} A.$$

Take $\mathfrak{X} = \mathfrak{M}^\mathcal{C}$, the category of right $\mathcal{C}$-comodules, and $\bar{\mathfrak{X}} = {}_R\mathfrak{M}$. Define a functor

$$u^*: \mathfrak{M}^\mathcal{C} \to {}_R\mathfrak{M}, \qquad M \mapsto M,$$

where right $\mathcal{C}$-comodule M is given a left R-module structure by $rm = \sum m_{(0)} r(m_{(1)})$. Since $\mathcal{C}$ is a locally projective left A-module, the functor u^* has a right adjoint, the rational functor (see [5, 20.1]),

$$u_* = \operatorname{Rat}^\mathcal{C} : {}_R\mathfrak{M} \to \mathfrak{M}^\mathcal{C}, \qquad \operatorname{Rat}^\mathcal{C}(M) = \{n \in M \mid n \text{ is rational}\},$$

where an element $n \in M$ is said to be rational provided there exists $\sum_i m_i \otimes_A c_i \in M \otimes_A \mathcal{C}$ such that, for all $r \in R$, $rm = \sum_i m_i r(c_i)$. Here, the left R-module M is seen as a right A-module via the anti-algebra map $A \to R$, $a \mapsto \varepsilon_\mathcal{C}(-a)$.

Example 3.4 (Coseparable corings). Recall that an A-coring $(\mathcal{C}, \Delta_\mathcal{C}, \varepsilon_\mathcal{C})$ is said to be *coseparable* [10] if there exists a $(\mathcal{C}, \mathcal{C})$-bicomodule retraction of the coproduct $\Delta_\mathcal{C}$. This is equivalent to the existence of a *cointegral* defined as an (A, A)-bimodule map $\delta: \mathcal{C} \otimes_A \mathcal{C} \to A$ such that $\delta \circ \Delta_\mathcal{C} = \varepsilon_\mathcal{C}$, and

$$(\mathcal{C} \otimes_A \delta) \circ (\Delta_\mathcal{C} \otimes_A \mathcal{C}) = (\delta \otimes_A \mathcal{C}) \circ (\mathcal{C} \otimes_A \Delta_\mathcal{C}).$$

Furthermore, this is equivalent to the separability of the forgetful functor $(-)_A : \mathfrak{M}^\mathcal{C} \to \mathfrak{M}_A$ [4, Theorem 3.5]). Since this forgetful functor is a left adjoint to $- \otimes_A \mathcal{C} : \mathfrak{M}_A \to \mathfrak{M}^\mathcal{C}$, a coseparable coring $\mathcal{C}$ gives rise to an S-category $\mathbb{X}$ with

$$\mathfrak{X} = \mathfrak{M}^\mathcal{C}, \qquad \bar{\mathfrak{X}} = \mathfrak{M}_A, \qquad u^* = (-)_A, \qquad u_* = - \otimes_A \mathcal{C}.$$

This S-category is denoted by $\mathbb{X}_\delta^\mathcal{C}$. By [4, Theorem 3.5], the retraction ν of the unit of the adjunction is given explicitly, for all $M \in \mathfrak{M}^\mathcal{C}$,

$$\nu_M : M \otimes_A \mathcal{C} \to M, \qquad m \otimes_A c \mapsto \sum m_{(0)} \delta(m_{(1)} \otimes_A c).$$

In general, $\mathbb{X}_\delta^\mathcal{C}$ need not to be supplemented. However, if there exists

$$e \in \mathcal{C}^A := \{c \in \mathcal{C} \mid \forall a \in A, \ ac = ca\},$$

then $\mathbb{X}_\delta^{\mathcal{C}}$ can be supplemented with

$$u_! = -\otimes_A \mathcal{C}, \qquad \bar\eta_M : M \to M \otimes_A \mathcal{C}, \qquad m \mapsto m \otimes_A e.$$

This supplemented S-category is denoted by $\mathbb{X}_{\delta,e}^{\mathcal{C}}$.

Recall that an A-coring $\mathcal{C}$ is said to be *cosplit* if there exists an A-central element $e \in \mathcal{C}^A$ such that $\varepsilon_{\mathcal{C}}(e) = 1$. By [4, Theorem 3.3] this is equivalent to the separability of the functor $-\otimes_A \mathcal{C}$, and thus a cosplit coring gives rise to an S°-category. Therefore, a coring which is both cosplit and coseparable induces a self-dual, supplemented S-category.

In addition to the defining adjunction of an A-coring, $(-)_A \dashv -\otimes_A \mathcal{C}$, for any right $\mathcal{C}$-comodule P, there is a pair of adjoint functors

$$-\otimes_B P : \mathfrak{M}_B \to \mathfrak{M}^{\mathcal{C}}, \qquad \mathrm{Hom}^{\mathcal{C}}(P, -) : \mathfrak{M}^{\mathcal{C}} \to \mathfrak{M}_B,$$

where B is any subring of the endomorphism ring $S = \mathrm{End}^{\mathcal{C}}(P)$ (cf. [5, 18.21]). Depending on the choice of $\mathcal{C}$, P and B this adjunction provides a number of examples of Q-categories.

Example 3.5 (Comatrix corings). Take a (B, A)-bimodule P that is finitely generated and projective as a right A-module. Let $\mathbf{e} \in P \otimes_A P^*$ be the dual basis (where $P^* = \mathrm{Hom}_A(P, A)$), and let $\mathcal{C} = P^* \otimes_B P$ be the comatrix coring associated to P [9]. The coproduct and counit in $\mathcal{C}$ are given by

$$\Delta_{\mathcal{C}}(\xi \otimes_B p) = \xi \otimes_B \mathbf{e} \otimes_B p, \qquad \varepsilon_{\mathcal{C}}(\xi \otimes_B p) = \xi(p),$$

for all $p \in P$ and $\xi \in P^*$. P is a right $\mathcal{C}$-comodule with the coaction $\varrho^P : p \mapsto \mathbf{e} \otimes_B p$. Let

$$\mathfrak{X} = \mathfrak{M}_B, \qquad \bar{\mathfrak{X}} = \mathfrak{M}^{\mathcal{C}}, \qquad u^* = -\otimes_B P, \qquad u_* = \mathrm{Hom}^{\mathcal{C}}(P, -).$$

In view of [6, Proposition 2.3], $\mathbb{X} = (\bar{\mathfrak{X}} \underset{u^*}{\overset{u_*}{\rightleftarrows}} \mathfrak{X})$ is a Q-category if and only if the map

$$B \to P \otimes_A P^*, \qquad b \mapsto b\mathbf{e},$$

is pure as a morphism of left B-modules (equivalently, P is a totally faithful left B-module).

Example 3.6 (Strongly $(\mathcal{C}, A)$-injective comodules). Let $\mathcal{C}$ be an A-coring, let P be a right $\mathcal{C}$-comodule and $S = \mathrm{End}^{\mathcal{C}}(P)$. Following [18, 2.9], P is said to be *strongly* $(\mathcal{C}, A)$-*injective* if the coaction $\varrho^P : P \to P \otimes_A \mathcal{C}$ has a left S-module right $\mathcal{C}$-comodule retraction. For such a comodule, define

$$\mathfrak{X} = \mathfrak{M}_S, \qquad \bar{\mathfrak{X}} = \mathfrak{M}^{\mathcal{C}}, \qquad u^* = -\otimes_S P, \qquad u_* = \mathrm{Hom}^{\mathcal{C}}(P, -).$$

In view of [18, 3.2], if P is a finitely generated and projective as a right A-module, then $\mathbb{X} = (\bar{\mathfrak{X}} \underset{u^*}{\overset{u_*}{\rightleftarrows}} \mathfrak{X})$ is a Q-category.

Example 3.7 (($\mathcal{C}, A$)-injective Galois comodules). Recall that a right $\mathcal{C}$-comodule is said to be ($\mathcal{C}, A$)-*injective*, provided there is a right $\mathcal{C}$-colinear retraction of the coaction. The full subcategory of $\mathfrak{M}^{\mathcal{C}}$ consisting of all ($\mathcal{C}, A$)-injective comodules is denoted by $\mathfrak{I}^{\mathcal{C}}$.

Let P be a right comodule of an A-coring $\mathcal{C}$, and let $S = End^{\mathcal{C}}(P)$ and $T = End_A(P)$. Following [18, 4.1], P is said to be a *Galois comodule* if, for all $N \in \mathfrak{I}^{\mathcal{C}}$, the evaluation map

$$Hom^{\mathcal{C}}(P, N) \otimes_S P \to N, \qquad f \otimes_S p \to f(p),$$

is an isomorphism of right $\mathcal{C}$-comodules.

Let P be a Galois comodule, and assume that the inclusion $S \to T$ has a right S-module retraction. By [18, 4.3] this is equivalent to say that P is a ($\mathcal{C}, A$)-injective comodule, and hence one can consider the following pair of categories and adjoint functors:

$$\bar{\mathfrak{X}} = \mathfrak{M}_S, \qquad \mathfrak{X} = \mathfrak{I}^{\mathcal{C}}, \qquad u_* = - \otimes_S P : \bar{\mathfrak{X}} \to \mathfrak{X}, \qquad u^* = Hom^{\mathcal{C}}(P, -) : \mathfrak{X} \to \bar{\mathfrak{X}}.$$

Since the evaluation map is the counit of the adjunction $u_* \dashv u^*$, the Galois property of P means that the functor u^* is fully faithful. Thus $\mathbb{X} = (\bar{\mathfrak{X}} \underset{u^*}{\overset{u_*}{\rightleftarrows}} \mathfrak{X})$ is a Q°-category.

4. Examples of smooth and cosmooth objects

Let $\mathcal{C}$ be an A-coring, set $\mathfrak{X} = \mathfrak{M}^{\mathcal{C}}$, and consider the full subcategory of $\mathfrak{X}^2$ consisting of all monomorphisms in $\mathfrak{M}^{\mathcal{C}}$ with an A-module retraction. With these data one constructs a Q-category as in Example 3.1. This Q-category is denoted by $\mathbb{X}^{\mathcal{C}}$.

Theorem 4.1. *A right $\mathcal{C}$-comodule M is ($\mathcal{C}, A$)-injective if and only if M is a formally $\mathbb{X}^{\mathcal{C}}$-cosmooth object.*

Proof. In view of the discussion at the end of Example 3.1, an object $M \in \mathfrak{X} = \mathfrak{M}^{\mathcal{C}}$ if formally $\mathbb{X}^{\mathcal{C}}$-cosmooth if and only if, for all morphisms $f : N \to N'$ in $\mathfrak{M}^{\mathcal{C}}$ with right A-module retraction, the maps

$$\vartheta_f : Hom^{\mathcal{C}}(N', M) \to Hom^{\mathcal{C}}(N, M), \qquad g \mapsto g \circ f,$$

are surjective. This means that, for all $h \in Hom^{\mathcal{C}}(N, M)$, there is $g \in Hom^{\mathcal{C}}(N', M)$ completing the following diagram

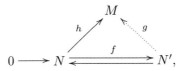

where the arrow $N' \to N$ is in $\mathfrak{M}_A$, and thus is equivalent to M being ($\mathcal{C}, A$)-injective, see [5, 18.18]. $\square$

The arguments used in the proof of Theorem 4.1, in particular, the identification of (co)smooth objects as object with a (co)splitting property, apply to all Q-categories of the type described in Example 3.1. This leads to reinterpretation of smooth algebras and coalgebras in abelian monoidal categories studied in [3].

Example 4.1. Let $(V, \otimes)$ be an abelian monoidal category, i.e., a monoidal category which is abelian and such that the tensor functors $- \otimes v$, $v \otimes -$ are additive and right exact, for all objects v of V. Let $\mathfrak{X}$ be the category of algebras in V, and let $\bar{\mathfrak{X}}$ be a full subcategory of $\mathfrak{X}^2$, consisting of *Hochschild algebra extensions*, i.e., of all surjective algebra morphisms split as morphisms in V and with a square-zero kernel. Denote the resulting Q-category by $\mathbb{HAE}$. In view of [3, Theorem 3.8], an algebra in V is formally smooth in the sense of [3, Definition 3.9], i.e., it has the Hochschild dimension at most 1, if and only if it is a formally $\mathbb{HAE}$-smooth object.

In particular if $(V, \otimes)$ is the category of vector spaces (with the usual tensor product), we obtain the characterisation of smooth algebras [16] (or semi-free algebras in the sense of [8]), described in [12, Proposition 4.3].

Example 4.2. Let $(V, \otimes)$ be an abelian monoidal category. Let $\mathfrak{X}$ be the category of coalgebras in V, and let $\bar{\mathfrak{X}}$ be a full subcategory of $\mathfrak{X}^2$, consisting of *Hochschild coalgebra extensions*, i.e., of all injective coalgebra morphisms $\sigma : C \to E$ split as morphisms in V and with the property $(p \otimes p) \circ \Delta_E = 0$, where $p : E \to$ cokerσ is the cokernel of σ. Denote the resulting Q-category by $\mathbb{HCE}$. In view of [3, Theorem 4.16], a coalgebra in V is formally smooth in the sense of [3, Definition 4.17] if and only if it is a formally $\mathbb{HCE}$-cosmooth object.

The following example is taken from [2].

Example 4.3. Let A and B be rings, and let M be a (B, A)-bimodule. Denote by $\mathcal{E}_M$ the class of all (B, B)-bilinear maps f such that $\mathrm{Hom}_B(M, f)$ splits as an (A, B)-bimodule map. A B-bimodule P is said to be $\mathcal{E}_M$-*projective*, provided every morphism $N \to P$ in $\mathcal{E}_M$ has a section. By the argument dual to that in the proof of Theorem 4.1 one can reinterpret $\mathcal{E}_M$-projectivity as formal smoothness as follows.

Take $\mathfrak{X}$ to be the category of B-bimodules and $\bar{\mathfrak{X}} = \mathcal{E}_M$, a full subcategory of $\mathfrak{X}^2$. Denote the resulting Q-category by $\mathbb{E}$. A B-bimodule P is formally $\mathbb{E}$-smooth if and only if, for all $f : N \to N' \in \mathcal{E}_M$, the function

$$\Theta(f) : \mathrm{Hom}_{B,B}(P, N) \to \mathrm{Hom}_{B,B}(P, N'), \qquad g \mapsto f \circ g,$$

is surjective. In terminology of [11, Chapter X], $\mathbb{E}$-smoothness of P is equivalent to the $\mathcal{E}_M$-projectivity of P.

A (B, A)-bimodule M is said to be *formally smooth* provided the kernel of the evaluation map

$$\mathrm{ev}_M : M \otimes_A \mathrm{Hom}_B(M, B) \to B, \qquad \mathrm{ev}_M \, (m \otimes_A f) = f(m).$$

is an $\mathcal{E}_M$-projective B-bimodule. Thus M is formally smooth if and only if $\ker \mathrm{ev}_M$ is formally $\mathbb{E}$-smooth.

Next we characterise all smooth and cosmooth objects in the supplemented S-category $\mathbb{X}^{\mathcal{C}}_{\delta,e}$ associated to a coseparable A-coring $\mathcal{C}$ with an A-central element e as in Example 3.4.

Proposition 4.1. *Let $\mathcal{C}$ be a coseparable A-coring with a cointegral δ and an A-central element e, and let $\mathbb{X}^{\mathcal{C}}_{\delta,e}$ be the associated supplemented S-category. A right $\mathcal{C}$-comodule M is formally $\mathbb{X}^{\mathcal{C}}_{\delta,e}$-smooth if and only if the map*

$$\kappa_M : M \to M, \qquad m \mapsto \sum m_{(0)}\delta(e\otimes_A m_{(1)}),$$

is a right A-linear section (i.e., κ_M has a left inverse in $End_A(M)$).

Proof. In this case, for all $N \in \mathfrak{M}_A$, the canonical morphisms r_N read

$$r_N : N\otimes_A\mathcal{C} \to N\otimes_A\mathcal{C}, \qquad n\otimes_A c \mapsto \sum n\otimes_A e_{(1)}\delta(e_{(2)}\otimes_A c).$$

Using the (defining adjunction) isomorphisms $\mathrm{Hom}^{\mathcal{C}}(M, N\otimes_A\mathcal{C}) \simeq \mathrm{Hom}_A(M,N)$, the maps

$$\mathrm{Hom}^{\mathcal{C}}(M,r_N) : \mathrm{Hom}^{\mathcal{C}}(M, N\otimes_A\mathcal{C}) \to \mathrm{Hom}^{\mathcal{C}}(M, N\otimes_A\mathcal{C}),$$

can be identified with

$$\vartheta_{M,N} : \mathrm{Hom}_A(M,N) \to \mathrm{Hom}_A(M,N), \qquad f \mapsto (N\otimes_A\varepsilon_{\mathcal{C}}) \circ r_N \circ (f\otimes_A\mathcal{C}) \circ \varrho^M,$$

where $\varrho^M : M \to M\otimes_A\mathcal{C}$ is the coaction. Hence $\mathrm{Hom}^{\mathcal{C}}(M,r_N)$ are surjective for all N if and only if $\vartheta_{M,N}$ are surjective for all N. These can be computed further, for all $m \in M$, $f \in \mathrm{Hom}_A(M,N)$,

$$\vartheta_{M,N}(f)(m) = \sum(N\otimes_A\varepsilon_{\mathcal{C}}) \circ r_N(f(m_{(0)})\otimes_A m_{(1)})$$

$$= \sum(N\otimes_A\varepsilon_{\mathcal{C}})(f(m_{(0)})\otimes_A e_{(1)}\delta(e_{(2)}\otimes_A m_{(1)}))$$

$$= \sum f(m_{(0)})\delta(e\otimes_A m_{(1)}) = \sum f(m_{(0)}\delta(e\otimes_A m_{(1)})) = f(\kappa_M(m)),$$

by the right A-linearity of f. Hence

$$\vartheta_{M,N}(f) = f \circ \kappa_M.$$

If κ_M has a retraction $\lambda_M \in End_A(M)$, then for all $f \in \mathrm{Hom}_A(M,N)$,

$$\vartheta_{M,N}(f \circ \lambda_M) = f \circ \lambda_M \circ \kappa_M = f,$$

i.e., the $\vartheta_{M,N}$ are surjective. If, on the other hand, all the $\vartheta_{M,N}$ are surjective, choose $N = M$ and take any $\lambda_M \in \vartheta_{M,M}^{-1}(M)$. Then

$$M = \vartheta_{M,M}(\lambda_M) = \lambda_M \circ \kappa_M,$$

so λ_M is a retraction of κ_M as required. $\qquad\square$

Example 4.4 (Modules graded by G-sets). Let G be a group, X be a (right) G-set and let $A = \oplus_{\sigma\in G}$ be a G-graded k-algebra. Following [13], a kX-graded right A-module $M = \oplus_{x\in X}M_x$ is said to be *graded by G-set X* provided, for all $x \in X$, $\sigma \in G$,

$$M_x A_\sigma \subseteq M_{x\sigma}.$$

A morphism of such modules is an A-linear map which preserves the X-grading. The resulting category is denoted by gr-(G, A, X). It is shown in [7, Section 4.6] that gr-(G, A, X) is isomorphic to the category of right comodules of the following coring $\mathcal{C}$. As a left A-module $\mathcal{C} = A \otimes kX$. The right A-multiplication is given by

$$(a \otimes x) a_\sigma = a a_\sigma \otimes x\sigma, \qquad \forall a \in A, x \in X, a_\sigma \in A_\sigma.$$

The coproduct and counit are defined by

$$\Delta_{\mathcal{C}}(a \otimes x) = (a \otimes x) \otimes_A (1_A \otimes x), \qquad \varepsilon_{\mathcal{C}}(a \otimes x) = a.$$

An object $M = \oplus_{x \in X} M_x$ in gr-(G, A, X) is a right $\mathcal{C}$-comodule with the coaction $\varrho^M : M \to M \otimes_A \mathcal{C}, m_x \mapsto m_x \otimes_A 1_A \otimes x$, where $m_x \in M_x$. Also in [7, Section 4.6] it is shown that $\mathcal{C}$ is a coseparable coring with a cointegral (cf. [19, Proposition 2.5.3])

$$\delta : \mathcal{C} \otimes_A \mathcal{C} \simeq A \otimes kX \otimes kX \to A, \qquad a \otimes x \otimes y \mapsto a \delta_{x,y}.$$

Thus gr-(G, A, X) gives rise to an S-category as in Example 3.4.

Let $X^G := \{x \in X \mid \forall \sigma \in G, \ x\sigma = x\}$ be the set of one-point orbits of G in X. If $X^G \neq \emptyset$, the above S-category can be supplemented as in Example 3.4 by

$$e := 1_A \otimes z, \qquad z \in X^G.$$

In this case, for any $M \in$ gr-(G, A, X), the map κ_M in Proposition 4.1 comes out as

$$\kappa_M(m_x) = m_x \delta_{x,z}, \qquad \forall m_x \in M_x.$$

Thus a graded module $M \in$ gr-(G, A, X) is formally $\mathbb{X}_{\delta,e}^{A \otimes kX}$-smooth if and only if it is concentrated in degree z, i.e., $M = M_z$.

Given an A-coring $\mathcal{C}$, the set of right A-module maps $\mathcal{C} \to A$, $\mathcal{C}^*$, is a ring with the unit $\varepsilon_{\mathcal{C}}$ and the product, for all $\xi, \xi' \in \mathcal{C}^*$,

$$\xi \xi' : \mathcal{C} \xrightarrow{\Delta_{\mathcal{C}}} \mathcal{C} \otimes_A \mathcal{C} \xrightarrow{\xi' \otimes_A \mathcal{C}} \mathcal{C} \xrightarrow{\xi} A.$$

Proposition 4.2. *Let $\mathcal{C}$ be a coseparable A-coring with a cointegral δ and an A-central element e, and let $\mathbb{X}_{\delta,e}^{\mathcal{C}}$ be the associated supplemented S-category. Then the following statements are equivalent:*

(1) *All right $\mathcal{C}$-comodules are formally $\mathbb{X}_{\delta,e}^{\mathcal{C}}$-cosmooth.*

(2) *The right A-linear map*

$$\lambda : \mathcal{C} \to A, \qquad c \mapsto \delta(e \otimes_A c),$$

   has a left inverse in the dual ring $\mathcal{C}^*$.

(3) *The regular right $\mathcal{C}$-comodule $\mathcal{C}$ is formally $\mathbb{X}_{\delta,e}^{\mathcal{C}}$-cosmooth.*

Proof. Note that $\mathcal{C}^*$ can be identified with $End^{\mathcal{C}}(\mathcal{C})$ via the map $\xi \mapsto (\xi \otimes_A \mathcal{C}) \circ \Delta_{\mathcal{C}}$ (with the inverse $f \mapsto \varepsilon_{\mathcal{C}} \circ f$). Under this identification the product in $\mathcal{C}^*$ coincides with the composition in $End^{\mathcal{C}}(\mathcal{C})$. Hence (2) is equivalent to saying that the map $r_A = (\lambda \otimes_A \mathcal{C}) \circ \Delta_{\mathcal{C}}$ has a retraction in $\mathfrak{M}^{\mathcal{C}}$. Denote this retraction by s_A. Note further that, since δ is a cointegral, the r_N defined in the proof of Proposition 4.1 can be written as $r_N = N \otimes_A r_A$. This implies that s_A is a section of r_A if and only if $s_N = N \otimes_A s_A$ is a retraction of $r_N = N \otimes_A r_A$, for all right A-modules N. Finally observe that for all $M \in \mathfrak{M}^{\mathcal{C}}$ and $N \in \mathfrak{M}_A$, the maps $\varphi_{M,N} := \mathrm{Hom}^{\mathcal{C}}(r_N, M)$ come out explicitly as

$$\varphi_{M,N} : \mathrm{Hom}^{\mathcal{C}}(N \otimes_A \mathcal{C}, M) \ni f \mapsto f \circ r_N \in \mathrm{Hom}^{\mathcal{C}}(N \otimes_A \mathcal{C}, M).$$

$(2) \Rightarrow (1)$ The property $s_N \circ r_N = N \otimes_A \mathcal{C}$, implies that, for all right $\mathcal{C}$-comodules M and right A-modules N, the maps $\varphi_{M,N}$ are surjective. Hence all right $\mathcal{C}$-comodules are formally $\mathbb{X}^{\mathcal{C}}_{\delta,e}$-cosmooth.

The implication $(1) \Rightarrow (3)$ is obvious.

$(3) \Rightarrow (2)$ If $\mathcal{C}$ is formally $\mathbb{X}^{\mathcal{C}}_{\delta,e}$-cosmooth, then $\varphi_{\mathcal{C},A} : \mathrm{End}^{\mathcal{C}}(\mathcal{C}) \to \mathrm{End}^{\mathcal{C}}(\mathcal{C})$ is surjective. Hence there exists $s_A \in \mathrm{End}^{\mathcal{C}}(\mathcal{C})$ such that

$$\mathcal{C} = \varphi_{\mathcal{C},A}(s_A) = s_A \circ r_A.$$

This completes the proof. $\qquad\square$

A coseparable A-coring $\mathcal{C}$ with a cointegral δ is said to be *Frobenius-coseparable* if there exists $e \in \mathcal{C}^A$ such that, for all $c \in \mathcal{C}$, $\delta(c \otimes_A e) = \delta(e \otimes_A c) = \varepsilon_{\mathcal{C}}(c)$. The element e is called a *Frobenius element*. In particular a Frobenius-coseparable coring is a Frobenius coring, see [5, 27.5].

Corollary 4.1. *Let $\mathcal{C}$ be a Frobenius-coseparable A-coring with cointegral δ and Frobenius element e. Then any right $\mathcal{C}$-comodule is formally $\mathbb{X}^{\mathcal{C}}_{\delta,e}$-cosmooth and $\mathbb{X}^{\mathcal{C}}_{\delta,e}$-smooth.*

Proof. The maps κ_M in Proposition 4.1 are all identity morphisms, hence they are sections and thus every right $\mathcal{C}$-comodule is formally $\mathbb{X}^{\mathcal{C}}_{\delta,e}$-smooth. The map λ in Proposition 4.2 coincides with the counit $\varepsilon_{\mathcal{C}}$. Since $\varepsilon_{\mathcal{C}}$ is a unit in $\mathcal{C}^*$, it has a left inverse, and thus every right $\mathcal{C}$-comodule is formally $\mathbb{X}^{\mathcal{C}}_{\delta,e}$-cosmooth. $\qquad\square$

References

[1] A. Ardizzoni, *Separable functors and formal smoothness*, J. K-Theory, in press (arXiv:math.QA/0407095).

[2] A. Ardizzoni, T. Brzeziński and C. Menini, *Formally smooth bimodules*, J. Pure Appl. Algebra 212 (2008) 1072–1085.

[3] A. Ardizzoni, C. Menini and D. Ştefan, *Hochschild cohomology and 'smoothness' in monoidal categories*, J. Pure Appl. Algebra 208 (2007), 297–330.

[4] T. Brzeziński, *The structure of corings. Induction functors, Maschke-type theorem, and Frobenius and Galois-type properties*, Algebr. Represent. Theory 5 (2002), 389–410.

[5] T. Brzeziński and R. Wisbauer, *Corings and Comodules*, Cambridge University Press, Cambridge, 2003.

[6] S. Caenepeel, E. De Groot and J. Vercruysse, *Galois theory for comatrix corings: Descent theory, Morita theory, Frobenius and separability properties*, Trans. Amer. Math. Soc. 359 (2007), 185–226.

[7] S. Caenepeel, G. Militaru and S. Zhu, *Frobenius and Separable Functors for Generalized Hopf Modules and Nonlinear Equations*, Springer, Berlin (2002)

[8] J. Cuntz and D. Quillen, *Algebra extensions and nonsingularity*, J. Amer. Math. Soc. 8 (1995), 251–289.

[9] L. El Kaoutit and J. Gómez-Torrecillas, *Comatrix corings: Galois corings, descent theory, and a structure theorem for cosemisimple corings*, Math. Z. 244 (2003), 887–906.

[10] F. Guzman, *Cointegrations, relative cohomology for comodules and coseparable corings*, J. Algebra 126 (1989), 211–224.

[11] P.J. Hilton and U. Stammbach, *A Course in Homological Algebra*, Springer, New York, (1971).

[12] M. Kontsevich and A. Rosenberg, *Noncommutative spaces*, Preprint MPIM2004-35, 2004.

[13] C. Năstăsescu, S. Raianu and F. Van Oystaeyen, *Modules graded by G-sets*, Math. Z. 203 (1990), 605–627.

[14] C. Năstăsescu, M. Van den Bergh and F. Van Oystaeyen, *Separable functors applied to graded rings*, J. Algebra 123 (1989), 397–413.

[15] M.D. Rafael, *Separable functors revisited*, Comm. Algebra 18 (1990), 1445–1459.

[16] W.F. Schelter, *Smooth algebras*, J. Algebra 103 (1986), 677–685.

[17] R. Wisbauer, *Foundations of Module and Ring Theory*, Gordon and Breach, Philadelphia, PA, 1991.

[18] R. Wisbauer, *On Galois comodules,* Comm. Algebra 34 (2006), 2683–2711.

[19] M. Zarouali Darkaoui, *Adjoint and Frobenius Pairs of Functors, Equivalences and the Picard Group for Corings*, PhD Thesis, University of Granada, 2007.

Tomasz Brzeziński
Department of Mathematics
Swansea University
Singleton Park
Swansea SA2 8PP, UK
e-mail: T.Brzezinski@swansea.ac.uk

Modules and Comodules
Trends in Mathematics, 125–141
© 2008 Birkhäuser Verlag Basel/Switzerland

Certain Chain Conditions in Modules over Dedekind Domains and Related Rings

Esperanza Sanchez Campos and Patrick F. Smith

For Robert Wisbauer on the occasion of his 65th birthday

Abstract. Necessary and sufficient conditions are given for a module over a Dedekind domain to satisfy the ascending chain condition on n-generated submodules for every positive integer n or on submodules with uniform dimension at most n for every positive integer n. These results are then extended to modules over commutative Noetherian domains which need not be Dedekind.

Mathematics Subject Classification (2000). 13E15.

Keywords. Chain conditions; Dedekind domain; Uniform dimension.

1. Introduction

In this paper all rings are commutative with identity and all modules are unitary. Let R be any ring. We shall assume that if S is a subring of the ring R then S and R have the same identity element. Given a positive integer n, an R-module satisfies *n-acc* provided every ascending chain of n-generated submodules terminates. We shall say that an R-module M satisfies *pan-acc* provided M satisfies n-acc for every positive integer n. Clearly Noetherian modules satisfy pan-acc. However, Renault [18] shows that if R is a Noetherian ring then every free R-module satisfies pan-acc. Frohn [9] improved Renault's Theorem by proving that if R is a Noetherian ring then every direct product of copies of R is an R-module satisfying pan-acc. Modules satisfying these chain conditions over not-necessarily Noetherian rings have been extensively studied in recent years (see, for example, [1]–[5], [8]–[9], [11]–[12] and [17]–[18]).

Before we proceed we introduce one item of notation. Let R be a ring and let M be an R-module. For each maximal ideal P of R we set $M(P) = \{m \in M : P^k m = 0$ for some positive integer $k\}$. Then $M(P)$ is clearly a submodule of M. Note that if R is a one-dimensional Noetherian domain then every torsion

R-module is a direct sum of submodules of the form $M(P)$, for some maximal ideals P of R.

In [4, Theorem 3], Baumslag and Baumslag characterize which $\mathbb{Z}$-modules satisfy pan-acc. Nicolas [17, Théorème 2.3] characterizes which torsion-free modules over a Dedekind domain satisfy pan-acc. In this paper, we shall show that if R is a Dedekind domain then an R-module M with torsion submodule T satisfies pan-acc if and only if M satisfies the following two conditions:

(i) T is reduced and $M(P) \neq 0$ for at most a finite number of maximal ideals P of R, and

(ii) every countably generated torsion-free submodule of M is projective.

Heinzer and Lantz [11, p. 272] point out that Fuchs [10, p. 125] has shown that, for each positive integer n, there exist a torsion-free $\mathbb{Z}$-module A_n such that A_n satisfies n-acc but not $(n+1)$-acc.

We shall also consider modules with a somewhat different chain condition. Let R be any ring. Given a positive integer n, we shall say that an R-module M satisfies *nd-acc* provided every ascending chain of submodules with uniform dimension at most n terminates. In addition, an R-module M satisfies *pand-acc* provided M satisfies nd-acc for every positive integer n. Recall that a non-zero R-module X has *uniform dimension* n, for some positive integer n, provided X contains a direct sum $X_1 \oplus \cdots \oplus X_n$ of non-zero submodules $X_i\,(1 \leq i \leq n)$ but X does not contain a direct sum of $n+1$ non-zero submodules. A zero module has uniform dimension 0. For more information about uniform dimension see [16, Section 2.2]. Note that every semisimple R-module satisfies pand-acc. However, by [18, Lemme 1.1], in general not every semisimple R-module satisfies 1-acc. Note that the $\mathbb{Z}$-module A_n mentioned above satisfies nd-acc but not $(n+1)$d-acc, for each positive integer n (see [11, p. 272]).

We shall prove a companion theorem to our result above. Let R be a Dedekind domain and let M be an R-module with torsion submodule T. Then M satisfies pand-acc if and only if

(i) T is reduced, and

(ii) every countably generated torsion-free submodule of M is projective.

The proofs of these two theorems occupy the next section. In Section 3, we consider modules over domains which are related to Dedekind domains but need not themselves be Dedekind. Finally, in Section 4, we prove that if R is a non-local Noetherian domain and P any maximal ideal of R then the R-module $\oplus_{k=1}^{\infty}(R/P^k)$ satisfies pan-acc but the R-module $\prod_{k=1}^{\infty}(R/P^k)$ does not satisfy 1-acc.

2. Modules over Dedekind domains

To prove our results for modules over Dedekind domains we need some more or less well-known results. We shall include some proofs for completeness. Note that we shall always assume that the Dedekind domains we consider are not fields.

Lemma 2.1. *Let R be a Dedekind domain. Then the following statements are equivalent for a torsion-free R-module M.*

(i) *M satisfies pan-acc.*
(ii) *M satisfies pand-acc.*
(iii) *Every countably generated submodule of M is projective.*

Proof. By [17, Théorème 2.3]. □

Let R be a domain. An R-module M is called *divisible* in case $M = cM$ for every non-zero element c of R. Recall that the domain R is Dedekind if and only if every divisible R-module is injective (see, for example, [19, Theorem 4.25]). For a general domain R, an R-module M is called *reduced* if 0 is the only divisible submodule of M.

Lemma 2.2. *Let R be domain which is not a field and let M be a non-zero divisible R-module. Then M does not satisfy 1-acc.*

Proof. Suppose that M satisfies 1-acc. Let $m \in M$. Suppose that $cm = 0$ for some $0 \neq c \in R$. There exists $m_1 \in M$ such that $m = cm_1$. Similarly there exists $m_2 \in M$ such that $m_1 = cm_2$. Then $Rm \subseteq Rm_1 \subseteq Rm_2 \subseteq \ldots$ is an ascending chain of cyclic submodules. There exists a positive integer k such that $Rm_k = Rm_{k+1}$. Then $m_{k+1} = am_k$ for some $a \in R$ and hence $m_k = cm_{k+1} = cam_k$ so that $(1 - ca)m_k = 0$. But $c^2 m_1 = cm = 0$, $c^3 m_2 = c^2 m_1 = 0$ and, by induction, $c^{k+1} m_k = 0$. It follows that $m_k = 0$. Therefore, $m = 0$. Thus M is torsion-free.

Let $0 \neq c \in R$ such that c is not a unit in R. Let $0 \neq m \in M$. By the above argument there exists $m_k \in M$, for some positive integer k, such that $0 \neq Rm \subseteq Rm_k$ and $(1 - ca)m_k = 0$ for some $a \in R$. In this case $ca = 1$, a contradiction. The result follows. □

Lemma 2.3. *Let P_i ($i \in I$) be an infinite collection of distinct maximal ideals of a ring R and let M denote the R-module $\bigoplus_{i \in I}(R/P_i)$. Then M does not satisfies 1-acc.*

Proof. If P_i ($1 \leq i \leq n$) are distinct maximal ideals of R, for some positive integer n, then $(R/P_1) \oplus \cdots \oplus (R/P_n) \cong R/(P_1 \cap \cdots \cap P_n)$ and so is cyclic. The result follows. □

Corollary 2.1. *Let R be a domain and let M be an R-module such that M satisfies 1-acc. Then $M(P) \neq 0$ for at most a finite number of maximal ideals P of R.*

Proof. Let P be a maximal ideal of R such that $M(P) \neq 0$. Then the R-module R/P embeds in $M(P)$. Apply Lemma 2.3. □

Lemma 2.4. *Let R be a Dedekind domain and let n be a positive integer. Then an R-module M with torsion submodule T satisfies n-acc if and only if*

(i) *T satisfies n-acc, and*
(ii) *every countably generated torsion-free submodule of M satisfies n-acc.*

Proof. The necessity is clear. Conversely, suppose that M satisfies (i) and (ii). Let K be any n-generated submodule of M. Then $K/(K \cap T) \cong (K + T)/T$ which is a finitely generated torsion-free R-module and hence is projective. It follows that $K \cap T$ is a direct summand of K so that $K \cap T$ is also n-generated.

Let $L_1 \subseteq L_2 \subseteq L_3 \subseteq \cdots$ be any ascending chain of n-generated submodules of M. By the above remark, $L_1 \cap T \subseteq L_2 \cap T \subseteq \cdots$ is an ascending chain of n-generated submodules of T. There exists a positive integer k such that $L_k \cap T = L_{k+1} \cap T = \cdots$. Let $L = \bigcup_{i \geq 1} L_i$. Then $L \cap T = L_k \cap T$. It follows that $L \cap T$ is an n-generated torsion R-module and so $c(L \cap T) = 0$ for some $0 \neq c \in R$. Note that $T \cap cL = 0$. For, let $y \in T \cap cL$. Then $y = cx$ for some $x \in L$ and $ay = 0$ for some $0 \neq a \in R$. This implies that $(ac)x = 0$ so that $x \in L \cap T$ and hence $cx = 0$, i.e., $y = 0$. It follows that cL is a countably generated torsion-free submodule of M. Moreover, $cL_k \subseteq cL_{k+1} \subseteq \cdots$ is an ascending chain of n-generated submodules of cL. By (ii) there exists a positive integer $t \geq k$ such that $cL_t = cL_{t+1} = \cdots$. Let $i \geq t$. Let $u \in L_{i+1}$. Then there exists $v \in L_i$ such that $cu = cv$ so that $u - v \in L_{i+1} \cap T \subseteq L_i$. Thus $u \in L_i$. It follows that $L_t = L_{t+1} = \cdots$, as required. $\qquad \square$

In view of Lemmas 2.1 and 2.4 we now consider torsion modules over a Dedekind domain R. The next lemma is a special case of [4, p. 686 Lemma 1] but we include a proof for completeness.

Lemma 2.5. *Let R be a domain and let n be a positive integer. Let M be a torsion R-module such that $M = M(P_1) \oplus \cdots \oplus M(P_k)$ for some positive integer k and distinct maximal ideals P_i $(1 \leq i \leq k)$. Then M satisfies n-acc if and only if $M(P_i)$ satisfies n-acc for all $1 \leq i \leq k$.*

Proof. The necessity is clear. Conversely, suppose that $M(P_i)$ satisfies n-acc for all $1 \leq i \leq k$. Let L be any n-generated submodule of M. Then $L = (L \cap M(P_1)) \oplus \cdots \oplus (L \cap M(P_k))$ and $L \cap M(P_i)$ is an n-generated submodule of $M(P_i)$ for each $1 \leq i \leq k$. Clearly it follows that M satisfies n-acc $\qquad \square$

In view of Lemma 2.5 we now consider a reduced module M over a Dedekind domain R such that $M = M(P)$ for some maximal ideal P. Note that the localization of R at P is a DVR (discrete valuation ring) and that M is an R_P-module such that for any submodule L of M and any positive integer n, L is an n-generated R-submodule of M if and only if L is an n-generated R_P-submodule of M. Thus the R-module M satisfies n-acc if and only if the R_P-module M satisfies n-acc.

Lemma 2.6. *Let R be a DVR and let M be an n-generated torsion R-module for some positive integer n. Then $M = M_1 \oplus \cdots \oplus M_k$ is a direct sum of cyclic submodules M_i $(1 \leq i \leq k)$ for some positive integer $k \leq n$.*

Proof. Suppose that $M = Rm_1 + \cdots + Rm_n$ where $m_i \in M$ $(1 \leq i \leq n)$ and suppose that $M \neq 0$. Let $p \in R$ such that Rp is the unique maximal ideal of R. There exists a positive integer t such that $p^t M = 0$ but $p^{t-1} M \neq 0$. Without loss of generality $p^t m_1 = 0$ but $p^{t-1} m_1 \neq 0$. Now M/Rm_1 is a direct sum of $(k - 1)$

cyclic submodules for some positive integer $k \leq n$, by induction on n. Also by [13, Lemma 4 and Theorem 5 (see the remark on p. 36)], Rm_1 is a direct summand of M. Thus M is a direct sum of k cyclic submodules. □

Lemma 2.7. *Let R be a DVR and let M be a reduced torsion R-module. Then there exists a submodule N of M such that N is a direct sum of cyclic submodules and M/N is a divisible module.*

Proof. By [13, Theorem 9] there exists $0 \neq m \in M$ such that Rm is a direct summand of M. Next Zorn's Lemma gives a maximal subset $\{x_i : i \in I\}$ of M such that $\sum_{i \in I} Rx_i$ is a direct sum and $\oplus_{i \in I} Rx_i$ is a pure submodule of M (see [13, p. 18]). Let $N = \oplus_{i \in I} Rx_i$. Suppose that M/N is not divisible. By [13, Theorems 3 and 9], there exists a submodule L of M containing N such that L/N is a non-zero cyclic direct summand of M/N. Next [13, Lemma 2] gives that L is a pure submodule of M. Moreover, by [13, Theorem 5], $L = N \oplus Rx$ for some $x \in L$. Thus $\{x_i : i \in I\} \subset \{x\} \cup \{x_i : i \in I\}$ contradicts the choice of the set $\{x_i : i \in I\}$. It follows that M/N is divisible. □

Lemma 2.8. *Let R be a DVR. Then every reduced torsion R-module satisfies pan-acc.*

Proof. Let M be a reduced torsion R-module. Let $L_1 \subseteq L_2 \subseteq L_3 \subseteq \cdots$ be any ascending chain of n-generated submodules of M. Let $L = \bigcup_{i \geq 1} L_i$. Then L is a submodule of M. By Lemma 2.7, there exists a submodule N of M such that N is a direct sum of cyclic submodules and L/N is a divisible module.

Suppose that N contains a direct sum $Rx_1 \oplus \cdots \oplus Rx_{n+1}$ for some non-zero elements $x_i \in N$ ($1 \leq i \leq n+1$). Then $Rx_1 \oplus \cdots \oplus Rx_{n+1}$ is contained in an n-generated submodule L_k, for some positive integer k. This contradicts Lemma 2.6 because every non-zero cyclic submodule of M is uniform. It follows that N is a finite direct sum of cyclic submodules and hence $cN = 0$ for some $0 \neq c \in R$.

Define a mapping $\varphi : L \to L$ by $\varphi(x) = cx$ for all $x \in L$. Then $\mathrm{im}\varphi \cong L/\ker\varphi$ and $N \subseteq \ker\varphi$, so that $\mathrm{im}\varphi$ is a divisible submodule of M. It follows that $\mathrm{im}\varphi = 0$, $cL = 0$ and hence $L = N$. Thus $L_k = L_{k+1} = \cdots$ for some positive integer k. □

Theorem 2.2. *Let R be a Dedekind domain and let M be an R-module with torsion submodule T. Then M satisfies pan-acc if and only if*

(i) *T is reduced and $M(P) \neq 0$ for at most a finite number of maximal ideals P of R, and*

(ii) *every countably generated torsion-free submodule of M is projective.*

Proof. The necessity follows by Lemmas 2.1 and 2.2 and Corollary 2.1. The sufficiency follows by Lemmas 2.1, 2.4, 2.5 and 2.8 (see the remarks preceding Lemma 2.6). □

We now turn our attention to finding which modules over a Dedekind domain satisfy pand-acc. Let M be any module. A submodule K of M is called *closed (in M)* if K has no proper essential extension in M. For example, every direct summand of M is closed in M. Given any submodule N of M, by Zorn's Lemma there exists a closed submodule K of M such that N is an essential submodule of K. If the module M has finite uniform dimension then we shall denote the uniform dimension of M by $u(M)$. If $u(M) = n$, for some positive integer n, then we shall call M n-dimensional.

Lemma 2.9. *Let N and K be submodules of a module M.*

(a) *If N and M/N both have finite uniform dimension then so too does M and*
    $$u(M) \leq u(N) + u(M/N).$$
(b) *If M is a module with finite uniform dimension and K is a closed submodule of M then the modules K and M/K both have finite uniform dimension and*
    $$u(M) = u(K) + u(M/K).$$

Proof. See [7, 5.10]. □

Note that if R is a domain and K a submodule of an R-module M such that M/K is a torsion-free R-module then K is a closed submodule of M and part (b) of Lemma 2.9 applies if M has finite uniform dimension.

Lemma 2.10. *Let R be any ring and let n be a positive integer. Then the following statements are equivalent for an R-module M.*

(i) *M satisfies nd-acc.*
(ii) *Every countably generated submodule of M satisfies nd-acc.*
(iii) *Every submodule N of M with uniform dimension at most n is Noetherian.*

Proof. (i) ⇒ (ii) Clear.

(ii) ⇒ (iii) Let L be any submodule of M with uniform dimension at most n. Let $L_1 \subseteq L_2 \subseteq L_3 \subseteq \cdots$ be any ascending chain of finitely generated submodules of L and let $L' = \bigcup_{i \geq 1} L_i$. Clearly L' is countably generated and L_i has uniform dimension at most n for all $i \geq 1$. By (ii), $L_k = L_{k+1} = \cdots$ for some positive integer k. It follows that L satisfies the ascending chain condition on finitely generated submodules and hence L is Noetherian.

(iii) ⇒ (i) Let $H_1 \subseteq H_2 \subseteq H_3 \subseteq \cdots$ be any ascending chain of submodules of M, each with uniform dimension at most n. If $H = \bigcup_{i \geq 1} H_i$ then H is a submodule of M with uniform dimension at most n. By (iii), there exists a positive integer t such that $H_t = H_{t+1} = \cdots$. It follows that M satisfies nd-acc. □

Lemma 2.11. *Let R be a domain. Let M be an R-module with torsion submodule T and let n be a positive integer. Then M satisfies nd-acc if and only if*

(i) *T satisfies nd-acc, and*
(ii) *every countably generated torsion-free submodule of M satisfies nd-acc.*

Proof. The necessity is clear. Conversely, suppose that M satisfies (i) and (ii). Let N be an n-dimensional submodule of M. Then $N \cap T$ is a k-dimensional submodule of T, for some $k \leq n$, so that $N \cap T$ is Noetherian. There exists a non-zero element c of R such that $c(N \cap T) = 0$. Then $T \cap cN = 0$ (see the proof of Lemma 2.4) and

cN is a torsion-free submodule of M. The mapping that sends x to cx $(x \in N)$ is an epimorphism from N to cN with kernel $N \cap T$. Because T is a closed submodule of M and hence also of $N + T$, $(N + T)/T$ has uniform dimension h for some $h \leq n$ (Lemma 2.9) and thus cN is h-dimensional. By (ii) and Lemma 2.10, cN is Noetherian. Thus N is Noetherian. The result follows by applying Lemma 2.10 again. $\qquad \square$

Lemma 2.12. *Let R be a DVR and let M be a reduced torsion R-module. Then M satisfies pand-acc.*

Proof. Let N be a finite-dimensional submodule of M. By [13, Theorem 9] there exists a cyclic direct summand K of N (because N is reduced and not torsion-free). Then $N = K \oplus K'$ for some submodule K'. By induction on the dimension of N, K' is Noetherian. Thus N is Noetherian. Now apply Lemma 2.10. $\qquad \square$

Theorem 2.3. *Let R be a Dedekind domain and let M be an R-module with torsion submodule T. Then M satisfies pand-acc if and only if*

(i) *T is reduced, and*

(ii) *every countably generated torsion-free submodule of M is projective.*

Proof. Suppose that M satisfies pand-acc. By Lemma 2.1, every countably generated torsion-free submodule of M is projective. Moreover, T is clearly reduced (see [19, Proposition 2.10]). Conversely, suppose that T is reduced and that every countably generated torsion-free submodule of M is projective. Let n be any positive integer. By Lemma 2.1, every torsion-free submodule of M satisfies nd-acc. Thus to prove that M satisfies nd-acc it is sufficient to prove that T satisfies nd-acc by Lemma 2.11. Let X be any submodule of T with uniform dimension at most n. Then $X = X(P_1) \oplus \cdots \oplus X(P_k)$ for some positive integer k and distinct maximal ideals P_i $(1 \leq i \leq k)$ of R. For each $1 \leq i \leq k$, $X(P_i)$ is a reduced torsion module over the DVR R_{P_i} and $X(P_i)$ has uniform dimension at most n, so that $X(P_i)$ is Noetherian by Lemma 2.12. Thus X is Noetherian. By Lemma 2.10, T satisfies nd-acc and this completes the proof. $\qquad \square$

Corollary 2.4. *Let R be a Dedekind domain and let M be any R-module such that M satisfies pan-acc. Then M satisfies pand-acc.*

Proof. By Theorems 2.2 and 2.3. $\qquad \square$

Note that Nicolas proves that the converse of Corollary 2.4 holds in case the module M is torsion-free.

3. Related rings

Given the classification of Baumslag and Baumslag for modules over $\mathbb{Z}$ which satisfy pan-acc, it is not too surprising that this classification can be extended to modules over a Dedekind domain (Theorem 2.2) and that there is a corresponding theorem for modules over a Dedekind domain satisfying pand-acc (Theorem 2.3).

Now we consider modules over domains that are not themselves Dedekind domains but are closely related to Dedekind domains in some way. This represents a first attempt to explore what happens more generally. We immediately lose the fact that for a Dedekind domain R, every finitely generated torsion-free R-module is projective and consequently the torsion submodule of every finitely generated R-module is a direct summand. We begin with an elementary result. Recall that all rings considered are commutative.

Lemma 3.1. *Let R be a subring of a ring S such that S is a finitely generated R-module. Let M be an S-module such that, as an R-module, M satisfies pan-acc. Then the S-module M satisfies pan-acc.*

Proof. There exist a positive integer k and elements $s_i \in S$ $(1 \leq i \leq k)$ such that $S = Rs_1 + \cdots + Rs_k$. Let n be any positive integer. Then every n-generated S-submodule of M is an nk-generated R-submodule of M. The result follows. $\square$

Lemma 3.2. *Let R be a subring of a domain S such that S is a finitely generated R-module and let M be any S-module. Then*

(i) *S is integral over R.*
(ii) *$A \cap R \neq 0$ for every non-zero ideal A of S.*
(iii) *The torsion submodule of the S-module M coincides with the torsion submodule of the R-module M. In particular, M is a torsion-free S-module if and only if M is a torsion-free R-module.*
(iv) *The S-module M is reduced if and only if the R-module M is reduced.*

Moreover, if R is a Dedekind domain then S is a one-dimensional Noetherian domain.

Proof. (i) By [14, Theorem 12].

(ii) By (i).

(iii) By (ii).

(iv) Suppose that the S-module M is not reduced. Then there exists a non-zero S-submodule N of M such that $N = aN$ for all non-zero $a \in S$. Now N is an R-submodule of M and $N = bN$ for all non-zero $b \in R$. Thus the R-module M is not reduced. Conversely, suppose that the R-module M is not reduced and let L be a non-zero R-submodule of M such that $L = bL$ for all non-zero $b \in R$. Let $H = SL$. Then H is a non-zero S-submodule of M. Let $0 \neq a \in S$. By (ii) there exists $c \in S$ such that $0 \neq ca \in R$. Now $H = SL = S(caL) \subseteq aSL = aH \subseteq H$, so that $H = aH$. It follows that H is a divisible S-submodule of M and hence the S-module M is not reduced.

The last part follows by [14, Theorem 48]. $\square$

We now consider when modules over certain rings related to Dedekind domains satisfy pan-acc. Note that in the following result the torsion submodule T of M is both the torsion submodule of the R-module M and the torsion submodule of the S-module M by Lemma 3.2(iii).

Theorem 3.1. *Let R be a subring of a commutative domain S such that R is a Dedekind domain and S is a finitely generated R-module. Let M be an S-module with torsion submodule T. Then the S-module M satisfies pan-acc if and only if*

(i) *T satisfies pan-acc, and*

(ii) *every countably generated torsion-free submodule of the S-module M satisfies pan-acc.*

Proof. The necessity is clear. For the sufficiency, first of all note that $S = Rs_1 + \cdots + Rs_k$ for some positive integer k and $s_i \in S$ $(1 \le i \le k)$. Let L be a finitely generated S-module. Then $(L + T)/T \cong L/(L \cap T)$ is a finitely generated torsion-free R-module and hence is projective. It follows that $L = (L \cap T) \oplus K$ for some R-submodule K of L.

Let $L_1 \subseteq L_2 \subseteq L_3 \subseteq \cdots$ be any ascending chain of n-generated S-submodules of M. Note that this chain is also an ascending chain of nk-generated R-submodules of M. Hence $L_1 \cap T \subseteq L_2 \cap T \subseteq \cdots$ is an ascending chain of nk-generated R-submodules of T and so an ascending chain of nk-generated S-submodules of T. There exist a positive integer $h \ge 1$ such that $L_h \cap T = L_{h+1} \cap T = \cdots$. Let $L = \bigcup_{i \ge 1} L_i$. Then $L_h \cap T = L \cap T$. It follows that there exists $0 \ne a \in R$ such that $a(L \cap T) = 0$. Now aL is a countably generated torsion-free S-submodule of M (see the proof of Lemma 2.4). Next $aL_h \subseteq aL_{h+1} \subseteq \cdots$ is an ascending chain of n-generated S-submodules of aL. By (ii) there exists a positive integer $t \ge h$ such that $aL_t = aL_{t+1} = \cdots$. Let $y \in L_{t+1}$. There exists $x \in L_t$ such that $ay = ax$ so that $y - x \in L_{t+1} \cap T \subseteq L_t$. Thus $y \in L_t$. It follows that $L_t = L_{t+1} = \cdots$, as required. $\qquad\square$

Let R be a subring of a ring S. For each prime ideal Q of S, $Q \cap R$ is clearly a prime ideal of R. Given any prime ideal P of R we let $\pi_S(P)$ denote the (possibly empty) set of prime ideals Q of S such that $P = Q \cap R$.

Lemma 3.3. *Let R be a subring of a domain S such that S is a finitely generated R-module and let M be any S-module. Let P be any maximal ideal of R. Then $\pi_S(P)$ is a non-empty finite collection of maximal ideals of S. Moreover $M(P) \ne 0$ if and only if $M(Q) \ne 0$ for some $Q \in \pi_S(P)$.*

Proof. By [14, Theorem 44], $\pi_S(P)$ is non-empty. Note that R/P is a field and hence S/SP is an Artinian ring. In particular, S/SP is semilocal. This proves that $\pi_S(P)$ is a finite collection of maximal ideals of S. If $M(Q) \ne 0$ for some $Q \in \pi_S(P)$ then clearly $M(P) \ne 0$. Conversely, suppose that $M(P) \ne 0$. There exist a positive integer k and maximal ideals Q_i $(1 \le i \le k)$ of S such that $\pi_S(P) = \{Q_1, \ldots, Q_k\}$. Then Q_i/SP $(1 \le i \le k)$ are the minimal prime ideals of the Artinian ring S/SP. There exists a positive integer t such that $(Q_1 \ldots Q_k)^t \subseteq SP$. If $0 \ne m \in M(P)$ then $P^h m = 0$ for some positive integer h and hence $(Q_1 \ldots Q_k)^{th} m = 0$. It follows that $M(Q_i) \ne 0$ for some $1 \le i \le k$. $\qquad\square$

Theorem 3.1 shows that in certain cases a module over a domain satisfies pan-acc precisely when its torsion submodule and its countably generated torsion-free submodules satisfy pan-acc. Our next result deals with the case of torsion modules.

Theorem 3.2. *Let R be a subring of a commutative domain S such that R is a Dedekind domain and S is a finitely generated R-module. Then the following statements are equivalent for a torsion S-module M.*

(i) *M satisfies pan-acc.*

(ii) *M satisfies 1-acc.*

(iii) *M is a reduced S-module such that $M(Q) \neq 0$ for at most a finite number of maximal ideals Q of S.*

Proof. (i) $\Rightarrow$ (ii) Clear.

(ii) $\Rightarrow$ (iii) By Lemma 2.2 and Corollary 2.1.

(iii) $\Rightarrow$ (i) Suppose that (iii) holds. By Lemmas 3.2(iv) and 3.3, the R-module M is reduced and $M(P) \neq 0$ for at most a finite number of maximal ideals P of R. By Theorem 2.2, the R-module M satisfies pan-acc. Finally by Lemma 3.1, the S-module M satisfies pan-acc. $\quad\square$

Next we recall a well-known result.

Lemma 3.4. *A ring R is perfect if and only if every R-module satisfies pan-acc.*

Proof. See [12] or [18, Proposition 1.2]. $\quad\square$

Lemma 3.5. *Let S be a subring of a ring R such that R is a finitely generated S-module and there exists a finitely generated ideal A of R with $A \subseteq S$ and the ring S/A perfect. Let M be an R-module and let X be an S-submodule of M such that the S-module X satisfies pan-acc. Then the R-module RX satisfies pan-acc.*

Proof. There exist a positive integer k and elements $r_i \in R$ $(1 \leq i \leq k)$ such that $R = Sr_1 + \cdots + Sr_k$ and there exist a positive integer t and elements a_i $(1 \leq i \leq t)$ of A such that $A = Ra_1 + \cdots + Ra_t$. Let n be any positive integer. Let $L_1 \subseteq L_2 \subseteq L_3 \subseteq \cdots$ be any ascending chain of n-generated R-submodules of the R-module RX. Then $AL_1 \subseteq AL_2 \subseteq AL_3 \subseteq \cdots$ is an ascending chain of nkt-generated S-submodules of the S-module X. By hypothesis, we can suppose without loss of generality that $AL_1 = AL_2 = \cdots$. Let $L = \bigcup_{i \geq 1} L_i$. Note that $A(L/L_1) = 0$, so that $L_1/L_1 \subseteq L_2/L_1 \subseteq L_3/L_1 \subseteq \cdots$ is an ascending chain of nk-generated submodules of the (S/A)-module L/L_1. By Lemma 3.4, there exists a positive integer h such that $L_h/L_1 = L_{h+1}/L_1 = \cdots$ and hence $L_h = L_{h+1} = \cdots$. $\quad\square$

Lemma 3.6. *Let R be a Noetherian domain. Then every free R-module satisfies pand-acc.*

Proof. Let F be any free R-module and let n be any positive integer. Let L be a submodule of F of uniform dimension n. Then there exists an n-generated essential submodule K of L. There exist submodules F_1 and F_2 of F such that $F = F_1 \bigoplus F_2$, F_1 is finitely generated and $K \subseteq F_1$. Note that $(L + F_1)/F_1 \cong L/(L \cap F_1)$ and so is a torsion module. It follows that $L \subseteq F_1$ and hence L is Noetherian. The result follows by Lemma 2.10. $\square$

Lemma 3.7. *Let R be a domain and let M be a torsion-free R-module such that M satisfies nd-acc for some positive integer n. Then M satisfies n-acc.*

Proof. Let $L_1 \subseteq L_2 \subseteq L_3 \subseteq \cdots$ be any ascending chain of n-generated submodules of M. For each $i \geq 1$, L_i is a torsion-free homomorphic image of the free R-module $R^{(n)}$, so that L_i has uniform dimension at most n. By hypothesis, there exists $k \geq 1$ such that $L_k = L_{k+1} = \cdots$. $\square$

We next prove a result for torsion-free modules over certain domains. Note that this result generalizes Nicolas' Theorem (see Lemma 2.1).

Theorem 3.3. *Let S be a subring of a Dedekind domain R such that R is a finitely generated S-module and there exists a non-zero ideal A of R with $A \subseteq S$. Then the following statements are equivalent for a torsion-free S-module M.*

(i) *M satisfies pand-acc.*
(ii) *M satisfies pan-acc.*
(iii) *Every countably generated S-submodule of M can be embedded in a free S-module.*

Proof. (i) $\Rightarrow$ (ii) By Lemma 3.7.

(ii) $\Rightarrow$ (iii) Let X be any countably generated torsion-free S-module which satisfies pan-acc. Because the Dedekind domain R is a finitely generated module over its subring S, Eakin's Theorem gives that S is a Noetherian domain (see, for example, [15, Theorem 3.6]). By [14, Theorem 48], S is one-dimensional and hence S/A is an Artinian, and hence perfect, ring. Let M denote the R-module $R \otimes_S X$. We can identify X with the S-submodule $S \otimes_S X$ of M. By Lemma 3.5, the R-module RX satisfies pan-acc. Note that the R-module RX is countably generated and, by Lemma 3.2(iii), is torsion-free. Lemma 2.1 gives an R-monomorphism $\phi : RX \to R^{(I)}$, for some index set I. Let $0 \neq a \in A$. Then $aR^{(I)} \subseteq S^{(I)}$ and $a\phi : X \to S^{(I)}$ is an S-monomorphism. Thus X embeds in the free S-module $S^{(I)}$. Clearly (iii) follows.

(iii) $\Rightarrow$ (i) By Lemmas 2.10 and 3.6. $\square$

This brings us to the following generalization of Theorem 2.2.

Theorem 3.4. *Let R be a subring of a Dedekind domain R' such that R is also a Dedekind domain and R' is a finitely generated R-module. Let A be any non-zero ideal of R' and let S be any subring of R' such that $R + A \subseteq S$. Then an S-module M with torsion submodule T satisfies pan-acc if and only if*

(i) *T is reduced and $T(Q) \neq 0$ for at most a finite number of maximal ideals Q of S, and*
(ii) *every countably generated torsion-free submodule of M can be embedded in a free S-module.*

Proof. By Theorems 3.1, 3.2 and 3.3. □

With the notation of Theorem 3.4, we have been unable to find a corresponding characterization for S-modules M which satisfy pand-acc. Our problem is finding an analogue for Theorem 3.2 (see Lemma 2.11 and Theorem 3.4).

We now show that the situation described in Theorem 3.4 arises naturally. Let R be *any* Dedekind domain and let K denote the field of fractions of R. Let K' be any finite separable extension of K and let R' denote the integral closure of R in K'. By [20, p. 264 Theorem 7], R' is a finitely generated R-module and, by [20, p. 281 Theorem 19], R' is a Dedekind domain. Let A be any non-zero ideal of R'. Then any subring S of R' such that $R + A \subseteq S$ satisfies the hypotheses of Theorem 3.4. For example, if $R = \mathbb{Z}$, $K = \mathbb{Q}$ and K' is the field $\mathbb{Q}[\sqrt{d}]$, for any square-free integer d in $\mathbb{Z}$, then $R' = \mathbb{Z}[\sqrt{d}]$ or $\mathbb{Z}[(1 + \sqrt{d})/2]$ (see, for example, [6, p. 86 Theorem 1.3]). It follows that the domain $S = \mathbb{Z}[\sqrt{d}]$ satisfies the hypotheses of Theorem 3.4 for every square-free rational integer d. Note that the ring $\mathbb{Z}[\sqrt{d}]$ is not a Dedekind domain if $d \equiv 1 \pmod 4$ (see [6, p. 86 Theorem 1.3]).

4. Direct sums and direct products

Let R be a (commutative) ring and let P be a maximal ideal of R. Let I be an index set and let k_i be a positive integer for each $i \in I$. We shall show that, in case R is Noetherian, the direct sum of the modules R/P^{k_i} satisfies pan-acc but, in general, the direct product of the modules R/P^{k_i} does not satisfy 1-acc unless the integers k_i $(i \in I)$ are bounded above. We also characterize when the direct sum of cyclic modules over a one-dimensional Noetherian domain satisfies pan-acc. Finally in this section we shall show that if M_i $(i \in I)$ is any collection of modules each satisfying pand-acc then the module $M = \oplus_{i \in I} M_i$ also satisfies pand-acc.

Let P be a maximal ideal of a ring R. Then an ideal A of R will be called *P-primary* provided there exists a positive integer k such that $P^k \subseteq A \subseteq P$. Note that if c is an element of R such that $c \notin P$ then $R = Rc + A$ for every P-primary ideal A of R. Note further that if A is a P-primary ideal of R then $P^k(R/A) = 0$ for some positive integer k. It follows that if A_i $(i \in I)$ is any collection of P-primary ideals of R and M is the R-module $\oplus_{i \in I}(R/A_i)$ then $\cap_{k=1}^{\infty} P^k M = 0$.

Lemma 4.1. *Let P be a maximal ideal of any ring R and let A_i $(i \in I)$ be any non-empty collection of P-primary ideals of R. Then the R-module $M = \oplus_{i \in I}(R/A_i)$ satisfies 1-acc.*

Proof. Suppose that M does not satisfy 1-acc. Then there exist $0 \neq m_1 \in M$ and a properly ascending chain $Rm_1 \subset Rm_2 \subset \cdots$ of cyclic submodules of M. For each $j \geq 1$, $m_j = c_j m_{j+1}$ for some $c_j \in R$. Suppose that there exists $j \geq 1$ such that $c_j \notin P$. There exists t such that $P^t m_{j+1} = 0$. Moreover $1 \in Rc_j + P^t$ so that $m_{j+1} \in Rc_j m_{j+1} + P^t m_{j+1} \subseteq Rm_j$, and this implies that $Rm_j = Rm_{j+1}$, a contradiction. Thus $c_j \in P$ for all $j \geq 1$. Then

$$m_1 = c_1 m_2 = c_1 c_2 m_3 = \cdots = c_1 \ldots c_k m_{k+1} \in P^k M$$

for all $k \geq 1$. Thus $m_1 \in \cap_{k=1}^{\infty} P^k M$. But this implies that $m_1 = 0$, a contradiction. It follows that M satisfies 1-acc. $\qquad\square$

Theorem 4.1. *Let P be a maximal ideal of a Noetherian ring R and let A_i ($i \in I$) be any non-empty collection of P-primary ideals of R. Then the R-module $M = \oplus_{i \in I}(R/A_i)$ satisfies pan-acc.*

Proof. We prove the result by induction on n. The case $n = 1$ has been dealt with in Lemma 4.1. Suppose that $n \geq 2$. Let $L_1 \subseteq L_2 \subseteq \cdots$ be any ascending chain of non-zero n-generated submodules of M. Suppose that for each $s \geq 1$ there exists $t > s$ such that $L_s \subseteq PL_t$. Then there exists an increasing sequence $s_1 < s_2 < \cdots$ such that $L_1 \subseteq PL_{s_1} \subseteq P^2 L_{s_2} \subseteq \cdots$ so that $L_1 \subseteq \cap_{k=1}^{\infty} P^k M = 0$, a contradiction. Thus we can suppose without loss of generality that $L_1 \not\subseteq PL_s$ for all $s \geq 2$. For each $i \geq 1$ there exist elements x_{ij} ($1 \leq j \leq n$) such that $L_i = Rx_{i1} + \cdots + Rx_{in}$. Let $x = x_{11}$. Without loss of generality we can suppose that $x \notin PL_i$ for all $i \geq 2$. For each $i \geq 2$,

$$x = a_{i1}x_{i1} + \cdots + a_{in}x_{in}.$$

Without loss of generality we can suppose that $a_{i1} \notin P$ for all $i \geq 2$. Because every element of M is annihilated by some power of the maximal ideal P, it is not difficult to see that $x_{i1} \in Rx + Rx_{i2} + \cdots + Rx_{in}$ and thus $L_i = Rx + Rx_{i2} + \cdots + Rx_{in}$ for all $i \geq 2$. There exists a finite subset J of I such that $x \in N = \oplus_{j \in J}(R/A_j)$. Note that by induction on n, the module M/N satisfies $(n\text{-}1)$-acc. It follows that the ascending chain $(L_1 + N)/N \subseteq (L_2 + N)/N \subseteq \cdots$ of $(n\text{-}1)$-generated submodules $(L_i + N)/N$ ($i \geq 1$) of M/N terminates. Thus there exists a positive integer k such that $(L_k + N)/N = (L_{k+1} + N)/N = \cdots$. However, N is a Noetherian R-module so that, without loss of generality, we can suppose that $L_k \cap N = L_{k+1} \cap N \cdots$. It follows that $L_k = L_{k+1} = \cdots$. $\qquad\square$

We drop the assumption just for a moment that all rings be commutative in order to prove the following result.

Theorem 4.2. *Let R be a right Noetherian ring and let n be any positive integer. Then every finite direct sum of right R-modules satisfying n-acc also satisfies n-acc.*

Proof. By induction on the number of summands it is sufficient to prove that if an R-module $M = M_1 \oplus M_2$ is a direct sum of submodules M_1, M_2 such that M_1 and M_2 both satisfy n-acc then M also satisfies n-acc. Let $L_1 \subseteq L_2 \subseteq \cdots$ be any ascending chain of n-generated submodules of M. Let $L = \bigcup_{i=1}^{\infty} L_i$ and note that L is a submodule of M. Note that $(L_1 + M_1)/M_1 \subseteq (L_2 + M_1)/M_1 \subseteq \cdots$ is an ascending chain of n-generated submodules of the module $M/M_1 \cong M_2$. By hypothesis, there exists a positive integer k such that $(L_k + M_1)/M_1 = (L_{k+1} + M_1)/M_1 = \cdots$ and it follows that $L + M_1 = L_k + M_1$. But $L/(L \cap M_1) \cong (L + M_1)/M_1 = (L_k + M_1)/M_1$ which is a finitely generated module. It follows that $L/(L \cap M_1)$ is a Noetherian module. Similarly, $L/(L \cap M_2)$ is a Noetherian module. But the mapping $\theta : L \rightarrow (L/(L \cap M_1)) \oplus (L/(L \cap M_2))$ defined by

$\theta(x) = (x + (L \cap M_1), x + (L \cap M_2))$, for all x in L, is a monomorphism. Thus L is a Noetherian R-module. Hence the ascending chain $L_1 \subseteq L_2 \subseteq \cdots$ of submodules of L terminates. It follows that M satisfies n-acc. $\quad\square$

Theorem 4.2 allows us to prove the following generalization of [4, p. 694 Corollary].

Corollary 4.3. *Let R be a right Noetherian ring, let n be a positive integer and let $M_i (i \in I)$ be any collection of nonsingular right R-modules which each satisfy n-acc. Then the direct product $\prod_{i \in I} M_i$ also satisfies n-acc.*

Proof. By Theorem 4.2 and [1, Theorem 1.5]. $\quad\square$

We now return to commutative rings. We aim to show that certain modules over (commutative) one-dimensional Noetherian domains satisfy pan-acc.

Let R be a one-dimensional Noetherian domain. (Note that Dedekind domains are particular examples of one-dimensional Noetherian domains.) Every non-zero prime ideal of R is maximal and hence the ring R/A is Artinian for every non-zero ideal A of R by [19, Theorem 4.6]. Let M be any cyclic R-module. Either $M \cong R$ or M is Artinian. If M is a non-zero Artinian cyclic module then there exist a positive integer k, maximal ideals P_i $(1 \leq i \leq k)$ and a P_i-primary ideal A_i of R for each $1 \leq i \leq k$ such that $M \cong (R/A_1) \oplus \cdots \oplus (R/A_k)$ by [19, Theorem 4.28]

Theorem 4.4. *Let R be a one-dimensional Noetherian domain and let M be an R-module which is a direct sum of cyclic submodules. Then the following statements are equivalent.*

  (i) *M satisfies pan-acc.*
  (ii) *M satisfies 1-acc.*
  (iii) *$M(P) \neq 0$ for at most a finite number of maximal ideals P of R.*

Proof. (i) $\Rightarrow$ (ii) Clear.

(ii) $\Rightarrow$ (iii) By Corollary 2.1.

(iii) $\Rightarrow$ (i) By the above remarks there exist a positive integer k and submodules M_i $(1 \leq i \leq k)$ of M such that $M = M_1 \oplus \cdots \oplus M_k$ where M_1 is zero or free and for each $2 \leq i \leq k$ there exist a maximal ideal P_i of R, an index set J_i and P_i-primary ideals A_j $(j \in J_i)$ such that $M_i \cong \oplus_{j \in J_i} (R/A_j)$. By Theorem 4.1, for each $2 \leq i \leq k$, M_i satisfies pan-acc. Also by [4, Theorem 8] M_1 satisfies pan-acc. Finally, by Theorem 4.2, M satisfies pan-acc. $\quad\square$

Note that the proof of Theorem 4.4 shows that if R is any Noetherian ring and an R-module M is a direct sum of cyclic Artinian modules then M satisfies pan-acc if and only if $M(P) \neq 0$ for at most a finite number of maximal ideals P of R, by [19, Theorem 4.28]. We now consider direct products. As we have remarked above, under certain circumstances direct products of modules do satisfy pan-acc (see [1], [9], [18]). In contrast we have the following result.

Theorem 4.5. *Let P be a maximal ideal of a non-local ring R such that $\cap_{k=1}^{\infty} P^k = 0$ and let k_i $(i \in I)$ be a non-empty collection of positive integers. Then the following statements are equivalent for the R-module $M = \prod_{i \in I}(R/P^{k_i})$.*

(i) *M satisfies pan-acc.*
(ii) *M satisfies 1-acc.*
(iii) *There exists a positive integer N such that $k_i \leq N$ for all $i \in I$.*

Proof. (i) $\Rightarrow$ (ii) Clear.

(ii) $\Rightarrow$ (iii) Suppose that there does not exist a positive integer N such that $k_i \leq N$ for all $i \in I$. Then $\cap_{i \in I} P^{k_i} = 0$. Let c be any non-unit in R such that $c \notin P$. Let $a_i \in R \backslash P$ $(i \in I)$. For each $i \in I$, $P^{k_i}(a_i + P^{k_i}) = 0$ and $R = Rc + P^{k_i}$ so that $a_i + P^{k_i} = cb_i + P^{k_i}$ for some $b_i \in R$. Thus $(a_i + P^{k_i}) = c(b_i + P^{k_i})$. Note that $b_i \in R \backslash P$ for each $i \in I$. If $x = (a_i + P^{k_i})$ and $y = (b_i + P^{k_i})$ then $x = cy$. Suppose that $Rx = Ry$. Then $y = dx$ for some $d \in R$ and hence $x = cy = cdx$. Thus $(1 - cd)x = 0$ and hence $(1 - cd)a_i \in P^{k_i}$ for all $i \in I$. Because $a_i \notin P$ it follows that $1 - cd \in P^{k_i}$ for all $i \in I$. Hence $1 - cd \in \cap_{i \in I} P^{k_i} = 0$ and $cd = 1$, a contradiction. Thus $Rx \subset Ry$. Starting with the element $z_1 = (1 + P^{k_i})$ of M, in this way we can form an infinite properly ascending chain $Rz_1 \subset Rz_2 \subset \cdots$ of cyclic submodules of M. Thus M does not satisfy 1-acc.

(iii) $\Rightarrow$ (i) Suppose that there exists a positive integer N such that $k_i \leq N$ for all $i \in I$. Then $P^N M = 0$ and M is a module over the perfect ring R/P^N. By Lemma 3.4 M satisfies pan-acc. $\qquad \square$

In particular, let p be any prime in the ring $\mathbb{Z}$ and let M denote the $\mathbb{Z}$-module $\oplus_{k \geq 1}(\mathbb{Z}/\mathbb{Z}p^k)$. Then M satisfies pan-acc but the $\mathbb{Z}$-module $\prod_{k \geq 1}(\mathbb{Z}/\mathbb{Z}p^k)$ does not satisfy 1-acc, nor does the $\mathbb{Z}$-module $M^{\mathbb{N}}$.

Finally we look again at the pand-acc condition. First recall the following well-known fact.

Lemma 4.2. *Let $K \subseteq L$ be submodules of a module M such that K is a closed submodule of L and L is a closed submodule of M. Then K is a closed submodule of M.*

Proof. See [7, p. 6 (4)]. $\qquad \square$

Lemma 4.3. *Let K be a closed submodule of a module M such that K and M/K both satisfy pand-acc. Then M satisfies pand-acc.*

Proof. Let N be any submodule of M such that N has finite uniform dimension. There exists a closed submodule L of K such that $N \cap K$ is an essential submodule of L. By Lemma 4.2 L is a closed submodule of M. Note that $u(L) = u(N \cap K) \leq u(N)$. By hypothesis, L is Noetherian. Now $(N + L)/N \cong L/(N \cap L)$ so that $(N + L)/N$ is Noetherian and hence has finite uniform dimension. Thus $N + L$ has finite uniform dimension by Lemma 2.9(a) and hence so too does $(N + L)/L$ by Lemma 2.9(b) because L is a closed submodule of $N + L$. But

$$(N + K)/K \cong (N + L)/[(N + L) \cap K] = (N + L)/[L + (N \cap K)] = (N + L)/L.$$

Thus $(N + K)/K$ has finite uniform dimension. By hypothesis, $(N + K)/K$ is Noetherian and hence so too is $(N + L)/L$. Because L is Noetherian, $N + L$ is also Noetherian and it follows that N is Noetherian. Now apply Lemma 2.10. □

The next result is due to J. Jenkins (unpublished). Note that it holds for any ring whether or not the ring is commutative.

Theorem 4.6. *Let a module* $M = \oplus_{i \in I} M_i$ *be a direct sum of submodules* M_i $(i \in I)$. *Then* M *satisfies pand-acc if and only if for each* i *in* I, M_i *satisfies pand-acc.*

Proof. The necessity is clear. Conversely, suppose that, for each $i \in I$, M_i satisfies pand-acc. Let N be any submodule of M such that N has finite uniform dimension. Then N contains a finitely generated essential submodule L. There exists a finite subset J of I such that $L \subseteq \oplus_{j \in J} M_j$. Note that $L \cap \oplus_{i \in I \setminus J} M_i = 0$ and hence $N \cap \oplus_{i \in I \setminus J} M_i = 0$. It follows that N embeds in $\oplus_{j \in J} M_j$. But by Lemma 4.3 the module $\oplus_{j \in J} M_j$ satisfies pand-acc. Thus N is Noetherian. The result follows by Lemma 2.10. □

Acknowledgement

The first author is partially supported by Ministerio de Ciencia y Tecnología (MTM2004-08115C0404), FEDER, and PAI (FQM-0125). This work was done during a visit of the second author to the Universidad de Málaga in 2006. He would like to thank the Departamento de Álgebra, Geometría y Topología and the project MTM2004-08115C040 for their hospitality and financial support.

References

[1] M.E. Antunes Simões and P.F. Smith, Direct products satisfying the ascending chain condition for submodules with a bounded number of generators, Comm. Algebra 23, 3525–3540 (1995).

[2] M.E. Antunes Simões and P.F. Smith, On the ascending chain condition for submodules with a bounded number of generators, Comm. Algebra 24, 1713–1721 (1996).

[3] M.E. Antunes Simões and P.F. Smith, Rings whose free modules satisfy the ascending chain condition on submodules with a bounded number of generators, J. Pure Appl. Algebra 123, 51–66 (1998).

[4] B. Baumslag and G. Baumslag, On ascending chain conditions, Proc. London Math. Soc. (3) 22, 681–704 (1971).

[5] P.M. Cohn, *Free Rings and their Relations* (Academic Press, London 1971).

[6] P.M. Cohn, *Algebraic Numbers and Algebraic Functions* (Chapman and Hall, London 1991).

[7] N.V. Dung, D.V. Huynh, P.F. Smith and R. Wisbauer, *Extending Modules* (Longman, Harlow 1994).

[8] D. Frohn, A counterexample concerning ACCP in power series rings, Comm. Algebra 30, 2961–2966 (2002).

[9] D. Frohn, Modules with n-acc and the acc on certain types of annihilators, J. Algebra 256, 467–483 (2002).

[10] L. Fuchs, *Infinite Abelian Groups Vol II* (Academic Press, New York 1973).

[11] W. Heinzer and D. Lantz, Commutative rings with ACC on n-generated ideals, J. Algebra 80, 261–278 (1983).

[12] D. Jonah, Rings with the minimum condition for principal right ideals have the maximum condition for principal left ideals, Math. Z. 113, 106–112 (1970).

[13] I. Kaplansky, *Infinite Abelian Groups* (Univ. Michigan, Ann Arbor 1954).

[14] I. Kaplansky, *Commutative Rings* (Allyn and Bacon, Boston 1970).

[15] H. Matsumura, *Commutative Ring Theory*, Cambridge Studies in Adv. Math. 8 (Cambridge Univ. Press, Cambridge 1986).

[16] J.C. McConnell and J.C. Robson, *Noncommutative Noetherian Rings* (John Wiley and Sons, Chichester 1987).

[17] A.-M. Nicolas, Sur les modules tels que toute suite croissante de sous-modules engendrés par n générateurs soit stationnaire, J. Algebra 60, 249–260 (1979).

[18] G. Renault, Sur des conditions de chaînes ascendantes dans des modules libres, J. Algebra 47, 268–275 (1977).

[19] D.W. Sharpe and P. Vamos, *Injective Modules*, Cambridge Tracts in Mathematics and Mathematical Physics, No. 62 (Cambridge University Press, London 1972).

[20] O. Zariski and P. Samuel, *Commutative Algebra Vol. I* (Van Nostrand, Princeton 1958).

Esperanza Sanchez Campos
Departamento de Álgebra, Geometría y Topología
Universidad de Málaga
29080 Málaga, Spain
e-mail: esperanz@agt.cie.uma.es

Patrick F. Smith
Department of Mathematics
University of Glasgow
Glasgow G12 8QW
Scotland UK
e-mail: pfs@maths.gla.ac.uk

Modules and Comodules
Trends in Mathematics, 143–168
© 2008 Birkhäuser Verlag Basel/Switzerland

τ-Injective Modules

Stelios Charalambides and John Clark

To Robert Wisbauer on the occasion of his 65th birthday.

Abstract. In this article we consider injective modules relative to a torsion theory τ. We introduce τ-M-injective and s-τ-M-injective modules, relatively τ-injective modules, the τ-M-injective hull and Σ-τ-M-injective and Σ-s-τ-M-injective modules. We then examine the relationship between these new and known concepts.

Some of the new results obtained include a Generalized Fuchs Criterion characterizing s-τ-M-injective modules, a Generalized Azumaya's Lemma characterizing τ-$\bigoplus_I M_i$-injective modules, the proof of the existence and uniqueness up to isomorphism of the τ-M-injective hull and generalizations of results by Albu and Năstăsescu, Faith, and Cailleau on necessary and sufficient conditions for a module to be Σ-s-τ-M-injective, Σ-τ-injective and for a direct sum of Σ-s-τ-M-injective modules to be Σ-s-τ-M-injective.

1. Introduction and preliminaries

After the introduction and preliminary sections, the article has three main sections, on the topics of τ-injectivity, relative τ-injective hulls and relative Σ-τ-injective modules.

Injective modules relative to a torsion theory τ can be defined in a variety of ways. We begin with the usual definition of τ-injective modules and the Generalized Baer Criterion for determining τ-injectivity by looking at τ-dense left ideals of the ring. We then extend this definition by looking at τ-injectivity relative to a fixed module M, thus introducing τ-M-injective modules. A slightly different flavour of

This paper is part of the first author's University of Otago Ph.D. thesis, written under the supervision of the second author. The first author gratefully acknowledges the support of the Commonwealth Scholarship and Fellowship Committee of New Zealand. Both authors are very grateful to the referee for their careful reading of the article, their suggestions, and for supplying reference [PRY].

this is also introduced, the s-τ-M-injective modules which generalize the τ^f-quasi-injective modules given in [B, 4.1.16]. Subsequently the relationship between these definitions is explored. In particular we show how one version of relative injectivity can be used to characterize the others.

Next we modify the Generalized Fuchs Criterion, characterizing s-τ-quasi-injective modules, and characterize s-τ-M-injectivity using certain types of τ-dense left ideals.

The section on relative injective hulls is motivated by a flaw in an argument in [B] on the existence and uniqueness of τ-quasi-injective hulls. We repair the proof and extend the concept to that of a τ-M-injective hull, generalizing the M-injective hull found in [W, 17.8]. The existence and uniqueness up to isomorphism of τ-quasi-injective hulls is a consequence. The section ends with a short proof of a torsion-theoretic version of Azumaya's Lemma using a characterization of τ-M-injectivity via the concept of a trace.

The last section, on Σ-τ-injective modules, is motivated by a result of Faith and another in [AN] which characterize Σ-injectivity by an ascending chain condition on annihilators. We obtain torsion-theoretic versions of both results, and of some of their corollaries, most notably a result found in [AN] which is based on a method of [Cai] and characterizes when a direct sum of Σ-s-τ-M-injectives is Σ-s-τ-M-injective.

Unless otherwise stated, all modules will be left R-modules for some unitary ring R, and all homomorphisms will be R-module homomorphisms. We denote the class of left R-modules by R-Mod. If $N, M \in R$-Mod, we write $N \leq M$ to denote that N is a submodule of M while if N is an essential submodule of M we write $N \leq_e M$. If X is a subset of the module M and $n \in M$, we let $(X : n)$ denote the set $\{\, r \in R \,|\, rn \in X \,\}$.

In what follows, $\tau = (\mathcal{T}, \mathcal{F})$ denotes a hereditary torsion theory on R-Mod.

A submodule N of M is called τ-*dense* (τ-*pure*) if M/N is τ-torsion (τ-torsionfree), in which case we write $N \leq_{\tau\text{-d}} M$ ($N \leq_{\tau\text{-p}} M$). We let $\mathcal{D}_\tau(M)$ denote the set of all τ-dense submodules of M and $\mathcal{P}_\tau(M)$ denote the set of all τ-pure submodules of M. The τ-*pure closure* of N in M is the intersection of all the τ-pure submodules of M that contain N. It is denoted by $\mathrm{Cl}_\tau^M(N)$, or, when M and τ are understood, by N^c. Thus:

$$\mathrm{Cl}_\tau^M(N) = N^c = \bigcap \{\, K \in \mathcal{P}_\tau(M) \,|\, N \leq K \,\}.$$

A submodule N of a module M is called τ-*essential* in M if it is τ-dense and essential in M, i.e., $N \in \mathcal{D}_\tau(M)$ and $N \leq_e M$. In this case we write $N \leq_{\tau\text{-e}} M$ and say M is a τ-*essential extension* of N. If $N \leq M$, we say that N is τ-*essentially closed* in M if N has no proper τ-essential extensions in M, i.e., if $N \leq_{\tau\text{-e}} E \leq M$ then $N = E$. If all its nonzero submodules are τ-dense, M is called τ-*compact* (vacuously 0 is τ-compact).

The following collects together some basic torsion theory results for later use.

Proposition 1.1. *Let τ be a torsion theory and $M \in R$-Mod.*

(1) *If $K \in \mathcal{D}_\tau(M)$ and $K \leqslant L \leqslant M$ then $L \in \mathcal{D}_\tau(M)$.*

(2) *If $K, L \in \mathcal{D}_\tau(M)$ then $K \cap L \in \mathcal{D}_\tau(M)$.*

(3) *Given $K \leqslant L \leqslant M$, we have $K \leqslant_{\tau\text{-}d} M$ if and only if $K \leqslant_{\tau\text{-}d} L \leqslant_{\tau\text{-}d} M$.*

(4) *If $N \leqslant M$ then $t_\tau(M/N) = N^c/N$ and $N^c = \{\, m \in M \mid (N : m) \in \mathcal{D}_\tau(R) \,\}$, and so N is τ-dense in N^c.*

(5) *If $m \in M$ and $I \in \mathcal{D}_\tau(R)$ then $Im \in \mathcal{D}_\tau(Rm)$.*

(6) *Given $K \leqslant L \leqslant M$, we have $K \leqslant_{\tau\text{-}e} M$ if and only if $K \leqslant_{\tau\text{-}e} L$ and $L \leqslant_{\tau\text{-}e} M$.*

(7) *If $N \leqslant M$ then there is a τ-essentially closed submodule N^* of M with $N \leqslant_{\tau\text{-}e} N^*$.*

Proof. Items (1), (2), and (4) appear in [B, 1.1.11 and 2.1.7] while the proofs of (3) and (5) are routine. Item (6) follows from (3) and [DHSW, 1.5(2)]. Using (6) and the property that the union of an ascending chain of torsion modules is also torsion, the proof of (7) is a straightforward modification of the well-known argument showing that every submodule N of M has an essential closure $\overline{N}$ in M (see [DHSW, p. 6]). □

2. τ-injectivity and related concepts

Recall that a module M is called *injective* if, for any module L, every homomorphism from a submodule K of L to M extends to a homomorphism from L to M. A torsion-theoretic version of this, as found in [B], now follows.

Definition 2.1. *A module M is τ-injective if, for every module L, every homomorphism from a τ-dense submodule of L to M extends to a homomorphism from L to M.*

Before detailing the properties of τ-injectivity we give an example, from Crivei [C98].

A module is called *semi-Artinian* if it is either 0 or each of its nonzero factor modules has a nonzero socle (equivalently, a simple submodule). With $\mathcal{T}$ as the class of all semi-Artinian R-modules and $\mathcal{F}$ as the class of all R-modules with zero socle, $\tau_D = (\mathcal{T}, \mathcal{F})$ is a hereditary torsion theory called the *Dickson torsion theory*. In [C98, Theorem 6] Crivei shows that an R-module A is τ_D-injective if and only if, for any maximal left ideal $\mathcal{M}$ of R, any homomorphism $f : \mathcal{M} \to A$ can be extended to a homomorphism $g : R \to A$. In [C98, Example 16] he then shows that if $R = F[[x,y]]$, the power series in indeterminates x, y over a field F, the R-module R is τ_D-injective but not injective.

We now continue with the theory of τ-injectivity. The following is a torsion-theoretic analogue of Baer's Criterion for injectivity. It is essentially given as [B, 4.1.2].

Proposition 2.1 (Generalized Baer Criterion). *Let M be an R-module. Then the following are equivalent:*

(1) *M is τ-injective.*
(2) *Every homomorphism from a τ-dense left ideal of R to M extends to a homomorphism from R to M.*
(3) *For each $I \in \mathcal{D}_\tau(R)$ and each $f \in \mathrm{Hom}_R(I, M)$, there exists an $m \in M$ such that $f(a) = am$ for all $a \in I$.*

Remark 2.1. The τ-dense left ideals of R in Proposition 2.1 can be replaced by τ-essential left ideals of R. (See Bland [B, 4.1.3])

Let $M, E \in R$-Mod. Then E is called M-*injective* if every homomorphism from a submodule of M to E extends to a homomorphism from M to E. This is equivalent to the seemingly stronger requirement that if $K \leqslant L \leqslant M$ then every homomorphism from K to E extends to one from L to E, i.e., E is L-injective for every $L \leqslant M$ (see [W, 16.3] for details). This results in the following two torsion-theoretic definitions.

Definition 2.2. *Let $M, E \in R$-Mod. Then E is called τ-M-injective if any homomorphism from a τ-dense submodule of M to E extends to a homomorphism from M to E. We say that E is s-τ-M-injective if, for any $N \leqslant M$, any homomorphism from a τ-dense submodule of N to E extends to a homomorphism from N to E. [Here the "s" in "s-τ" is short for "strongly".]*

We note that s-τ-M-injective modules are referred to as M-τ-*injective* modules in [PRY][1].

Remark 2.2. From the definitions of τ-injectivity and s-τ-M-injectivity we get:

(1) Every s-τ-M-injective module is τ-M-injective.
(2) A module is s-τ-M-injective if and only if it is (s-)τ-N-injective for every $N \leqslant M$.

A module Q is said to be *quasi-injective* if it is Q-injective. The following definition combines two torsion-theoretic analogues of this found in [B, 4.1.16, 4.1.19]. Note that we have changed Bland's notation from $\tau^{\underline{f}}$-quasi-injective to s-τ-quasi-injective.

Definition 2.3. *A module Q is said to be s-τ-quasi-injective if, given $N \leq Q$ and $K \in \mathcal{D}_\tau(N)$, any $f \in \mathrm{Hom}_R(K, Q)$ can be extended to a $g \in \mathrm{Hom}_R(N, Q)$. A module Q is called τ-quasi-injective if, given $K \in \mathcal{D}_\tau(Q)$, any $f \in \mathrm{Hom}_R(K, Q)$ can be extended to a $g \in \mathrm{Hom}_R(Q, Q)$.*

---

[1] Reference [PRY] provides an alternative to our Proposition 2.3, the Generalized Fuchs Criterion, which is then used to prove our Lemma 4.2, but there is no other obvious overlap with the results presented here.

Remark 2.3. Using the definition of s-τ-M-injectivity and the Generalized Baer Criterion we get the following reformulations.

(1) A module E is τ-injective if and only if it is (s-)τ-M-injective for all $M \in$ R-Mod.

(2) A module E is τ-injective if and only if it is (s-)τ-$_RR$-injective.

(3) A module Q is s-τ-quasi-injective if and only if it is s-τ-Q-injective if and only if it is (s-)τ-K-injective for every $K \leqslant Q$.

(4) A module Q is τ-quasi-injective if and only if it is τ-Q-injective.

In [Fu], Fuchs obtained a condition similar to Baer's Criterion characterizing quasi-injectivity, and Bland in [B90] generalized this to s-τ-quasi-injective modules. Our next aim is to further generalize the condition so as to characterize s-τ-M-injective modules.

Definition 2.4. *For any module M, let $\Omega(M)$ denote the set of all left ideals of R which contain the left annihilator of an element of M. Thus a left ideal I of R is in $\Omega(M)$ if and only if there is an $m \in M$ such that $(0 : m) \leqslant I$. Similarly we denote the set of all left ideals of R which contain the left annihilator of a finite subset Y of M by $\bar{\Omega}(M)$.*

The following extends [B, 4.1.17] and is used in the proof of Proposition 2.3.

Proposition 2.2. *Let E, M be R-modules. Then the following conditions are equivalent.*

(1) *E is s-τ-M-injective.*

(2) *If $m \in M$ and $K \in \mathcal{D}_\tau(Rm)$, any map $f \in \mathrm{Hom}_R(K, E)$ can be extended to a map $g \in \mathrm{Hom}_R(Rm, E)$.*

(3) *If K and N are R-modules, not necessarily submodules of M, with $K \in \mathcal{D}_\tau(N)$ and $\Omega(N) \subseteq \Omega(M)$, any $f \in \mathrm{Hom}_R(K, E)$ can be extended to a $g \in \mathrm{Hom}_R(N, E)$.*

Proof. (1) $\Rightarrow$ (2) is obvious and taking N to be a submodule of M gives (3) $\Rightarrow$ (1).

(2) $\Rightarrow$ (3). Suppose that K, N and $f \in \mathrm{Hom}_R(K, E)$ are as given in (3). Let $\mathcal{S} = \{ (g, L) \mid K \leqslant L \leqslant N, \ g : L \to E, \ g|_K = f \}$. Partially order $\mathcal{S}$ by setting $(g_1, L_1) \leqslant (g_2, L_2)$ if $L_1 \leqslant L_2$ and $g_1 = g_2|_{L_1}$. Then $\mathcal{S}$ is inductive and so, by Zorn's Lemma, has a maximal element, (h, X) say. It suffices to show that $X = N$.

Let $n \in N$. We construct a pair $(h^*, X + Rn)$ that contains (h, X). Since $(0 : n) \in \Omega(N) \subseteq \Omega(M)$, there exists an $m \in M$ such that $(0 : m) \leqslant (0 : n)$. Define a map $\varphi : (X : n)m \to E$ by $\varphi(am) = h(an)$ for any $a \in (X : n)$. This is a well-defined homomorphism and since $X \in \mathcal{D}_\tau(N)$ we have $N/X \in \mathcal{T}$ and so $(X : n) \in \mathcal{D}_\tau(R)$ by Proposition 1.1 (4).

Hence, by Proposition 1.1 (5), we get $(X : n)m \in \mathcal{D}_\tau(Rm)$. Thus, by (2), φ extends to a homomorphism $\varphi^* : Rm \to E$. This allows us to define a map $h^* : X + Rn \to E$ which extends $h : X \to E$ as follows. For any $x \in X$ and $r \in R$, let $h^*(x + rn) = h(x) + \varphi^*(rm)$. It is easy to check that h^* is well defined. Clearly

$h^*|_X = h$ and so $h^*|_K = h|_K = f$. Thus $(h^*, X + Rn) \in \mathcal{S}$ and $(h, X) \leqslant (h^*, X + Rn)$. The maximality of (h, X) now gives $X + Rn = X$ and so $n \in X$. Hence $X = N$ as required. $\qquad\square$

The following generalizes [B, 4.1.18], where it is shown for $E = M$, thereby characterizing s-τ-quasi-injective modules. It is our version of the Generalized Fuchs Criterion.

Proposition 2.3 (Generalized Fuchs Criterion). *Let E, M be R-modules. Then the following conditions are equivalent.*

(1) *E is s-τ-M-injective.*
(2) *For each $I \in \mathcal{D}_\tau(R)$, any map $f \in \operatorname{Hom}_R(I, E)$ with $\operatorname{Ker} f \in \Omega(M)$ can be extended to a map $g \in \operatorname{Hom}_R(R, E)$.*
(3) *For each $I \in \mathcal{D}_\tau(R)$, and each $f \in \operatorname{Hom}_R(I, E)$ with $\operatorname{Ker} f \in \Omega(M)$, there exists an $x \in E$ such that $f(a) = ax$ for all $a \in I$.*

Proof. $(1) \Rightarrow (2)$. The following commutative diagram illustrates the proof.

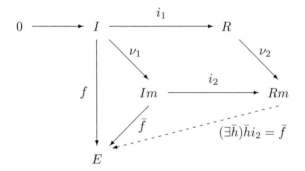

Suppose that $I \in \mathcal{D}_\tau(R)$, E is an s-τ-M-injective module and $f \in \operatorname{Hom}_R(I, E)$ with $\operatorname{Ker} f \in \Omega(M)$. Then $(0 : m) \leqslant \operatorname{Ker} f$ for some $m \in M$. Let $i_1 : I \to R$ and $i_2 : Im \to Rm$ be the inclusion maps. Define $\nu_1 : I \to Im$ by $a \mapsto am$ for any $a \in I$ and $\nu_2 : R \to Rm$ by $r \mapsto rm$ for any $r \in R$. Next define $\bar{f} : Im \to E$ by $am \mapsto f(a)$ for any $a \in I$. This is a good definition since if $a_1, a_2 \in I$ are such that $a_1 m = a_2 m$ then $a_1 - a_2 \in (0 : m) \leqslant \operatorname{Ker} f$. Since $I \in \mathcal{D}_\tau(R)$, Proposition 1.1 (5) gives $Im \in \mathcal{D}_\tau(Rm)$. Hence, by Proposition 2.2, there is a homomorphism $\bar{h} : Rm \to E$, such that $\bar{h} i_2 = \bar{f}$. To finish, define $g : R \to E$ by $g = \bar{h} \nu_2$. Then $g i_1 = \bar{h} \nu_2 i_1 = \bar{h} i_2 \nu_1 = \bar{f} \nu_1 = f$.

$(2) \Rightarrow (1)$. Let K and N be submodules of M with $K \in \mathcal{D}_\tau(N)$ and $f \in \operatorname{Hom}_R(K, E)$. Let $\mathcal{S} = \{ (g, L) \mid K \leqslant L \leqslant N, \; g : L \to E, \; g|_K = f \}$. First note that $\mathcal{S} \neq \emptyset$ since $(f, K) \in \mathcal{S}$. Then we may partially order $\mathcal{S}$ as in the proof of Proposition 2.2 and, with respect to this, $\mathcal{S}$ has a maximal element, (h, X) say. It suffices to show that $X = N$.

Let $n \in N$. We construct a pair $(h^*, X + Rn)$ that contains (h, X). Since $X \in \mathcal{D}_\tau(N)$ we have $N/X \in \mathcal{T}$ and so $(0 : n + X) = (X : n) \in \mathcal{D}_\tau(R)$. Define $\varphi : (X : n) \to E$ by $\varphi(a) = h(an)$ for any $a \in (X : n)$. Then, since $(0 : n) \leqslant \operatorname{Ker} \varphi$, we have $\operatorname{Ker} \varphi \in \Omega(M)$. Thus, by (2), φ extends to a homomorphism $\varphi^* : R \to E$. This allows us to define a map $h^* : X + Rn \to E$ which extends $h : X \to E$ as follows. For any $x \in X$ and $r \in R$, let $h^*(x + rn) = h(x) + \varphi^*(r)$. It is easy to see that h^* is well defined. Clearly $h^*|_X = h$ and so $h^*|_K = h|_K = f$. Thus $(h^*, X + Rn) \in \mathcal{S}$ and $(h, X) \leqslant (h^*, X + Rn)$. Therefore the maximality of (h, X) gives $X + Rn = X$ and so $n \in X$. Hence $X = N$ as required.

$(2) \Leftrightarrow (3)$. This is trivial. $\qquad\qquad\qquad\qquad\qquad\qquad\qquad\qquad\qquad\qquad\quad\square$

Lemma 2.1. *If M is a τ-compact module, then a τ-M-injective module is M-injective.*

Proof. This follows since every nonzero submodule of M is τ-dense. $\qquad\quad\square$

We now state a torsion-theoretic version of Azumaya's Lemma ([AF, 16.13(2)]). (The original can be recovered when τ is the torsion theory in which every module is τ-torsion.) While a direct proof of this is possible, to save room we instead give a short proof in the next section using techniques developed there (see Lemma 3.2).

Lemma 2.2 (Generalized Azumaya's Lemma). *If E and $M = \bigoplus_I M_i$ are modules, then E is τ-M-injective if and only if E is τ-M_i-injective for each $i \in I$.*

By Remark 2.2, given modules E and $M = \bigoplus_I M_i$ such that E is s-τ-M-injective, then E is s-τ-M_i-injective for each $i \in I$. However the converse of this remains open.

The next result is based on [G, 8.4] where it is established in the special case of τ-injectivity. The proof is by standard arguments and so we omit it.

Proposition 2.4. *The following statements are true for any module M.*

(1) *Every direct summand of a $(s\text{-})\tau$-M-injective module is $(s\text{-})\tau$-M-injective.*
(2) *Let $\{ E_i \,|\, i \in I \}$ be a family of modules. Then $\prod_{i \in I} E_i$ is $(s\text{-})\tau$-M-injective if and only if each E_i is $(s\text{-})\tau$-M-injective.*

The following lemma can be found in [DHSW, 7.5].

Lemma 2.3. *Let M_1 and M_2 be R-modules and let $M = M_1 \oplus M_2$. Then M_1 is M_2-injective if and only if, for every submodule N of M such that $N \cap M_1 = 0$, there exists a submodule M_2' of M such that $N \leqslant M_2'$ and $M = M_1 \oplus M_2'$.*

An investigation of when a finite direct sum of self-τ-divisible modules is self-τ-divisible is given in [CC]. There a module D is called *self-τ-divisible* if, for any $P \in \mathcal{P}_\tau(D)$ and any $f \in \operatorname{Hom}_R(P, D)$, there exists a $g \in \operatorname{Hom}_R(D, D)$ such that $g|_P = f$. We now adapt some of their arguments to obtain analogous results for τ-quasi-injective modules.

The first result is inspired by [CC, 3.3] and is a torsion-theoretic analogue of Lemma 2.3. The proof requires only minor modifications of the proof of the original.

Proposition 2.5. *Let A_1 and A_2 be modules, $A = A_1 \oplus A_2$, and $\pi_1 : A \to A_1$ and $\pi_2 : A \to A_2$ be the canonical projections. Then A_1 is τ-A_2-injective if and only if, for every submodule B of A such that $B \cap A_1 = 0$ and $\pi_2(B) \in \mathcal{D}_\tau(A_2)$, there exists a submodule C of A such that $A = A_1 \oplus C$ and $B \leqslant C$.*

Proof. ($\Rightarrow$) Suppose that A_1 is τ-A_2-injective and let $B \leqslant A$ such that $B \cap A_1 = 0$ and $\pi_2(B) \in \mathcal{D}_\tau(A_2)$. The homomorphism $f = \pi_2|_B : B \to \pi_2(B)$ is an isomorphism, because $B \cap A_1 = 0$. Since A_1 is τ-A_2-injective and $\pi_2(B) \in \mathcal{D}_\tau(A_2)$, $\pi_1 f^{-1} : \pi_2(B) \to A_1$ extends to a homomorphism $g : A_2 \to A_1$. Let $C = \{\, g(a) + a \,|\, a \in A_2 \,\}$. It is routine to check that C is a submodule of A and $A = A_1 \oplus C$. Furthermore, for any $b \in B$ we have $b = \pi_1(b) + \pi_2(b) = \pi_1 f^{-1}\pi_2(b) + \pi_2(b) = g\pi_2(b) + \pi_2(b) \in C$, and so $B \leqslant C$, as required.

($\Leftarrow$) Let $D \in \mathcal{D}_\tau(A_2)$ and $f \in \operatorname{Hom}_R(D, A_1)$. Define $B = \{\, d - f(d) \,|\, d \in D \,\}$. It is easily seen that $B \leqslant A$ and $B \cap A_1 = 0$. Furthermore $\pi_2(B) = D$ and so $\pi_2(B) \in \mathcal{D}_\tau(A_2)$. Hence, by assumption, there is a submodule C of A such that $B \leqslant C$ and $A = A_1 \oplus C$. Let $\pi : A = A_1 \oplus C \to A_1$ be the canonical projection with kernel C. Let $i_D : D \to A_2$ and $i_2 : A_2 \to A$ denote the inclusion homomorphisms. If we define $g : A_2 \to A_1$ by $g = \pi i_2$ then for any $d \in D$ we have $g i_D(d) = \pi i_2 i_D(d) = \pi(d - f(d) + f(d)) = f(d)$, since $d - f(d) \in B \leqslant C$, and so $g i_D = f$. Thus A_1 is τ-A_2-injective as required. $\square$

The next result is inspired by [CC, 3.4] and again the proof requires only minor modifications of the proof of the original result.

Proposition 2.6. *Let $A \in R$-Mod, D be a τ-A-injective module and $B \leqslant A$. Then:*
(1) *D is τ-A/B-injective.*
(2) *If $B \in \mathcal{D}_\tau(A)$ then D is τ-B-injective.*

Proof. (1) Let $C/B \in \mathcal{D}_\tau(A/B)$ and $f \in \operatorname{Hom}_R(C/B, D)$. Let $i_1 : C/B \to A/B$ and $i_2 : C \to A$ be the inclusion maps and $\nu_1 : C \to C/B$ and $\nu_2 : A \to A/B$ be the natural epimorphisms. Since $C \in \mathcal{D}_\tau(A)$, there is an $h \in \operatorname{Hom}_R(A, D)$ such that $h i_2 = f \nu_1$. Define $g : A/B \to D$ by $g(a+B) = h(a)$, so that $g \nu_2 = h$. This is a good definition since, for any $a_1, a_2 \in A$, if $a_1 + B = a_2 + B$ then $a_1 - a_2 \in B$, and so $h(a_1 - a_2) = h i_2(a_1 - a_2) = f\nu_1(a_1 - a_2) = f((a_1 - a_2) + B) = 0$. Furthermore, $g i_1 \nu_1 = g \nu_2 i_2 = h i_2 = f \nu_1$. Since ν_1 is an epimorphism we have $g i_1 = f$ and so D is τ-A/B-injective, as required.

(2) Suppose that $B \leqslant_{\tau\text{-d}} A$. Let $C \leqslant_{\tau\text{-d}} B$ and let $f : C \to D$ be a homomorphism. Then, by the transitivity of $\leqslant_{\tau\text{-d}}$, we have $C \leqslant_{\tau\text{-d}} A$, and so f extends to a homomorphism $g : A \to D$. Clearly $g|_B$ is an extension of f and so D is τ-B-injective, as required. $\square$

The following definition describes injectivity relationships in a family of modules.

Definition 2.5. *The members of a family $\{M_i\}_I$ of modules are said to be relatively $(\tau$-$)$injective if M_i is $(\tau$-$)M_j$-injective for any $i, j \in I$ such that $i \neq j$.*

By Propositions 2.6 and 2.4 we have the following corollary.

Corollary 2.7. *Let A_1 and A_2 be modules such that $A_1 \oplus A_2$ is τ-quasi-injective. Then A_1 and A_2 are both τ-quasi-injective and relatively τ-injective.*

We now extend the main result in [CC, 3.7] for τ-injective modules. This time the proof does not follow that of [CC, 3.6] but uses instead our version of Azumaya's Lemma.

Theorem 2.1. *Let $A_1, \ldots, A_n$ be relatively τ-injective modules. Then $A_1 \oplus \cdots \oplus A_n$ is τ-quasi-injective if and only if A_i is τ-quasi-injective for each $1 \leqslant i \leqslant n$.*

Proof. ($\Rightarrow$) For each $1 \leqslant i \leqslant n$, A_i is τ-$(A_1 \oplus \cdots \oplus A_n)$-injective by Proposition 2.4 and so, by Lemma 2.2, A_i is τ-A_i-injective. Thus A_i is τ-quasi-injective, as required.

($\Leftarrow$) Since $A_1, \ldots, A_n$ are relatively τ-injective, if each A_i is τ-quasi-injective, then A_i is τ-A_j-injective for any $1 \leqslant i, j \leqslant n$. Thus, by Lemma 2.2, A_i is τ-$(A_1 \oplus \cdots \oplus A_n)$-injective for each i and so, by Proposition 2.4, $A_1 \oplus \cdots \oplus A_n$ is τ-$(A_1 \oplus \cdots \oplus A_n)$-injective. $\square$

From the proof of Theorem 2.1, it is clear that the following corollary holds.

Corollary 2.8. *Let $A_1, \ldots, A_n$ be modules. Then $A_1 \oplus \cdots \oplus A_n$ is τ-quasi-injective if and only if A_i is τ-A_j-injective for any $1 \leqslant i, j \leqslant n$.*

3. Relative τ-injective hulls

Here we explore various types of injective hulls relative to a torsion theory τ. We begin with some useful properties of τ-injectivity which we use to prove a known result on the existence and uniqueness of τ-injective hulls. We then define τ-M-injective hulls, generalizing M-injective hulls, and prove their existence and uniqueness up to isomorphism for M-generated modules. This is used to simplify a proof of a result in [B], asserting the existence and uniqueness of a τ-quasi-injective hull, whose proof is flawed. We end the section with a streamlined proof of a relative version of Azumaya's Lemma.

The following is a generalization of a result found in [G, 8.4]. We will use it to define the torsion-theoretic analogue of an injective hull.

Proposition 3.1. *Let M be an R-module. Then every τ-pure submodule of an s-τ-M-injective R-module is s-τ-M-injective.*

Proof. Let E be an s-τ-M-injective R-module and $N \in \mathcal{P}_\tau(E)$. We wish to show that N is s-τ-M-injective. We proceed using our Generalized Fuchs Criterion. Let $f : I \to N$ be an R-homomorphism with $I \in \mathcal{D}_\tau(R)$ and $\operatorname{Ker} f \in \Omega(M)$. Consider the following diagram, where i_I, i_N denote the inclusion mappings and ν_I, ν_N the canonical epimorphisms.

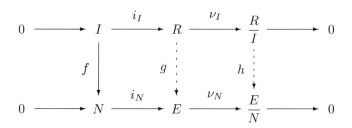

Note that $f : I \to N$ induces a homomorphism $i_N f : I \to E$, which extends to a homomorphism $g : R \to E$, because clearly $\operatorname{Ker} f \leqslant \operatorname{Ker} i_N f \in \Omega(M)$ and E is s-τ-M-injective. Thus we have $g i_I = i_N f$. This induces a homomorphism $h : R/I \to E/N$, defined by $h(r + I) = g(r) + N$. It is easily seen that this is a well-defined homomorphism.

Also note that $R/I \in \mathcal{T}$, because $I \in \mathcal{D}_\tau(R)$, and $E/N \in \mathcal{F}$, because $N \in \mathcal{P}_\tau(E)$. Thus $h : R/I \to E/N$ is a homomorphism from a τ-torsion to a τ-torsionfree module and so zero. Hence $h(R/I) = (g(R) + N)/N = N/N$ and so $g(R) \leqslant N$. Thus we have a homomorphism $g : R \to N$ with $g i_I = f$ and so N is s-τ-M-injective. $\qquad\square$

Corollary 3.2. *The following statements are true:*

(1) *Every τ-pure submodule of a τ-injective module is τ-injective.*
(2) *Every τ-pure submodule of a s-τ-quasi-injective module is s-τ-quasi-injective.*

Proof. These follow from Remarks 2.3 and 2.2. $\qquad\square$

Similarly, by replacing the short exact sequence $0 \to I \to R \to R/I \to 0$ in the proof of Proposition 3.1 with $0 \to D \to M \to M/D \to 0$, where $D \in \mathcal{D}_\tau(M)$, we get:

Proposition 3.3. *Every τ-pure submodule of a τ-M-injective module is τ-M-injective.*

We recall the following characterization of the injective hull $E = E(M)$ of a module M (see, e.g., [L, 3.30]).

Theorem 3.1. *For modules $M \leqslant E$, the following are equivalent:*

(1) *E is injective and an essential extension of M.*
(2) *E is a maximal essential extension of M.*
(3) *E is a minimal injective extension of M.*

We will see later that this result still holds if injectivity is replaced by τ-injectivity, and essential by τ-essential. The direction (2) $\Rightarrow$ (3) in the relative version is stated in [G, p. 84] and proved in [B, p. 104]. To prove the other implications we first investigate the relationship between τ-injectivity, direct sums and τ-essential extensions.

It is worth mentioning here that a module X is M-injective if and only if $f(M) \leqslant X$ for any $f \in \operatorname{Hom}_R(E(M), E(X))$ (see [DHSW, 2.1]). We provide a torsion-theoretic analogue of this in the next section and use it to obtain a relative M-injective hull.

But let us first give a torsion-theoretic version of the injective hull which appears in [G, p. 84]. Note that, by Corollary 3.2, since every injective module is τ-injective we immediately get that τ-injective hulls as defined below are indeed τ-injective.

Definition 3.1. *A τ-injective hull of a module M is the τ-pure closure of M in an injective hull $E(M)$ and denoted by $E_\tau(M)$, i.e., $E_\tau(M) = Cl_\tau^{E(M)}(M)$. (Since the injective hull $E(M)$ is unique up to isomorphism, $E_\tau(M)$ is also likewise unique.)*

Since $M \leqslant_e E(M)$ and, by Proposition 1.1 (4), every module is τ-dense in its τ-pure closure, we have $M \leqslant_{\tau\text{-e}} E_\tau(M)$.

An alternative definition of the τ-injective hull appears in [B, 5.1.1]. There a τ-injective module E is called a τ-injective hull of a module M if there is a monomorphism $\varphi : M \to E$ with $\varphi(M) \leqslant_{\tau\text{-e}} E$. In [B, 5.1.2] Bland states that every module has a τ-injective hull, unique up to isomorphism. He shows that $Cl_\tau^{E(M)}(M)$ is indeed a τ-injective hull by an argument similar to that of the proof of Proposition 3.1.

We next record for easy reference a result of Nishida, [N]. (See also [G, Exercise 8.11].)

Remark 3.1. The following are equivalent for a module M:

(1) M is τ-injective.
(2) M is a direct summand of its τ-pure closure in any module containing it.
(3) M is a direct summand of its τ-pure closure in $E(M)$, i.e., of $E_\tau(M)$.

The following corollary is a variation of the above result and appears in [S, 1.2].

Corollary 3.4. *A module M is τ-injective if and only if M is a direct summand of every module N in which it is τ-dense.*

The next two results generalize two well-known characterizations of injective modules. These results are probably known but we have not found them explicitly in the literature.

Proposition 3.5. *A module M is τ-injective if and only if M has no proper τ-essential extensions.*

Proof. ($\Rightarrow$) Suppose that $M \leqslant_{\tau\text{-e}} X$ for some module X. Then $M \in \mathcal{D}_\tau(X)$ and so, by Corollary 3.4, $X = M \oplus Y$ for some $Y \leqslant X$. But, as $M \leqslant_e X$ and $M \cap Y = 0$, we must have $Y = 0$ and thus $M = X$. Therefore M has no proper τ-essential extensions.

($\Leftarrow$) As noted earlier $M \leqslant_{\tau\text{-e}} E_\tau(M)$ and so $M = E_\tau(M)$ and thus it is τ-injective. $\qquad\square$

The next result characterizes when a submodule of a τ-injective module is τ-injective.

Proposition 3.6. *Let* $M \leqslant E$, *where* E *is* τ-*injective. Then* M *is* τ-*injective if and only if* M *is* τ-*essentially closed in* E.

Proof. ($\Rightarrow$) By Proposition 3.5, if M is τ-injective then it has no proper τ-essential extensions in E and thus M is τ-essentially closed in E.

($\Leftarrow$) Let N be *any* τ-essential extension of M and $i_N : M \to N$ and $i_E : M \to E$ be the inclusion maps. Since $M \in \mathcal{D}_\tau(N)$ and E is τ-injective, $i_E = f i_N$ for some $f : N \to E$. Then $\ker f = 0$ since $\ker f i_N = 0$ and i_N is an essential monomorphism. Thus $f(N) \cong N$. But $M \leqslant f(N)$ and, as $M \leqslant_e N$ and $f : N \to f(N)$ is an isomorphism, we have $f(M) \leqslant_e f(N)$. Also, since $M \leqslant_{\tau\text{-d}} N$, we have $f(M) \leqslant_{\tau\text{-d}} f(N)$. Thus $f(M) \leqslant_{\tau\text{-e}} f(N) \leqslant E$. But $f(M) = M$ and M is τ-essentially closed in E, so $f(M) = f(N)$. Hence $M = N$ and so M has no proper τ-essential extensions, and so, by Proposition 3.5, is τ-injective. $\square$

We are now ready to state and prove the torsion-theoretic version of Theorem 3.1, which forms the basis for a very useful characterization of the τ-injective hull of a module.

Proposition 3.7. *For modules* $M \leqslant E$ *the following are equivalent:*

(1) E *is* τ-*injective and a* τ-*essential extension of* M.
(2) E *is a maximal* τ-*essential extension of* M.
(3) E *is a minimal* τ-*injective extension of* M.

Proof. (1) $\Rightarrow$ (2). Suppose that F is a τ-essential extension of M that contains E. Then, by Proposition 1.1 (6), we get $M \leqslant_{\tau\text{-e}} E \leqslant_{\tau\text{-e}} F$. Thus, by Corollary 3.4, $F = E \oplus X$ for some module X. Since $M \leqslant E$ and $M \leqslant_{\tau\text{-e}} F$ we have $X = 0$ and so $F = E$.

(2) $\Rightarrow$ (3). If E is a maximal τ-essential extension of M, then E is τ-injective by Proposition 3.5. Let D be a τ-injective module such that $M \leqslant D \leqslant E$. As $M \leqslant_{\tau\text{-e}} E$ we have $D \leqslant_{\tau\text{-e}} E$ and so, by Proposition 3.5, $D = E$. Thus E is a minimal τ-injective extension of M.

(3) $\Rightarrow$ (1). Since $M \leqslant E$, by Proposition 1.1 (7), there is a τ-essentially closed submodule M^* of E with $M \leqslant_{\tau\text{-e}} M^*$. By Proposition 3.6, M^* is τ-injective and hence the minimality of E gives $M^* = E$ and so $M \leqslant_{\tau\text{-e}} E$. $\square$

Clearly from Definition 3.1 we have the following result.

Proposition 3.8. *A module* M *and its* τ-*injective hull,* $E = E_\tau(M)$, *satisfy the (equivalent) properties of Proposition 3.7.*

The following is a torsion-theoretic analogue of the result that every injective module containing a module M contains an injective hull of M.

Proposition 3.9. *A* τ-*injective module* E *contains a* τ-*injective hull of each* $M \leq E$.

Proof. By Proposition 1.1 (7), E has a τ-essentially closed submodule M^* with $M \leqslant_{\tau\text{-e}} M^*$. By Proposition 3.6, M^* is τ-injective and so $E_\tau(M) = M^*$ by Proposition 3.8. □

As already mentioned, given modules X and M, X is M-injective if and only if $f(M) \leqslant X$ for any $f \in \operatorname{Hom}_R(E(M), E(X))$. The next result generalizes this.

Proposition 3.10. *Let X and M be modules. Then the following are equivalent:*

(1) X *is* τ-M-*injective.*
(2) *For any* $f \in \operatorname{Hom}_R(M, E_\tau(X))$, *we have* $f(M) \leqslant X$.
(3) *For any* $f \in \operatorname{Hom}_R(E_\tau(M), E_\tau(X))$, *we have* $f(M) \leqslant X$.

Proof. $(1) \Rightarrow (2)$. Let $f : M \to E_\tau(X)$ and $L = f^{-1}(X)$. We claim that $L \in \mathcal{D}_\tau(M)$ and so the restriction $f|_L : L \to X$ extends to a homomorphism $g : M \to X$. To prove our claim we consider the following isomorphism

$$\frac{M}{L} = \frac{M}{f^{-1}(X)} \cong \frac{f(M) + X}{X} \leqslant \frac{E_\tau(X)}{X} \in \mathcal{T},$$

given by $m + f^{-1}(X) \mapsto f(m) + X$ for all $m \in M$. To finish the proof it suffices to show that $(f - g)(M) = 0$, since then $f(M) = g(M) \leqslant X$. As $X \leqslant_{\tau\text{-e}} E_\tau(X)$, we need only to show that $X \cap (f - g)(M) = 0$. Given $x \in X \cap (f - g)(M)$ there is an $m \in M$ with $x = (f - g)(m)$ and so $f(m) = x + g(m) \in X$. Thus $m \in f^{-1}(X) = L$ and so $f(m) = g(m)$ and hence $x = 0$.

$(2) \Rightarrow (3)$. For any $f : E_\tau(M) \to E_\tau(X)$ we have $f(M) = f|_M(M) \leqslant X$ (by (2)).

$(3) \Rightarrow (1)$. Let $N \in \mathcal{D}_\tau(M)$ and $f : N \to X$ be a homomorphism. We can consider f as a homomorphism from N to $E_\tau(X)$ which extends to a homomorphism $g : E_\tau(M) \to E_\tau(X)$ since $N \leqslant_{\tau\text{-d}} M \leqslant_{\tau\text{-e}} E_\tau(M)$ and $E_\tau(X)$ is τ-injective. By (3), $g(M) \leqslant X$ and so the restriction of g to M is a homomorphism from M to X which extends f. □

As a corollary we get a torsion-theoretic analogue of a well-known characterization of quasi-injectivity (see [MM, 1.15]). This has also been shown directly in Crivei [C01, 2.1].

Corollary 3.11. *A module M is τ-quasi-injective if and only if it is invariant under all endomorphisms of its τ-injective hull, i.e., $f(M) \leqslant M$ for any $f \in \operatorname{End}_R(E_\tau(M))$.*

We use the following definition to prove the existence of τ-M-injective hulls.

Definition 3.2. *For modules M and X let $\Delta_{M,X} = \operatorname{Hom}_R(E_\tau(M), E_\tau(X))$. For any $K \leqslant E_\tau(M)$, let $\Delta_{M,X}K = \{ \sum_{i=1}^n f_i(k_i) \mid f_i \in \Delta_{M,X}, \; k_i \in K \; \text{for} \; i = 1, \ldots, n; \; n \in \mathbb{N} \}$.*

It is clear from the above definition that $\Delta_{M,X}K$ is a submodule of $E_\tau(X)$.

Recall that, given modules X and M, we say that X is M-*generated* if there is an epimorphism $f : M^{(I)} \to X$ for some index set I, where $M^{(I)} = \bigoplus_I M$.

Proposition 3.12. *The following statements are true for modules X and M:*

(1) *X is τ-M-injective if and only if $\Delta_{M,X}M \leqslant X$.*

(2) *$\Delta_{M,X}M$ is τ-M-injective.*

(3) *If X is M-generated then $X \leqslant \Delta_{M,X}M$ and $\Delta_{M,X}M$ is the intersection of all the τ-M-injective submodules of $E_\tau(X)$ containing X.*

Proof. (1) This is clear from Proposition 3.10 and the definition of $\Delta_{M,X}M$.

(2) To simplify our notation let $Q = \Delta_{M,X}M$. By (1), it suffices to show that $\Delta_{M,Q}M \leqslant Q$. Note that by definition $Q \leqslant E_\tau(X)$ and so, by Proposition 3.9, $E_\tau(Q) \leqslant E_\tau(X)$. Thus $\Delta_{M,Q} \subseteq \Delta_{M,X}$ and so $\Delta_{M,Q}M \leqslant \Delta_{M,X}M = Q$ as required.

(3) We first show that if X is M-generated, then $X \leqslant \Delta_{M,X}M$. Let $f : M^{(I)} \to X$ be an epimorphism, for some index set I, let $i_X : X \to E_\tau(X)$ and $i_M : M \to E_\tau(M)$ be the inclusion mappings, and let $\varphi_i : M \to M^{(I)}$ be the ith canonical injection, for any $i \in I$. Then, since $M \leqslant_{\tau\text{-d}} E_\tau(M)$, for each $i \in I$ there is a $g_i : E_\tau(M) \to E_\tau(X)$ such that $g_i i_M = i_X f \varphi_i$, and thus for any $m \in M$ we have $g_i(m) = f\varphi_i(m)$.

   Given $x \in X$ there is a $y = (m_i)_{i \in I} \in M^{(I)}$ such that $x = f(y)$. We may write $y = \sum_{j=1}^t \varphi_{i_j}(m_{i_j})$ where $i_1, \ldots, i_t \in I$ and $\varphi_{i_j} : M \to M^{(I)}$ is the i_jth injection. Then $x = f(\sum_{j=1}^t \varphi_{i_j}(m_{i_j})) = \sum_{j=1}^t f\varphi_{i_j}(m_{i_j}) = \sum_{j=1}^t g_{i_j}(m_{i_j}) \in \Delta_{M,X}M$, as required.

   Now let $\mathcal{B} = \{B \leqslant E_\tau(X) \,|\, X \leqslant B,\ \Delta_{M,B}M \leqslant B\}$. Note that, by (1), $\Delta_{M,B}M \leqslant B$ simply means that B is τ-M-injective. To prove our claim it suffices to show that $\bigcap \mathcal{B} = \bigcap_{B \in \mathcal{B}} B = \Delta_{M,X}M$. By (2), $\Delta_{M,X}M$ is τ-M-injective and so, as it contains X, it belongs to $\mathcal{B}$ and so $\bigcap \mathcal{B} \leqslant \Delta_{M,X}M$. Take an arbitrary $B \in \mathcal{B}$. To finish it suffices to show that $\Delta_{M,X}M \leqslant B$. Note that $X \leqslant B \leqslant E_\tau(X)$ and, as $X \leqslant_{\tau\text{-e}} E_\tau(X)$, we have $E_\tau(B) = E_\tau(X)$. Thus $\Delta_{M,B} = \Delta_{M,X}$ and so $\Delta_{M,X}M = \Delta_{M,B}M \leqslant B$, as required. $\square$

   The above result allows us to define the τ-M-injective hull of a module X and show that this exists and is unique up to isomorphism when X is M-generated.

Definition 3.3. *Let X, M and E be modules. Then E is said to be a τ-M-injective hull of X, and denoted by $E_{\tau\text{-}M}(X)$, if there is a monomorphism $f : X \to E$ such that:*

(i) *$f(X) \leqslant_{\tau\text{-e}} E$,*

(ii) *E is a minimal τ-M-injective extension of $f(X)$, i.e., $f(X) \leqslant E$, E is τ-M-injective, and if Q is τ-M-injective such that $f(X) \leqslant Q \leqslant E$ then $Q = E$.*

Proposition 3.13. *Let M be a module. Then every M-generated module X has a τ-M-injective hull, namely $\Delta_{M,X}M$, and this is unique up to isomorphism.*

Proof. For existence, set $E_{\tau\text{-}M}(X) = \Delta_{M,X}M$ and let $\varphi : X \to E_{\tau\text{-}M}(X)$ be the canonical monomorphism given by Proposition 3.12 (3). Then, by the definition of $\Delta_{M,X}M$, we have $\varphi(X) = X \leqslant_{\tau\text{-e}} E_{\tau\text{-}M}(X) \leqslant_{\tau\text{-e}} E_\tau(X)$. Moreover, by

Proposition 3.12 (3), $E_{\tau\text{-}M}(X)$ is the smallest τ-M-injective submodule of $E_\tau(X)$ containing $\varphi(X)$, and so minimal.

For uniqueness, suppose that A is also a τ-M-injective hull of X, and that $f : X \to A$ is the associated monomorphism. We will construct an isomorphism $\psi : E_{\tau\text{-}M}(X) \to A$. The following diagram illustrates the rest of the proof:

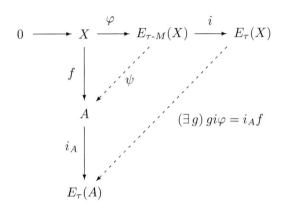

Here i and i_A are the inclusion maps. We can extend f to a monomorphism $i_A f : X \to E_\tau(A)$, and φ to a monomorphism $i\varphi : X \to E_\tau(X)$. As $E_\tau(A)$ is τ-injective, there is a homomorphism $g : E_\tau(X) \to E_\tau(A)$ such that $gi\varphi = i_A f$. We claim that g is an isomorphism. Our goal is to show that $g(E_{\tau\text{-}M}(X)) = A$ and thus taking $\psi = g|_{E_{\tau\text{-}M}(X)}$ we get an isomorphism $\psi : E_{\tau\text{-}M}(X) \to A$.

First note that g is a monomorphism since $gi\varphi = i_A f$ is a monomorphism and $i\varphi$ is an essential monomorphism. Before we show that g is onto we make and prove a few more claims.

We simplify our notation by taking $E_M = g(E_{\tau\text{-}M}(X))$ and $E = g(E_\tau(X))$. By the minimality of A as a τ-M-injective τ-essential extension of $f(X)$ it suffices to show that:

(1) E_M is τ-M-injective,
(2) $f(X) \leqslant E_M$,
(3) $E_M \leqslant A$ or, equivalently, $A \cap E_M = E_M$.

(1) This is obvious since $E_M = g(E_{\tau\text{-}M}(X)) \cong E_{\tau\text{-}M}(X)$, a τ-M-injective module.

(2) We have $f(X) = i_A f(X) = gi\varphi(X) = g(\varphi(X)) \leqslant g(E_{\tau\text{-}M}(X)) = E_M \leqslant g(E_\tau(X)) = E \leqslant E_\tau(A)$. That is $f(X) \leqslant E_M \leqslant E \leqslant E_\tau(A)$.

(3) Using the above observation and the fact that $f(X) \leqslant_{\tau\text{-e}} A \leqslant_{\tau\text{-e}} E_\tau(A)$ we get $f(X) \leqslant_{\tau\text{-e}} E_M \leqslant_{\tau\text{-e}} E \leqslant_{\tau\text{-e}} E_\tau(A)$. Hence $E_\tau(E_M) = E_\tau(E) = E_\tau(A)$. But $E = g(E_\tau(X)) \cong E_\tau(X)$ and so E is τ-injective. Then, by Proposition 3.6, E is τ-essentially closed in $E_\tau(A)$ and so $g(E_\tau(X)) = E_\tau(A)$. Hence g is onto and so an isomorphism.

Similarly $f(X) \leqslant_{\tau\text{-e}} A \cap E_M \leqslant_{\tau\text{-e}} E_\tau(A)$ and so $E_\tau(A \cap E_M) = E_\tau(A) = E$. These observations give $\Delta_{M,E_M} = \Delta_{M,A} = \Delta_{M,A\cap E_M}$ and so $\Delta_{M,A\cap E_M}M = \Delta_{M,A}M \cap \Delta_{M,E_M}M \leqslant A \cap E_M$. As a result, $A \cap E_M$ and its isomorphic copy $g^{-1}(A \cap E_M)$ are τ-M-injective. Now $g(\varphi(X)) = gi\varphi(X) = i_A f(X) \leqslant A$. As already seen, $g(\varphi(X)) \leqslant g(E_{\tau\text{-}M}(X)) = E_M$. Thus $g(\varphi(X)) \leqslant A \cap E_M$ and so $\varphi(X) \leqslant g^{-1}(A \cap E_M) \leqslant g^{-1}(E_M) = g^{-1}g(E_{\tau\text{-}M}(X)) = E_{\tau\text{-}M}(X)$. Hence, by the minimality of $E_{\tau\text{-}M}(X)$ as a τ-M-injective submodule of $E_\tau(X)$ containing $\varphi(X)$, we have $g^{-1}(A \cap E_M) = g^{-1}(E_M)$. Thus $A \cap E_M = E_M$, as required.   □

If E is a τ-M-injective hull of X, we will write $E = E_{\tau\text{-}M}(X)$. The following remark gives a useful property of τ-M-injective hulls.

Remark 3.2. Every τ-M-injective τ-essential extension of a module X contains a τ-M-injective hull of X. That is, if A is τ-M-injective and $X \leqslant_{\tau\text{-e}} A$ then $E_{\tau\text{-}M}(X) \leqslant A$.

Proof. Since $X \leqslant_{\tau\text{-e}} A \leqslant_{\tau\text{-e}} E_\tau(A)$ we have $E_\tau(A) = E_\tau(X)$ and so $E_{\tau\text{-}M}(X) = \Delta_{M,X}M = \Delta_{M,A}M \leqslant A$.   □

We now show that the τ-M-injective hull generalizes the concept of a τ-injective hull.

Remark 3.3. For any R-module M we have $E_\tau(M) = E_{\tau\text{-}R}(M)$.

Proof. By Propositions 3.13 and 3.12 (3), $M \leqslant_{\tau\text{-e}} E_{\tau\text{-}R}(M) = \Delta_{R,M}R \leqslant E_\tau(M)$. But, by Remark 2.3 (2), $E_{\tau\text{-}R}(M)$ is τ-injective since it is τ-R-injective. Thus, by Proposition 3.8, $E_\tau(M) = E_{\tau\text{-}R}(M)$, as required.   □

The τ-M-injective hull generalizes the concept of an M-injective hull found in [W, 17.8]. To see this, we first recall a few definitions from [W, p. 118 and p. 141].

For any module M, let $\sigma[M]$ denote as usual the Wisbauer category of M, i.e., the full subcategory of R-Mod consisting of submodules of M-generated modules.

Definition 3.4. *Let $M \in R$-Mod and E, X be modules in $\sigma[M]$. Then E is called an M-injective hull of X (or an injective hull of X in $\sigma[M]$) if E is A-injective for every module $A \in \sigma[M]$ and there is a monomorphism $\varphi : X \to E$ such that $\varphi(X) \leqslant_e E$. We denote an M-injective hull of X by $E_M(X)$.*

The following definition can be found in [W, 13.4, 13.5].

Definition 3.5. *Given R-modules U and N, the trace of U in N is defined as*

$$\mathrm{Tr}(U, N) = \left\{ \sum_{i=1}^{t} f_i(u_i) \mid f_i \in \mathrm{Hom}_R(U, N),\ u_i \in U \ \text{for } i = 1, \ldots, t;\ t \in \mathbb{N} \right\}.$$

We now state a result from [W, 17.9] on the existence of M-injective hulls.

Proposition 3.14. *Let M be an R-module. Then the following are true.*

(1) *Every module X in $\sigma[M]$ has an injective hull $E_M(X)$ in $\sigma[M]$.*
(2) *If $X \in \sigma[M]$ then $E_M(X) \cong \mathrm{Tr}(M, E(X))$ where $E(X)$ is any injective hull of X in R-Mod.*
(3) *The injective hull of a module in $\sigma[M]$ is unique up to isomorphism.*

Remark 3.4. For any modules M and X we have $\mathrm{Tr}(M, E_\tau(X)) = \Delta_{M,X} M = E_{\tau\text{-}M}(X)$.

Proof. Since $E_\tau(X)$ is τ-injective, any $f \in \mathrm{Hom}_R(M, E_\tau(X))$ can be extended to an $\bar{f} \in \mathrm{Hom}_R(E_\tau(M), E_\tau(X))$ and so $\mathrm{Tr}(M, E_\tau(X)) \leqslant \Delta_{M,X} M$. Conversely, for any $f \in \mathrm{Hom}_R(E_\tau(M), E_\tau(X))$, $f|_M \in \mathrm{Hom}_R(M, E_\tau(X))$ and so $\Delta_{M,X} M \leqslant \mathrm{Tr}(M, E_\tau(X))$. $\qquad\square$

Note that when all modules are τ-torsion, we have $E_\tau(X) = E(X)$ and so, by the two preceding results, $E_M(X) \cong \mathrm{Tr}(M, E(X)) = \mathrm{Tr}(M, E_\tau(X)) = \Delta_{M,X} M = E_{\tau\text{-}M}(X)$.

The following definition is motivated by [B, 5.1.7].

Definition 3.6. *For any module M let $\Delta_M = \mathrm{End}_R(E_\tau(M)) = \Delta_{M,M}$. For any $X \leqslant E_\tau(M)$ let $\Delta_M X = \{ \sum_{i=1}^n f_i(x_i) \mid f_i \in \Delta_M, x_i \in X \text{ for } i = 1, \ldots, n; n \in \mathbb{N} \}$.*

Clearly if $X \leqslant E_\tau(M)$ then $X \leqslant \Delta_M X \leqslant E_\tau(M)$ and it is routine to show that

$$\Delta_M(\Delta_M X) = \Delta_M X. \tag{$\star$}$$

The following proposition appears in [B, 5.1.7]. We note that there is a flaw in the proof, namely that $M/(M \cap f^{-1}(M)) \cong (M + f^{-1}(M))/M$. This is remedied by using the isomorphism $M/(M \cap f^{-1}(M)) \cong (M + f(M))/M$ instead. We provide a simpler alternative below, using Proposition 3.12.

Proposition 3.15. *The following statements are true for any module M:*

(1) *M is τ-quasi-injective if and only if $\Delta_M M = M$.*
(2) *$\Delta_M M$ is τ-quasi-injective.*
(3) *$\Delta_M M$ is the intersection of all the τ-quasi-injective submodules of $E_\tau(M)$ containing M.*

Proof. (1) We always have $M \leqslant \Delta_M M$. By Proposition 3.12, M is τ-quasi-injective if and only if $\Delta_{M,M} M \leqslant M$ if and only if $\Delta_M M = M$.

(2) Let $Q = \Delta_M M$. Then by (1) it suffices to show that $\Delta_Q Q = Q$. Note that, since $M \leqslant Q \leqslant E_\tau(M)$ and $M \leqslant_{\tau\text{-e}} E_\tau(M)$, it follows from Proposition 1.1 (6) that $M \leqslant_{\tau\text{-e}} Q \leqslant_{\tau\text{-e}} E_\tau(M)$. Hence, by Proposition 3.8, $E_\tau(Q) = E_\tau(M)$ and so $\Delta_Q = \Delta_M$. Therefore, using $(\star)$, we get $\Delta_Q Q = \Delta_M Q = \Delta_M(\Delta_M M) = \Delta_M M = Q$, as required.

(3) Let $\mathcal{B} = \{ B \leqslant E_\tau(M) \mid M \leqslant B, \Delta_B B = B \}$. It suffices to show that $\bigcap \mathcal{B} = \bigcap_{B \in \mathcal{B}} B = \Delta_M M$. By (2), $\Delta_M M$ is τ-quasi-injective and, as it contains M, belongs to $\mathcal{B}$. Thus $\bigcap \mathcal{B} \leqslant \Delta_M M$. Take an arbitrary $B \in \mathcal{B}$. To finish it suffices to show that $\Delta_M M \leqslant B$. Note that $M \leqslant B \leqslant E_\tau(M)$, and as $M \leqslant_{\tau\text{-e}} E_\tau(M)$, it follows that $E_\tau(B) = E_\tau(M)$. Thus $\Delta_B = \Delta_M$ and so $\Delta_M M \leqslant \Delta_M B = \Delta_B B = B$, as required. $\qquad\square$

Since $E_{\tau\text{-}M}(M) = \Delta_{M,M}M = \Delta_M M$, we have the following useful corollary.

Corollary 3.16. $E_{\tau\text{-}M}(M)$ *is* τ*-quasi-injective.*

Definition 3.7. *Let M and Q be modules. Then Q is said to be a τ-quasi-injective hull of M if there is a monomorphism $f : M \to Q$ such that:*

(i) $f(M) \leqslant_{\tau\text{-}e} Q$,

(ii) Q *is a minimal τ-quasi-injective extension of $f(M)$, i.e., $f(M) \leqslant Q$, Q is τ-quasi-injective, and if E is τ-quasi-injective such that $f(M) \leqslant E \leqslant Q$ then $E = Q$.*

We denote a τ-quasi-injective hull of M by $Q_\tau(M)$.

We use the next result to show that $Q_\tau(M)$ exists and is unique up to isomorphism.

Lemma 3.1. *If Q is a τ-quasi-injective τ-essential extension of M then:*

(1) Q *is τ-M-injective,*

(2) $E_{\tau\text{-}M}(M) \leqslant Q$.

Proof. (1) We have $M \leqslant_{\tau\text{-}e} Q \leqslant_{\tau\text{-}e} E_\tau(Q)$ and so $E_\tau(Q) = E_\tau(M)$. Hence $\Delta_{M,Q}M \leqslant \Delta_{M,Q}Q = \Delta_{Q,Q}Q = \Delta_Q Q = Q$. Therefore, by Proposition 3.12, Q is τ-M-injective.

(2) This follows from Remark 3.2.   □

We now give a simple proof of [B, 5.1.8] on the existence and uniqueness of $Q_\tau(M)$.

Proposition 3.17. *Every module M has a τ-quasi-injective hull which is unique up to isomorphism. In fact $Q_\tau(M) = E_{\tau\text{-}M}(M)$.*

Proof. For existence, note that $E_{\tau\text{-}M}(M) = \Delta_{M,M}M = \Delta_M M$ and so, by Proposition 3.15 (3), $E_{\tau\text{-}M}(M)$ is a τ-quasi-injective hull for M.

Next suppose that Q is also a τ-quasi-injective hull of M. Then $E_{\tau\text{-}M}(M) \leqslant Q$ by Lemma 3.1. But, by Corollary 3.16, $E_{\tau\text{-}M}(M)$ is τ-quasi-injective and so the minimality of Q gives $Q = E_{\tau\text{-}M}(M)$. As $E_{\tau\text{-}M}(M)$ is unique up to isomorphism so is $Q_\tau(M)$.   □

We now provide the short proof promised earlier for Lemma 2.2. Before we proceed, note that, by Proposition 3.12 (1) and Remark 3.4, X is τ-M-injective if and only if $\mathrm{Tr}(M, E_\tau(X)) \leqslant X$.

Lemma 3.2 (Generalized Azumaya's Lemma). *If X and $M = \bigoplus_I M_\beta$ are R-modules, then X is τ-M-injective if and only if X is τ-M_α-injective for each $\alpha \in I$.*

Proof. ($\Rightarrow$) Let $\alpha \in I$ and $f : M_\alpha \to E_\tau(X)$. If $\varphi_\alpha : M_\alpha \to M$ and $\pi_\alpha : M \to M_\alpha$ denote the canonical injection and projection respectively, then we have $f(M_\alpha) = f\pi_\alpha\varphi_\alpha(M_\alpha) = f\pi_\alpha(\varphi_\alpha(M_\alpha)) \leqslant f\pi_\alpha(M) \leqslant X$, since $f\pi_\alpha : M \to E_\tau(X)$ and X is τ-M-injective. Hence $\mathrm{Tr}(M_\alpha, E_\tau(X)) \leqslant X$ and so X is τ-M_α-injective.

($\Leftarrow$) Let $f : M \rightarrow E_\tau(X)$ and $y \in M$. Then $y = \sum_{j=1}^t m_{\alpha_j}$ where $m_{\alpha_j} \in M_{\alpha_j}$, $\alpha_j \in I$ and $t \in \mathbb{N}$. Thus, with φ_α and π_α as above, $f(y) = \sum_{j=1}^t f(m_{\alpha_j}) = \sum_{j=1}^t f\varphi_{\alpha_j}\pi_{\alpha_j}(y) \in X$, since $f\varphi_{\alpha_j} : M_{\alpha_j} \rightarrow E_\tau(X)$ and X is τ-M_α-injective for each α. Thus $\mathrm{Tr}(M, E_\tau(X)) \leqslant X$ and so X is τ-M-injective. $\qquad\square$

4. Relative Σ-τ-injective modules

In this section we seek conditions for a direct sum of copies of a module E to be (s-)τ-M-injective. As a result we derive torsion-theoretic analogues of results by Faith, Albu and Năstăsescu, and Cailleau.

Definition 4.1. *If P is a property of a module M then M is said to be Σ-P if $M^{(A)}$ is (has) P for any index set A, i.e., the direct sum of $|A|$ copies of M is (has) P. On the other hand, if $M^{(A)}$ is P for any countable index set A, we say that M is countably Σ-P.*

To simplify our notation in this section, given a subset X of R, we let $^\perp X$ denote the left annihilator of X in R, and $X^\perp$ the right annihilator of X in R. In other words $^\perp X = \{\, r \in R \,|\, rx = 0, \forall x \in X \,\}$, and $X^\perp = \{\, r \in R \,|\, xr = 0, \forall x \in X \,\}$.

We will also use the following notation to distinguish between left and right annihilators as well as the ring or module from which the elements of the annihilator come.

Definition 4.2. *Let X be a nonempty subset of a ring R and Y be a nonempty subset of a left R-module M. We let*

$$l_R(Y) = {}^{\perp_R}Y = \{\, r \in R \,|\, rY = 0 \,\}, \quad r_M(X) = X^{\perp_M} = \{\, m \in M \,|\, Xm = 0 \,\}.$$

When the context is clear we will simply write $^\perp Y$ and $X^\perp$. Recall that:

$$\Omega(M) = \{\, I \leqslant R \,|\, (\exists m \in M)\,(0 : m) \leqslant I \,\},$$
$$\bar{\Omega}(M) = \{\, I \leqslant R \,|\, (\exists \text{ finite } Y \subseteq M)\,(0 : Y) \leqslant I \,\}.$$

If E is also a left R-module then we follow [AN, p. 118] and use the notation:

$$\mathcal{A}_M(E) = \{\, l_R(D) \,|\, \emptyset \neq D \subseteq E, (\exists m \in M)\, l_R(m) \leqslant l_R(D) \,\}$$
$$= \{\, l_R(D) \in \Omega(M) \,|\, \emptyset \neq D \subseteq E \,\},$$
$$\bar{\mathcal{A}}_M(E) = \{\, l_R(D) \,|\, \emptyset \neq D \subseteq E, (\exists \text{ finite } Y \subseteq M)\, l_R(Y) \leqslant l_R(D) \,\}$$
$$= \{\, l_R(D) \in \bar{\Omega}(M) \,|\, \emptyset \neq D \subseteq E \,\}.$$

In particular, if $M = {}_R R$ then for any $\emptyset \neq D \subseteq E$ we have $l_R(1) = \{0\} \subseteq l_R(D)$ and so

$$\mathcal{A}_R(E) = \{\, l_R(D) \,|\, \emptyset \neq D \subseteq E \,\}.$$

The next lemma allows us to construct a *properly* descending chain from a *properly* ascending chain of left annihilators.

Lemma 4.1. *If I and J are left ideals of R which are left annihilators with $I \subset J$ then $J^\perp \subset I^\perp$.*

Proof. Obviously $J^\perp \subseteq I^\perp$. If $J^\perp = I^\perp$ then we would have $^\perp(J^\perp) = {}^\perp(I^\perp)$ and so, by the double annihilator condition (see [Fa, p. 65]), $J = I$ which contradicts our assumption. Therefore $J^\perp \neq I^\perp$ and thus $J^\perp \subset I^\perp$. $\qquad\square$

Before we state the next lemma we need the following definition.

Definition 4.3. *Let $\{\, M_\alpha \mid \alpha \in A \,\}$ be a family of modules and $M = \bigoplus_{\alpha \in A} M_\alpha$. For any $x = (x_\alpha)_{\alpha \in A} \in M$ we define the* support *of x to be the set $\{\, \alpha \in A \mid x_\alpha \neq 0 \,\}$, and denote it by $\operatorname{supp}(x)$. For any $X \subseteq M$ we define the* support *of X to be the set $\bigcup_{x \in X} \operatorname{supp}(x) = \{\, \alpha \in A \mid (\exists\, x \in X)\ x_\alpha \neq 0 \,\}$, and denote it by $\operatorname{supp}(X)$.*

The following lemma generalizes results found in [G, 8.13] and [AN, 10.2].

Lemma 4.2. *Let $\{\, E_\alpha \mid \alpha \in A \,\}$ be an infinite family of s-τ-M-injective modules. If $\bigoplus_{\alpha \in C} E_\alpha$ is s-τ-M-injective for any countable subset C of A, then $\bigoplus_{\alpha \in A} E_\alpha$ is s-τ-M-injective.*

Proof. Let $\pi_\beta : \bigoplus_{\alpha \in A} E_\alpha \to E_\beta$ be the natural projection. Note that, given $I \in \mathcal{D}_\tau(R)$ and a homomorphism $f : I \to \bigoplus_{\alpha \in A} E_\alpha$ with $\operatorname{Ker} f \in \Omega(M)$ for which $\operatorname{supp}(\operatorname{Im}(f))$ is finite, then f is a map into $\bigoplus_{\alpha \in F} E_\alpha$, where F is a finite subset of A. Also, by Proposition 2.4, $\bigoplus_{\alpha \in F} E_\alpha$ is s-τ-M-injective and so, by Proposition 2.3, f extends to a map from R to $\bigoplus_{\alpha \in F} E_\alpha$. Hence, if we suppose to the contrary that $\bigoplus_{\alpha \in A} E_\alpha$ is not s-τ-M-injective, then there is a τ-dense left ideal I of R and a map $f : I \to \bigoplus_{\alpha \in A} E_\alpha$ for which $\operatorname{Ker} f \in \Omega(M)$, such that $\operatorname{supp}(\operatorname{Im}(f))$ is not finite. Then there is a countably infinite subset C of A such that for any $\alpha \in C$, $\pi_\alpha(\operatorname{Im}(f)) \neq 0$, i.e., $C \subseteq \operatorname{supp}(\operatorname{Im}(f))$. Define $\pi_C : \bigoplus_{\alpha \in A} E_\alpha \to \bigoplus_{\alpha \in C} E_\alpha$ as follows: for $x \in \bigoplus_{\alpha \in A} E_\alpha$, let $\pi_C(x) = x_C$ where

$$\pi_\alpha(x_C) = \pi_\alpha(\pi_C(x)) = \begin{cases} \pi_\alpha(x), & \text{if } \alpha \in C \\ 0, & \text{otherwise.} \end{cases}$$

We claim that $C = \operatorname{supp}(\operatorname{Im}(\pi_C f))$. First let $\alpha \in \operatorname{supp}(\operatorname{Im}(\pi_C f))$. Then there exists an $r \in I$ such that $\pi_\alpha(\pi_C f(r)) \neq 0$. But this implies that $\alpha \in C$, by our definition of π_C. Therefore we have $\operatorname{supp}(\operatorname{Im}(\pi_C f)) \subseteq C$.

For the reverse containment, let $\alpha \in C$. Then there is an $r \in I$ such that $\pi_\alpha(f(r)) \neq 0$. Hence we have $\pi_\alpha(\pi_C f(r)) = \pi_\alpha(f(r)) \neq 0$. Therefore $\alpha \in \operatorname{supp}(\operatorname{Im}(\pi_C f))$, as required.

Consider the map $\pi_C f : I \to \bigoplus_{\alpha \in C} E_\alpha$. By assumption $\bigoplus_{\alpha \in C} E_\alpha$ is s-τ-M-injective. Furthermore, since $\operatorname{Ker} f \in \Omega(M)$, there is an $m \in M$ such that $(0 : m) \leqslant \operatorname{Ker} f \leqslant \operatorname{Ker} \pi_C f$ and so $\operatorname{Ker} \pi_C f \in \Omega(M)$. Thus, by Proposition 2.3 (3), there exists a $y \in \bigoplus_{\alpha \in C} E_\alpha$ such that $\pi_C f(r) = ry$ for all $r \in I$ and thus $C = \operatorname{supp}(\operatorname{Im}(\pi_C f)) \subseteq \operatorname{supp}(y)$. This is a contradiction since $\operatorname{supp}(y)$ is finite, whereas C is countably infinite. $\qquad\square$

Remark 2.3(2) now gives the following, found in [G, 8.13], as a corollary.

Corollary 4.1. *Let* $\{\,E_\alpha \,|\, \alpha \in A\,\}$ *be an* infinite *family of τ-injective modules. If* $\bigoplus_{\alpha \in C} E_\alpha$ *is τ-injective for any countable subset C of A, then* $\bigoplus_{\alpha \in A} E_\alpha$ *is τ-injective.*

With a little more work one gets the following.

Corollary 4.2. *Let* $\{\,E_\alpha \,|\, \alpha \in A\,\}$ *be an* infinite *family of s-τ-quasi-injective modules. If* $\bigoplus_{\alpha \in C} E_\alpha$ *is s-τ-quasi-injective for any countable subset C of A, then* $\bigoplus_{\alpha \in A} E_\alpha$ *is s-τ-quasi-injective.*

Proof. Since $\bigoplus_{\alpha \in A} E_\alpha$ is s-τ-quasi-injective if and only if $\bigoplus_{\alpha \in A} E_\alpha$ is s-τ-$\bigoplus_{\alpha \in A} E_\alpha$-injective, by Lemma 4.2 it suffices to show that $\bigoplus_{\alpha \in C} E_\alpha$ is s-τ-$\bigoplus_{\alpha \in A} E_\alpha$-injective for any countable subset C of A. Using the Generalized Fuchs Criterion, we need only show that, given $I \in \mathcal{D}_\tau(R)$, every homomorphism $f : I \to \bigoplus_{\alpha \in C} E_\alpha$ with $\operatorname{Ker} f \in \Omega(\bigoplus_{\alpha \in A} E_\alpha)$ extends to a homomorphism $g : R \to \bigoplus_{\alpha \in C} E_\alpha$. Given any $m \in \bigoplus_{\alpha \in A} E_\alpha$, there is a finite subset F of A such that $m \in \bigoplus_{\alpha \in F} E_\alpha$. Therefore, if $(0 : m) \leqslant \operatorname{Ker} f$, we have $\operatorname{Ker} f \in \Omega(\bigoplus_{\alpha \in F} E_\alpha)$. Thus it remains to show that $\bigoplus_{\alpha \in C} E_\alpha$ is s-τ-$\bigoplus_{\alpha \in F} E_\alpha$-injective.

We prove the slightly more general statement that, for any countable subsets C_1 and C_2 of A, $\bigoplus_{\alpha \in C_1} E_\alpha$ is s-τ-$\bigoplus_{\alpha \in C_2} E_\alpha$-injective. Since $C_1 \cup C_2$ is countable, by assumption $\bigoplus_{\alpha \in C_1 \cup C_2} E_\alpha$ is s-τ-$\bigoplus_{\alpha \in C_1 \cup C_2} E_\alpha$-injective. Hence, by Proposition 2.4, $\bigoplus_{\alpha \in C_1} E_\alpha$ is s-τ-$\bigoplus_{\alpha \in C_1 \cup C_2} E_\alpha$-injective and so, by Remark 2.2, $\bigoplus_{\alpha \in C_1} E_\alpha$ is s-τ-$\bigoplus_{\alpha \in C_2} E_\alpha$-injective. $\square$

Remark 4.1. Let M be a left R module. Then we have:
(1) M is Σ-τ-injective if and only if M is Σ-(s)-τ-$_R R$-injective.
(2) If M is Σ-s-τ-quasi-injective then M is Σ-s-τ-M-injective.
(3) M is Σ-τ-quasi-injective if and only if M is Σ-τ-M-injective.

Proof. (1) By Remark 2.3, M is Σ-τ-injective if and only if $M^{(I)}$ is (s)-τ-$_R R$-injective for any index set I if and only if M is Σ-(s)-τ-$_R R$-injective.

(2) If M is Σ-s-τ-quasi-injective then, by Remark 2.3, $M^{(I)}$ is s-τ-$M^{(I)}$-injective for any index set I. Thus, by Remark 2.2, $M^{(I)}$ is s-τ-M-injective for any I and hence M is Σ-s-τ-M-injective.

(3) By Lemma 2.2, M is Σ-τ-quasi-injective if and only if $M^{(I)}$ is τ-$M^{(I)}$-injective for any I if and only if $M^{(I)}$ is τ-M-injective for any I if and only if M is Σ-τ-M-injective. $\square$

Taking $E_\alpha = E$ for every $\alpha \in A$ in Lemma 4.2 and Corollary 4.2 give the following corollaries.

Corollary 4.3. *If a module E is countably Σ-s-τ-M-injective then it is Σ-s-τ-M-injective.*

Corollary 4.4. *Every countably Σ-s-τ-quasi-injective module is Σ-s-τ-quasi-injective.*

In order to state the next two theorems we need the following definition.

Definition 4.4. *Let M be an R-module. A left ideal I of R is called an M-annihilator if there is an $N \subseteq M$ such that $I = (0 : N)$, i.e., I is the annihilator of a subset of M.*

The motivation for our main result in this section, Theorem 4.2, comes from a theorem by Albu and Năstăsescu, [AN, 10.4], which we state below. (Taking $M = R$ gives a well-known theorem of Faith, [AF, 25.1].)

Theorem 4.1. *The following statements are equivalent for an M-injective module E:*

(1) E *is* Σ-M*-injective.*
(2) *The set* $\bar{\mathcal{A}}_M(E)$ *of left ideals of R satisfies the ACC.*
(3) $\mathcal{A}_M(E)$ *satisfies the ACC.*

We now state the main result of this section which characterizes Σ-s-τ-M-injective modules, and generalizes the above theorem.

Theorem 4.2. *Let E be an s-τ-M-injective module. Then the following statements are equivalent.*

(1) E *is countably* Σ-s-τ-M*-injective, i.e.,* $E^{(\mathbb{N})}$ *is s-τ-M-injective.*
(2) *If* $I_1 \subseteq I_2 \subseteq \cdots$ *is an ascending chain in* $\mathcal{A}_M(E)$ *such that* $I_\infty = \bigcup_{i=1}^\infty I_i$ *is τ-dense in R then there is a positive integer n such that* $I_n = I_{n+k}$ *for all* $k \in \mathbb{N}$.
(3) *The following two conditions hold:*
    (a) $\mathcal{A}_M(E) \cap \mathcal{D}_\tau(R)$ *satisfies the ACC.*
    (b) *If* $I_1 \subseteq I_2 \subseteq \cdots$ *is an ascending chain in* $\mathcal{A}_M(E)$ *such that* $I_\infty = \bigcup_{i=1}^\infty I_i$ *is τ-dense in R, then I_n is τ-dense in R for some $n \in \mathbb{N}$.*
(4) E *is* Σ-s-τ-M*-injective, i.e.,* $E^{(A)}$ *is s-τ-M-injective for any set A.*

Proof. $(1) \Rightarrow (2)$. Assume to the contrary that (2) does not hold. Then there exist left ideals $I_1, I_2, \ldots$ in $\mathcal{A}_M(E)$ and an $m \in M$ such that

$$(0 : m) \subseteq I_1 \subset I_2 \subset \cdots$$

and $I_\infty = \bigcup_{i=1}^\infty I_i$ is τ-dense in R. By Lemma 4.1, this yields the strictly descending chain $(0 : m)^{\perp_E} \supseteq I_1^{\perp_E} \supset I_2^{\perp_E} \supset \cdots$. For every $n \in \mathbb{N}$ choose an $x_n \in I_n^{\perp_E} \setminus I_{n+1}^{\perp_E}$. Moreover, for a fixed $a \in I_\infty$, let n be the smallest positive integer such that $a \in I_n$. Then, for every $k \geq 0$, $a \in I_n \subset I_{n+k}$ and $x_{n+k} \in I_{n+k}^{\perp_E}$. Hence $ax_{n+k} = 0$. This ensures that the map $f : I_\infty \to E^{(\mathbb{N})}$ sending a to $(ax_1, ax_2, \ldots)$ is a well-defined homomorphism.

Moreover, since $\{x_1, x_2, \ldots\} \subseteq (0 : m)^{\perp_E}$, for any $r \in (0 : m)$ we have $f(r) = (rx_1, rx_2, \ldots) = 0$. Hence $(0 : m) \leqslant \operatorname{Ker} f$. But, by assumption, $E^{(\mathbb{N})}$ is s-τ-M-injective and as I_∞ is τ-dense, we can infer, using the Generalized Fuchs Criterion, that there exists an element $y = (y_1, \ldots, y_t, 0, \ldots) \in E^{(\mathbb{N})}$ such that $f(a) = ay = (ay_1, \ldots, ay_t, 0, \ldots)$ for all $a \in I_\infty$. We then have $(ax_1, ax_2, \ldots) = (ay_1, \ldots, ay_t, 0, \ldots)$. This implies that $ax_{t+1} = 0$ for all $a \in I_\infty$ and thus $x_{t+1} \in$

$I_\infty{}^{\perp_E}$. But as $I_{t+2} \subset I_\infty$ we have $I_\infty{}^{\perp_E} \subseteq I_{t+2}{}^{\perp_E}$ and so $x_{t+1} \in I_{t+2}{}^{\perp_E}$. This contradicts the fact that $x_{t+1} \in I_{t+1}{}^{\perp_E} \setminus I_{t+2}{}^{\perp_E}$.

$(2) \Rightarrow (3)$. In (3a), for any ascending chain of τ-dense left ideals $I_1 \subseteq I_2 \subseteq \cdots$, I_∞ is also τ-dense since it contains each I_k and so by (2) the chain becomes stationary. In (3b), condition (2) gives $I_n = I_\infty$ for some $n \geqslant 1$, and so I_n is τ-dense.

$(3) \Rightarrow (1)$. Assume (3). Let I be a τ-dense left ideal of R and $f : I \to E^{(\mathbb{N})}$ be a homomorphism such that $(0 : m) \leqslant \operatorname{Ker} f$ for some $m \in M$. Since E is s-τ-M-injective, by Proposition 2.4 the direct product $E^{\mathbb{N}}$ is likewise. Since $E^{(\mathbb{N})}$ is a submodule of $E^{\mathbb{N}}$, we can consider f as a homomorphism from I to $E^{\mathbb{N}}$ and so, by the Generalized Fuchs Criterion, there is an element $x = (x_1, x_2, \dots) \in E^{\mathbb{N}}$ such that $f(a) = ax$ for all $a \in I$.

Let $X = \{x_1, x_2, \dots\}$ and $X_k = X \setminus \{x_1, x_2, \dots, x_k\} = \{x_{k+1}, x_{k+2}, \dots\}$ for any $k \geqslant 1$. Then the descending chain $X \supseteq X_1 \supseteq X_2 \supseteq \cdots$ yields an ascending chain ${}^{\perp}X \subseteq {}^{\perp}X_1 \subseteq {}^{\perp}X_2 \subseteq \cdots$ of E-annihilators in R.

To simplify our notation, let $J_{k+1} = {}^{\perp}X_k$ for all $k \geqslant 0$, where $X_0 = X$, and $J_\infty = \bigcup_{i=1}^{\infty} J_i$. Note that for any $r \in (0 : m) \leqslant \operatorname{Ker} f$ we have $0 = f(r) = rx = (rx_1, rx_2, \dots)$ and so $r \in {}^{\perp_R}\{x_1, x_2, \dots\} = {}^{\perp}X = J_1$. Hence we have $J_i \in \mathcal{A}_M(E)$ for any $i \in \mathbb{N}$.

Since $f(I) \leqslant E^{(\mathbb{N})}$, for any $a \in I$, either $ax_k = 0$ for all $k \in \mathbb{N}$, or there is a largest integer $n \in \mathbb{N}$ such that $ax_n \neq 0$. In either case there is an $n \in \mathbb{N}$ such that $ax_{n+k} = 0$ for all $k \geqslant 1$. Therefore $a \in {}^{\perp}X_n = J_{n+1} \subseteq J_\infty$ and so $I \subseteq J_\infty$. But I is τ-dense in R so J_∞ is likewise. Then, by (3b), there is an $\ell \geq 1$ such that J_ℓ is τ-dense in R. This yields an ascending chain $J_\ell \subseteq J_{\ell+1} \subseteq \cdots$ in $\mathcal{A}_M(E) \cap \mathcal{D}_\tau(R)$ which becomes stationary by (3a), say at $J_t = {}^{\perp}X_{t-1}$. This implies that $I \subseteq J_\infty = J_t = {}^{\perp}X_{t-1}$. Consequently, for any $a \in I$ we have $ax_{t+k} = 0$ for all $k \geqslant 0$ and so $f(a) = ay$ where $y = (x_1, x_2, \dots, x_t, 0, \dots)$. Since $y \in E^{(\mathbb{N})}$ then, by the Generalized Fuchs Criterion, $E^{(\mathbb{N})}$ is τ-injective.

$(1) \Rightarrow (4)$ follows from Corollary 4.3 while $(4) \Rightarrow (1)$ is obvious. $\square$

Note that, by Remark 4.1, the module E is Σ-τ-injective if and only if it is Σ-s-τ-${}_R R$-injective and $\mathcal{A}_R(E)$ is the set of all E-annihilators in R. Thus, we can use Theorem 4.2 with $M = {}_R R$ to get the following characterization of Σ-τ-injectivity.

Theorem 4.3. *The following statements are equivalent for a τ-injective module E.*

(1) *E is countably Σ-τ-injective, i.e., $E^{(\mathbb{N})}$ is τ-injective.*

(2) *If $I_1 \subseteq I_2 \subseteq \cdots$ is an ascending chain of E-annihilators in R such that $I_\infty = \bigcup_{i=1}^{\infty} I_i$ is τ-dense in R, there exists an $n \in \mathbb{N}$ such that $I_n = I_{n+k}$ for all $k \in \mathbb{N}$.*

(3) *The τ-dense E-annihilator left ideals in R satisfy the ACC and, if $I_1 \subseteq I_2 \subseteq \cdots$ is an ascending chain of E-annihilators in R such that $I_\infty = \bigcup_{i=1}^{\infty} I_i$ is τ-dense in R, then I_n is τ-dense in R for some $n \in \mathbb{N}$.*

(4) *E is Σ-τ-injective, i.e., $E^{(A)}$ is τ-injective for any set A.*

We also have the following corollaries which generalize [AN, 10.6, 10.7].

Corollary 4.5. *Let $\{\, E_i \mid 1 \leqslant i \leqslant n \,\}$ be a family of s-τ-M-injective modules. If E_i is Σ-s-τ-M-injective for each i, then $\bigoplus_{i=1}^{n} E_i$ is Σ-s-τ-M-injective.*

Proof. Note that $(\bigoplus_{i=1}^{n} E_i)^{(\mathbb{N})} = \bigoplus_{i=1}^{n} E_i^{(\mathbb{N})}$. Thus, by Proposition 2.4 (2), $\bigoplus_{i=1}^{n} E_i^{(\mathbb{N})}$ is s-τ-M-injective and so, by Theorem 4.2, $\bigoplus_{i=1}^{n} E_i$ is Σ-s-τ-M-injective, as required. $\qquad\square$

Corollary 4.6. *If $E_1, \ldots, E_n$ are Σ-τ-injective modules, then $\bigoplus_{i=1}^{n} E_i$ is Σ-τ-injective.*

Proof. By Remark 4.1 it suffices to take $_R M = {}_R R$ in Corollary 4.5. $\qquad\square$

The following theorem generalizes [AN, 10.10]. The proof we give here requires only minor modifications to that in [AN], which in turn is based on that in [Cai].

Theorem 4.4. *Let M be a module, $\{\, E_\lambda \mid \lambda \in \Lambda \,\}$ be a family of modules and $E = \bigoplus_{\lambda \in \Lambda} E_\lambda$. Suppose that E is s-τ-M-injective. Then E is Σ-s-τ-M-injective if and only if E_λ is Σ-s-τ-M-injective for all $\lambda \in \Lambda$.*

Proof. Clearly for any $\lambda \in \Lambda$, $\mathcal{A}_M(E_\lambda) \subseteq \mathcal{A}_M(E)$. Hence, by Theorem 4.2, if E is Σ-s-τ-M-injective then E_λ is Σ-s-τ-M-injective for every $\lambda \in \Lambda$.

The following diagram illustrates the proof of the converse.

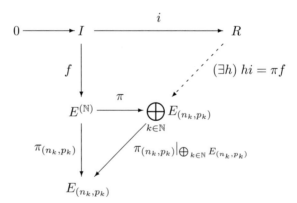

Corollary 4.5 takes care of the case when Λ is finite. Now suppose that Λ is countably infinite, i.e., $E = \bigoplus_{n \in \mathbb{N}} E_n$. Then, setting $E_{(n,p)} = E_n$ for any $(n,p) \in \mathbb{N} \times \mathbb{N}$, we have $E^{(\mathbb{N})} = \bigoplus_{(n,p) \in \mathbb{N} \times \mathbb{N}} E_{(n,p)}$. Let I be a τ-dense left ideal of R and $f : I \to E^{(\mathbb{N})}$ be a homomorphism such that $(0 : m) \leqslant \operatorname{Ker} f$ for some $m \in M$. Furthermore, for any $(n,p) \in \mathbb{N} \times \mathbb{N}$, let $\pi_{(n,p)} : E^{(\mathbb{N})} \to E_{(n,p)}$ denote the canonical projection and $A_{(n,p)}$ be the image of I under $\pi_{(n,p)} f$. Clearly

$\operatorname{Ker} \pi_{(n,p)} = \bigoplus_{(m,q) \neq (n,p)} E_{(m,q)}$. Suppose that the following statement holds:

$$(\forall n \in \mathbb{N}), (\exists (n', p) \in \mathbb{N} \times \mathbb{N}) \text{ with } n' > n \text{ and } A_{(n',p)} \neq 0. \qquad (\star)$$

Then we can construct a sequence (n_k, p_k) of elements of $\mathbb{N} \times \mathbb{N}$ such that for each $k \in \mathbb{N}$ we have $n_{k+1} > n_k$ and $A_{(n_k, p_k)} \neq 0$. Since $\bigoplus_{k \in \mathbb{N}} E_{(n_k, p_k)} = \bigoplus_{k \in \mathbb{N}} E_{n_k}$ is a direct summand of E, it is s-τ-M-injective.

Now consider the canonical projection $\pi : E^{(\mathbb{N})} \to \bigoplus_{k \in \mathbb{N}} E_{(n_k, p_k)}$. Then $(0 : m) \leqslant \operatorname{Ker} f \leqslant \operatorname{Ker} \pi f$ and so, by Proposition 2.3, there is an $x \in \bigoplus_{k \in \mathbb{N}} E_{(n_k, p_k)}$ such that $\pi f(a) = ax$ for all $a \in I$. Therefore there is a $t \in \mathbb{N}$ such that $\operatorname{Im} \pi f \leqslant \bigoplus_{k=1}^{t} E_{(n_k, p_k)}$. Thus, for any $k > t$, we have $A_{(n_k, p_k)} = \operatorname{Im} \pi_{(n_k, p_k)} f = \operatorname{Im} \pi_{(n_k, p_k)}|_{\bigoplus_{k \in \mathbb{N}} E_{(n_k, p_k)}} \pi f = 0$. Consequently, statement $(\star)$ fails and so there is an $n \in \mathbb{N}$ such that $A_{(n',p)} = 0$ whenever $(n', p) \in \mathbb{N} \times \mathbb{N}$ with $n' > n$. Thus $\operatorname{Im} f \subseteq (\bigoplus_{p \in \mathbb{N}} E_{(1,p)}) \oplus \cdots \oplus (\bigoplus_{p \in \mathbb{N}} E_{(n,p)}) = E_1^{(\mathbb{N})} \oplus \cdots \oplus E_n^{(\mathbb{N})}$ which, by Proposition 2.4, is s-τ-M-injective and so f extends to R, i.e., $E^{(\mathbb{N})}$ is s-τ-M-injective.

Now let Λ be an arbitrary set. Then $E^{(\mathbb{N})} = \bigoplus_{(\lambda, p) \in \Lambda \times \mathbb{N}} E_{(\lambda, p)}$, where $E_{(\lambda, p)} = E_\lambda$ for all $(\lambda, p) \in \Lambda \times \mathbb{N}$. Suppose to the contrary that $E^{(\mathbb{N})}$ is not s-τ-M-injective. Then, by Lemma 4.2, there is a countable subfamily $\mathcal{C}$ of $\{ E_{(\lambda, p)} \mid (\lambda, p) \in \Lambda \times \mathbb{N} \}$ such that $\bigoplus_{C \in \mathcal{C}} C$ is not s-τ-M-injective. Set $\Delta = \{ \lambda \in \Lambda \mid \exists p \in \mathbb{N} \text{ such that } E_{(\lambda, p)} \in \mathcal{C} \}$. Since Δ is countable, by our proof above $D = \bigoplus_{\delta \in \Delta} E_\delta$ is Σ-s-τ-M-injective. But $\bigoplus_{C \in \mathcal{C}} C$ is a direct summand of $D^{(\mathbb{N})} = \bigoplus_{(\delta, p) \in \Delta \times \mathbb{N}} E_{(\delta, p)}$ and so s-τ-M-injective, a contradiction. $\square$

Note that there appears to be a slight error in the last three lines of the proof in [AN, 10.10]. The argument in the last paragraph of our proof above rectifies this. We end with a corollary that generalizes [AN, 10.11].

Corollary 4.7. *Let M be a module, $\{ E_\lambda \mid \lambda \in \Lambda \}$ be a family of modules and $E = \bigoplus_{\lambda \in \Lambda} E_\lambda$. Suppose that E is τ-injective. Then E is Σ-τ-injective if and only if E_λ is Σ-τ-injective for all $\lambda \in \Lambda$.*

Proof. Take $M = {}_R R$ in Theorem 4.4. $\square$

References

[AF] F.W. Anderson and K.R. Fuller, *Rings and Categories of Modules*, 2nd ed., Graduate Texts in Mathematics, 13, Springer-Verlag, New York, 1992.

[AN] T. Albu and C. Năstăsescu, *Relative Finiteness in Module Theory*, Pure and Applied Mathematics, 84, Marcel Dekker, Inc., New York and Basel, 1984.

[B90] P.E. Bland, *A note on quasi-divisible modules*, Comm. Algebra **18** (1990), 1953–1959.

[B] P.E. Bland, *Topics in Torsion Theory*, Mathematical Research, 103, Wiley-VCH, Berlin, 1998.

[Cai] A. Cailleau, *Une caractérisation des modules Σ-injectifs*, C.R. Acad. Sci. Sér. A-B **269** (1969), 997–999.

[CC]   I. Crivei and S. Crivei, *Divisible modules with respect to a torsion theory*, Algebras, Rings and their Representations, Proceedings of the International Conference on Algebras, Modules and Rings, Lisbon, Portugal, 14–18 July 2003 (A. Facchini, K. Fuller, C.M. Ringel, and C. Santa-Clara, eds.), World Scientific, Singapore, 2006, pp. 25–36.

[C98]   S. Crivei, *m-injective modules*, Mathematica (Cluj) **40 (63)** (1998), 71–78.

[C01]   S. Crivei, *A note on τ-quasi-injective modules*, Studia Univ. Babeş-Bolai Math. **46** (2001), 33–39.

[DHSW]   N.V. Dung, D.V. Huynh, P.F. Smith, and R. Wisbauer, *Extending modules*, Pitman Research Notes in Mathematics, 313, Longman Scientific & Technical, Essex, 1994.

[Fa]   C. Faith, *Algebra II Ring Theory*, Grundlehren der mathematischen Wissenschaften, 191, Springer-Verlag, New York, 1976.

[Fu]   L. Fuchs, *On quasi-injective modules*, Ann. Scuola Norm. Sup. Pisa **23** (1969), 541–546.

[G]   J.S. Golan, *Torsion Theories*, Pitman Monographs and Surveys in Pure and Applied Mathematics, 29, Longman Scientific and Technical, New York, 1986.

[L]   T.Y. Lam, *Lectures on Modules and Rings*, Graduate Texts in Mathematics, 189, Springer-Verlag, New York, 1999.

[MM]   S.H. Mohamed and B.J. Müller, *Continuous and Discrete Modules*, London Math. Soc. Lecture Notes, 147, Cambridge University Press, Cambridge, 1990.

[N]   K. Nishida, *Divisible modules, codivisible modules, and quasi-divisible modules*, Comm. Algebra **5** (1977), 591–610.

[PRY]   Y.S. Park, S.H. Rim, and S. Young-Su, *On relative τ-injective modules*, Far East J. Math. Sci. **5** (1997), 19–27.

[S]   P.F. Smith, *Injective dimension relative to a torsion theory*, Algebras, Rings and their Representations, Proceedings of the International Conference on Algebras, Modules and Rings, Lisbon, Portugal, 14-18 July 2003 (A. Facchini, K. Fuller, C.M. Ringel and C. Santa-Clara, eds.), World Scientific, Singapore, 2006, pp. 343–356.

[W]   R. Wisbauer, *Foundations of Module and Ring Theory*, Algebra, Logic and Applications, 3, Gordon and Breach, Reading, 1991.

Stelios Charalambides and John Clark
University of Otago
Department of Mathematics and Statistics
PO Box 56
Dunedin, New Zealand
e-mail: stelios@maths.otago.ac.nz
e-mail: jclark@maths.otago.ac.nz

Modules and Comodules
Trends in Mathematics, 169–172
© 2008 Birkhäuser Verlag Basel/Switzerland

A Note on Polynomial Rings over Nil Rings

M.A. Chebotar, W.-F. Ke, P.-H. Lee and E.R. Puczyłowski

To Robert Wisbauer

Abstract. Let R be a nil ring with $pR = 0$ for some prime number p. We show that the polynomial ring $R[x, y]$ in two commuting indeterminates x, y over R cannot be homomorphically mapped onto a ring with identity. This extends, in finite characteristic case, a result obtained by Smoktunowicz [8] and gives a new approximation, in that case, of a positive solution of Köthe's problem.

Mathematics Subject Classification (2000). 16N40, 16N80.

Keywords. Nil ring, polynomial ring, Brown-McCoy radical, Köthe's problem.

It is shown in [6] that the polynomial ring $R[x]$ in the indeterminate x over a nil ring R cannot be homomorphically mapped onto a ring with identity or, shortly, $R[x]$ is Brown-McCoy radical. A natural question is whether this is also the case for the polynomial ring $R[x, y]$ in two commuting indeterminates x, y over a nil ring R. The problem was posed in the paper [4] and then studied in many papers (e.g., [1], [2] [5], [7]), but still remains open. Smoktunowicz [8] showed that if $R[x]$ is Jacobson radical for a ring R, then $R[x, y]$ is Brown-McCoy radical. It is well known that if $R[x]$ is Jacobson radical, then R is nil and that the converse implication is equivalent to Köthe's problem [3]. Hence Smoktunowicz's result shows that if there should exist an example of a nil ring R such that $R[x, y]$ can be homomorphically mapped onto a ring with identity, then R would give a counterexample to Köthe's problem. Consequently the result proved in this paper can be viewed as a step towards a positive solution of Köthe's problem at least in case of finite characteristic p. To our best knowledge there are no results in the literature concerning Köthe's problem in which the characteristic plays a substantial role. Thus

Acknowledgment: The fourth author was supported by KBN Grant No. 1 P03A 032 27. This work was finished when he was a visiting scholar at the National Center for Theoretical Sciences, Taipei Office, Taiwan.

Corresponding author: P.-H. Lee, Department of Mathematics, National Taiwan University, Taipei 106, Taiwan. E-mail: `phlee@math.ntu.edu.tw`.

maybe the result of this paper indicates that the Köthe's problem in the finite characteristic case is "easier" than that in the general case.

The following result was proved in [6, Theorem 1].

Theorem 1. *The polynomial ring $R[x]$ over a ring R is Brown-McCoy radical if, and only if, R cannot be homomorphically mapped onto a ring containing non-nilpotent central elements.*

From this result it follows that to prove that $R[x, y]$ is Brown-McCoy radical for a ring R, it suffices to show that $R[x]$ cannot be homomorphically mapped onto a ring containing non-nilpotent central elements. We will show that the latter holds if R is a nil ring with $pR = 0$ for some prime p. The heart of its proof lies on some observation, which we illustrate in the case of $p = 2$.

Suppose that $\varphi : R[x] \to P$ is a homomorphism onto a ring P containing non-nilpotent central elements. Let $u = a_0 + a_1 x + a_2 x^2 + \cdots + a_n x^n$ be such that $\varphi(u)$ is a non-nilpotent central element of P. Write $u = u_0 + u_1$ where $u_0 = a_0 + a_2 x^2 + \cdots$ and $u_1 = a_1 x + a_3 x^3 + \cdots$. Then $\varphi(u_0) + \varphi(u_1) = \varphi(u)$ is central in P, and so $\varphi(u_0)\varphi(u_1) = \varphi(u_1)\varphi(u_0)$. Thus $\varphi(u)^2 = (\varphi(u_0) + \varphi(u_1))^2 = \varphi(u_0)^2 + \varphi(u_1)^2 = \varphi(u_0^2) + \varphi(u_1^2) = \varphi(u_0^2 + u_1^2)$. Note that $u' = u_0^2 + u_1^2$ is a polynomial of the form (setting $a_k = 0$ for $k > n$):

$$u' = a_0^2 + (a_0 a_2 + a_1^2 + a_2 a_0)x^2 + \cdots + \left(\sum_{i=0}^{2k} a_i a_{2k-i} \right) x^{2k} + \cdots + a_n^2 x^{2n}.$$

Denote $x_1 = x^2$ and $a'_k = \sum_{i=0}^{2k} a_i a_{2k-i}$. We have

$$u' = a'_0 + a'_1 x_1 + \cdots + a'_k x_1^k + \cdots + a'_n x_1^n,$$

where we shall keep in mind that $a'_n = a_n^2$.

Now, we write again $u' = u'_0 + u'_1$ where $u'_1 = a'_0 + a'_2 x_1^2 + \ldots$ and $u'_2 = a'_1 x_1 + a'_3 x_1^3 + \ldots$. Since $\varphi(u'_0) + \varphi(u'_1) = \varphi(u') = \varphi(u)^2$ is also central in P, we have $\varphi(u'_0)\varphi(u'_1) = \varphi(u'_1)\varphi(u'_0)$ and so $\varphi(u')^2 = \varphi(u_0'^2 + u_1'^2)$. Then $u'' = u_0'^2 + u_1'^2$ is a polynomial of the form (setting $a'_k = 0$ for $k > n$):

$$u'' = a_0'^2 + (a'_0 a'_2 + a_1'^2 + a'_2 a'_0)x_1^2 + \cdots + \left(\sum_{i=0}^{2k} a'_i a'_{2k-i} \right) x_1^{2k} + \cdots + a_n'^2 x_1^{2n}.$$

Denote $x_2 = x_1^2 = x^{2^2}$ and $a''_k = \sum_{i=0}^{2k} a'_i a'_{2k-i}$. We have

$$u'' = a''_0 + a''_1 x_2 + \cdots + a''_k x_2^k + \cdots + a''_n x_2^n,$$

where we shall keep in mind that $a''_n = a_n'^2 = a_n^{2^2}$.

Continuing in this manner l times, we get that $\varphi(u^{(l)}) = \varphi(u)^{2^l}$ is a nonzero central element of P, where $u^{(l)}$ is an element of the form

$$u^{(l)} = a_0^{(l)} + a_1^{(l)} x_l + \cdots + a_k^{(l)} x_l^k + \cdots + a_n^{(l)} x_l^n,$$

with $x_l = x^{2^l}$ and $a_n^{(l)} = a_n^{2^l}$. When l is large enough, $a_n^{(l)} = 0$ since R is nil. We can continue to reach a contradiction that $\varphi(0)$ is a nonzero central element in P.

Now we consider the case that p is an arbitrary prime. The general result is a consequence of the following lemma, which is of independent interest. We denote by $\mathbb{Z}_p$ the Galois field of p elements. Clearly a ring R is a $\mathbb{Z}_p$-algebra if and only if $pR = 0$.

Lemma 2. *Let A be a $\mathbb{Z}_p$-algebra and $b_0, b_1, \ldots, b_n \in A$. If $b = b_0 + b_1 + \cdots + b_n$ is a central element of A, then b^p is the sum of elements of the form $b_{i_1} b_{i_2} \ldots b_{i_p}$ with $i_1 + i_2 + \cdots + i_p \equiv 0 \pmod{p}$.*

Proof. For every $i = 1, \ldots, p - 1$, let a_i be the sum of all b_j such that $j \equiv i \pmod{p}$. It is clear that $b = a_0 + a_1 + \cdots + a_{p-1}$ and it suffices to show that b^p is the sum of elements of the form $a_{i_1} a_{i_2} \cdots a_{i_p}$ with $i_1 + i_2 + \cdots + i_p \equiv 0 \pmod{p}$.

Consider A as a $\mathbb{Z}_p$-subalgebra of the group algebra $A^{\#}[G]$, where $A^{\#}$ is the $\mathbb{Z}_p$-algebra with identity adjoined to A, and G the cyclic group of order p generated by the element g. Let $u = a_0 + a_1 g + \cdots + a_{p-1} g^{p-1}$; then $u^p = c_0 + c_1 g + \cdots + c_{p-1} g^{p-1}$ where each c_j is the sum of elements of the form $a_{i_1} a_{i_2} \cdots a_{i_p}$ with $i_1 + i_2 + \cdots + i_p \equiv j \pmod{p}$. Let $v = b - u$; then $v = w(1 - g)$ where

$$w = a_1 + a_2(1 + g) + \cdots + a_{p-1}(1 + g + \cdots + g^{p-2}).$$

Thus, $v^p = w^p(1 - g)^p = w^p(1 - g^p) = 0$ since $1 - g$ is a central element of $A^{\#}[G]$. As $u = b - v$ and b is a central element of $A^{\#}[G]$, we have

$$c_0 + c_1 g + \cdots + c_{p-1} g^{p-1} = u^p = b^p - v^p = b^p \in A.$$

Hence, $c_i = 0$ for $i = 1, \cdots, p - 1$ and so $b^p = u^p = c_0$ is the sum of elements of the form $a_{i_1} a_{i_2} \cdots a_{i_p}$ with $i_1 + i_2 + \cdots + i_p \equiv 0 \pmod{p}$. $\square$

Now we prove the key result of this note.

Theorem 3. *If R is a nil $\mathbb{Z}_p$-algebra, then $R[x]$ cannot be homomorphically mapped onto a ring with non-nilpotent central elements.*

Proof. Assume the contrary. Let n be the smallest integer such that some polynomial $u = a_0 + a_1 x + \cdots + a_n x^n \in R[x]$ of degree n can be mapped via a homomorphism onto a non-nilpotent central element of some ring. We may assume further that the nilpotency index k of a_n is the smallest among such polynomials.

Let $\varphi : R[x] \to P$ be such a homomorphism. Set $b = \varphi(u)$ and $b_i = \varphi(a_i x^i)$ for each $i = 0, 1, \ldots, n$. Then $b = b_0 + b_1 + \cdots + b_n$ is a central element of P, so applying the above lemma, we get that $b^p = \varphi(v)$, where

$$v = \sum_{i_1 + i_2 + \cdots + i_p \equiv 0 \,(\mathrm{mod}\, p)} a_{i_1} a_{i_2} \cdots a_{i_p} x^{i_1 + i_2 + \cdots + i_p}.$$

Write $v = c_0 + c_1 x^p + c_2 x^{2p} + \cdots + c_n x^{np}$ for some $c_0, c_1, \ldots, c_n \in R$ and set $w = c_0 + c_1 x + c_2 x^2 + \cdots + c_n x^n$. Note that $c_n = a_n^p$. Let $\psi : R[x] \to R[x]$ be the homomorphism defined by $\psi(r_0 + r_1 x + \cdots r_k x^k) = r_0 + r_1 x^p + \cdots + r_k x^{pk}$. Clearly $\varphi \circ \psi : R[x] \to P$ is a homomorphism and w is a polynomial in $R[x]$ of degree n such that $(\varphi \circ \psi)(w) = \varphi(\psi(w)) = \varphi(v) = b^p$ is a non-nilpotent central element of P. However, the leading coefficient $c_n = a_n^p$ of w has nilpotency index $\leq [k/p] + 1 < k$. This contradiction completes the proof. $\square$

Combining Theorems 1 and 3, we are able to conclude the main theorem.

Theorem 4. *If R is a nil ring with $pR = 0$ for some prime p, then $R[x, y]$ cannot be homomorphically mapped onto a ring with identity.*

References

[1] Ferrero, M. *Unitary strongly prime rings and ideals.* Proceedings of the 35th Symposium on Ring Theory and Representation Theory (Okayama, 2002), 101–111, Symp. Ring Theory Representation Theory Organ. Comm., Okayama, 2003.

[2] Ferrero, M.; Wisbauer, R. *Unitary strongly prime rings and related radicals.* J. Pure Appl. Algebra **181** (2003), 209–226.

[3] Krempa, J. *Logical connections among some open problems in non-commutative rings.* Fund. Math. **76** (1972), 121–130.

[4] Puczyłowski, E.R. *Some questions concerning radicals of associative rings.* Theory of radicals (Szekszard, 1991), 209–227, Colloq. Math. Soc. Janos Bolyai, 61, North-Holland, Amsterdam, 1993.

[5] Puczyłowski, E.R. *Some results and questions on nil rings.* 15th School of Algebra (Portuguese) (Canela, 1998). Mat. Contemp. **16** (1999), 265–280.

[6] Puczyłowski, E.R.; Smoktunowicz, A. *On maximal ideals and the Brown-McCoy radical of polynomial rings.* Comm. Algebra **26** (1998), 2473–2482.

[7] Smoktunowicz, A. *On some results related to Köthe's conjecture.* Serdica Math. J. **27** (2001), 159–170.

[8] Smoktunowicz, A. *R[x, y] is Brown-McCoy radical if R[x] is Jacobson radical.* Proc. 3rd Int. Algebra Conf. (Tainan, 2002), 235–240, Kluwer Acad. Publ., Dordrecht, 2003.

M.A. Chebotar
Department of Mathematical Sciences, Kent State University
Kent OH 44242, USA
e-mail: chebotar@math.kent.edu

W.-F. Ke
Department of Mathematics, National Cheng Kung University
and National Center for Theoretical Sciences, Tainan Office
Tainan 701, Taiwan
e-mail: wfke@mail.ncku.edu.tw

P.-H. Lee
Department of Mathematics, National Taiwan University
and National Center for Theoretical Sciences, Taipei Office
Taipei 106, Taiwan
e-mail: phlee@math.ntu.edu.tw

E.R. Puczyłowski
Institute of Mathematics, University of Warsaw
Warsaw, Poland
e-mail: edmundp@mimuw.edu.pl

Modules and Comodules

Trends in Mathematics, 173–183

© 2008 Birkhäuser Verlag Basel/Switzerland

QI-modules

John Dauns and Yiqiang Zhou

Abstract. This paper is about QI-modules M and the full subcategory $\sigma[M]$ of Mod-R subgenerated by M. A ring R is a right QI-ring if every quasi-injective right R-module is injective. The module M is QI if all quasi-injective modules in $\sigma[M]$ are M-injective. Three classes of rings are presented for each of which the QI-modules are precisely the semisimple modules. The QI-modules M are characterized in terms of properties of some lattices of classes of modules in $\sigma[M]$.

1. Introduction

Throughout R is an associative ring with identity, modules are right unital. Let Mod-R be the category of the right R-modules. For an arbitrary right R-module M, $\sigma[M]$ denotes the full subcategory of Mod-R subgenerated by M, which is systematically studied in Wisbauer [14]. Following Wisbauer [13], a module M is called a QI-module if every quasi-injective module in $\sigma[M]$ is M-injective. It is easy to see that every semisimple module is QI and it is known that every QI-module is locally noetherian. It is beyond the scope of this article to explore when "M is QI" relative to some other categories built out of M, such as Gen $M \subseteq \sigma[M] \subseteq \pi[M]$, where Gen M is generated by M, while $\pi[M]$ is just like $\sigma[M]$, except that sums have been replaced by products, and which also was introduced by Wisbauer [16].

We first give some easy characterizations of a QI-module and present a few classes of rings R for which the QI-modules are precisely the semisimple modules. The main results Theorem 3.4 and Theorem 3.9 of this paper are the diverse equivalent characterizations of QI-modules M in terms of properties of some lattices of classes of modules in $\sigma[M]$. In order to decide what are all QI-modules M, we have two problems. What are all quasi-injective modules? And for specific modules M, what are all quasi-injective modules inside $\sigma[M]$? These questions can be answered in terms of certain right ideals of the ring R, and for the readers benefit we give an outline in Section 4 at the end.

For two right R-modules N and M, let $E_M(N)$ be the M-injective hull of N and $E(N)$ the injective hull. For a subset X of M, $X^\perp = \{r \in R : Xr = 0\}$ is the right annihilator. For $x \in M$, let $x^\perp = \{x\}^\perp$.

2. The class of QI-modules

A module M is called strongly prime if M is contained in every nonzero fully invariant submodule of $E(M)$ (see [1]). It is easy to see that a uniform module M is strongly prime iff M is contained in every nonzero quasi-injective submodule of $E(M)$. The equivalence (1) $\Leftrightarrow$ (4) in the next proposition is known (see [17]).

Proposition 2.1. *The following are equivalent:*

1) M *is a QI-module.*
2) *Every module in $\sigma[M]$ is QI.*
3) *Every M-injective module in $\sigma[M]$ is QI.*
4) *M is a locally noetherian module and every cyclic uniform module in $\sigma[M]$ is strongly prime.*

Proof. (1) $\Rightarrow$ (2). Let $N \in \sigma[M]$ and let $P \in \sigma[N]$ be quasi-injective. Then $P \in \sigma[M]$, so P is M-injective. Thus, P is X-injective for all $X \in \sigma[M]$. In particular, P is N-injective. So N is QI.

(2) $\Rightarrow$ (3). It is obvious.

(3) $\Rightarrow$ (1). Let $N \in \sigma[M]$ be quasi-injective. Then $N \subseteq E_M(N) \in \sigma[M]$. So $E_M(N)$ is QI by (3). But, $N \in \sigma[E_M(N)]$, so N is $E_M(N)$-injective. Thus, $N = E_M(N)$ is M-injective. So M is QI. □

Thus, the class $\mathcal{K}_q$ of all QI-modules is closed under submodules and factor modules. But in general, $\mathcal{K}_q$ is not closed under injective hulls, nor under direct products, nor under extensions of modules.

Example 2.1. Let $R = \mathbb{Z}$. If $N = \mathbb{Z}_{p^\infty}$ where p is a prime, then N is the injective hull of soc(N). Since soc(N) is not a direct summand of N, N is not QI (though soc(N) is QI). If $N = \mathbb{Z}_4$, then soc(N) $\cong \mathbb{Z}_2$ and $N/$soc(N) $\cong \mathbb{Z}_2$. So both soc(N) and $N/$soc(N) are QI. But N is not QI because soc(N) is not a direct summand of N. Moreover, $\mathbb{Z}$ is a subdirect product of simple $\mathbb{Z}$-modules, but $\mathbb{Z}$ is not QI.

A ring R is right semiartinian if every nonzero R-module has nonzero socle.

Proposition 2.2. *$\mathcal{K}_q$ is precisely the class of all semisimple R-modules, in either of the following cases:*

1) *R is right semiartinian.*
2) *R is a principal ideal domain.*
3) *R has finitely many maximal right ideals (in particular, R is local).*

Proof. (1) For any $0 \neq N \in \mathcal{K}_q$, soc($N$) is essential in N since R is right semiartinian. Because N is QI, soc(N) is N-injective and so soc(N) is a direct summand. Thus, $N = $ soc(N) is semisimple.

(2) We can assume that R is not a field. Let $0 \neq N \in \mathcal{K}_q$ and let $T(N)$ be the torsion submodule of N. If $N \neq T(N)$, then, since $N/T(N) \in \mathcal{K}_q$ and $\mathcal{K}_q$ is closed under submodules, there exists a cyclic torsionfree QI-module, say xR. Then $R_R \cong (xR)_R$ is QI. Since R is not a field, take $a \in R$ such that a is a

prime element in R. Thus, R/a^2R is QI. But aR/a^2R is simple, so it is R/a^2R-injective. This shows that aR/a^2R is a direct summand of R/a^2R. It follows that $aR/a^2R = R/a^2R$. That is $aR = R$, a contradiction. So $N = T(N)$ is torsion. For any $x \in N$, $xR \cong x_1R \oplus \cdots \oplus x_kR$ where $x_i^\perp = a_i^{m_i}R$ with $m_i > 0$ and with a_i a prime element of R (for each i). For the same reason that R/a^2R is not QI, x_iR is not QI unless $m_i = 1$. Hence each x_iR is simple and hence xR is semisimple. This shows that N is semisimple.

(3) Let $0 \neq N \in \mathcal{K}_q$. Because $\mathcal{K}_q$ is closed under submodules, to show that N is semisimple, we may assume that N is cyclic. So let $N = xR$ and hence $N \cong R/x^\perp$. Let $I_1, \ldots, I_m$ be the maximal right ideals of R. We can assume that, for some k with $1 \leq k \leq m$, $x^\perp \subseteq \cap_{i=1}^k I_i$ but $x^\perp \not\subseteq I_i$ for $i = k+1, \ldots, m$. We claim that $x^\perp = I := \cap_{i=1}^k I_i$. In fact, if $x^\perp \subset I$, then $0 \neq I/x^\perp \subseteq R/x^\perp$. Because N is QI, N is noetherian and so is $I/x^\perp$. Hence $I/x^\perp$ contains a maximal submodule, say $K/x^\perp$, where K is a right ideal of R. Thus, I/K is a simple submodule of R/K. But R/K is an image of N, so R/K is QI because $\mathcal{K}_q$ is closed under factor modules. Thus, I/K is R/K-injective. Hence $R/K = I/K \oplus J/K$ where J is a right ideal of R. It follows that $K = I \cap J$ and J is a maximal right ideal of R. Since $x^\perp \subseteq K \subseteq J$, $J = I_i$ for some $1 \leq i \leq k$. Hence $K = I \cap J = I$. This is a contradiction. So it must be that $x^\perp = I$. It follows that $N \cong R/I \subseteq \oplus_{i=1}^k R/I_i$. This shows that N is a submodule of a semisimple module. So N is semisimple. $\qquad\square$

Corollary 2.3. *Let R be a local ring with Jacobson radical $J(R)$. Then an R-module N is a QI-module iff N is a direct sum of copies of $R/J(R)$.*

Proposition 2.4. *The following hold:*

   1) *Every QI-module is a subdirect product of simple modules.*
   2) *$\mathcal{K}_q$ is closed under direct products if and only if every direct product of simple modules is QI.*

Proof. (1) This follows from the fact that every QI-module is co-semisimple ([14, p. 190, 23.1]). For the reader's convenience, we give a direct proof. Let N be a QI-module. By Birkhoff's Theorem, N is a subdirect product of modules $\{N_\alpha : \alpha \in \Gamma\}$ where each N_α has an essential socle which is simple. Since N_α is again QI, $\mathrm{soc}(N_\alpha)$ is N_α-injective. It follows that $N_\alpha = \mathrm{soc}(N_\alpha)$ is simple.

(2) The implication '$\Rightarrow$' is clear, and the implication '$\Leftarrow$' follows because of (1) and the fact that $\mathcal{K}_q$ is closed under submodules. $\qquad\square$

Thus, if every direct product of R-simple modules is semisimple (e.g., $R = \mathbb{Z}_{p^2}$ with p a prime), then $\mathcal{K}_q$ is closed under direct products.

3. QI-modules and lattices of module classes

In this section, we discuss the relationship between a QI-module M and properties of some lattices of classes of modules in $\sigma[M]$. We recall a few notions.

Definition 3.1. [3] *A module class $\mathcal{K}$ is called a natural class if $\mathcal{K}$ is closed under submodules, direct sums and injective hulls. For a module class $\mathcal{F}$, define $c(\mathcal{F}) = \{N \in \text{Mod-}R : \forall 0 \neq X \subseteq N, X \nrightarrow Y$ for any $Y \in \mathcal{F}\}$ and $d(\mathcal{F}) = \{N \in \text{Mod-}R : \forall 0 \neq X \subseteq N, \exists 0 \neq A \subseteq X$ with $A \hookrightarrow Y$ for some $Y \in \mathcal{F}\}$. For a module N, let $c(N) = c(\{N\})$ and $d(N) = d(\{N\})$. It is always true that $c(\mathcal{F})$ and $d(\mathcal{F})$ are natural classes. More generally, for a module M, a subclass $\mathcal{K}$ of $\sigma[M]$ is called an M-natural class if $\mathcal{K}$ is closed under submodules, direct sums and M-injective hulls. The collection of the M-natural classes (resp., the natural classes) is denoted by $\mathcal{N}(R, M)$ (resp., $\mathcal{N}(R)$). Notice that both $\mathcal{N}(R, M)$ and $\mathcal{N}(R)$ are sets, and $\mathcal{K} \in \mathcal{N}(R, M)$ iff $\mathcal{K} = \mathcal{L} \cap \sigma[M]$ for some $\mathcal{L} \in \mathcal{N}(R)$.*

A module class $\mathcal{T}$ is called a *hereditary pretorsion class* if $\mathcal{T}$ is closed under submodules, direct sums and factor modules. A module class $\mathcal{T}$ is a hereditary pretorsion class iff $\mathcal{K} = \sigma[M]$ for some module M (see [5]). The collection (or set) of the hereditary pretorsion classes is denoted by $\mathcal{T}^p(R)$.

Definition 3.2. [18, 3] *A module class $\mathcal{K}$ is called a pre-natural class if $\mathcal{K}$ is closed under submodules, direct sums, and $\text{tr}(\mathcal{K}, E(N)) \in \mathcal{K}$ for every $N \in \mathcal{K}$, where $\text{tr}(\mathcal{K}, E(N)) = \sum\{f(X) : f \in \text{Hom}(X, E(N)), X \in \mathcal{K}\}$. A module class $\mathcal{K}$ is a pre-natural class iff $\mathcal{K} = \mathcal{L} \cap \mathcal{T}$ where $\mathcal{L} \in \mathcal{N}(R)$ and $\mathcal{T} \in \mathcal{T}^p(R)$. The collection of the pre-natural classes is denoted by $\mathcal{N}^p(R)$. Note that $\mathcal{N}^p(R)$ is a set. For a module M, we let $\mathcal{N}^p(R, M)$ be the set of all pre-natural classes contained in $\sigma[M]$. That is $\mathcal{N}^p(R, M) = \{\mathcal{K} \cap \sigma[M] : \mathcal{K} \in \mathcal{N}^p(R)\}$.*

Lemma 3.3. [18, 3] *$\mathcal{N}^p(R)$ is a complete lattice with the least element $\mathbf{0} = \{0\}$ and the greatest element $\mathbf{1} = \text{Mod-}R$ under the following partial ordering and lattice operations:*

1) *For $\mathcal{K}_1, \mathcal{K}_2 \in \mathcal{N}^p(R)$, $\mathcal{K}_1 \leq \mathcal{K}_2 \Longleftrightarrow \mathcal{K}_1 \subseteq \mathcal{K}_2$.*
2) *For a set of pre-natural classes $\mathcal{K}_i$, $\wedge \mathcal{K}_i = \cap \mathcal{K}_i$ and $\vee \mathcal{K}_i = d(\mathcal{K}) \cap \sigma[M_\mathcal{K}]$ where $\mathcal{K} = \cup \mathcal{K}_i$ and $M_\mathcal{K} = \oplus \{X_\alpha : \alpha \in \Lambda\}$ with $\{X_\alpha : \alpha \in \Lambda\}$ a set of representatives of isomorphy classes of cyclic submodules of modules in $\mathcal{K}$.*

The set of all hereditary pretorsion classes contained in $\sigma[M]$ is denoted by M-ptors in [10]. That is M-ptors $= \{\mathcal{K} \cap \sigma[M] : \mathcal{K} \in \mathcal{T}^p(R)\}$. Notice that $\mathcal{N}(R)$, $\mathcal{N}(R, M)$, $\mathcal{N}^p(R, M)$, $\mathcal{T}^p(R)$ and M-ptors all are sublattices of $\mathcal{N}^p(R)$.

Theorem 3.4. *The following are equivalent for a module M:*

1) *M is a QI-module.*
2) *$\mathcal{N}^p(R, M) = \mathcal{N}(R, M)$.*
3) *M-ptors $\subseteq \mathcal{N}(R, M)$.*
4) *$\mathcal{N}^p(R, M)$ is a uniquely complemented lattice.*

Proof. (1) $\Rightarrow$ (2). Let $\mathcal{L} \in \mathcal{N}^p(R, M)$. Write $\mathcal{L} = \mathcal{K} \cap \sigma[M]$, where $\mathcal{K} \in \mathcal{N}^p(R)$. There exists a module N such that $\mathcal{K} = d(N) \cap \sigma[N]$ by [3, 2.5.4]. Since $\sigma[N] \cap \sigma[M]$ is a hereditary pretorsion class, $\sigma[N] \cap \sigma[M] = \sigma[P]$ for a module P. Hence $\mathcal{L} = d(N) \cap \sigma[P]$ is a P-natural class. Now we show that $\mathcal{L}$ is closed under M-injective hulls. Let $X \in \mathcal{L}$. Then $X \subseteq E_P(X) \subseteq E_M(X) \subseteq E(X)$. Since $E_P(X)$ is P-injective, it is quasi-injective, so $E_P(X)$ is M-injective by (1). Thus, $E_P(X) = E_M(X)$. But $E_P(X) \in \mathcal{L}$, since $\mathcal{L}$ is a P-natural class. So $E_M(X) \in \mathcal{L}$.

(2) $\Rightarrow$ (4). The implication follows from a result of Dauns that $\mathcal{N}(R, M)$ is a Boolean lattice (see [3, 6.2.16]).

(4) $\Rightarrow$ (3). Let $\mathcal{K}$ be a hereditary pretorsion class contained in $\sigma[M]$. Notice that $\mathcal{K} \in \mathcal{N}^p(R, M)$. So, by (4), there exists $\mathcal{L} \in \mathcal{N}^p(R, M)$ such that $\mathcal{K} \wedge \mathcal{L} = \mathbf{0}$ and $\mathcal{K} \vee \mathcal{L} = \mathbf{1}$. This shows that $\mathcal{K} \subseteq c(\mathcal{L}) \cap \sigma[M]$. Let $\mathcal{J} = c(\mathcal{L}) \cap \sigma[M]$. Then $\mathcal{J} \wedge \mathcal{L} = \mathbf{0}$ and $\mathcal{J} \vee \mathcal{L} = \mathbf{1}$ because $\mathcal{J} \vee \mathcal{L} \geq \mathcal{K} \vee \mathcal{L} = \mathbf{1}$. By (4), $\mathcal{K} = \mathcal{J} = c(\mathcal{L}) \cap \sigma[M]$ is an M-natural class.

(3) $\Rightarrow$ (1). Let $N \in \sigma[M]$ be a quasi-injective module. Then $\sigma[N] \subseteq \sigma[M]$. By (3), $\sigma[N]$ is an M-natural class. Hence $\sigma[N]$ is closed under M-injective hulls. In particular, $E_M(N) \in \sigma[N]$. Since N is quasi-injective, N is X-injective for each $X \in \sigma[N]$. Thus, N is $E_M(N)$-injective. This shows that N is a direct summand of $E_M(N)$. It must be that $N = E_M(N)$, so N is M-injective. $\square$

Definition 3.5. [15, 1.2] *For any subclasses $\mathcal{K}$ and $\mathcal{L}$ of $\sigma[M]$, let*

$$E_M(\mathcal{K}, \mathcal{L}) = \big\{ N \in \sigma[M] : \exists X \subseteq N \text{ such that } X \in \mathcal{K}, N/X \in \mathcal{L} \big\}.$$

A subclass $\mathcal{K}$ of $\sigma[M]$ is said to be closed under extensions in $\sigma[M]$ if $\mathcal{K} = E_M(\mathcal{K}, \mathcal{K})$.

Lemma 3.6. $E_M(\mathcal{K}, \mathcal{L}) = \mathcal{K} \vee \mathcal{L}$ *for all $\mathcal{K}, \mathcal{L} \in \mathcal{N}(R, M)$.*

Proof. Let $\mathcal{K}, \mathcal{L} \in \mathcal{N}(R, M)$. Then $\mathcal{K} \vee \mathcal{L} \in \mathcal{N}(R, M)$ by [3, 6.1.4]. So $\mathcal{K} \vee \mathcal{L}$ is closed under extensions in $\sigma[M]$ by [3, 2.4.5], and hence $E_M(\mathcal{K}, \mathcal{L}) \subseteq \mathcal{K} \vee \mathcal{L}$. Suppose $N \in \mathcal{K} \vee \mathcal{L}$. Let X be a submodule of N maximal with respect to $X \in \mathcal{L}$ and let Y be a submodule of N maximal with respect to $X \cap Y = 0$. (X and Y exist by Zorn's Lemma.) Then X is essentially embeddable in N/Y. So $N/Y \in \mathcal{L}$ by [3, 2.4.2]. The maximality of X shows that $Y \in \mathcal{J} := c(\mathcal{L}) \cap \sigma[M]$. But, by [3, 6.2.16], $\mathcal{N}(R, M)$ is a Boolean lattice. Hence we obtain that $Y \in \mathcal{J} \wedge (\mathcal{K} \vee \mathcal{L}) = (\mathcal{J} \wedge \mathcal{K}) \vee (\mathcal{J} \wedge \mathcal{L}) = (\mathcal{J} \wedge \mathcal{K}) \vee \mathbf{0} = \mathcal{J} \wedge \mathcal{K} \leq \mathcal{K}$. So $N \in E_M(\mathcal{K}, \mathcal{L})$. $\square$

Let M-tors be the set of all hereditary torsion classes contained in $\sigma[M]$. That is M-tors $= \big\{ \mathcal{K} \in M\text{-ptors} : \mathcal{K} = E_M(\mathcal{K}, \mathcal{K}) \big\}$.

Lemma 3.7. *The following are equivalent for a module M:*

1) M-ptor $= M$-tor.
2) $E_M(\mathcal{K}, \mathcal{L}) = \mathcal{K} \vee \mathcal{L}$ *for all $\mathcal{K}, \mathcal{L} \in M$-ptor.*

Proof. (1) $\Rightarrow$ (2). For $\mathcal{K}, \mathcal{L} \in M$-ptor, $\mathcal{K} \cup \mathcal{L} \subseteq \mathcal{K} \vee \mathcal{L} \in M$-ptor by [3, 6.1.10]. So, by (1), $\mathcal{K} \vee \mathcal{L} \in M$-tor and hence $E_M(\mathcal{K}, \mathcal{L}) \subseteq \mathcal{K} \vee \mathcal{L}$. But, by [15, 1.4, p. 75], $E_M(\mathcal{K}, \mathcal{L}) \in M$-ptor. This shows that $\mathcal{K} \vee \mathcal{L} \leq E_M(\mathcal{K}, \mathcal{L})$ since $\mathcal{K} \vee \mathcal{L}$ is the smallest pre-natural class containing $\mathcal{K} \cup \mathcal{L}$. So $E_M(\mathcal{K}, \mathcal{L}) = \mathcal{K} \vee \mathcal{L}$.

(2) $\Rightarrow$ (1). For $\mathcal{K} \in M$-ptor, $\mathcal{K} = E_M(\mathcal{K}, \mathcal{K})$ by (2). So $\mathcal{K} \in M$-tor. $\square$

For the Gabriel dimension of modules, we refer to [6, 7]. The Gabriel dimension of a module N, when exists, is denoted by $G\dim(N)$. Let $\mathcal{U}$ be the class of modules that have Gabriel dimension. Then $\mathcal{U} \in \mathcal{T}^p(R)$ and we let μ denote the left exact preradical corresponding to the hereditary pretorsion class $\mathcal{U}$. For a module M, $\sigma[M] \cap \mathcal{U}$ is again a hereditary pretorsion class, and we let μ_M be the left exact preradical corresponding to the hereditary pretorsion class $\sigma[M] \cap \mathcal{U}$. The next lemma can be proved as in the proof of [3, 6.4.15, p. 182] line by line. For the definition of an β-simple module (where β is an ordinal), see [3, p. 181].

Lemma 3.8. *Let M be a module such that M-ptor $= M$-tor and let β be an ordinal.*

1) *If $N \in \sigma[M]$ is a β-simple module, then $\mu_M(E_M(N))$ is β-simple.*
2) *If $N \in \sigma[M]$ is a semisimple module, then $N = \mu_M(E_M(N))$.*

Theorem 3.9. *The following are equivalent for a module M:*

1) *M is QI.*
2) *M is a locally noetherian module and M-ptor $= M$-tor.*
3) *M has Gabriel dimension and M-ptor $= M$-tor.*
4) *$\mathcal{K}_1 \vee \mathcal{K}_2 = E_M(\mathcal{K}_1, \mathcal{K}_2)$ for all $\mathcal{K}_1, \mathcal{K}_2 \in \mathcal{N}^p(R, M)$.*
5) *M-ptor $= M$-tor and $E_M(\mathcal{K}_1, \mathcal{K}_2) \in \mathcal{N}_r^p(R, M)$ for all $\mathcal{K}_1, \mathcal{K}_2 \in \mathcal{N}^p(R, M)$.*
6) *M-ptor $= M$-tor and every M-singular quasi-injective module in $\sigma[M]$ is M-injective.*

Proof. (1) $\Rightarrow$ (2). Suppose that (1) holds. Then M is locally noetherian. By Theorem 3.4, every $\mathcal{K}$ in M-ptor is an M-natural class, so $\mathcal{K}$ is closed under extensions in $\sigma[M]$ by [3, 2.4.5, p. 25].

(2) $\Rightarrow$ (3). If $x \in M$, then $(xR + \mu(M))/\mu(M)$ is noetherian, so it has Gabriel dimension (see [7, Proposition 2.3]). Thus, $xR + \mu(M) \in \mathcal{U}$ by [7, Lemma 1.3]. Thus $xR \subseteq \mu(M)$. So $M = \mu(M)$ has Gabriel dimension.

(3) $\Rightarrow$ (1). Since M has Gabriel dimension, $M \in \mathcal{U}$, so $\sigma[M] \subseteq \mathcal{U}$. Then μ_M is just the left exact preradical corresponding to the hereditary pretorsion class $\sigma[M]$. Let N be a semisimple module in $\sigma[M]$. Then $E_M(N) \in \sigma[M] \cap \mathcal{U}$. Hence $E_M(N) = \mu_M(E_M(N))$. But, by Lemma 3.8, $N = \mu_M(E_M(N))$. Hence $N = E_M(N)$ is M-injective. Thus, every semisimple module in $\sigma[M]$ is M-injective. This shows that M is locally noetherian. In fact, if $A_1 \subset A_2 \subset \cdots \subseteq xR \subseteq M$ is a strictly ascending chain of submodules, then for each $i \geq 1$, there exist submodules B_i and C_i of xR such that $A_i \subseteq B_i \subset C_i \subseteq A_{i+1}$ and C_i/B_i is simple. Let $A = \cup_i A_i$. For each $i \geq 1$, let $f_i : C_i \longrightarrow C_i/B_i$ be the canonical homomorphism. Since C_i/B_i is M-injective, there is a homomorphism $g_i : A \longrightarrow C_i/B_i$ which extends f_i. Let $S = \oplus_i(C_i/B_i)$. Define a map $f : A \longrightarrow S = \oplus_i(C_i/B_i)$ by $\pi_i \circ f(a) = g_i(a)$, where π_i is the projection of S onto C_i/B_i. It is easily seen that f is a well-defined homomorphism. Since S is M-injective, it is xR-injective. So there exists a homomorphism $g : xR \longrightarrow S$ which extends f. Since $g(xR)$ is cyclic, there exists some $n \geq 1$ such that $f(A) \subseteq g(xR) \subseteq \oplus_{j<n}(C_j/B_j)$. For any $a \in C_n$, we have $0 = \pi_n \circ f(a) = g_n(a) = a + B_n$, implying that $a \in B_n$. So, $C_n = B_n$, a contradiction. So M is locally noetherian. To prove that M is QI, it suffices to show that each $\mathcal{T}$ in

M-ptor is an M-natural class by Theorem 3.4, i.e., to show that $\mathcal{T}$ is closed under injective hulls. Now let $0 \neq N \in \mathcal{T}$ and we need to show that $E_M(N) \in \mathcal{T}$. Since M is locally noetherian, by [3, 3.1.25, p. 48], $E_M(N) = \oplus_t N_t$ where each N_t is uniform M-injective. Thus, it suffices to show that each $N_t \in \mathcal{T}$. Let $X = N \cap E_t$ and $Y = E_t$. Since X has Gabriel dimension (as $X \in \mathcal{U}$), X contains a nonzero β-simple submodule P for some ordinal β. Since $Y \in \sigma[M] \cap \mathcal{U}$, $Y = \mu_M(Y) = \mu_M(E_M(P))$ is β-simple by Lemma 3.8. Let τ be the left exact preradical corresponding to the hereditary pretorsion class $\sigma[N]$ and let $\mathcal{L} = \sigma[\tau(Y) \oplus Y/\tau(Y)]$. Thus, $\mathcal{L}$ is closed under extensions in $\sigma[M]$. So $Y \in \mathcal{L}$. Hence there exists an epimorphism $\theta : K \to Y$ where $K \subseteq A \oplus B$, $A \oplus \tau(Y)$ and $B = \oplus(Y/\tau(Y))$. Let $L = K \cap B$. Because Y is β-simple and because $\tau(Y)$ is essential in Y, $G \dim(Y/\tau(Y)) < \beta$ and hence $G \dim L < \beta$. Thus, $G \dim(\theta(L)) < \beta$. As Y is β-simple, it must be that $\theta(L) = 0$. Hence θ induces an epimorphism $K/L \to Y$. This shows that $K/L = K/(K \cap B) \cong (K + B)/B \hookrightarrow A$, so $K/L \in \sigma[X]$ since $A \in \sigma[X]$. Hence $Y \in \sigma[X] \subseteq \mathcal{T}$.

$(1) \Rightarrow (4)$. Suppose that M is QI. Then $\mathcal{N}^p(R, M) = \mathcal{N}(R, M)$ by Theorem 3.4. Thus, (4) follows by Lemma 3.6.

$(4) \Rightarrow (5)$. For $\mathcal{K} \in M$-ptor, $\mathcal{K} \vee \mathcal{K} = E_M(\mathcal{K}, \mathcal{K})$ by (4). So $\mathcal{K}$ is closed under extensions in $\sigma[M]$. The rest of (5) follows clearly by (4).

$(5) \Rightarrow (6)$. Let N be an M-singular quasi-injective module. Then $N \in \sigma[M]$. Let $\mathcal{K} = \sigma[M]$. Then $\mathcal{K}$ is closed under extensions in $\sigma[M]$ by (5). So $\mathcal{K}$ is a hereditary torsion class in $\sigma[M]$ (note that $\mathcal{K}$ need not be a hereditary torsion class in Mod-R). Let τ be the left exact radical in $\sigma[M]$ corresponding to $\mathcal{K}$. Since N is quasi-injective, $N = E_N(N) = \tau(E_M(N))$. Thus, $\tau(E_M(N)/N) = \tau(E_M(N)/\tau(E_M(N))) = \bar{0}$. Let $\mathcal{L} = d(E_M(N)/N) \cap \sigma[M]$. By (5), every $\mathcal{F}$ in M-ptor is closed under extensions in $\sigma[M]$, so every $\mathcal{F}$ in $\mathcal{N}^p(R, M)$ is closed under extensions in $\sigma[M]$ because of [3, 2.3.5]. So $E_M(N) \in E_M(\mathcal{L}, \mathcal{K})$. Thus, there exists $X \subseteq E_M(N)$ such that $X \in \mathcal{L}$ and $E_M(N)/X \in \mathcal{K}$. Because $\tau(E_M(N)/N) = \bar{0}$, $X \in \mathcal{L}$ shows that $\tau(X) = 0$. But $X \cap N \subseteq \tau(X)$, so $X \cap N = 0$. This shows that $X = 0$ because N is essential in $E_M(N)$. So, $E_M(N) \in \mathcal{K} = \sigma[N]$. Since N is quasi-injective, it follows that N is $E_M(N)$-injective. So $N = E_M(N)$ is M-injective.

$(6) \Rightarrow (1)$. Let $N \in \sigma[M]$ be quasi-injective. Let $\mathcal{T}$ be the class of the M-singular modules. Then $\mathcal{T} \in M$-ptor, so $\mathcal{T}$ is closed under extensions in $\sigma[M]$ by (6). Let τ be the left exact radical in $\sigma[M]$ corresponding to $\mathcal{T}$. Then $\tau(N/\tau(N)) = \bar{0}$. Thus, $\tau(N)$ is a closed submodule of N. Since N is quasi-injective, $N = \tau(N) \oplus P$. So $\tau(N)$ is an M-singular quasi-injective module and hence M-injective by (6). To show N is M-injective, it suffices to show that P is M-injective. Notice that P is non M-singular quasi-injective. Let $\mathcal{K} = \sigma[P]$. By Lemma 3.7, $E_M(\mathcal{K}, \mathcal{T}) = \mathcal{K} \vee \mathcal{T} = \mathcal{T} \vee \mathcal{K} = E_M(\mathcal{T}, \mathcal{K})$. From $0 \to P \to E_M(P) \to E_M(P)/P \to 0$, we have $E_M(P) \in E_M(\mathcal{K}, \mathcal{T}) = E_M(\mathcal{T}, \mathcal{K})$, so there exists $X \subseteq E_M(P)$ such that $X \in \mathcal{T}$ and $E_M(P)/X \in \mathcal{K}$. Since P is non M-singular, $P \cap X = 0$. So $X = 0$ since P is essential in $E_M(P)$. Thus, $E_M(P) \in \mathcal{K} = \sigma[P]$. Since P is quasi-injective, P is $E_M(P)$-injective. So $P = E_M(P)$ is M-injective. $\qquad \square$

4. Quasi-injective modules inside $\sigma[M]$

This section gives an overview of quasi-injective modules. The next lemma gives necessary and sufficient conditions entirely inside R in terms of right ideals in order for a module to be quasi-injective, in order for a module in $\sigma[M]$ to be quasi-injective. Its proof follows from [4], and in particular, (2) and (3) of Lemma 4.1 follow from [4, Lemma 2]. To show the contrast between quasi-injectivity and M-injectivity we also list (4) and (5).

Lemma 4.1. *Let* $M, N \in$ Mod-R *and* $P \in \sigma[M]$. *Then the following hold:*

1) N *is quasi-injective* $\Longleftrightarrow \forall x \in N$ *and* $I \leq R_R$, *any homomorphism* $\phi : xI \longrightarrow xR$ *extends to a homomorphism* $xR \longrightarrow xR$.

2) N *is quasi-injective* $\Longleftrightarrow \forall I \leq R_R$ *with* $x^{\perp} \subseteq I$ *for some* $x \in N$, *any homomorphism* $\phi : I \longrightarrow N$ *with* $x^{\perp} \subseteq \text{Ker}(\phi)$ *extends to a homomorphism* $R \longrightarrow N$.

3) P *is quasi-injective* $\Longleftrightarrow \forall y \in P$ *and* $I \leq R_R$ *with* $y^{\perp} \subseteq I$, *any homomorphism* $\phi : I \longrightarrow P$ *with* $y^{\perp} \subseteq \text{Ker}(\phi)$ *extends to a homomorphism* $R \longrightarrow P$. *(Here* yR *is a typical cyclic module of* $\sigma[M]$.)

4) P *is* M-*injective* $\Longleftrightarrow \forall x^{\perp} \subseteq I \leq R_R$, *where* $x \in M$, *any homomorphism* $\phi : I \longrightarrow P$ *with* $x^{\perp} \subseteq \text{Ker}(\phi)$ *extends to a homomorphism* $R \longrightarrow P$.

5) N *is* M-*injective* $\Longleftrightarrow \forall x^{\perp} \subseteq I \leq R_R$, *where* $x \in M$, *any homomorphism* $\phi : I \longrightarrow N$ *with* $x^{\perp} \subseteq \text{Ker}(\phi)$ *is of the form* $\phi(i) = yi$ *for all* $i \in I$, *for some* $y \in N$.

6) N *is quasi-injective* $\Longleftrightarrow$ *above (5) holds for* $M = N$, *and* $x \in N$.

Notation 4.2. *For any modules* M_R *and* N_R, *define* $\Lambda = \text{End}_R(E(N))$, $S = \text{End}_R(E_M(N))$ *and* $\widetilde{N} = \Lambda N$. *Then* $N \subseteq \widetilde{N} \subseteq E(N)$. *Note that* $\widetilde{N}$ *and* $E_M(N)$ *are fully invariant in* $E(N)$, *and hence quasi-injective.*

Remark 4.3. *From now on,* $N \in \sigma[M]$.

1) $\widetilde{N} = \Lambda N = SN \subseteq E_M(N)$.

2) *Hence* N *is fully invariant in* $E(N) \Longleftrightarrow N$ *is fully invariant in* $E_M(N)$.

Remark 4.4. *The following are all equivalent:*

1) M *is a QI-module.*

2) $\forall$ *fully invariant submodule* N *of* $E(N)$ *with* $N \in \sigma[M]$, *there does not exist a fully invariant submodule* P *of* $E(N)$ *with* $N \subset P \in \sigma[M]$.

3) *Same as (2), but with 'E' replaced by 'E_M'.*

Proof. Note that if M is QI and $N, P \in \sigma[M]$ such that N is fully invariant in $E(N)$ and P is fully invariant in $E(P)$, then $N = E_M(N)$ and $P = E_M(P)$.

$(1) \Longrightarrow (2)$. Suppose that $N \subset P \subseteq E(N)$. Since $N \leq_e P$ is essential, we get that $E_M(N) = E_M(P)$, or $N = P$, a contradiction.

$(1) \Longleftarrow (2)$. Let $N \in \sigma[M]$ be quasi-injective. Since $N \in \sigma[M]$, $N \subseteq E_M(N)$. If $N \subset E_M(N)$ with $N \neq E_M(N)$, let $P = E_M(N)$ which contradicts (1).

Clearly, (2) and (3) are equivalent. $\qquad\square$

Proposition 4.5. *For any module N, $\widetilde{N} \in \sigma[N]$. Furthermore, $\widetilde{N}^\perp = N^\perp$.*

Proof. From $\widetilde{N} = \Lambda N = \Sigma\{fN : f \in \mathrm{Hom}_R(N, E(N))\}$, and from $fN \cong N/\mathrm{Ker}(f)$, it follows that $\widetilde{N}$ is isomorphic to a quotient of some direct sum of N's, and $\widetilde{N} \in \sigma[N]$. As a consequence of the latter, $\widetilde{N}N^\perp = 0$, and $N^\perp \subseteq \widetilde{N}^\perp$. From $N \subseteq \widetilde{N}$ it follows that $\widetilde{N}^\perp \subseteq N^\perp$. Thus $N^\perp = \widetilde{N}^\perp$. $\qquad\square$

We give a description of quasi-injective modules in $\sigma[M]$.

Remark 4.6. *Let $N \in \sigma[M]$, and let $\mathcal{F} = \{L \leq R : L \supseteq \cap_{i=1}^n x_i^\perp$ for some n and some $x_i \in N\}$. Notice that if $L \supseteq \cap_{i=1}^n x_i^\perp$ then $a^{-1}L := (a+L)^\perp \supseteq \cap_{i=1}^n (x_i a)^\perp$ for each $a \in R$. Thus, $\mathcal{F}$ is a prefilter (see [3, 2.1.8]). Let τ be the left exact preradical determined by $\mathcal{F}$ (see [3, 2.1.3]), and hence $\tau(E_M(N)) = \{x \in E_M(N) : x^\perp \in \mathcal{F}\}$. Then $\tau(E_M(N))$ is a submodule of $E_M(N)$. In fact, if $x_1, x_2, x \in \tau(E_M(N))$ and $r \in R$, then $(x_1 - x_2)^\perp \supseteq x_1^\perp \cap x_2^\perp \in \mathcal{F}$. Since $x^\perp \in \mathcal{F}$, for some $x_i \in N$, $x^\perp \supseteq \cap_{i=1}^n x_i^\perp$, and thus $(xr)^\perp = r^{-1}x^\perp := (r + x^\perp)^\perp \supseteq r^{-1}[\cap_{i=1}^n x_i^\perp] = \cap_{i=1}^n (x_i r)^\perp$, where $x_i r \in N$. Consequently, $r^{-1}x^\perp \in \mathcal{F}$. So $x_1 - x_2$, $xr \in \tau(E_M(N))$.*

In general the inclusion $\{x^\perp : x \in N\} \subset \mathcal{F}$ is proper (see [11, p. 1282]). The following hold:

1) $\widetilde{N} = \{x \in E_M(N) : x^\perp \supseteq y^\perp$ for some $y \in N\}$.
2) $\widetilde{N} = \tau(E_M(N))$.
3) $\mathcal{F}$ *is the prefilter generated by $\{x^\perp : x \in N\}$.*

Proof. Conclusion (3) holds by definition of $\mathcal{F}$.

(1) The starting point is the result of de la Rosa and Viljoen [11, Lemma (1), p. 1281]), which is proved by use of Fuchs [4, Lemma 2, p. 542], which says that (1) above holds if $E_M(N)$ is replaced by $E(N)$. Since each $x \in \widetilde{N}$ is a quotient of $yR \in \sigma[M]$, $xR \subseteq \mathrm{trace}(\sigma[M], E(N)) = E_M(N)$. Thus (1) holds.

(2) Clearly, by (1), $\widetilde{N} \subseteq \tau(E_M(N))$. For any n, let $z \in \tau(E_M(N))$ with $z^\perp \supseteq \cap_{i=1}^n y_i^\perp$ for $y_i \in N$. We now apply conclusion (1) to the quasi-injective hull of n copies of N which equals $\widetilde{N} \oplus \cdots \oplus \widetilde{N}$. Take the diagonal element $x = (z, \ldots, z) \in E_M(N) \oplus \cdots \oplus E_M(N)$ and let $y = (y_1, \ldots, y_n) \in N \oplus \cdots \oplus N$. Then $z^\perp = x^\perp \supseteq y^\perp = (y_1, \ldots, y_n)^\perp = \cap_{i=1}^n y_i^\perp$. Thus $(z, \ldots, z) \in \widetilde{N} \oplus \cdots \oplus \widetilde{N}$, and hence $z \in \widetilde{N}$. Since we can do this argument for any finite n, $\widetilde{N} = \tau(E_M(N))$. $\qquad\square$

A ring R is called a right V-ring if every simple R-module is injective.

Lemma 4.7. *If R is not a right V-ring, and N is a QI-module, then $x^\perp \neq 0$ for each $x \in N$.*

Proof. If not, let $R_R \cong xR \in \sigma[N] = \mathrm{Mod}\text{-}R$. Let $P \in \sigma[N]$ be a simple non-injective module. Then P is quasi-injective, hence N-injective, hence xR-injective, and thus injective, a contradiction. $\qquad\square$

Lemma 4.8. *The following domains R are not V-rings. Consequently, any QI-module N over any of these rings is as in the previous lemma, i.e., N does not contain a free submodule.*

  1) *R has a non-zero primitive ideal.*
  2) *R is commutative and not a field.*

Proof. (1) In this case, $\exists\, 0 \neq a \in (R/L)^{\perp} \subseteq L < R$, where $L < R$ is a maximal right ideal. Map $aR \longrightarrow R/L$ by $ar \mapsto r + L$, and if R/L is injective, extend this to R by $1 \mapsto b + L$. Then $ba \in (R/L)^{\perp} \subseteq L$, while $L = ba + L = 1 + L$ is a contradiction. So R/L is a simple non-injective module.

(2) In this case, R has a non-zero maximal ideal and (1) applies. $\qquad\square$

Proposition 4.9. *Assume that R is any ring such that for any uniform cyclic modules $0 \neq yR < xR$, $(xR)^{\perp} \subset (yR)^{\perp}$ is a proper inclusion. Then any QI-module N which contains no free submodules is semisimple.*

Proof. By Proposition 2.1(4), the module N contains an essential direct sum of non-zero uniform cyclic modules $xR \leq N$. If xR is not simple, there exists $0 \neq yR \subset xR$. By the hypothesis on R, $(xR)^{\perp} \subset (yR)^{\perp}$. Then $\widetilde{xR} \in \sigma[xR]$ by Proposition 4.5. Since xR is QI, and $\widetilde{yR} \subseteq \widetilde{xR}$, it follows that the quasi-injective module $\widetilde{yR}$ is $\widetilde{xR}$-injective. So $\widetilde{yR}$ is a direct summand of $\widetilde{xR}$. Thus, $\widetilde{yR} = \widetilde{xR}$ because xR is uniform. So $(yR)^{\perp} = \widetilde{yR}^{\perp} = \widetilde{xR}^{\perp} = (xR)^{\perp}$ by Proposition 4.5. This contradicts the assumption. So each xR is simple. Thus N contains an essential submodule P that is semisimple. Since N is QI, P is N-injective and hence a direct summand of N. Thus, $N = P$ is semisimple. $\qquad\square$

All commutative rings satisfy the hypothesis of the last proposition.

Corollary 4.10. *If R is a commutative ring, then any torsion (i.e., $\forall 0 \neq x \in N$, $x^{\perp} \neq 0$) QI-module N is semisimple.*

Acknowledgment

The work of the second author was supported by NSERC Grant OGP0194196.

References

[1] J. Beachy and W.D. Blair, Rings whose faithful left ideals are cofaithful, *Pacific J. Math.* **58**(1975), 1–13.

[2] J. Dauns, Unsaturated classes of modules, *Abelian groups and modules* (Colorado Springs, CO, 1995), pp. 211–225, Lecture Notes in Pure and Appl. Math., **182**, Dekker, New York, 1996.

[3] J. Dauns and Y. Zhou, Classes of Modules, *Pure and Applied Mathematics (Boca Raton)*, **281**, Chapman & Hall/CRC, Boca Raton, FL, 2006.

[4] L. Fuchs, On quasi-injective modules, *Annali dela Scuola Norm. Sup. Pisa* **23**(1969), 541–546.

[5] J.L. García Hernández and J.L. Gómez Pardo, V-rings relative to Gabriel topologies, *Comm. Alg.* **13**(1985), 58–83.

[6] R. Gordon and J.C. Robson, Krull Dimension, *Mem. Amer. Math. Soc.* **133**, 1973.

[7] R. Gordon and J.C. Robson, The Gabriel dimension of a module, *J. Algebra* **29**(1974), 459–473.

[8] S.H. Mohamed and B.J. Müller, Continuous and Discrete Modules, *London Math. Soc. Lecture Note Series* **147**, Cambridge University Press, New York, 1990.

[9] S. Page and Y. Zhou, On direct sums of injective modules and chain conditions, *Canad. J. Math.* **46**(1994), 634–647.

[10] F. Raggi, J.R. Montes and R. Wisbauer, The lattice structure of hereditary pretorsion classes, *Comm. Algebra* **29**(2001), 131–140.

[11] B. de la Rosa, and G. Viljoen, A note on quasi-injective modules, *Comm. Alg.*, **15**(6)(1987), 1279–1286.

[12] M.L. Teply, On the idempotence and stability of kernel functors, *Glasgow Math. J.* **37**(1995), 37–43.

[13] R. Wisbauer, Generalized co-semisimple modules, *Comm. Algebra* **18**(12)(1990), 4235–4253.

[14] R. Wisbauer, Foundations of Module and Ring Theory, *Gordon and Breach Science Publishers*, 1991.

[15] R. Wisbauer, On module classes closed under extensions, in: *Rings and Radicals*, pp. 73–97, Pitman Res. Notes Math. Ser., **346**, Longman, Harlow, 1996.

[16] R. Wisbauer, Cotilting objects and dualities, *Representations of algebras (Sao Paulo, 1999)*, 215–233, Lecture Notes in Pure and Appl. Math., **224**, Dekker, New York, 2002.

[17] Y. Zhou, Direct sums of M-injective modules and module classes, *Comm. Algebra*, **23**(3)1995, 927–940.

[18] Y. Zhou, The lattice of pre-natural classes of modules, *J. Pure Appl. Algebra* **140**(2)(1999), 191–207.

John Dauns
Tulane University
New Orleans
Louisiana 70118-5698, USA
e-mail: dauns@tulane.edu

Yiqiang Zhou
Memorial University of Newfoundland
St.John's
Newfoundland A1C 5S7, Canada
e-mail: zhou@math.mun.ca

Modules and Comodules

Trends in Mathematics, 185–201

© 2008 Birkhäuser Verlag Basel/Switzerland

Corings with Exact Rational Functors and Injective Objects

L. El Kaoutit and J. Gómez-Torrecillas

Professor Robert Wisbauer gewidmet

Abstract. We describe how some aspects of abstract localization on module categories have applications to the study of injective comodules over some special types of corings. We specialize the general results to the case of Doi-Koppinen modules, generalizing previous results in this setting.

Introduction

The Wisbauer category $\sigma[M]$ subgenerated by a module M [20] is a flexible and useful tool when applied to some at a first look unrelated situations. This has been the case of the categories of comodules over corings, which, under suitable conditions, become Wisbauer's categories [2, 5, 12]. On the other hand, as it was explained in [6], the categories of entwined modules and, henceforth, of Doi-Koppinen modules, are instances of categories of comodules over certain corings, which ultimately enlarges the field of influence of the methods from Module Theory developed in [20]. The present paper has been deliberately written from this point of view, although with a necessarily different style. To illustrate how abstract results on modules may successfully be applied to more concrete situations, we have chosen a topic from the theory of Doi-Koppinen modules with roots in the theory of graded rings and modules, namely, the transfer of the injectivity from relative modules (Doi-Koppinen, graded) to the underlying modules over the ground ring (comodule algebra, graded algebra). This was studied at the level of Doi-Koppinen modules in [10], giving versions in this framework of results on graded modules from [9]. The methods developed in [10] rest on the exactness of the rational functor

This research is supported by the grants MTM2004-01406 and MTM2007-61673 from the Ministerio de Educación y Ciencia of Spain, and P06-FQM-01889 from the Consejería de Innovación, Ciencia y Empresa of Andalucía, with funds from FEDER (Unión Europea).

for semiperfect coalgebras over fields [17], which allows the construction of a suitable adjoint pair between the category of Doi-Koppinen modules and the category of modules over the smash product [10, Theorem 3.5]. The pertinent observation here, from the point of view of corings, is that one of the functors in that adjoint pair is already a rational functor for the coring associated to the comodule algebra [2, Proposition 3.21]. Thus, a relevant ingredient in [10] is, under this interpretation, the exactness of the trace functor defined by a Wisbauer category of modules or, equivalently, the exactness of the preradical associated to a closed subcategory of a category of modules. Here, we make explicit the fact that the exactness of such a preradical is equivalent to the property of being, up to an equivalence of categories, the canonical functor of a localization (Theorem 1.1), and, henceforth, it has a right adjoint, which is explicitly described. This right adjoint will preserve injective envelopes, since it is a section functor (Proposition 1.3). We then deduce the general form of the transfer of injective objects stated in [10].

In the rest of this paper, we specialize the former general scheme to corings with exact rational functors and, even more, to Doi-Koppinen modules where the coacting coalgebra has an exact rational functor.

The results of this paper should not be considered as completely new. In fact, most of them could be gathered, with suitable adaptations (not always obvious), from other sources. Thus, our text resembles a mini-survey. However, we believe that the reader will not find elsewhere the statements made here, nor the applications to the transfer of injectivity, since they do not intend to be reproductions of previously published results. We hope we have presented a study of some aspects of the theory of corings and their comodules in a new light.

Notations and basic notions. Throughout this paper the word ring will refer to an associative unital algebra over a commutative ring K. The category of all left modules over a ring R will be denoted by $_R\mathsf{Mod}$, being Mod_R the notation for the category of all right R-modules. The notation $X \in \mathcal{A}$ for a category $\mathcal{A}$ means that X is an object of $\mathcal{A}$, and the identity morphism attached to any object X will be denoted by the same character X.

Recall from [19] that an A-coring is a three-tuple $(\mathfrak{C}, \Delta_{\mathfrak{C}}, \varepsilon_{\mathfrak{C}})$ consisting of an A-bimodule $\mathfrak{C}$ and two homomorphisms of A-bimodules (the comultiplication and the counity)

$$\mathfrak{C} \xrightarrow{\ \Delta_{\mathfrak{C}}\ } \mathfrak{C} \otimes_A \mathfrak{C} \ , \quad \mathfrak{C} \xrightarrow{\ \varepsilon_{\mathfrak{C}}\ } A$$

such that $(\Delta_{\mathfrak{C}} \otimes_A \mathfrak{C}) \circ \Delta_{\mathfrak{C}} = (\mathfrak{C} \otimes_A \Delta_{\mathfrak{C}}) \circ \Delta_{\mathfrak{C}}$ and $(\varepsilon_{\mathfrak{C}} \otimes_A \mathfrak{C}) \circ \Delta_{\mathfrak{C}} = (\mathfrak{C} \otimes_A \varepsilon_{\mathfrak{C}}) \circ \Delta_{\mathfrak{C}} = \mathfrak{C}$.

A right $\mathfrak{C}$-comodule is a pair (M, ρ_M) consisting of a right A-module M and a right A-linear map $\rho_M : M \to M \otimes_A \mathfrak{C}$, called right $\mathfrak{C}$-coaction, such that $(M \otimes_A \Delta_{\mathfrak{C}}) \circ \rho_M = (\rho_M \otimes_A \mathfrak{C}) \circ \rho_M$ and $(M \otimes_A \varepsilon_{\mathfrak{C}}) \circ \rho_M = M$. A morphism of right $\mathfrak{C}$-comodules (or a right $\mathfrak{C}$-colinear map) is a right A-linear map $f : M \to M'$ satisfying $\rho_{M'} \circ f = (f \otimes_A \mathfrak{C}) \circ \rho_M$. The K-module of all right $\mathfrak{C}$-colinear maps between two right comodules $M_{\mathfrak{C}}$ and $M'_{\mathfrak{C}}$ is denoted by $\mathrm{Hom}^{\mathfrak{C}}(M, M')$. Right $\mathfrak{C}$-comodules and their morphisms form a K-linear category $\mathsf{Comod}_{\mathfrak{C}}$. Although not abelian in general, $\mathsf{Comod}_{\mathfrak{C}}$ is a Grothendieck category provided $_A\mathfrak{C}$ is a flat module, see [12, Section 1]. The category $_{\mathfrak{C}}\mathsf{Comod}$ of left $\mathfrak{C}$-comodules is symmetrically defined.

For more information on corings and comodules, the reader is referred to [5] and its bibliography.

1. Exactness of a preradical, localization, and injective objects

In this section we will derive from [14] some facts on quotient categories that will be useful in the sequel. Recall that a full subcategory $\mathcal{C}$ of a Grothendieck category $\mathcal{G}$ is said to be *closed* if any subobject and any quotient object of an object belonging to $\mathcal{C}$ is in $\mathcal{C}$, and any direct sum of objects of $\mathcal{C}$ is in $\mathcal{C}$. A closed subcategory $\mathcal{C}$ of $\mathcal{G}$ defines a preradical $\mathfrak{r} : \mathcal{G} \to \mathcal{G}$, which sends an object X of $\mathcal{G}$ to its largest subobject $\mathfrak{r}(X)$ belonging to $\mathcal{C}$. This preradical is left exact, since it is right adjoint to the inclusion functor $\mathcal{C} \subseteq \mathcal{G}$. By $\mathsf{Ker}(\mathfrak{r})$ we denote the full subcategory of $\mathcal{G}$ with objects defined by the condition $\mathfrak{r}(X) = 0$.

A full subcategory $\mathcal{L}$ of $\mathcal{G}$ is *dense* if for any short exact sequence in $\mathcal{G}$

$$0 \longrightarrow X \longrightarrow Y \longrightarrow Z \longrightarrow 0 \,,$$

Y is in $\mathcal{L}$ if, and only if, both X and Z are in $\mathcal{L}$. From [13, 15.11] we know that a dense subcategory $\mathcal{L}$ is *localizing* in the sense of [14] if and only if it is stable under coproducts. Following [14, Chapter III], every localizing subcategory $\mathcal{L}$ of $\mathcal{G}$ defines a new Grothendieck category $\mathcal{G}/\mathcal{L}$ (the *quotient* category), and an exact functor $\mathbf{T} : \mathcal{G} \to \mathcal{G}/\mathcal{L}$ (the *canonical functor*) that admits a right adjoint $\mathbf{S} : \mathcal{G}/\mathcal{L} \to \mathcal{G}$. The counit $\phi_- : \mathbf{T} \circ \mathbf{S} \to \mathrm{id}_{\mathcal{G}/\mathcal{L}}$ of this adjunction is a natural isomorphism. The unit $\psi_- : \mathrm{id}_{\mathcal{G}} \to \mathbf{S} \circ \mathbf{T}$ satisfies the property that both the kernel and the cokernel of $\psi_X : X \to (\mathbf{S} \circ \mathbf{T})(X)$ belong to $\mathcal{L}$ for every object X of $\mathcal{G}$.

The exactness of a preradical $\mathfrak{r}$ can be expressed in terms of quotient categories, as the following proposition shows. The underlying ideas of its proof can be traced back to [17, Theorem 2.3].

Proposition 1.1. *Let $\mathcal{C}$ be a closed subcategory of a Grothendieck category $\mathcal{G}$ with associated preradical $\mathfrak{r} : \mathcal{G} \to \mathcal{C}$, and inclusion functor $\mathfrak{l} : \mathcal{C} \to \mathcal{G}$. The following statements are equivalent:*

(i) *$\mathfrak{r}$ is an exact functor;*
(ii) *$\mathcal{K} = \mathsf{Ker}(\mathfrak{r})$ is a localizing subcategory of $\mathcal{C}$ with canonical functor $\mathbf{T}$, and there exists an equivalence of categories $H : \mathcal{G}/\mathcal{K} \to \mathcal{C}$ such that $H \circ \mathbf{T} = \mathfrak{r}$.*

Proof. (i) $\Rightarrow$ (ii) Since $\mathfrak{r}$ is exact and preserves coproducts, we easily get that $\mathcal{K} = \mathsf{Ker}(\mathfrak{r})$ is a localizing subcategory. Consider the canonical adjunctions

$$\mathcal{C} \xrightarrow[\mathfrak{r}]{\mathfrak{l}} \mathcal{G} \,, \qquad \mathcal{G} \xrightarrow[\mathbf{S}]{\mathbf{T}} \mathcal{G}/\mathcal{K}$$

where $\mathbf{S}$ is right adjoint to $\mathbf{T}$, and $\mathfrak{r}$ is right adjoint to the inclusion functor $\mathfrak{l}$. Composing we get a new adjoint pair $\mathbf{T} \circ \mathfrak{l} : \mathcal{C} \leftrightarrows \mathcal{G}/\mathcal{K} : \mathfrak{r} \circ \mathbf{S}$, which we claim to provide an equivalence of categories. The unit of this new adjunction is given by

$$\mathrm{id}_{\mathcal{C}} = \mathfrak{r}\mathfrak{l} \xrightarrow{\ \mathfrak{r}\,\psi_{\mathfrak{l}}\ } \mathfrak{r}\,\mathbf{S}\,\mathbf{T}\,\mathfrak{l}$$

where ψ_- is the unit of the adjunction $\mathbf{T} \dashv \mathbf{S}$. For any object M of $\mathcal{C}$, there is an exact sequence

$$0 \longrightarrow X \longrightarrow \mathfrak{l}(M) \xrightarrow{\ \psi_{\mathfrak{l}(M)}\ } \mathbf{S}\mathbf{T}\mathfrak{l}(M) \longrightarrow Y \longrightarrow 0$$

with X and Y in $\mathcal{K}$. Apply the exact functor $\mathfrak{r}$ to obtain an isomorphism $\mathfrak{r}(\psi_{\mathfrak{l}(M)})$: $M = \mathfrak{r}\mathfrak{l}(M) \cong \mathfrak{r}\mathbf{S}\mathbf{T}\mathfrak{l}(M)$. Therefore, $\mathfrak{r}\psi_{\mathfrak{l}(-)}$ is a natural isomorphism. The counit of the adjunction $\mathbf{T} \circ \mathfrak{l} \dashv \mathfrak{r} \circ \mathbf{S}$ is given by the following composition

$$\mathbf{T}\,\mathfrak{l}\,\mathfrak{r}\,\mathbf{S} \xrightarrow{\ \mathbf{T}\,\lambda_{\mathbf{S}}\ } \mathbf{T}\,\mathbf{S} \xrightarrow[\cong]{\ \phi_-\ } \mathrm{id}_{\mathcal{G}/\mathcal{K}}$$

where λ_- is the counit of the adjunction $\mathfrak{l} \dashv \mathfrak{r}$, and ϕ_- is the counit of the adjunction $\mathbf{T} \dashv \mathbf{S}$. For any object N of $\mathcal{G}/\mathcal{K}$, $\lambda_{\mathbf{S}(N)}$ is a monomorphism with cokernel in $\mathcal{K}$ since $\mathfrak{r}$ is exact. Thus, [14, Lemme 2, p. 366] implies that $\mathbf{T}(\lambda_{\mathbf{S}(N)})$ is an isomorphism. Therefore, $\phi_N\,\mathbf{T}(\lambda_{\mathbf{S}(N)})$ is an isomorphism. Therefore, $\mathbf{T} \circ \mathfrak{l}$ is an equivalence of categories. On the other hand, by [14, Corollaire 3, p. 368], there exists a functor $H : \mathcal{G}/\mathcal{K} \to \mathcal{C}$ such that $H \circ \mathbf{T} = \mathfrak{r}$. By composing on the right with $\mathfrak{l}$ we get $H \circ \mathbf{T} \circ \mathfrak{l} = \mathfrak{r} \circ \mathfrak{l} = \mathrm{id}_{\mathcal{C}}$. From this, and using that $\mathbf{T} \circ \mathfrak{l}$ is an equivalence, we get that H is an equivalence.(ii) $\Rightarrow$ (i) This is obvious, since $\mathbf{T}$ is always exact. □

In the rest of this section we consider $\mathcal{G} = \mathsf{Mod}_B$, the category of right modules over a ring B. We fix the following notation: $\mathcal{C}$ is a closed subcategory of Mod_B, with preradical $\mathfrak{r} : \mathsf{Mod}_B \to \mathcal{C}$, and inclusion functor $\mathfrak{l} : \mathcal{C} \to \mathsf{Mod}_B$. We will consider the twosided ideal $\mathfrak{a} = \mathfrak{r}(B_B)$, and $\mathcal{K} = \mathsf{Ker}(\mathfrak{r})$.

The following proposition collects a number of well-known consequences of assuming that $\mathfrak{r}$ is exact. A short proof is included.

Proposition 1.2. *If $\mathfrak{r}$ is exact then $\mathfrak{a}$ is an idempotent ideal of B such that $_B(B/\mathfrak{a})$ is flat, and $\mathfrak{r}(M) = M\mathfrak{a}$ for every right B-module M. In this way, $\mathcal{K} = \mathsf{Ker}(\mathfrak{r})$ becomes a localizing subcategory of Mod_B stable under direct products and injective envelopes.*

Proof. Since $\mathfrak{r}$ preserves epimorphisms it follows easily that $\mathfrak{r}(M) = M\mathfrak{a}$, for any right B-module M. In particular, we get that $\mathcal{K} = \{M \in \mathsf{Mod}_B \mid M\mathfrak{a} = 0\}$. This easily implies that $\mathcal{K}$ is a localizing subcategory stable under direct products and essential extensions. Finally, the flatness of $_B(B/\mathfrak{a})$ can be proved as follows. We know that $\mathcal{K}$ is isomorphic to $\mathsf{Mod}_{B/\mathfrak{a}}$. Let $\pi : B \to B/\mathfrak{a}$ be the canonical projection; the functor $- \otimes_B (B/\mathfrak{a}) : \mathsf{Mod}_B \to \mathsf{Mod}_{B/\mathfrak{a}}$ is left adjoint to the restriction of scalars functor $\pi_* : \mathsf{Mod}_{B/\mathfrak{a}} \to \mathsf{Mod}_B$. Up to the isomorphism $\mathcal{K} \cong \mathsf{Mod}_{B/\mathfrak{a}}$, π_* is nothing but the inclusion functor $j : \mathcal{K} \to \mathsf{Mod}_B$. Since $\mathcal{K}$ is stable under injective envelopes, the functor $- \otimes_B (B/\mathfrak{a})$ has to be exact, that is, $_B(B/\mathfrak{a})$ is a flat module. □

If $\mathfrak{a}$ is any idempotent ideal of B such that $_B(B/\mathfrak{a})$ is flat, then there is a canonical isomorphism of B-bimodules $\mathfrak{a} \cong \mathfrak{a} \otimes_B \mathfrak{a}$. This isomorphism makes $\mathfrak{a}$ a B-coring with counit given by the inclusion $\mathfrak{a} \subseteq B$. We say that $\mathfrak{a}$ is a *left idempotent B-coring* to refer to this situation. The forgetful functor $U : \mathsf{Comod}_{\mathfrak{a}} \to \mathsf{Mod}_B$ induces then an isomorphism of categories between $\mathsf{Comod}_{\mathfrak{a}}$ and the full subcategory of Mod_B whose objects are the modules M_B such that $M\mathfrak{a} = M$.

Corollary 1.1. *Assume that $\mathfrak{r}$ is an exact functor. Then*

(i) *The ideal $\mathfrak{a} = \mathfrak{r}(B_B)$ is a left idempotent B-coring whose category of all right comodules $\mathsf{Comod}_{\mathfrak{a}}$ is isomorphic to the quotient category $\mathsf{Mod}_B/\mathcal{K}$. In particular $\mathfrak{a}$ is a generator of $\mathsf{Mod}_B/\mathcal{K}$.*

(ii) *The functor $F = \mathrm{Hom}_B(\mathfrak{a}, -) \circ \mathfrak{l} : \mathcal{C} \to \mathsf{Mod}_B$ is right adjoint to $\mathfrak{r}$, where $\mathfrak{l} : \mathcal{C} \to \mathsf{Mod}_B$ is the inclusion functor. In particular if E is an injective object of $\mathcal{C}$, then $F(E)_B$ is an injective right module.*

Proof. (i) By Proposition 1.2, $\mathfrak{a}$ is a left idempotent B-coring. Its category of right comodules clearly coincides with the torsion class $\mathcal{C}$, and the stated isomorphism of categories follows by Proposition 1.1.

(ii) Given any object (M, M') in $\mathsf{Mod}_B \times \mathcal{C}$, we get natural isomorphisms

$$\mathrm{Hom}_{\mathcal{C}}(\mathfrak{r}(M), M') \cong \mathrm{Hom}_B(M \otimes_B \mathfrak{a}, \mathfrak{l}(M')) \cong \mathrm{Hom}_B(M, \mathrm{Hom}_B(\mathfrak{a}, \mathfrak{l}(M'))),$$

since $\mathfrak{r}(M) = M\mathfrak{a} \cong M \otimes_B \mathfrak{a}$. This means that F is right adjoint to $\mathfrak{r}$. In particular, F preserves injectives since $\mathfrak{r}$ is exact. $\square$

Given a module M in Mod_B, the *Wisbauer category* $\sigma[M]$ associated to M is the full subcategory of Mod_B whose objects are all M-subgenerated modules (see [20]). By definition, it is a closed subcategory and, in fact, it is easy to prove that every closed subcategory of Mod_B is of the form $\sigma[M]$. Therefore, the following theorem, that summarizes some of the previous results, complements [5, 42.16].

Theorem 1.1. *Let $\mathcal{C}$ be a closed subcategory of a category of modules Mod_B with associated preradical $\mathfrak{r} : \mathsf{Mod}_B \to \mathcal{C}$. Let $\mathfrak{l} : \mathcal{C} \to \mathsf{Mod}_B$ be the inclusion functor, and $\mathfrak{a} = \mathfrak{r}(B)$. The following statements are equivalent.*

(i) *$\mathfrak{r} : \mathsf{Mod}_B \to \mathcal{C}$ is an exact functor;*

(ii) *$\mathcal{K} = \mathrm{Ker}(\mathfrak{r})$ is a localizing subcategory of Mod_B with canonical functor $\mathbf{T}$, and there exists an equivalence $H : \mathsf{Mod}_B/\mathcal{K} \to \mathcal{C}$ such that $\mathfrak{r} = H \circ \mathbf{T}$;*

(iii) *$F = \mathrm{Hom}_B(\mathfrak{a}_B, -) \circ \mathfrak{l} : \mathcal{C} \to \mathsf{Mod}_B$ is right adjoint to $\mathfrak{r}$;*

(iv) *$\mathfrak{a}$ is a left idempotent B-coring and the forgetful functor $U : \mathsf{Comod}_{\mathfrak{a}} \to \mathsf{Mod}_B$ induces an isomorphism of categories $\mathsf{Comod}_{\mathfrak{a}} \cong \mathcal{C}$;*

(v) *$\mathfrak{a}^2 = \mathfrak{a}$ and $M\mathfrak{a} = M$ for every M in $\mathcal{C}$.*

Proof. The equivalences (i) $\Leftrightarrow$ (ii) and (i) $\Leftrightarrow$ (iii) are immediate from Proposition 1.1 and Corollary 1.1.

(i) $\Rightarrow$ (iv) is a consequence of Proposition 1.1 and Corollary 1.1(i).

(iv) $\Rightarrow$ (v) Obvious.

(v) $\Rightarrow$ (i) We have easily that $\mathfrak{r}(M) = M\mathfrak{a}$, for every right B-module M. From this we get immediately that $\mathfrak{r}$ is a right exact functor. $\square$

Given a right B-module $M \in \mathcal{C}$, by $E_{\mathcal{C}}(M)$ we denote its injective hull in the Grothendieck category $\mathcal{C}$. According to Theorem 1.1, if $\mathfrak{r}$ is exact, then it becomes essentially the canonical functor associated to a localization with a section functor (the terminology is taken from [14]). As a section functor, $\operatorname{Hom}_B(\mathfrak{a}, -)$ will preserve injective envelopes, as stated in Proposition 1.3. We give a detailed proof of this fact, suitable for the forthcoming applications to more concrete situations.

Proposition 1.3. *Assume that* $\mathfrak{r} : \operatorname{Mod}_B \to \mathcal{C}$ *is exact, and let* $M \in \mathcal{C}$. *The map*

$$\zeta_M : M \to \operatorname{Hom}_B(\mathfrak{a}, E_{\mathcal{C}}(M)) \qquad (m \mapsto \zeta_M(m)(a) = ma, \ m \in M, a \in \mathfrak{a})$$

gives an injective envelope of M *in* Mod_B. *As a consequence,* M *is injective in* Mod_B *if and only if* M *is injective in* $\mathcal{C}$ *and* ζ_M *is an isomorphism.*

Proof. By Theorem 1.1, the functor $F = \operatorname{Hom}_B(\mathfrak{a}, -) \circ \iota : \mathcal{C} \to \operatorname{Mod}_B$ is right adjoint to the exact functor $\mathfrak{r}$. Therefore, $F(E_{\mathcal{C}}(M)) = \operatorname{Hom}_B(\mathfrak{a}, E_{\mathcal{C}}(M))$ is injective in Mod_B. On the other hand, ζ_M is obviously a right B-linear map. Let us show that it is injective. Let $m \in M$ such that $\zeta_M(m) = 0$, that is, $m\mathfrak{a} = 0$. By Theorem 1.1, we have $mB = m\mathfrak{a}$, which implies $m = 0$. Let us prove that ζ_M is essential. Pick a non zero element $f \in \operatorname{Hom}_B(\mathfrak{a}, E_{\mathcal{C}}(M))$, so there exists $0 \neq u \in \mathfrak{a}$ such that $0 \neq f(u) \in E_{\mathcal{C}}(M)$. Since M is essential in $E_{\mathcal{C}}(M)$, there exists a non zero element $b \in B$ such that $0 \neq f(u)b \in M$. Since $B/\mathfrak{a}$ is flat as a left B-module (Proposition 1.2), there exists $w \in \mathfrak{a}$ with $ub = ubw$ (see, e.g., [5, 42.5]). If we consider the map $g = fub$, then $g(x) = (fub)(x) = f(ubx)$, for all $x \in \mathfrak{a}$, that is $g = fub = \zeta_M(f(ub))$ is a non zero element of $\zeta_M(M)$, as $g(w) = f(ub) \neq 0$. $\square$

Definition 1.1. *Assume that* $\mathfrak{a}$ *has a set of local units in the sense of* [1], *that is,* $\mathfrak{a}$ *contains a set* E *of commuting idempotents such that for every* $x \in \mathfrak{a}$ *there exits* $e \in E$ *such that* $xe = ex = x$. *A right* B-module M *is said to be of* finite support *if there exits a finite subset* $F \subseteq E$ *such that* $m\left(\sum_{e \in F} e\right) = m$ *for every* $m \in M$.

A straightforward argument proves that if M_B is of finite support, then every $f \in \operatorname{Hom}_B(\mathfrak{a}, M)$ is of the form $f(x) = mx$ for some $m \in M$. Therefore, we deduce from Proposition 1.3:

Corollary 1.2. *Assume that* $\mathfrak{a}$ *has a set of local units, and let* $M \in \mathcal{C}$ *of finite support. Then* M *is injective in* Mod_B *if and only if* M *is injective in* $\mathcal{C}$. *As a consequence, given a homomorphism of rings* $A \to B$ *with* ${}_A B$ *flat, we deduce that if* M *is injective in* $\mathcal{C}$, *then* M *is injective in* Mod_A.

Remark 1.1. In Definition 1.1 and Corollary 1.2, it suffices to assume that $\mathfrak{a}$ contains a set of commuting idempotents E such that $\mathfrak{a} = \sum_{e \in E} eB$.

In what follows we specialize our results to the case where the subcategory $\mathcal{C}$ is isomorphic to the category of right comodules over a given A-coring $\mathfrak{C}$. This is the case when $\mathfrak{C}$ is member of a rational pairing $\mathcal{T} = (\mathfrak{C}, B, \langle -, - \rangle)$. Rational pairings for coalgebras over commutative rings were introduced in [15] and used in [3] to study the category of right comodules over the finite dual coalgebra associated to certain algebras over Noetherian commutative rings. This development was adapted for corings in [12], see also [2].

Recall from [12, Section 2] that a three-tuple $\mathcal{T} = (\mathcal{C}, B, \langle -, - \rangle)$ consisting of an A-coring $\mathcal{C}$, an A-ring B (i.e., B is an algebra extension of A) and a balanced A-bilinear form $\langle -, - \rangle : \mathcal{C} \times B \to A$, is said to be *a right rational pairing over* A provided

(1) $\beta_A : B \to {}^*\mathcal{C}$ is a ring anti-homomorphism, where ${}^*\mathcal{C}$ is the left dual convolution ring of $\mathcal{C}$ defined in [19, Proposition 3.2], and

(2) α_M is an injective map, for each right A-module M,

where α_- and β_- are the following natural transformations

$$\beta_N : B \otimes_A N \longrightarrow \operatorname{Hom}({}_A\mathcal{C}, {}_AN), \qquad \alpha_M : M \otimes_A \mathcal{C} \longrightarrow \operatorname{Hom}(B_A, M_A)$$
$$b \otimes_A n \longrightarrow [c \mapsto \langle c, b \rangle n] \qquad\qquad m \otimes_A c \longrightarrow [b \mapsto m \langle c, b \rangle].$$

Given a right rational pairing $\mathcal{T} = (\mathcal{C}, B, \langle -, - \rangle)$ over A, we can define a functor called the *right rational functor* as follows. An element m of a right B-module M is called *rational* if there exists a set of *right rational parameters* $\{(c_i, m_i)\} \subseteq \mathcal{C} \times M$ such that $mb = \sum_i m_i \langle c_i, b \rangle$, for all $b \in B$. The set of all rational elements in M is denoted by $\operatorname{Rat}^{\mathcal{T}}(M)$. As it was explained in [12, Section 2], the proofs detailed in [15, Section 2] can be adapted in a straightforward way in order to get that $\operatorname{Rat}^{\mathcal{T}}(M)$ is a B-submodule of M and the assignment $M \mapsto \operatorname{Rat}^{\mathcal{T}}(M)$ is a well-defined functor $\operatorname{Rat}^{\mathcal{T}} : \mathsf{Mod}_B \to \mathsf{Mod}_B$, which is in fact a left exact preradical. Therefore, the full subcategory $\operatorname{Rat}^{\mathcal{T}}(\mathsf{Mod}_B)$ of Mod_B whose objects are those B-modules M such that $\operatorname{Rat}^{\mathcal{T}}(M) = M$ is a closed subcategory. Furthermore, $\operatorname{Rat}^{\mathcal{T}}(\mathsf{Mod}_B)$ is a Grothendieck category which is shown to be isomorphic to the category of right comodules $\mathsf{Comod}_\mathcal{C}$ as [12, Theorem 2.6'] asserts (see also [2, Proposition 2.8]).

Example 1.1. Let $\mathcal{C}$ be an A-coring such that ${}_A\mathcal{C}$ is a locally projective left module (see [21, Theorem 2.1] and [2, Lemma 1.29]). Consider the endomorphism ring $\operatorname{End}({}_\mathcal{C}\mathcal{C})$ as a subring of the endomorphism ring $\operatorname{End}({}_A\mathcal{C})$, that is, with multiplication opposite to the composition of maps. Since $\Delta_\mathcal{C}$ is a left $\mathcal{C}$-colinear and a right A-linear map, the canonical ring extension $A \to \operatorname{End}({}_A\mathcal{C})$ factors throughout the extension $\operatorname{End}({}_\mathcal{C}\mathcal{C}) \hookrightarrow \operatorname{End}({}_A\mathcal{C})$. Therefore, the three-tuple $\mathcal{T} = (\mathcal{C}, \operatorname{End}({}_\mathcal{C}\mathcal{C}), \langle -, - \rangle)$, where the balanced A-bilinear $\langle -, - \rangle$ map is defined by $\langle c, f \rangle = \varepsilon_\mathcal{C}(f(c))$, for $(c, f) \in \mathcal{C} \times \operatorname{End}({}_\mathcal{C}\mathcal{C})$ is a rational pairing since $\operatorname{End}({}_\mathcal{C}\mathcal{C})$ is already a ring anti-isomorphic to ${}^*\mathcal{C}$ via the beta map associated to $\langle -, - \rangle$. We refer to $\mathcal{T}$ as *the right canonical pairing* associated to $\mathcal{C}$.

The following theorem complements [5, 20.8].

Theorem 1.2. *Let* $\mathcal{T} = (\mathcal{C}, B, \langle -, - \rangle)$ *be a right rational pairing with rational functor* $\operatorname{Rat}^{\mathcal{T}} : \mathsf{Mod}_B \to \mathsf{Mod}_B$, *and put* $\mathfrak{a} = \operatorname{Rat}^{\mathcal{T}}(B_B)$, $\mathcal{K} = \mathsf{Ker}(\operatorname{Rat}^{\mathcal{T}})$. *The following* *statements are equivalent:*

(i) $\mathrm{Rat}^{\mathcal{T}} : \mathsf{Mod}_B \to \mathsf{Mod}_B$ *is an exact functor;*

(ii) $\mathcal{K}$ *is a localizing subcategory of* Mod_B *with canonical functor* $\mathbf{T}$*, and there exists an equivalence* $H : \mathsf{Mod}_B/\mathcal{K} \to \mathrm{Rat}^{\mathcal{T}}(\mathsf{Mod}_B)$ *such that* $\mathrm{Rat}^{\mathcal{T}} = H \circ \mathbf{T}$*;*

(iii) $F = \mathrm{Hom}_B\,(\mathfrak{a}_B\,, -) \circ \mathfrak{l} : \mathrm{Rat}^{\mathcal{T}}(\mathsf{Mod}_B) \to \mathsf{Mod}_B$ *is right adjoint to* $\mathrm{Rat}^{\mathcal{T}}$*;*

(iv) $\mathfrak{a}$ *is a left idempotent* B*-coring and the forgetful functor* $U : \mathsf{Comod}_{\mathfrak{a}} \to \mathsf{Mod}_B$ *induces an isomorphism of categories* $\mathsf{Comod}_{\mathfrak{a}} \cong \mathrm{Rat}^{\mathcal{T}}(\mathsf{Mod}_B) \cong \mathsf{Comod}_{\mathfrak{C}}$*;*

(v) $_B\mathfrak{a}$ *is a pure submodule of* $_B B$*,* $\mathfrak{a}^2 = \mathfrak{a}$*, and* $\mathfrak{C}\mathfrak{a} = \mathfrak{C}$*.*

Proof. By Theorem 1.1 we only need to show that (v) $\Rightarrow$ (iv) since (iv) $\Rightarrow$ (v) is clear. We have that $\mathfrak{a}$ is a left idempotent B-coring and $\mathfrak{C} \cong \mathfrak{C} \otimes_B \mathfrak{a}$ as right B-modules. Given any rational right B-module X with its canonical structure of right $\mathfrak{C}$-comodule, we obtain a B-linear isomorphism $X \cong X\square_{\mathfrak{C}}(\mathfrak{C} \otimes_B \mathfrak{a})$ (recall that the comultiplication is a right B-linear map), where the symbol $-\square_{\mathfrak{C}}-$ refers to the cotensor bifunctor over $\mathfrak{C}$. Using the left version of [16, Lemma 2.2], we get

$$X \cong X\square_{\mathfrak{C}}\mathfrak{C} \cong X\square_{\mathfrak{C}}(\mathfrak{C} \otimes_B \mathfrak{a}) \cong (X\square_{\mathfrak{C}}\mathfrak{C}) \otimes_B \mathfrak{a} \cong X \otimes_B \mathfrak{a}.$$

That is, X is in fact a right $\mathfrak{a}$-comodule. $\square$

Remark 1.2. Right rational pairings are instances of right coring measuring in the sense of [4]. In this way, given an exact rational functor $\mathrm{Rat}^{\mathcal{T}}$ the isomorphism of categories $\mathsf{Comod}_{\mathfrak{C}} \cong \mathsf{Comod}_{\mathfrak{a}}$ stated in Theorem 1.2 can be interpreted as an isomorphism of corings in an adequate category. Following to [4, Definition 2.1], a B-coring $\mathfrak{D}$ is called a *right extension* of an A-coring $\mathfrak{C}$ provided $\mathfrak{C}$ is a $(\mathfrak{C}, \mathfrak{D})$-bicomodule with the left regular coaction $\Delta_{\mathfrak{C}}$. Corings understood as pairs $(\mathfrak{C} : A)$ (i.e., $\mathfrak{C}$ is an A-coring) and morphisms understood as right coring extensions (i.e., a pairs consisting of an action and coaction) with their bullet composition form a category denoted by $\mathbf{CrgExt}^r_K$ (see [4] for more details). If we apply this to the setting of Theorem 1.2, then it can be easily checked that $(\mathfrak{C} : A)$ and $(\mathfrak{a} : B)$ become isomorphic objects in the category $\mathbf{CrgExt}^r_K$.

From Proposition 1.3 and Corollary 1.2, we obtain:

Proposition 1.4. *Let* $\mathcal{T} = (\mathfrak{C}, B, \langle -, - \rangle)$ *be a right rational pairing with rational functor* $\mathrm{Rat}^{\mathcal{T}} : \mathsf{Mod}_B \to \mathsf{Mod}_B$*, and put* $\mathfrak{a} = \mathrm{Rat}^{\mathcal{T}}(B_B)$*. Assume that* $\mathrm{Rat}^{\mathcal{T}}$ *is an exact functor. Let* M *be a right* $\mathfrak{C}$*-comodule, and* $E(M_{\mathfrak{C}})$ *its injective hull in* $\mathsf{Comod}_{\mathfrak{C}}$*.*

(a) *The map*

$$\zeta_M : M \to \mathrm{Hom}_B\,(\mathfrak{a}\,, E(M_{\mathfrak{C}})) \quad (m \mapsto \zeta_M(m)(a) = ma, \; m \in M, a \in \mathfrak{a})$$

gives an injective envelope of M *in* Mod_B*.*

(b) M *is injective in* Mod_B *if and only if* M *is injective in* $\mathsf{Comod}_{\mathfrak{C}}$ *and* ζ_M *is an isomorphism.*

(c) *If* $_A B$ *is flat,* ζ_M *is an isomorphism, and* M *is injective in* $\mathsf{Comod}_{\mathfrak{C}}$*, then* M *in injective in* Mod_A*.*

(d) *If $\mathfrak{a}$ has a set of local units and M is of finite support, then M is injective in* $\mathsf{Comod}_{\mathfrak{C}}$ *if and only if M in injective in* Mod_B.

(e) *Assume that $\mathfrak{a}$ has a set of local units, M is of finite support and $_AB$ is flat. If M is injective in* $\mathsf{Comod}_{\mathfrak{C}}$, *then M is injective in* Mod_A.

2. Rational functors for entwined and Doi-Koppinen modules

In this section, we shall study the exactness of the rational functors for the corings coming from entwining structures. When specialized to the entwining structures given by a comodule algebra, we will obtain a result from [2]. Most of results in [10] are deduced.

2.1. Entwining structures with rational functor

Recall from [7] that an entwining structure over K is a three-tuple $(A, C)_\psi$ consisting of a K-algebra A with multiplication μ and unity 1, a K-coalgebra C with comultiplication Δ and counity ε, and a K-module map $\psi : C \otimes_K A \to A \otimes_K C$ satisfying

$$\psi \circ (C \otimes_K \mu) = (\mu \otimes_K C) \circ (A \otimes_K \psi) \circ (\psi \otimes_K A),$$
$$(A \otimes_K \Delta) \circ \psi = (\psi \otimes_K C) \circ (C \otimes_K \psi) \circ (\Delta \otimes_K A), \qquad (2.1)$$
$$\psi \circ (C \otimes_K 1) = 1 \otimes_K C, \quad (A \otimes_K \varepsilon) \circ \psi = \varepsilon \otimes_K A.$$

By [6, Proposition 2.2] the corresponding A-coring is $\mathfrak{C} = A \otimes_K C$ with the A-bimodule structure given by $a''(a' \otimes_K c)a = a''a'\psi(c \otimes_K a)$, $a, a', a'' \in A$, $c \in C$, the comultiplication $\Delta_{\mathfrak{C}} = A \otimes_K \Delta$, and the counit $\varepsilon_{\mathfrak{C}} = A \otimes_K \varepsilon$. Furthermore, the category of right $\mathfrak{C}$-comodules is isomorphic to the category of right entwined modules.

The map $(\phi, \nu) : (C, K) \to (\mathfrak{C}, A)$ defined by $\nu(1) = 1$ and $\phi(c) = 1 \otimes_K c$, is a homomorphism of corings in the sense of [16]. As in [16] the associated induction and ad-induction functors to this morphism are, respectively, given by $\mathcal{O} : \mathsf{Comod}_{\mathfrak{C}} \to \mathsf{Comod}_C$ and $- \otimes_K A : \mathsf{Comod}_C \to \mathsf{Comod}_{\mathfrak{C}}$, where $\mathcal{O}$ is the cotensor functor $- \square_{\mathfrak{C}}(A \otimes_K C)$. When $\mathsf{Comod}_{\mathfrak{C}}$ is interpreted as the category of entwined modules, $\mathcal{O}$ is naturally isomorphic to the forgetful functor. Moreover, there is a natural isomorphism

$$\mathrm{Hom}_{\mathfrak{C}}(M \otimes_K A, N) \overset{\cong}{\longrightarrow} \mathrm{Hom}_C(M, \mathcal{O}(N))$$
$$f \longmapsto [m \mapsto f(m \otimes_K 1)]$$
$$[m \otimes_K a \mapsto g(m)a] \longleftarrow\!\shortmid\; g,$$

for every pair of comodules $(M_C, N_{\mathfrak{C}})$. Thus the functor $\mathcal{O}$ is a right adjoint functor of $- \otimes_K A$. If C_K is a flat module, then $\mathcal{O}$ is exact, since $U_A : \mathsf{Comod}_{\mathfrak{C}} \to \mathsf{Mod}_A$ is already an exact functor (see, [12, Proposition 1.2]).

We know from [6] that the left dual convolution ring $^*\mathfrak{C}$ is isomorphic as a K-module to $\mathrm{Hom}_K(C, A)$. Up to this isomorphism the convolution multiplication reads

$$f \cdot g = \mu \circ (A \otimes_K f) \circ \psi \circ (C \otimes_K g) \circ \Delta_C, \quad f, g \in \mathrm{Hom}_K (C, A). \qquad (2.2)$$

The connection between this convolution ring and the usual coalgebra convolution ring C^* is given by the following homomorphism of rings

$$\Phi : C^* \longrightarrow {}^*\mathfrak{C}, \qquad\qquad (x \longmapsto A \otimes_K x). \qquad (2.3)$$

Proposition 2.1. *Let* $(A, C)_\psi$ *be an entwining structure over* K *such that* C_K *is a locally projective module and consider its corresponding* A-*coring* $\mathfrak{C} = A \otimes_K C$. *Suppose that there is a right rational pairing* $\mathfrak{T} = (\mathfrak{C}, B, \langle -, - \rangle)$ *and an anti-morphism of* K-*algebras* $\varphi : C^* \to B$ *which satisfy the following two conditions:* (1) $\beta \circ \varphi = \Phi$, *where* $\beta : B \to {}^*\mathfrak{C}$ *is the anti-homomorphism of* K-*algebras associated to* $\mathfrak{T}$ *and* Φ *is the homomorphism of rings given in equation* (2.3); (2) *for every pair of elements* $(a, x) \in A \times C^*$, *there exists a finite subset of pairs* $\{(x_i, a_i)\}_i \subseteq C^* \times A$ *such that* $a\varphi(x) = \sum_i \varphi(x_i)a_i$. *Then, by restricting scalars we have*

$$\mathrm{Rat}^{\mathfrak{T}} (M_B) = \mathrm{Rat}^r_C ({}_{C^*}M)$$

for every right B-*module* M, *where* $\mathrm{Rat}^r_C(-)$ *is the canonical right rational functor associated to the* K-*coalgebra* C.

Proof. Start with an arbitrary element $m \in \mathrm{Rat}^{\mathfrak{T}}(M_B)$ with right rational system of parameters $\{(\sum_k a_{kj} \otimes_K c_{kj}, m_j)\}_j \subset \mathfrak{C} \times M$. Then for every $x \in C^*$, we have

$$xm = m\varphi(x) = \sum_{k,j} m_j \langle a_{kj} \otimes_K c_{kj}, \varphi(x) \rangle$$

$$= \sum_{k,j} m_j \beta(\varphi(x))(a_{kj} \otimes_K c_{kj})$$

$$= \sum_{k,j} m_j \Phi(x)(a_{kj} \otimes_K c_{kj}), \quad \Phi = \beta \circ \varphi,$$

$$= \sum_{k,j} m_j a_{kj} x(c_{kj}),$$

thus $\{c_{kj}, m_j a_{kj}\} \subset C \times M$ is a right rational system of parameters for $m \in {}_{C^*}M$; that is $m \in \mathrm{Rat}^r_C({}_{C^*}M)$. Therefore, $\mathrm{Rat}^{\mathfrak{T}}(M_B) \subseteq \mathrm{Rat}^r_C({}_{C^*}M)$. Conversely, start with a pair of elements $(a, x) \in A \times C^*$, and let $\{(x_i, a_i)\}_i \subset C^* \times A$ be the finite system given by hypothesis, that is, $a\varphi(x) = \sum_i \varphi(x_i)a_i$. So, for every element $m \in \mathrm{Rat}^r_C({}_{C^*}M)$ with right C-coaction $\rho_{\mathrm{Rat}^r_C({}_{C^*}M)}(m) = \sum_{(m)} m_{(0)} \otimes_K m_{(1)}$, we have

$$x(ma) = (ma)\varphi(x) = m(a\varphi(x))$$

$$= \sum m(\varphi(x_i)a_i), \ a\varphi(x) = \sum \varphi(x_i)a_i$$

$$= \sum (x_i m)a_i$$

$$= \sum (m_{(0)} x_i(m_{(1)}))a_i$$

$$= \sum m_{(0)} \left(\Phi(x_i)(1 \otimes_K m_{(1)})a_i \right)$$

$$= \sum m_{(0)} \left(\beta(\varphi(x_i))(1 \otimes_K m_{(1)})a_i \right)$$

$$= \sum m_{(0)} \left(\beta(\varphi(x_i)a_i)(1 \otimes_K m_{(1)}) \right), \; \beta \text{ is right } A\text{-linear}$$

$$= \sum m_{(0)} \langle 1 \otimes_K m_{(1)}, \varphi(x_i)a_i \rangle$$

$$= \sum m_{(0)} \langle 1 \otimes_K m_{(1)}, a\varphi(x) \rangle$$

$$= \sum m_{(0)} \langle a_\psi \otimes_K m_{(1)}^\psi, \varphi(x) \rangle, \; \psi(m_{(1)} \otimes_K a) = \sum a_\psi \otimes_K m_{(1)}^\psi$$

$$= \sum m_{(0)} a_\psi x(m_{(1)}^\psi).$$

We conclude that $ma \in \mathrm{Rat}_C^r({}_{C^*}M)$ with right C-coaction $\rho_{\mathrm{Rat}_C^r({}_{C^*}M)}(ma) = \sum m_{(0)}a_\psi \otimes_K m_{(1)}^\psi$. From which we conclude that $\mathrm{Rat}_C^r({}_{C^*}M)$ is an entwined module, and thus a right $\mathfrak{C}$-comodule or, equivalently, a right rational B-submodule of M_B. Therefore, $\mathrm{Rat}_C^r({}_{C^*}M) \subseteq \mathrm{Rat}^\mathcal{T}(M_B)$. $\qquad\square$

2.2. The category of Doi-Koppinen modules

We apply the results of Subsection 2.1 to the category of Doi-Koppinen modules. This category is identified with the category of right rational modules over a well-known ring. Some results of this section were proved by different methods for particular Hopf algebras in [8, Theorem 2.3], for algebras and coalgebras over a field in [10, Proposition 2.7], and more recently for bialgebras in [2, Theorem 3.18, Proposition 3.21].

Let $\mathbf{H}$ be a Hopf K-algebra, (A, ρ_A) a right $\mathbf{H}$-comodule K-algebra, and (C, ϱ_C) a left $\mathbf{H}$-module K-coalgebra. That is, $\rho_A : A \to A \otimes_K \mathbf{H}$ and $\varrho_C : \mathbf{H} \otimes_K C \to C$ are, respectively, a K-algebra and a K-coalgebra map. We will use Sweedler's notation, that is, $\Delta_C(c) = \sum_{(c)} c_{(1)} \otimes_K c_{(2)}$, $\Delta_\mathbf{H}(h) = \sum_{(h)} h_{(1)} \otimes_K h_{(2)}$, and $\rho_A(a) = \sum_{(a)} a_{(1)} \otimes_K a_{(2)}$, for every $c \in C$, $h \in \mathbf{H}$ and $a \in A$.

Following [11, 18], a *Doi-Koppinen module* is a left A-module M with a structure of right C-comodule ρ_M such that, for every $a \in A$, $m \in M$,

$$\rho_M(am) = \sum a_{(0)}m_{(0)} \otimes_K a_{(1)}m_{(1)}.$$

A morphism between two Doi-Koppinen modules is a left A-linear and right C-colinear map. Doi-Koppinen modules and their morphisms form the category ${}_A\mathcal{M}(\mathbf{H})^C$.

Consider the following K-map (A^o means the opposite ring of A)

$$\psi : C \otimes_K A^o \longrightarrow A^o \otimes_K C, \qquad (c \otimes_K a^o \longmapsto \sum_{(a)} a_{(0)}^o \otimes_K a_{(1)}c). \quad (2.4)$$

It is easily seen that the map ψ satisfies all identities of equation (2.1). That is, $(A^o, C)_\psi$ is an entwining structure over K. So, consider the associated A^o-coring

$\mathfrak{C} = A^o \otimes_K C$, the A^o-biactions are then given by

$$b^o(a^o \otimes_K c) = (ab)^o \otimes_K c, \quad \text{and} \quad (b^o \otimes_K c)a^o = b^o \psi(c \otimes_K a^o) = \sum (a_{(0)}b)^o \otimes_K a_{(1)}c,$$

for every $a^o, b^o \in A^o$ and $c \in C$. The convolution multiplication of $\operatorname{Hom}_K(C, A^o)$ comes out from the general equation (2.2), as

$$f.g(c) = \sum_{(c)} \left(f(g(c_{(2)})_{(1)} c_{(1)}) g(c_{(2)})_{(0)} \right)^o \in A^o, \tag{2.5}$$
$$f, g \in \operatorname{Hom}_K(C, A^o) \text{ and } c \in C.$$

This multiplication coincides with the generalized smash product of A by C, denoted by $\sharp(C, A)$ in [18, (2.1)].

Define the smash product $A\sharp C^*$ whose underling K-module is the tensor product $A \otimes_K C^*$ and internal multiplication is given by

$$(a\sharp x).(b\sharp y) = \sum ab_{(0)}\sharp(xb_{(1)})y,$$

for $a \otimes_K x$, $b \otimes_K y \in A \otimes_K C^*$, and where the left $\mathbf{H}$-action on C^* is induced by the right $\mathbf{H}$-action on C. The unit of this multiplication is $1\sharp\varepsilon_C$. Moreover, it is clear that the maps

$$-\sharp\varepsilon_C : A \longrightarrow A\sharp C^*, \quad 1\sharp- : C^* \longrightarrow A\sharp C^*$$
$$a \longmapsto a\sharp\varepsilon_C \qquad\qquad x \longmapsto 1\sharp x$$

are K-algebra maps, and an easy computation shows that

$$\alpha_A : A\sharp C^* \longrightarrow \operatorname{Hom}_K(C, A^o), \qquad (a\sharp x \longmapsto [c \mapsto a^o x(c)])$$

is also a K-algebra morphism where $\operatorname{Hom}_K(C, A^o)$ is endowed with the multiplication of equation (2.5).

Proposition 2.2. [2, Proposition 3.21] *Let $\mathbf{H}$ be a Hopf K-algebra, A a right $\mathbf{H}$-comodule K-algebra and C a left $\mathbf{H}$-module K-coalgebra. Consider $\mathfrak{C} = A^o \otimes_K C$ the A^o-coring associated to the entwining structure $(A^o, C)_\psi$ where ψ is defined by (2.4), and let $B = (A\sharp C^*)^o$. Suppose that C_K is a locally projective module. Then $\mathcal{T} = (\mathfrak{C}, B, \langle -, - \rangle)$ is right rational pairing over A^o with the bilinear form $\langle -, - \rangle$ defined by*

$$\mathfrak{C} \times B \longrightarrow A^o$$
$$(a^o \otimes_K c, (b\sharp x)^o) \longmapsto \langle a^o \otimes_K c, (b\sharp x)^o \rangle = a^o b^o x(c)$$

$a, b \in A$, $c \in C$, $x \in C^*$. Moreover, $(A^o, C)_\psi$, $\mathcal{T}$ and $\varphi = (1\sharp-)^o : C^* \to B$ satisfy the conditions (1) and (2) stated in Proposition 2.1, and using restriction of scalars, we obtain

$$\operatorname{Rat}^{\mathcal{T}}(M_B) = \operatorname{Rat}^r_C({}_{C^*}M),$$

for every right B-module M.

Proof. First we show that $\langle -, - \rangle$ is bilinear and balanced. For $a, b, e \in A$, $x \in C^*$ and $c \in C$, we compute

$$\langle a^o \otimes_K c, (b \sharp x)^o e^o \rangle = \langle a^o \otimes_K c, ((e \sharp \varepsilon_C)(b \sharp x))^o \rangle$$
$$= \sum \langle a^o \otimes_K c, (e b_{(0)} \sharp (\varepsilon_C b_{(1)}) x)^o \rangle$$
$$= \sum a^o b_{(0)}^o e^o ((\varepsilon_C b_{(1)}) x)(c)$$
$$= \sum a^o b_{(0)}^o e^o \varepsilon_{\mathbf{H}}(b_{(1)}) x(c)$$
$$= a^o b^o e^o x(c)$$
$$= \langle a^o \otimes_K c, (b \sharp x)^o \rangle e^o,$$

which shows that $\langle -, - \rangle$ is right A^o-linear, and

$$\langle (a^o \otimes_K c) e^o, (b \sharp x)^o \rangle = \langle a^o \psi(c \otimes_K e^o), (b \sharp x)^o \rangle$$
$$= \sum \langle a^o e_{(0)}^o \otimes_K e_{(1)} c, (b \sharp x)^o \rangle$$
$$= \sum a^o e_{(0)}^o b^o x(e_{(1)} c)$$
$$= \sum a^o e_{(0)}^o b^o (x e_{(1)})(c)$$
$$= \langle a^o \otimes_K c, \sum (b e_{(0)}) \sharp x e_{(1)} \rangle$$
$$= \langle a^o \otimes_K c, (b \sharp x)(e \sharp \varepsilon_C) \rangle$$
$$= \langle a^o \otimes_K c, e^o (b \sharp x)^o \rangle,$$

which proves that $\langle -, - \rangle$ is A^o-balanced. The pairing $\langle -, - \rangle$ is clearly left A^o-linear. Consider now the right natural transformation associated to $\langle -, - \rangle$:

$$\alpha_N : \quad N \otimes_{A^o} \mathfrak{C} \longrightarrow \mathrm{Hom}_{A^o}(B_{A^o}, N)$$
$$n \otimes_{A^o} a^o \otimes_K c \longmapsto [(b \sharp x)^o \mapsto n \langle a^o \otimes_K c, (b \sharp x)^o \rangle = n(a^o b^o x(c))].$$

We need to show that α_- is injective. So let $\sum_i n_i \otimes_{A^o} 1 \otimes_K c_i \in N \otimes_{A^o} \mathfrak{C}$ whose image by α_N is zero. Since C_K is locally projective, associated to the finite set $\{c_i\}_i$ there exists a finite set $\{(c_l, x_l)\} \subset C \times C^*$ such that $c_i = \sum_l c_l x_l(c_i)$. The condition

$$\alpha_N(\sum_i n_i \otimes_{A^o} 1 \otimes_K c_i)((1 \sharp x_l)^o) = \sum_i n_i x_l(c_i) = 0, \text{ for all the } l\text{'s},$$

implies that

$$\sum_i n_i \otimes_{A^o} 1 \otimes_K c_i = \sum_{i,l} n_i \otimes_{A^o} 1 \otimes_K x_l(c_i) c_l = \sum_l \left(\sum_i n_i x_l(c_i) \right) \otimes_{A^o} 1 \otimes_K c_l = 0.$$

That is, α_N is an injective map for every right A^o-module N. Therefore, $\mathcal{T}$ is a right rational system. Lastly, the map $\beta : B \to {}^*\mathfrak{C}$ sending $(b \sharp x)^o \mapsto [a^o \otimes_K c \mapsto \langle a^o \otimes_K c, (b \sharp x)^o \rangle]$ is an anti-homomorphism of K-algebras, and B is a K-algebra extension of A^o, thus $\mathcal{T}$ is actually a right rational pairing.

Let $a^o \in A^o$, $c \in C$ and $x \in C^*$, then

$$\beta(\varphi(x))(a^o \otimes_K c) = \langle a^o \otimes_K c, (1\sharp x)^o \rangle = a^o x(c) = \Phi(x)(a^o \otimes_K c)$$

which implies the condition (1) of Proposition 2.1. For the condition (2), it is easily seen that the set $\{a^o_{(0)}, xa_{(1)}\}$, where $\rho_A(a) = \sum a_{(0)} \otimes_K a_{(1)}$, satisfies this condition for the pair $(a^o, x) \in A^o \times C^*$. The last stated assertion is a consequence of Proposition 2.1, and this finishes the proof. □

Theorem 2.1. *Let $\mathbf{H}$ be a Hopf K-algebra, A a right $\mathbf{H}$-comodule K-algebra and C a left $\mathbf{H}$-module K-coalgebra. Consider $\mathfrak{C} = A^o \otimes_K C$ the A^o-coring associated to the entwining structure $(A^o, C)_\psi$ where ψ is defined by (2.4), and set $B = (A\sharp C^*)^o$. Suppose that C_K is a locally projective module and consider the right rational pairing $\mathcal{T} = (\mathfrak{C}, B, \langle -, - \rangle)$ over A^o of Proposition 2.2, and put $\mathfrak{a} = \mathrm{Rat}^{\mathcal{T}}(B_B)$. If A_K is a flat module and $\mathrm{Rat}^r_C(-)$ is an exact functor, then*

(a) *$\mathrm{Rat}^r_C(_{C^*}C^*)$ is a right $\mathbf{H}$-submodule of C^* and $\mathfrak{a} = A \otimes_K \mathrm{Rat}^r_C(_{C^*}C^*)$.*

(b) *For each right B-module M, the map*

$$\mathrm{Hom}_B\left(\mathfrak{a}_B, M\right) \longrightarrow \mathrm{Hom}_{C^*}\left(\mathrm{Rat}^r_C(_{C^*}C^*), M\right) \qquad (2.6)$$

sending f onto the morphism $\widehat{f}$ defined by $\widehat{f}(c^*) = f(1 \otimes c^*)$ for $c^* \in C^*$ is an isomorphism of K-modules.

(c) *If, for every left $A\sharp C^*$-module M, we endow $\mathrm{Hom}_{C^*}\left(\mathrm{Rat}^r_C(_{C^*}C^*), M\right)$ with the structure of a left $A\sharp C^*$-module transferred from that of*

$$\mathrm{Hom}_{A\sharp C^*}\left(_{A\sharp C^*}\mathfrak{a}, M\right)$$

via the isomorphism (2.6), then we obtain a functor

$$\mathrm{Hom}_{C^*}\left(\mathrm{Rat}^r_C(_{C^*}C^*), -\right) : {}_A\mathcal{M}(\mathbf{H})^C \longrightarrow {}_{A\sharp C^*}\mathrm{Mod},$$

which is right adjoint to the functor (see Proposition 2.2)

$$\mathrm{Rat}^r_C : {}_{A\sharp C^*}\mathrm{Mod} \longrightarrow {}_A\mathcal{M}(\mathbf{H})^C.$$

Proof. (a) Let $y \in \mathrm{Rat}^r_C(_{C^*}C^*)$ with rational system of parameters $\{(y_i, c_i)\}_i \subset C^* \times C$. For any $h \in \mathbf{H}$ and $x \in C^*$, we obtain as in [10, Lemma 3.1]:

$$(x(yh))(c) = \sum_{(c)} x(c_{(1)})y(hc_{(2)}) = \sum_{(c),(h)} x(\varepsilon_{\mathbf{H}}(h_{(1)})c_{(1)})y(h_{(2)}c_{(2)})$$

$$= \sum_{(c),(h)} x(S(h_{(1)})h_{(2)}c_{(1)})y(h_{(3)}c_{(2)}) = \sum_{(h)}(xS(h_{(1)})y)(h_{(2)}c)$$

$$= \sum_{(h),i} y_i(h_{(2)}c)(xS(h_{(1)}))(c_i) = \sum_{(h),i} y_i(h_{(2)}c)x(S(h_{(1)})c_i),$$

for every $c \in C$, where S is the antipode of $\mathbf{H}$. That is,

$$x(yh) = \sum_{(h),i}(y_i h_{(2)})\, x(S(h_{(1)})c_i).$$

Hence, $\{(y_i h_{(2)}, S(h_{(1)})c_i)\} \subset C^* \times C$ is a rational system of parameters for yh. Thus $yh \in \text{Rat}_C^r(_{C^*}C^*)$, and $\text{Rat}_C^r(_{C^*}C^*)$ is a right $\mathbf{H}$-submodule of C^*. Since A_K is a flat module, an easy computation shows now that $A \otimes_K \text{Rat}_C^r(_{C^*}C^*)$ is a two-sided ideal of $A \natural C^*$. Let $x \in C^*$, $a \otimes_K y \in A \otimes_K \text{Rat}_C^r(_{C^*}C^*)$, and $\{(y_i, c_i)\}_i \subset C^* \times C$ a rational system of parameters for y. Applying the smash product, we get

$$x(a \otimes_K y) = \sum_{(a)} a_{(0)} \otimes_K ((xa_{(1)})y) = \sum_{(a), i} (a_{(0)} \otimes_K y_i)x(a_{(1)}c_i);$$

this means that $\{(a_{(0)} \otimes_K y_i, a_{(1)}c_i)\}_{(a), i} \subset (A \otimes_K \text{Rat}_C^r(_{C^*}C^*)) \times C$ is a rational system of parameters for $a \otimes_K y \in {}_{C^*}\left(A \otimes_K \text{Rat}_C^r(_{C^*}C^*)\right)$. Proposition 2.2, implies now that $A \otimes_K \text{Rat}_C^r(_{C^*}C^*) \subseteq \mathfrak{a}$. Conversely, we know that $\mathfrak{a}$ is a right $\mathfrak{C}$-comodule, so the underlying K-module is a right C-comodule, and, since Rat_C^r is exact, $\mathfrak{a} = \text{Rat}_C^r(_{C^*}C^*)\mathfrak{a}$. From this equality, it is easy to see that $\mathfrak{a} \subseteq A \otimes_K \text{Rat}_C^r(_{C^*}C^*)$, and the desired equality is derived.

(b) We know that $B = (A \natural C^*)^o$ and, by (a), we have $\mathfrak{a} = A \otimes_K \text{Rat}^r(_{C^*}C^*)$. Consider the homomorphism of algebras $C^* \to A \natural C^*$, which gives, as usual, the induction functor $(A \natural C^*) \otimes_{C^*} - : {}_{C^*}\text{Mod} \to {}_{A \natural C^*}\text{Mod}$ which is left adjoint to the restriction of scalars functor ${}_{A \natural C^*}\text{Mod} \to {}_{C^*}\text{Mod}$. The mapping $f \mapsto \widehat{f}$ is then defined as the composition

$$\text{Hom}_{A \natural C^*}\left(A \otimes_K \text{Rat}_C^r(_{C^*}C^*), M\right)$$
$$\cong \text{Hom}_{A \natural C^*}\left((A \natural C^*) \otimes_{C^*} \text{Rat}_C^r(_{C^*}C^*), M\right) \cong \text{Hom}_{C^*}\left(\text{Rat}_C^r(_{C^*}C^*), M\right),$$

where the second is the adjointness isomorphism, and the first one comes from the obvious isomorphism $(A \natural C^*) \otimes_{C^*} \text{Rat}_C^r(_{C^*}C^*) \cong A \otimes_K \text{Rat}_C^r(_{C^*}C^*)$.

(c) This is a consequence of (b) and Theorem 1.2. □

Keep, in the following corollary, the hypotheses of Theorem 2.1.

Corollary 2.1. *If M is an object of ${}_A\mathcal{M}(\mathbf{H})^C$, and $E({}_A M^C)$ denotes its injective hull in the category ${}_A\mathcal{M}(\mathbf{H})^C$, then*

(a) *The map*

$$\zeta_M : M \to \text{Hom}_{C^*}\left(\text{Rat}_C^r(_{C^*}C^*), E({}_A M^C)\right)$$
$$(m \mapsto \zeta_M(m)(c^*) = c^*m, \; m \in M))$$

gives an injective envelope of M in ${}_{A \natural C^*}\text{Mod}$.

(b) *M is injective in ${}_{A \natural C^*}\text{Mod}$ if and only if M is injective in ${}_A\mathcal{M}(\mathbf{H})^C$ and ζ_M is an isomorphism.*

(c) *Assume that the antipode of $\mathbf{H}$ is bijective, and that C^* is flat as a K-module. Let M be injective in ${}_A\mathcal{M}(\mathbf{H})^C$. If M has finite support as a right C-comodule, then M is injective as a left A-module.*

Proof. The two first statements follow from Proposition 1.4 and Theorem 2.1. For the last statement, observe that B_{A° is a flat module. Now, the proof of [10, Lemma 2.6] runs here to prove that $_{A^\circ}B$ is flat. $\qquad\square$

Remark 2.1. We have proved that, under suitable conditions,

$$\mathrm{Rat}^{\mathcal{T}}(M_B) = \mathrm{Rat}^r_C(_{C^*}M) = M\mathfrak{a} = \mathrm{Rat}^r_C(_{C^*}C^*)M, \qquad (2.7)$$

for every right B-module M. Therefore, equation (2.7) establishes a radical functor: $t : {}_{A\sharp C^*}\mathsf{Mod} \to {}_A\mathcal{M}(\mathbf{H})^C$ which acts on objects by $M \to \mathrm{Rat}^r_C(_{C^*}C^*)M$. This radical was used in [10, Lemma 2.9] for left and right semiperfect coalgebras over a commutative field. In this way, if we apply our results and [17, Proposition 2.2] to this setting, then most part of the results stated in [10] become consequences of the results stated in this paper. In particular, let us mention that, for a semiperfect coalgebra over a field, any comodule of finite support in the sense of [10] becomes of finite support in the sense of Definition 1.1. Finally, let us note that the results from [10] are only applicable to group-graded algebras over a field. This restriction has been dropped by our approach, and we fully cover the case of graded rings (take $K = \mathbb{Z}$), since the $\mathbb{Z}$-coalgebra $\mathbb{Z}G$, where the elements of the group G are all group-like, is easily shown to have an exact rational functor. Of course, the category of comodules over this coalgebra is not semiperfect.

References

[1] G.D. Abrams, *Morita equivalence for rings with local units*, Comm. Algebra **11** (1983), 801–837.

[2] J.Y. Abuhlail, *Rational modules for corings*, Comm. Algebra **31** (2003), no. 12, 5793–5840.

[3] J.Y. Abuhlail, J. Gómez-Torrecillas, and F.J. Lobillo, *Duality and rational modules in Hopf algebras over commutative rings*, J. Algebra **240** (2001), 165–184.

[4] T. Brzeziński, *A note on coring extensions*, Ann. Univ. Ferrara, Sez. VII, Sc. Mat. **LI**(III) (2005), 15–27.

[5] T. Brzeziński and R. Wisbauer, *Corings and comodules*, LMS, vol. 309, Cambridge University Press, 2003.

[6] T. Brzeziński, *The structure of corings. Induction functors, Maschke-type theorem, and Frobenius and Galois-type properties*, Alg. Rep. Theory **5** (2002), 389–410.

[7] T. Brzeziński and S. Majid, *Coalgebra bundles*, Comm. Math. Phys. **191** (1998), 467–492.

[8] C. Cai and H. Chen, *Coactions, smash product, and Hopf modules*, J. Algebra **167** (1994), 85–99.

[9] S. Dăscălescu, C. Năstăsescu, A. del Rio, and F. van Oystaeyen, *Gradings of finite support. Applications to injective objects*, J. Pure Appl. Algebra **107** (1996), 193–206.

[10] S. Dăscălescu, C. Năstăsescu, and B. Torrecillas, *Co-Frobenius Hopf algebras: Integrals, Doi-Koppinen modules and injective objects*, J. Algebra **220** (1999), 542–560.

[11] Y. Doi, *Unifying Hopf modules*, J. Algebra. **153** (1992), 373–385.

[12] L. El Kaoutit, J. Gómez-Torrecillas, and F.J. Lobillo, *Semisimple corings*, Algebra Colloq. **11** (2004), 427–442.

[13] C. Faith, *Algebra: rings, modules and categories, I*, Grundlehren Math. Wiss., 190, Springer, New York, 1973.

[14] P. Gabriel, *Des catégories abéliennes*, Bull. Soc. Math. France **90** (1962), 323–448.

[15] J. Gómez-Torrecillas, *Coalgebras and comodules over commutative rings*, Rev. Roumaine Math. Pures Appl. **43** (1998), no. 5-6, 591–603.

[16] J. Gómez-Torrecillas, *Separable functors in corings*, Int. J. Math. Math. Sci. **30** (2002), no. 4, 203–225.

[17] J. Gómez-Torrecillas and C. Năstăsescu, *Quasi-co-Fronenius coalgebras*, J. Algebra, **174** (1995), 909–923.

[18] M. Koppinen, *Variations on the smash product with applications to group-graded rings*, J. Pure Appl. Algebra **104** (1995), 61–80.

[19] M. Sweedler, *The predual theorem to the Jacobson-Bourbaki theorem*, Trans. Amer. Math. Soc. **213** (1975), 391–406.

[20] R. Wisbauer, *Foundations of Module and Ring Theory*, Gordon and Breach, Reading-Paris, 1991.

[21] B. Zimmermann-Huisgen, *Pure submodules of direct products of free modules*, Math. Ann. **224** (1976), 233–245.

L. El Kaoutit
Departamento de Álgebra
Facultad de Educación y Humanidades
Universidad de Granada
El Greco N° 10
E-51002 Ceuta, España
e-mail: kaoutit@ugr.es

J. Gómez-Torrecillas
Departamento de Álgebra
Facultad de Ciencias
Universidad de Granada
E-18071 Granada, España
e-mail: gomezj@ugr.es

Modules and Comodules
Trends in Mathematics, 203–225
© 2008 Birkhäuser Verlag Basel/Switzerland

Preradicals of Associative Algebras and their Connections with Preradicals of Modules

M. Luísa Galvão

Abstract. We study preradicals on an universal class $\mathcal{D}$ of algebras and we present a process to construct preradicals over algebras from certain families of preradicals over modules. We also define a torsion Plotkin radical on the class of all associative algebras which satisfies dual properties of the Jacobson radical.

Mathematics Subject Classification (2000). Primary 16N20; Secondary 16S90, 16S99.

Keywords. Preradical, socle, radical.

Let A be the class of all associative algebras (over a commutative ring R with 1) not necessarily with identity and let $\mathcal{D}$ be an universal subclass (i.e., closed under ideal and homomorphic images) of $\mathcal{A}$. An assignment $\rho : \mathcal{D} \to \mathcal{D}$ is a preradical on $\mathcal{D}$ if, for every $A \in \mathcal{D}$, $\rho(A)$ is an ideal of A and, for every homomorphism $f : A \to B$, with $A, B \in \mathcal{D}$, $f(\rho(A)) \subseteq \rho(f(A))$ [10]. Denoting by $\mathcal{D}$-pr the class of all preradicals on $\mathcal{D}$, in $\mathcal{D}$-pr we can define a partial order by: $\rho_1 \leq \rho_2$ if $\rho_1(A) \subseteq \rho_2(A)$, $(A \in \mathcal{D})$ and, for this relation, the pair $(\mathcal{D}$-pr, $\leq)$ behaves like a complete lattice. Using this partial order we can define in $\mathcal{D}$-pr the two associative binary operations meet and join, and will consider another binary operation, the *coproduct* denoted : which is defined by $(\rho_1 : \rho_2)(A)/\rho_1(A) = \rho_2(A/\rho_1(A)), (A \in \mathcal{D})$. The composition of two preradicals, $\rho_1 \circ \rho_2$, defined in the usual way, in general, is not a preradical. However, when possible, it can be considered the dual of the coproduct.

   In Section 2 we analyze some properties of the class $\mathcal{D}$-pr, in connection with the partial order, the coproduct and the composition and in Section 3 we present a process to construct preradicals on $\mathcal{D}$ from certain families of preradicals over modules. In particular, supposing $A \in \mathcal{A}$, in A-*Mod*, the category of right A-modules, the concepts of radical and socle defined respectively by: for every

This work was developed within the Project POCTI-ISFL-1-143 "Fundamental and Applied Algebra" of Centro de Álgebra da Universidade de Lisboa, financed by FCT and FEDER.

$M \in A$-Mod,

$$J(M) = \bigcap \{N : M/N \in \Sigma\} \text{ and } Soc(M) = \sum \{N : N \leq M, N \in \Sigma\}.$$

(where Σ is the class of all simple A-modules), are preradicals of module and from the above construction we can conclude that the maps: $S_r : A \to Soc(A_A)$, $S_l : A \to Soc(_AA)$, $J_r : A \to J(A_A)$ and $J_l : A \to J(_AA)$, $(A \in \mathcal{A})$ are preradicals on $\mathcal{A}$ such that $J_r : J_l = J_l : J_r = (J_r \vee J_l) : (J_r \vee J_l) = J =$ the Jacobson radical. Also, supposing β the prime (or Baer) radical, we prove that the maps $\beta_r : A \to P(A_A)$ and $\beta_l : A \to P(_AA)$, $(A \in \mathcal{A})$ where $P(A_A)$ and $P(_AA)$ are the prime radical of A_A and $_AA$, respectively, are preradicals such that $\beta = \beta_r : \beta_l = \beta_l : \beta_r = (\beta_r \vee \beta_l) : (\beta_r \vee \beta_l)$.

One of our concerns in this paper is to define in $(\mathcal{A}$-pr, $\leq)$ a preradical which satisfies dual properties of the Jacobson radical. So, in Section 4 we define the preradical $Soc : \mathcal{A} \to \mathcal{A}$ which in $(\mathcal{A}$-pr, $\leq)$ satisfies dual properties of the radical J, in particular, $Soc = S_r \circ S_l = S_l \circ S_r = (S_r \wedge S_l) \circ (S_r \wedge S_l)$. This preradical $Soc : A \to Soc(A)$ is a torsion Plotkin radical but not a KA-radical. [See definitions in Section 1.]

1. Preliminaries

Let $\mathcal{A}$ be the class of all associative algebras. We will denote $I \trianglelefteq A$, (resp. $I \trianglelefteq_r A, I \trianglelefteq_l A$) if I is an ideal (resp. right ideal, left ideal) of the algebra $A \in \mathcal{A}$. A subclass $\mathcal{D}$ of $\mathcal{A}$ is called *abstract* if it contains the algebra $\{0\}$ and all isomorphic copies of algebras from $\mathcal{D}$. An abstract class is *hereditary* if it is closed under ideals and it is *universal* if it is hereditary and closed under homomorphic images. In the following, even if some results are valid under more general conditions, we will suppose that $\mathcal{D}$ is an universal subclass of $\mathcal{A}$ and every subclass of $\mathcal{D}$ is abstract.

An assignment $\rho : \mathcal{D} \to \mathcal{D}$ is an *ideal-mapping* if, for every $A \in \mathcal{D}$, $\rho(A) \trianglelefteq A$. With ρ we associate the classes,

$$R_\rho = \{A \in \mathcal{D} : \rho(A) = A\} \qquad S_\rho = \{A \in \mathcal{D} : \rho(A) = 0\}.$$

An ideal-mapping ρ is *idempotent* if, for every $A \in \mathcal{D}$, $\rho(\rho(A)) = \rho(A)$ and is *complete* if $I = \rho(I) \trianglelefteq A$ implies $I \subseteq \rho(A)$. We say that: ρ is *hereditary* if $I \trianglelefteq A = \rho(A)$, implies $\rho(I) = I$, i.e., if R_ρ is hereditary; ρ is *torsion* if, for every $I \trianglelefteq A$, $\rho(I) = I \cap \rho(A)$; ρ is *ADS*, (from Anderson-Divinsky-Sulinski) if, for every $I \trianglelefteq A \in \mathcal{D}$, $\rho(I) \trianglelefteq A$; ρ is *isotone* if, for every $I \trianglelefteq A \in \mathcal{D}$, $\rho(I) \subseteq \rho(A)$.

In particular if ρ is torsion, ρ is *ADS* and isotone. If ρ is isotone then ρ is complete and S_ρ is hereditary [13, Prop. 4.3]. The following properties follow directly from the respective definitions:

Proposition 1.1. *Let ρ be an ideal-mapping on $\mathcal{D}$. The following are equivalent:*

a) ρ *is torsion.*
b) ρ *is idempotent, isotone and hereditary.*
c) ρ *is isotone and, for every $I \trianglelefteq A \in \mathcal{D}$ such that $I \subseteq \rho(A)$, then $\rho(I) = I$.*

An ideal-mapping ρ is a *preradical* if for every homomorphism $f : A \to B$, with $A, B \in \mathcal{D}$, $f(\rho(A)) \subseteq \rho(f(A))$. It is easily checked that, for every preradical ρ, $\rho(0) = 0$ and that, if $f : A \to B$ is an isomorphism then $f(\rho(A)) = \rho(B)$. The *null preradical* denoted by 0, associates to each $A \in \mathcal{D}$ the zero ideal. The *identity preradical*, denoted by 1, associates to each $A \in \mathcal{D}$ the ideal A. If ρ is a preradical then R_ρ is homomorphically closed and S_ρ is closed under subdirect sums [13, Prop. 1.1].

A preradical ρ is called an *Hoehnke radical* (*H-radical*, for short) if, for every $A \in \mathcal{D}$ $\rho(A/\rho(A)) = 0$. A *Kurosh-Amitzur radical* (*KA-radical*, for short) is an *H*-radical, idempotent and complete. An idempotent and complete preradical is called a *Plotkin radical*. For more details about *H*-radicals, *KA*-radicals and Plotkin radicals the reader is referred to [1], [7], [11] and [13].

Proposition 1.2. *Let ρ be a preradical on $\mathcal{D}$. For every $A \in \mathcal{D}$ and $\{A_i : i \in I\} \subseteq \mathcal{D}$*

a) *If $A = \prod_{i \in I} A_i$ then $\rho(A) \subseteq \prod_{i \in I} \rho(A_i)$.*
b) *If $A_i \trianglelefteq A$, $(i \in I)$ and $A = \bigoplus_{i \in I} A_i$ then $\rho(A) \subseteq \bigoplus_{i \in I} \rho(A_i)$. If ρ is isotone the equality holds.*
c) *For every $I \trianglelefteq A$, $\frac{\rho(A)+I}{I} \subseteq \rho(A/I)$. In particular,*
    i) *$\rho(A/I) = 0$ implies $\rho(A) \subseteq I$.*
    ii) *$I \subseteq \rho(A)$ implies $\rho(A)/I \subseteq \rho(A/I)$.*
d) *ρ is H-radical if and only if for every $I \trianglelefteq A$ such that $I \subseteq \rho(A)$, then $\rho(A/I) = \frac{\rho(A)}{I}$.*

Proof. a) If $p_k : \prod_{i \in I} A_i \to A_k$, $(k \in I)$ are the projections, then $p_k(\rho(A)) \subseteq \rho(A_k)$ and $\rho(A) \subseteq \prod_{i \in I} p_i(\rho(A)) \subseteq \prod_{i \in I} \rho(A_i)$.
b) As in a), $A = \bigoplus_{i \in I} A_i$, implies $\rho(A) \subseteq \bigoplus_{i \in I} \rho(A_i)$. If ρ is isotone, as $\rho(A_i) \subseteq \rho(A)$, $(i \in I)$; then $\bigoplus_{i \in I} \rho(A_i) \subseteq \rho(A)$.
c) Straightforward.
d) Let ρ be an *H*-radical and $I \trianglelefteq A$ such that $I \subseteq \rho(A)$. By c), $\rho(A/I) \subseteq \frac{\rho(A)}{I}$ and, since $\rho\left(\frac{A/I}{\rho(A)/I}\right) \simeq \rho\left(\frac{A}{\rho(A)}\right) = 0$, by c) i), $\frac{\rho(A)}{I} \subseteq \rho(A/I)$. The converse is clear. $\qquad\square$

Proposition 1.3. *Let ρ be an ideal-mapping.*

a) *If ρ is idempotent and ADS then ρ is isotone if and only if it is complete.*
b) *If ρ is an H-radical then ρ is isotone if and only if S_ρ is hereditary.*

Proof. a) Straightforward.

b) Let ρ be an *H*-radical such that S_ρ is hereditary and $I \trianglelefteq A$. If $\pi : A \to A/\rho(A)$ is the canonical epimorphism then

$$\pi(\rho(I)) = \frac{\rho(I) + \rho(A)}{\rho(A)} \subseteq \rho(\pi(I)) = \rho\left(\frac{I + \rho(A)}{\rho(A)}\right) = 0.$$

So $\rho(I) \subseteq \rho(A)$. The converse is clear. $\qquad\square$

2. The big lattice of preradicals

Let $\mathcal{D}$-pr be the class of all preradicals on $\mathcal{D}$. In $\mathcal{D}$-pr we can define the partial order: if $\rho_1, \rho_2 \in \mathcal{D}$-pr, then $\rho_1 \leq \rho_2$ if $\rho_1(A) \subseteq \rho_2(A)$, $(A \in \mathcal{D})$. For this relation the pair $(\mathcal{D}\text{-pr}, \leq)$ behaves like a complete lattice except that, in general, it is not a set; hence it is called a *big lattice*. For every family $(\rho_i)_I$ of preradicals the meet and the join are given by:

$$\left(\bigwedge_{i \in I} \rho_i \right)(A) = \bigcap_{i \in I} \rho_i(A) \text{ and } \left(\bigvee_{i \in I} \rho_i \right)(A) = \sum_{i \in I} \rho_i(A), (A \in \mathcal{D}).$$

Using this partial order we can define in $\mathcal{D}$-pr the two associative binary operations $\wedge$ and $\vee$, and we will consider another binary operation, the *coproduct* denoted $:$ which is defined by: for every $\rho_1, \rho_2 \in \mathcal{D}$-pr, $(\rho_1 : \rho_2)(A)/\rho_1(A) = \rho_2(A/\rho_1(A))$, $(A \in \mathcal{D})$. It is clear that $(\rho_1 : \rho_2)(A) \trianglelefteq A$. If $f : A \to B$ is a homomorphism and $\bar{f} : A/\rho_1(A) \to B/\rho_1(f(A))$ the canonical homomorphism then

$$\bar{f}(\rho_2(A/\rho_1(A))) = \frac{f[(\rho_1:\rho_2)(A)]+\rho_1(f(A))}{\rho_1(f(A))} \subseteq \rho_2[f(A)/\rho_1(f(A))]$$
$$= (\rho_1 : \rho_2)(f(A))/\rho_1(f(A))$$

so, $f[(\rho_1 : \rho_2)(A)] \subseteq (\rho_1 : \rho_2)(f(A))$. It is straightforward to see that this operation is also associative.

For the composition of two preradicals, $\rho_1 \circ \rho_2 : \mathcal{D} \to \mathcal{D}$, defined in the usual way, in general, $\rho_1 \circ \rho_2$ is not an ideal mapping and then, not a preradical. However, if ρ_1 is ADS and isotone then for every preradical ρ_2, $\rho_1 \circ \rho_2$ is a preradical. Indeed, for every $A \in \mathcal{D}$, $\rho_1(\rho_2(A)) \trianglelefteq A$ and, if $f : A \to B$ is a homomorphism, then

$$f(\rho_1(\rho_2(A))) \subseteq \rho_1(f(\rho_2(A))) \subseteq \rho_1(\rho_2(f(A))) = (\rho_1 \circ \rho_2)(f(A)).$$

Remark 2.1.

I) For every $\rho_1, \rho_2 \in \mathcal{D}$-pr, $\rho_1 \wedge \rho_2 \leq \rho_1 \vee \rho_2 \leq \rho_1 : \rho_2$. The first inequality is clear. Considering the canonical epimorphism $\pi : A \to A/\rho_1(A)$ then $\pi(\rho_2(A)) = \frac{\rho_1(A)+\rho_2(A)}{\rho_1(A)} \subseteq \rho_2(A/\rho_1(A)) = (\rho_1 : \rho_2)(A)/\rho_1(A)$.

II) If $\rho_1, \rho_2 \in \mathcal{D}$-pr are such that $\rho_1 \circ \rho_2 \in \mathcal{D}$-pr then $\rho_1 \circ \rho_2 \leq \rho_2$. If ρ_1 is isotone, then $\rho_1 \circ \rho_2 \leq \rho_1 \wedge \rho_2$.

III) $1 \circ \rho = \rho \circ 1 = \rho$, $0 \circ \rho = \rho \circ 0 = 0$, $(\rho : 1) = (1 : \rho) = 1$, $(0 : \rho) = (\rho : 0) = \rho$, for every $\rho \in \mathcal{D}$-pr.

IV) For every $\sigma, \tau, \rho \in \mathcal{D}$-pr.
  1. σ is idempotent if and only if $\sigma \circ \sigma = \sigma$.
  2. σ is H-radical if and only if $\sigma : \sigma = \sigma$.
  3. If $\sigma \leq \tau$ then $\sigma : \rho \leq \tau : \rho$ and $\rho : \sigma \leq \rho : \tau$. Indeed, if $\pi : A/\sigma(A) \to A/\tau(A)$, $(A \in \mathcal{D})$, is the canonical epimorphism then

$$\pi[\rho(A/\sigma(A))] = (\sigma : \rho)(A)/\tau(A) \subseteq \rho(A/\tau(A)) = (\tau : \rho)(A)/\tau(A).$$

The second relation is clear.

Proposition 2.1. *Suppose ρ and θ preradicals such that $\rho \leq \theta$. Then*

a) *If ρ is idempotent, $\theta \circ \rho = \rho$. If ρ is idempotent and isotone, $\theta \circ \rho = \rho = \rho \circ \theta$.*

b) *If θ is H-radical, $\rho : \theta = \theta : \rho = \theta$.*

Proof. a) If ρ is idempotent then $\theta(\rho(A)) \subseteq \rho(A) = \rho(\rho(A)) \subseteq \theta(\rho(A)) = (\theta \circ \rho)(A)$, $(A \in \mathcal{D})$. If ρ is idempotent and isotone, then $\rho(A) = \rho(\rho(A)) \subseteq \rho(\theta(A)) \subseteq \rho(A)$.

b) If θ is H-radical, for every $A \in \mathcal{A}$, $\frac{(\theta:\rho)(A)}{\theta(A)} = \rho\left(\frac{A}{\theta(A)}\right) \subseteq \theta\left(\frac{A}{\theta(A)}\right) = 0$, thus $\theta(A) = (\theta : \rho)(A)$. For the canonical epimorphism $\pi : A/\rho(A) \to A/\theta(A)$ we have $\pi(\theta(A/\rho(A))) \subseteq \theta(A/\theta(A)) = 0$; therefore

$$\theta(A/\rho(A)) = \frac{(\rho : \theta)(A)}{\rho(A)} \subseteq \mathrm{Ker}(\pi) = \frac{\theta(A)}{\rho(A)}$$

hence $(\rho : \theta)(A) \subseteq \theta(A)$. Since $\theta \leq \rho : \theta$, the equality holds.   □

Corollary 2.2.

a) *If ρ and θ are ADS Plotkin radicals such that $\rho \circ \theta = \theta \circ \rho$, then $\rho \circ \theta$ is the greatest ADS Plotkin radical that is contained in $\rho \wedge \theta$.*

b) *If ρ and θ are H-radicals such that $\rho : \theta = \theta : \rho$ then $\rho : \theta$ is the least H-radical that contains $\rho \vee \theta$.*

Proof. a) By 1.3.a), ρ and θ are isotone and, $\rho \circ \theta = \theta \circ \rho$ is a preradical. Also it is easy to check that $\rho \circ \theta$ is isotone and ADS. By other hand,

$$(\rho \circ \theta) \circ (\rho \circ \theta) = (\rho \circ \rho) \circ (\theta \circ \theta) = \rho \circ \theta.$$

Supposing τ an ADS Plotkin radical such that $\tau \leq \rho \wedge \theta$, by 2.1.a), $\theta \circ \tau = \rho \circ \tau = \tau$, so $(\rho \circ \theta) \circ \tau = \rho \circ (\theta \circ \tau) = \rho \circ \tau = \tau$. As $\rho \circ \theta$ is isotone, by Remark 2.1.II), $\tau \leq \rho \circ \theta$.

b) As $(\rho : \theta) : (\rho : \theta) = (\rho : \rho) : (\theta : \theta) = \rho : \theta$ then $(\rho : \theta)$ is H-radical. Supposing τ is an H-radical such that $\rho \vee \theta \leq \tau$, by 2.1.b), $(\rho : \theta) : \tau = \rho : (\theta : \tau) = \rho : \tau = \tau$ and $\rho : \theta \leq \tau$.   □

As for preradicals of modules [cf. 12] we can state the following properties that relate the operations in $\mathcal{D}$-pr.

Proposition 2.3. *Let $\sigma, \tau, \rho \in \mathcal{D}$-pr and let $\{\rho_i\}_I \subseteq \mathcal{D}$-pr.*

a) 1) *(Modular law)* $\sigma \leq \rho \Rightarrow \sigma \vee (\tau \wedge \rho) = (\sigma \vee \tau) \wedge \rho$.

2) *If $\{\rho_i\}_I$ is a directed set[1], then $\tau \wedge \left(\bigvee_{i \in I} \rho_i\right) = \bigvee_{i \in I}(\tau \wedge \rho_i)$.*

3) $\left(\tau : \bigwedge_{i \in I} \rho_i\right) = \bigwedge_{i \in I}(\tau : \rho_i), \quad \left(\tau : \bigvee_{i \in I} \rho_i\right) = \bigvee_{i \in I}(\tau : \rho_i)$.

b) *When both sides of the following are defined,*

1) $\left(\bigwedge_{i \in I} \rho_i\right) \circ \tau = \bigwedge_{i \in I}(\rho_i \circ \tau), \left(\bigvee_{i \in I} \rho_i\right) \circ \tau = \bigvee_{i \in I}(\rho_i \circ \tau)$.

2) *If ρ is isotone, $(\tau : \rho) \circ \sigma \leq ((\tau \circ \sigma) : (\rho \circ \sigma))$ and σ is H-radical if and only if, for every $\tau, \rho \in \mathcal{D}$-pr, the equality holds.*

3) *If σ is isotone, $(\sigma : \tau) \circ (\sigma : \rho) \leq (\sigma : (\tau \circ \rho))$ and σ is idempotent if and only if the equality holds, for every $\tau, \rho \in \mathcal{D}$-pr.*

---

[1] $\{\rho_i\}_I$ is a directed set if for every $i, j \in I$ there exists $k \in I$ such that $\rho_i \vee \rho_j \leq \rho_k$.

c) *If, for every $i \in I$,*
1) *ρ_i is isotone, then $\bigvee_{i\in I} \rho_i$ and $\bigwedge_{i\in I} \rho_i$ are isotone.*
2) *ρ_i is hereditary, then $\bigwedge_{i\in I} \rho_i$ is hereditary.*
3) *ρ_i is complete, then $\bigwedge_{i\in I} \rho_i$ is complete.*
4) *ρ_i is idempotent and isotone, then $\bigvee_{i\in I} \rho_i$ is idempotent and isotone.*
5) *ρ_i is H-radical, then $\bigwedge_{i\in I} \rho_i$ is H-radical.*

d) *If ρ and σ are isotone, then $\rho : \sigma$ is isotone.*

Proof. a) and b). Similar to proofs for preradicals of modules in [12].

c) 1), 2) and 3). Clear.

4) By c) 1), $\bigvee_{i\in I} \rho_i$ is isotone.
Since, $\rho_j \leq \bigvee_{i\in I} \rho_i$, $(j \in I)$ and ρ_j is idempotent and isotone by 2.1.a), $\rho_j \circ (\bigvee_{i\in I} \rho_i) = \rho_j$. Then, by b)1), $(\bigvee_{i\in I} \rho_i)\circ(\bigvee_{i\in I} \rho_i) = \bigvee_{j\in I} [\rho_j \circ (\bigvee_{i\in I} \rho_i)] = \bigvee_{j\in I} \rho_j$.

5) Since $\bigwedge_{i\in I} \rho_i \leq \rho_j$, $(j \in I)$ and ρ_j is H-radical, by 2.1.b), $(\bigwedge_{i\in I}\rho_i:\rho_j)=\rho_j$. Thus, by 2), $(\bigwedge_{i\in I} \rho_i : \bigwedge_{i\in I} \rho_i) = \bigwedge_{j\in I} (\bigwedge_{i\in I} \rho_i : \rho_j) = \bigwedge_{j\in I} \rho_j$.

d) Let $I \trianglelefteq A$ and $\pi : I/\rho(I) \to A/\rho(A)$ the canonical homomorphism. It follows that

$$\pi(\sigma(I/\rho(I))) \subseteq \sigma(\pi(I/\rho(I))) = \sigma((I + \rho(A))/\rho(A))$$

$$\frac{(\rho : \sigma)(I) + \rho(A)}{\rho(A)} \subseteq \sigma\left(\frac{I + \rho(A)}{\rho(A)}\right) \subseteq \sigma\left(\frac{A}{\rho(A)}\right) = \frac{(\rho : \sigma)(A)}{\rho(A)}. \qquad \square$$

In [13] a *summable preradical* (*s-preradical*, for short) is defined as an ideal-mapping σ for which there exists a class $\mathcal{C} \subseteq \mathcal{D}$, homomorphically closed such that $\sigma(A) = \sum\{I \trianglelefteq A : I \in \mathcal{C}\}$, $(A \in \mathcal{D})$ and it is proved that σ is an idempotent preradical such that $\mathcal{C} \subseteq R_\sigma$. It is also remarked that, in general, the class $\mathcal{C}$ is not well determined [13, Props. 1.8 and 1.9]. Nevertheless we will denote it by $s_\mathcal{C}$. Dually we can state

Proposition 2.4. *For every abstract class $\mathcal{C} \subseteq \mathcal{D}$ the ideal-mapping ρ defined by $\rho(A) := \bigcap\{I \trianglelefteq A : A/I \in \mathcal{C}\}$, $(A \in \mathcal{D})$ is an H-radical, for which $\mathcal{C} \subseteq S_\rho$. We will denote it by $r_\mathcal{C}$.*

Proposition 2.5. *Let $\rho \in \mathcal{D}$-pr.*

1) *Suppose $\widehat{\rho} = s_{R_\rho}$. Then $R_\rho \subseteq R_{\widehat{\rho}}$ and $\widehat{\rho}$ is greater or equal to any idempotent preradical $\tau \leq \rho$. If ρ is idempotent then $\widehat{\rho} \geq \rho$.*
$\widehat{\rho} \leq \rho$ if and only if ρ is complete. In this case, $R_\rho = R_{\widehat{\rho}}$, $\widehat{\rho}$ is a Plotkin radical and it is the largest idempotent preradical contained in ρ. Thus the equality $\widehat{\rho} = \bigvee\{\tau \in \mathcal{D}\text{-}pr : \tau \leq \rho$ and τ idempotent$\}$ holds.
2) *Suppose $\bar{\rho} = r_{S_\rho}$. Then $\rho \leq \bar{\rho}$ and $S_\rho = S_{\bar{\rho}}$. $\bar{\rho}$ is the least H-radical that contains ρ, i.e., $\bar{\rho} = \bigwedge\{\tau \in \mathcal{D}\text{-}pr : \rho \leq \tau$ and τ is H-radical$\}$.*

Proof. 1) Since R_ρ is homomorphically closed, we can consider the *s*-radical $\widehat{\rho} = \rho_{R_\rho}$ and it is clear that $R_\rho \subseteq R_{\widehat{\rho}}$. Supposing τ an idempotent preradical such that

$\tau \leq \rho$, then $\tau(A) = \tau(\tau(A)) \subseteq \rho(\tau(A)) \subseteq \tau(A), (A \in \mathcal{D})$; thus $\rho(\tau(A)) = \tau(A) \subseteq \widehat{\rho}(A)$. Therefore $\tau \leq \widehat{\rho}$ and $\bigvee\{\tau \in \mathcal{D}\text{-pr} : \tau \leq \rho \text{ and } \tau \text{ is idempotent}\} \leq \widehat{\rho}$.

It is clear that ρ is complete, if and only if $\widehat{\rho}(A) \subseteq \rho(A)$, $(A \in \mathcal{D})$, or equivalently, if and only if $\widehat{\rho} \leq \rho$. Therefore, in this case, since $\widehat{\rho}$ is idempotent, $\widehat{\rho} \leq \bigvee\{\tau \in \mathcal{D}\text{-pr} : \tau \leq \rho \text{ and } \tau \text{ idempotent}\}$ and the equality holds. On the other hand if $A \in R_{\widehat{\rho}}$, then $A = \widehat{\rho}(A) \subseteq \rho(A) \subseteq A$, thus $A \in R_{\rho}$ and $R_{\rho} = R_{\widehat{\rho}}$. Supposing $I \trianglelefteq A$ such that $I = \widehat{\rho}(I)$, then $I \in R_{\widehat{\rho}} = R_{\rho}$ and therefore $I \subseteq \widehat{\rho}(A) = \sum\{I \trianglelefteq A : \rho(I) = I\}$. Thus $\widehat{\rho}$ is complete and consequently a Plotkin radical.

2) [13, prop. 1.20]. □

Corollary 2.6. *Let $\rho \in \mathcal{D}$-pr*

   a) *ρ is a Plotkin radical if and only if $\rho = \widehat{\rho}$.*
   b) *ρ is an H-radical if and only if $\rho = \bar{\rho}$.*
   c) *ρ is a KA-radical if and only if $\rho = \widehat{\rho} = \bar{\rho}$.*

Proof. a) [13, Prop. 1.12], b) clear, c) [13, Prop. 1.18]. □

With any abstract class $\mathcal{C} \subseteq \mathcal{D}$ we associate the operator $\mathcal{U}$ defined by

$$\mathcal{U}\mathcal{C} = \{A \in \mathcal{D} : A/I \notin \mathcal{C} \text{ for all } I \trianglelefteq A, I \neq A\}$$

Clearly $\mathcal{U}\mathcal{C}$ is homomorphically closed. The s-summable preradical $s_{\mathcal{U}\mathcal{C}}$ will be called the *upper preradical* of the class $\mathcal{C}$.

3. Modules and preradicals

Let $A, B \in \mathcal{D}$ and $f : A \to B$ be an epimorphism. It is well known that if M_B is a right B-module, then M is a right A-module that will be denoted $M_A^{f^{-1}}$ (in short M_A) under the action $xa = xf(a)$, for every $x \in M$ and $a \in A$. M_B and M_A have the same submodules, $\text{An}(M_A) = f^{-1}(\text{An}(M_B)) \supseteq \text{Ker}(f)$, (where $\text{An}(M_A)$ is the annihilator of M) and if $g : M_B \to T_B$ is a B-homomorphism then $g : M_A \to T_A$ is an A-homomorphism.

Conversely if M_A is a right A-module, such that $\text{Ker}(f) \subseteq \text{An}(M_A)$, M is a right B-module, that will be denoted M_B^f, (in short M_B), under the action $xb = xa$, if $b = f(a)$, for every $x \in M$ and $b \in B$. The modules M_B and M_A have the same submodules, $\text{An}(M_B) = f(\text{An}(M_A))$ and if $g : M_A \to T_A$ is an A-homomorphism, then $g : M_B \to T_B$ is a B-homomorphism. One can also note, that, for every B-module M_B, $\text{Ker}(f) \subseteq \text{An}(M_A)$ and $(M_A)_B = M_B$. In particular, if we consider the regular modules A_A and B_B,

   1) B is a right A-module under the action $b * a = bf(a)$ for every $b \in B$ and $a \in A$. One has $\text{Ker}(f) \subseteq \text{An}(B_A)$ and the map $f : A_A \to B_A$ is an A-module epimorphism.
   2) Supposing $\text{Ker}(f) \subseteq \text{An}(A_A)$, A is a right B-module, under the action $a * b = aa'$ if $b = f(a')$, for every $a \in A$ and $b \in A$ and the map $f : A_B \to B_B$ is a B-epimorphism.

For each $A \in \mathcal{D}$, let $A\text{-}Mod$ be the class of all right A-modules. Recall that a *preradical* σ on $A\text{-}Mod$ associates to each module $M \in A\text{-}Mod$ a submodule $\sigma(M)$ such that for each homomorphism $f : M \to N$, on has $f(\sigma(M)) \subseteq \sigma(N)$ and in the class A-pr of all preradicals in $A\text{-}Mod$ we can consider also a binary operation, called *coproduct* and denoted : and defined by: for every $\sigma_1, \sigma_2 \in A$-pr, $(\sigma_1 : \sigma_2)(M)/\sigma_1(M) = \sigma_2(M/\sigma_1(M))$, $(M \in A\text{-}Mod)$.

A preradical $\sigma : A\text{-}Mod \to A\text{-}Mod$ is an *idempotent* (resp. *radical*) if, for every $M \in A\text{-}Mod$, $\sigma(\sigma(M)) = \sigma(M)$(resp. $\sigma(M/\sigma(M)) = 0$). For every preradical $\sigma \in A$-pr the following is valid:

1. If $M \leq T$ then $\sigma(M) \leq \sigma(T)$.
2. If $M = \bigoplus_{i \in I} M_i$, with $M_i \leq M$, $(i \in I)$ then $\sigma(M) = \bigoplus_{i \in I} \sigma(M_i)$.
3. $\sigma(A_A) \trianglelefteq A$.
4. σ is a radical if and only if $\sigma : \sigma = \sigma$.
5. σ is idempotent if and only if $\sigma \circ \sigma = \sigma$.

For more details about preradical on $A\text{-}Mod$ see [6].

Supposing $(\sigma_A)_\mathcal{D}$ a family of assignments indexed in $\mathcal{D}$, where $\sigma_A \in A$-pr, $(a \in \mathcal{D})$, by 3. we can define an ideal-mapping $\sigma : \mathcal{D} \to \mathcal{D}$ by $\sigma(A) = \sigma_A(A_A)$, $(A \in \mathcal{D})$, that will be denoted $\sigma = \Delta_{A \in \mathcal{D}} \sigma_A$ (the *diagonal mapping* of $(\sigma_A)_\mathcal{D}$). We say that:

$(\sigma_A)_\mathcal{D}$ verifies the *condition* (P) (resp. *condition* (P^*)) if, for every $A, B \in \mathcal{D}$ and every epimorphism $f : A \to B$, $\sigma_A(B_B) \subseteq \sigma_B(B_B)$ (resp. $\sigma_A(B_A) = \sigma_B(B_B)$).

It is easy to check that, if $(\sigma_A)_\mathcal{D}$ verifies the condition (P), then $\sigma = \Delta_{A \in \mathcal{D}} \sigma_A$ is a preradical on $\mathcal{D}$. In fact, if $f : A \to B$ is an algebra epimorphism for the A-epimorphism, $f : A_A \to B_A$, we have, $f(\sigma(A)) = f(\sigma_A(A_A)) \subseteq \sigma_A(B_A) \subseteq \sigma_B(B_B) = \sigma(B)$.

Proposition 3.1. *Let $(\sigma_A)_\mathcal{D}$ and $(\beta_A)_\mathcal{D}$ be families of maps, where $\sigma_A, \beta_A \in A$-pr, $(A \in \mathcal{D})$. Supposing $\sigma = \Delta_{A \in \mathcal{D}} \sigma_A$ and $\beta = \Delta_{A \in \mathcal{D}} \beta_A$, if $(\sigma_A)_\mathcal{D}$ and $(\beta_A)_\mathcal{D}$ verify (P^*) then*

  a) *$(\sigma_A : \beta_A)_\mathcal{D}$ verifies (P^*) and $\sigma : \beta = \Delta_{A \in \mathcal{D}}(\sigma_A : \beta_A)$.*
  b) *If, for every $A \in \mathcal{D}$, σ_A is a radical in $A\text{-}Mod$, then $\sigma = \Delta_{A \in \mathcal{D}} \sigma_A$ is an H-radical on $\mathcal{D}$.*
  c) *If $A = \bigoplus_{i \in I} B_i$, with $B_i \trianglelefteq A$, $(i \in I)$ then $\sigma(A) = \bigoplus_{i \in I} \sigma(B_i)$.*

Proof. a) Let $f : A \to B$ be an algebra epimorphism. Supposing $I = \sigma_A(B_A) = \sigma_B(B_B) \trianglelefteq B$, $B \overset{\pi}{\to} B/I$ the canonical epimorphism and $g = \pi f : A \to B/I$, as $(\beta_A)_\mathcal{D}$ verifies (P^*), it follows, respectively, that

$$\beta_B(B_B/I) = \beta_B(B/I)_B = \beta_{B/I}\left[(B/I)_{B/I}\right]$$
$$\beta_A(B_A/I) = \beta_A(B/I)_A = \beta_{B/I}\left[(B/I)_{B/I}\right]$$

Then $\beta_A(B_A/I) = \beta_B(B_B/I)$ and so

$$\frac{(\sigma_A : \beta_A)(B_A)}{\sigma_A(B_A)} = \beta_A\left(\frac{B_A}{\sigma_A(B_A)}\right) = \beta_B(B_B/\sigma_B(B_B)) = \frac{(\sigma_B : \beta_B)(B_B)}{\sigma_B(B_B)}$$

i.e., $(\sigma_A : \beta_A)(B_A) = (\sigma_B : \beta_B)(B_B)$.

On the other hand, for every $A \in \mathcal{D}$, from the equality $\frac{(\sigma : \beta)(A)}{\sigma(A)} = \beta(A/\sigma(A))$ it follows:

$$\frac{(\sigma : \beta)(A)}{\sigma_A(A_A)} = \beta_{A/\sigma(A)} \left[(A/\sigma(A))_{A/\sigma(A)} \right] = \beta_A(A/\sigma_A(A_A)) = \frac{(\sigma_A : \beta_A)(A_A)}{\sigma_A(A_A)}$$

so $(\sigma : \beta)(A) = (\sigma_A : \beta_A)(A_A)$.

b) Consequence of 3.1.a).

c) If $A = \bigoplus_{i \in I} B_i$ with $B_i \trianglelefteq A$ also $A_A = \bigoplus_{i \in I} (B_i)_A$ with $(B_i)_A \leq A_A$. Then $\sigma(A) = \sigma_A(A_A) = \bigoplus_{i \in I} \sigma_A((B_i)_A)$. On the other hand one can easily check that, for every $i \in I$, the action of A on B_i defined by the projection $p_i : A \to B_i$ ($i \in I$), coincide with the multiplication on A, i.e., $b_i * a = b_i p_i(a) = b_i a$, for every $b_i \in B_i$ and $a \in A$. Thus, as $(\sigma_A)_{\mathcal{D}}$ verifies (P^*), we have $\sigma_A((B_i)_A) = \sigma_{B_i}(B_i) = \sigma(B_i)$. So, $\sigma(A) = \bigoplus_{i \in I} \sigma(B_i)$. $\qquad \square$

With the convention that the intersection and the sum of an empty family of submodules of a module M is equal to M and $\{0\}$, respectively, for every class $\mathcal{C} \subseteq A\text{-}Mod$ one can define the dual preradicals, *Reject* and *Trace* on A-*Mod*, denoted, respectively, $Re_{\mathcal{C}}$ and $Tr_{\mathcal{C}}$ as follows: ([4, 11-2.1.])

$$Re_{\mathcal{C}}(M) = \bigcap \{ \mathrm{Ker}(f) : f \in \mathrm{Hom}_A(M, C), C \in \mathcal{C} \}, \ (M \in A\text{-}Mod),$$
$$Tr_{\mathcal{C}}(M) = \sum \{ \mathrm{Im}(f) : f \in \mathrm{Hom}_A(C, M), C \in \mathcal{C} \}, \ (M \in A\text{-}Mod),$$

where the first is a radical and the second an idempotent.

It is easy to check that, if $\mathcal{C}$ is closed under non zero homomorphic image then, for every $M \in A\text{-}Mod, Tr_{\mathcal{C}}(M) = \sum \{ N : N \leq M, N \in \mathcal{C} \}$; if $\mathcal{C}$ is closed under nonzero submodules, $Re_{\mathcal{C}}(M) = \bigcap \{ N : M/N \in \mathcal{C} \}$; in this case, supposing $K \leq M$ we have

$$Re_{\mathcal{C}}(M/K) = \frac{\bigcap \{ N \leq M : K \subseteq N, M/N \in \mathcal{C} \}}{K}.$$

Let $(\mathcal{C}_A)_{\mathcal{D}}$ be a family of module classes indexed in $\mathcal{D}$ where, for every $A \in \mathcal{D}$, $\mathcal{C}_A \subseteq A\text{-}Mod$ and let $\mathcal{C} = \bigcup_{A \in \mathcal{D}} \mathcal{C}_A$. For every A we define $\mathrm{Ker}(\mathcal{C}_A) = \bigcap_{M \in \mathcal{C}_A} \mathrm{An}(M)$ and on $\mathcal{C}$ we will consider the two following conditions: For every epimorphism $f : A \to B$, where $A, B \in \mathcal{D}$

M$_1$) $M_B \in \mathcal{C}_B$ implies $M_A \in \mathcal{C}_A$

M$_2$) $M_A \in \mathcal{C}_A$ and $\mathrm{Ker}(f) \subseteq \mathrm{An}(M_A)$ implies $M_B \in \mathcal{C}_B$. See [7, page 118].

Proposition 3.2. *Let* $\mathcal{C} = \bigcup_{A \in \mathcal{D}} \mathcal{C}_A$ *where* $(\mathcal{C}_A)_{\mathcal{D}}$ *is a family of module classes indexed in* $\mathcal{D}$ *with,* $\mathcal{C}_A \subseteq A\text{-}Mod$, $(A \in \mathcal{D})$

1) *If* $\mathcal{C}$ *satisfies* M$_1$) *then the family of preradicals* $(Re_{\mathcal{C}_A})_{\mathcal{D}}$ *verifies* P; *so* $\sigma = \Delta_{A \in \mathcal{D}} Re_{\mathcal{C}_A}$ *is a preradical.*

2)   a) *If, for every* $A \in \mathcal{D}$, $\mathcal{C}_A$ *is closed under non zero submodules and* $\mathcal{C}$ *satisfies* M$_1$) *and* M$_2$) *then* $(Re_{\mathcal{C}_A})_{\mathcal{D}}$ *verifies* (P^*); *so* $\sigma = \Delta_{A \in \mathcal{D}} Re_{\mathcal{C}_A}$ *is an H-radical.*

b) *Let $\mathcal{C}$ such that, for every $A \in \mathcal{D}$, $\mathcal{C}_A$ is closed under non zero homomorphic images. If $\mathcal{C}$ satisfies M_2), the family of preradicals $(Tr_{\mathcal{C}_A})_\mathcal{D}$ verifies (P); so $\sigma = \Delta_{A\in\mathcal{D}}Tr_{\mathcal{C}_A}$ is a preradical; if $\mathcal{C}$ satisfies M_1) and M_2) then $(Tr_{\mathcal{C}_A})_\mathcal{D}$ verifies (P^*).*

Proof. 1) Clear.

2) Supposing $f : A \to B$ an epimorphism where $A, B \in \mathcal{D}$, since $\mathrm{Ker}(f) \subseteq \mathrm{An}(B_A)$ also $\mathrm{Ker}(f) \subseteq \mathrm{An}(B_A/N)$ and $\mathrm{Ker}(f) \subseteq \mathrm{An}(N_A)$, for every $N \le B_A$. So

a) If $\mathcal{C}$ satisfies M_1) and M_2), then $\{D \le B_A : B_A/D \in \mathcal{C}_A\} = \{T \le B_B : B_B/T \in \mathcal{C}_B\}$ and if, for every $A \in \mathcal{D}$, $\mathcal{C}_A$ is closed under non zero submodules then $Re_{\mathcal{C}_A}(B_A) = Re_{\mathcal{C}_B}(B_B)$.

b) Let $\mathcal{C}$ such that, for every $A \in \mathcal{D}$, $\mathcal{C}_A$ is closed under non zero homomorphic images. If $\mathcal{C}$ satisfies M_2), $\{N \le B_A : N_A \in \mathcal{C}_A\} \subseteq \{H \le B_B : H_B \in \mathcal{C}_B\}$ and $Tr_{\mathcal{C}_A}(B_A) = Tr_{\mathcal{C}_B}(B_B)$. If $\mathcal{C}$ satisfies M_1) and M_2) then $\{N \le B_A : N \in \mathcal{C}_A\} = \{N \le B_B : N \in \mathcal{C}_B\}$ and $Tr_{\mathcal{C}_A}(B_A) = Tr_{\mathcal{C}_B}(B_B)$.   □

Example 3.1. Supposing Σ_A^r the class of all simple right A-modules, (i.e., $MA \ne 0$ and 0 and M are its only submodules) for every $M_A \in A\text{-}Mod$, let $J(M_A)$ be its radical defined by $J(M_A) = Re_{\Sigma_A^r}(M_A)$ and dually let $Soc(M_A)$ be its socle defined by $Soc(M_A) = Tr_{\Sigma_A^r}(M)$ [4, 11-2.1.]. Since Σ_A^r is closed under non zero submodules and non zero homomorphic image, then

$$J(M_A) = \bigcap\{N \le M : M/N \in \Sigma_A^r\} \text{ and } Soc(M_A) = \sum\{N \le M : N \in \Sigma_A^r\}.$$

If we consider the family of module classes $(\Sigma_A^r)_A$, then $\Sigma_r = \bigcup_{A\in\mathcal{A}}\Sigma_A^r$ satisfies M_1) and M_2). [7, Prop. 3.14.6.]. Thus the families of preradicals $(Tr_{\Sigma_A^r})_A$ and $(Re_{\Sigma_A^r})$ where, for every $A \in \mathcal{A}$, $(Tr_{\Sigma_A^r})(M_A) = Soc(M_A)$ and $Re_{\Sigma_A^r}(M_A) = J(M_A)$, verify (P^*); then, by 3.2, $S_r = \Delta_{A\in\mathcal{A}}Tr_{\Sigma_A^r}$ where $S_r : A \to Soc(A_A)$, $(A \in \mathcal{A})$ is a preradical and $J_r = \Delta_{A\in\mathcal{A}}Re_{\Sigma_A^r}$ where $J_r : A \to J(A_A)$, $(A \in \mathcal{A})$ is an H-radical. Analogously, $S_l : A \to Soc(_A A) = S_l(A)$ $(A \in \mathcal{A})$ and $J_l : A \to J(_A A) = J_l(A)$ $(A \in \mathcal{A})$ are a preradical and an H-radical, respectively.

Recall that an ideal P of A is *prime* if $P \ne A$ and supposing $I, K \trianglelefteq A$ such that $IK \subseteq P$ then $I \subseteq P$ or $K \subseteq P$. A module $M \in A\text{-}Mod$ is a *prime module* if $MA \ne 0$, and $xI = 0$ implies $x = 0$ or $MI = 0$, for $x \in M$ and $I \trianglelefteq A$; or, equivalently, $xA \ne 0$ for every $x \in M\backslash 0$, and $\mathrm{An}(N) = \mathrm{An}(M)$ for every $0 \ne N \le M$ [7, p. 124]. It is clear that a non zero submodule of a prime module is a prime module. In particular a simple module M is a prime module. In [7, Props. 3.14.16 and 3.14.17] it is proved that

Proposition 3.3. *P is a prime ideal if and only if P is the annihilator of a prime module. In particular if P is prime then A_A/P is a prime module and $P = \mathrm{An}(A/P)$.*

Supposing $D \trianglelefteq_r A$, it is easy to conclude that A/D is a prime module if and only if $A^2 \not\subseteq D$ and $aI \subseteq D$ implies $a \in D$ or $AI \subseteq D$, for every $a \in A$ and $I \trianglelefteq A$. In particular, if $aA \subseteq D$ then $a \in D$. Such right ideal will be called a *prime right ideal.* Analogously, we define *prime left ideal*. From 3.3, it follows that every prime ideal is a prime right and left ideal.

Proposition 3.4.

   a) *The annihilator of every non zero element of a prime right (resp. left) module is a prime right (resp. left) ideal.*
   b) *Every prime ideal is an intersection of prime right (or left) ideals.*

Proof. a) Supposing M a prime module and $0 \neq x \in M$ let $\varphi : A_A \to xA$ be the epimorphism defined by $\varphi : a \to xa$, $(a \in A)$. As $xA \neq 0$, xA is a prime module and so $A_A/\mathrm{Ker}(\varphi) = A_A/An(x)$ is a prime module.
   b) Consequence of 3.3 and a). $\qquad\square$

Theorem 3.5. *Let σ be an ideal mapping on $\mathcal{D}$ and, for every $A \in \mathcal{D}$, let $\mathcal{C}_A^r$ (resp. $\mathcal{C}_A^l$) be an abstract class of prime right (resp. left) A-modules such that $\sigma(A) = \mathrm{Ker}(\mathcal{C}_A^r) = \mathrm{Ker}(\mathcal{C}_A^l)$. Supposing $\mathcal{C}_r = \bigcup \mathcal{C}_A^r$ and $\mathcal{C}_l = \bigcup \mathcal{C}_A^l$ closed under non zero submodules, then*

   a) *If $\mathcal{C}_r$ and $\mathcal{C}_l$ satisfy the $\mathrm{M}_1)$ condition, $\sigma_r = \Delta_{A \in \mathcal{D}} Re_{\mathcal{C}_A^r}$ and $\sigma_l = \Delta_{A \in \mathcal{D}} Re_{\mathcal{C}_A^l}$ are preradicals and $\sigma = \sigma_r : \sigma_l = \sigma_l : \sigma_r = (\sigma_r \vee \sigma_l) : (\sigma_r \vee \sigma_l)$. If A has identity then $\sigma(A) = \sigma_r(A) = \sigma_l(A)$.*
   b) *If $\mathcal{C}_r$ and $\mathcal{C}_l$ satisfy the $\mathrm{M}_1)$ and $\mathrm{M}_2)$ conditions then σ_r and σ_l are H-radicals and $\sigma = \overline{\sigma_r \vee \sigma_l}$ is the least H-radical which contains $\sigma_r \vee \sigma_l$.*

Proof. a) By 3.2.1) σ_r and σ_l are preradicals. As, for every $A \in \mathcal{D}$.

$$\sigma(A) = \bigcap\{An_A(M) : M \in \mathcal{C}_A^r\} = \bigcap\{An(x) : 0 \neq x \in M \in \mathcal{C}_A^r\} \qquad (1)$$

and, for every $0 \neq x \in M \in \mathcal{C}_A^r$, we have $A/An(x) \simeq xA \leq M$ then $A/An(x) \in \mathcal{C}_A^r$ and $\sigma(A) \supseteq \sigma_r(A) = \bigcap\{D \lhd_r A : A/D \in \mathcal{C}_A^r\}$. Analogously $\sigma(A) \supseteq \sigma_l(A)$. On the other hand, it is clear that $\sigma(A) = \bigcap\{D \lhd_r A : \sigma(A) \subseteq D, A/D \in \mathcal{C}_A^r\}$; so

$$\bigcap\{D \lhd_r A : \sigma_l(A) \subseteq D, A/D \in \mathcal{C}_A^r\} \subseteq \bigcap\{D \lhd_r A : \sigma(A) \subseteq D, A/D \in \mathcal{C}_A^r\} = \sigma(A).$$

Conversely, since

$$\sigma(A) = \bigcap\{An(M) : M \in \mathcal{C}_A^l\} \subseteq \bigcap\{An(A/E) : E \lhd_l A \text{ and } A/E \in \mathcal{C}_A^l\}$$

it follows that

$$\sigma(A)A \subseteq \bigcap\{An(A/E)A : E \lhd_l A \text{ and } A/E \in \mathcal{C}_A^l\}$$
$$\subseteq \bigcap\{E \lhd_l A : A/E \in \mathcal{C}_A^l\} = \sigma_l(A).$$

Then, for every $D \lhd_r A$ such that $\sigma_l(A) \subseteq D$ and $A/D \in \mathcal{C}_A^r$ we have $\sigma(A)A \subseteq D$. As A/D is a prime module, then $\sigma(A) \subseteq D$ and the equality $\sigma(A) = \bigcap\{D \lhd_r A : \sigma_l(A) \subseteq D, A/D \in \mathcal{C}_A^r\}$ holds. Then,

$$\frac{(\sigma_l : \sigma_r)(A)}{\sigma_l(A)} = \sigma_r(A/\sigma_l(A)) = \frac{\bigcap\{D \lhd_r A : \sigma_l(A) \subseteq D, A/D \in \mathcal{C}_A^r\}}{\sigma_l(A)} = \frac{\sigma(A)}{\sigma_l(A)}$$

and $(\sigma_l : \sigma_r)(A) = \sigma(A)$. So $\sigma_l : \sigma_r = \sigma$. Analogously, $\sigma_r : \sigma_l = \sigma$.

Since

$$\bigcap\{D \trianglelefteq_r A : (\sigma_r \vee \sigma_l)(A) \subseteq D, A/D \in \mathcal{C}_A^r\}$$
$$= \bigcap\{D \trianglelefteq_r A : \sigma_l(A) \subseteq D, A/D \in \mathcal{C}_A^r\}$$

then

$$\frac{(\sigma_r \vee \sigma_l : \sigma_r)(A)}{(\sigma_r : \sigma_l)(A)} = \sigma_r(A/(\sigma_r : \sigma_l)(A))$$

$$= \frac{\bigcap\{D \trianglelefteq_r A : (\sigma_r \vee \sigma_l)(A) \subseteq D, A/D \in \mathcal{C}_A^r\}}{(\sigma_r : \sigma_l)(A)} = \frac{\sigma(A)}{(\sigma_r : \sigma_l)(A)}.$$

So $(\sigma_r \vee \sigma_l : \sigma_r) = \sigma$ and analogously $(\sigma_r \vee \sigma_l : \sigma_l) = \sigma$. By 2.3.b), the equality $(\sigma_r \vee \sigma_l : \sigma_r \vee \sigma_l) = \sigma$ holds.

If A has identity then $\sigma(A) = \sigma(A)A \subseteq \sigma_l(A) \subseteq \sigma(A)$; so, $\sigma(A) = \sigma_l(A)$. Analogously $\sigma(A) = \sigma_r(A)$.

b) Consequence of 2.2.b), 2.5.2) and 3.2.1).   □

It is well known that, for every $A \in \mathcal{A}$, $J(A) = \mathrm{Ker}(\Sigma_A^r) = \mathrm{Ker}(\Sigma_A^l)$ where Σ_A^r and Σ_A^l are families of prime modules. As Σ^r and Σ^l are closed under submodules and satisfy M_1) and M_2), by 3.5.

Corollary 3.6.

a) $J = J_r : J_l = J_l : J_r = (J_r \vee J_l) : (J_r \vee J_l) = \overline{J_r \vee J_l}$.
b) J is the least H-radical that contains $J_r \vee J_l$.
c) If A has identity then $J(A) = J_r(A) = J_l(A)$.

In [2], Andrunakievich defines a *special class* as a class $\mathcal{E}$ of prime algebras, hereditary and closed under essential extensions (i.e., if $0 \neq I \trianglelefteq A$ is essential, $I \in \mathcal{E}$ and A is a prime algebra then $A \in \mathcal{E}$). By [7, Prop. 2.2.3], the upper preradical of a special class in a KA-radical which will be called a *special radical*. In particular, the prime and the Jacobson radical are special radicals [7, Props. 3.8.13 and 3.14.24].

If γ is a special radical, supposing $\mathcal{C}_A^r$ (resp. $\mathcal{C}_A^l$) the class of the right (resp. left) prime A-modules such that $\gamma(A/\mathrm{An}(M)) = 0$ it is easy to check that $\mathcal{C}_A^r$ and $\mathcal{C}_A^l$ are closed under submodules. From [7, Theorem 3.14.23] they satisfy the M_1) and M_2) conditions and, for every $A \in \mathcal{D}$, $\gamma(A) = \mathrm{Ker}(\mathcal{C}_A^r) = \mathrm{Ker}(\mathcal{C}_A^l)$. Then from 3.5, it follows

Proposition 3.7. *Let γ be a special radical and for every $A \in \mathcal{A}$, let $\mathcal{C}_A^r$ (resp. $\mathcal{C}_A^l$) be the class of prime right A-modules (resp. left) such that $\gamma(A/\mathrm{An}(M)) = 0$. Then*

a) $\gamma_r = \Delta_{A \in \mathcal{D}} Re_{\mathcal{C}_A^r}$ and $\gamma_l = \Delta_{A \in \mathcal{D}} Re_{\mathcal{C}_A^l}$ are H-radicals,
b) the equalities $\gamma = \gamma_r : \gamma_l = \gamma_l : \gamma_r = (\gamma_r \vee \gamma_l) : (\gamma_r \vee \gamma_l) = \overline{\gamma_r \vee \gamma_l}$ hold and γ is the least H-radical that contains $\gamma_r \vee \gamma_l$. If A has identity then $\gamma(A) = \gamma_r(A) = \gamma_l(A)$.

About the prime radical β we can observe that, by 3.3, the class of the right prime A-modules such that $\beta(A/\mathrm{An}(M)) = 0$ coincide with the class $\mathcal{P}_A^r$ of all prime right A-modules. As, for every $M \in A$-Mod, its prime radical $P(M)$ is defined by

$$P(M) = \mathrm{Re}_{\mathcal{P}_A^r}(M) = \bigcap \{N \leq M : M/N \in \mathcal{P}_A^r\}$$

then, supposing $\beta_r = \Delta_{A \in \mathcal{A}} \mathrm{Re}_{\mathcal{P}_A^r}$, β_r is an H-radical on $\mathcal{A}$ where, for every $A \in \mathcal{A}$, $\beta_r(A) = P(A_A) = \bigcap_{i \in I} \{D \trianglelefteq A : A/D \in \mathcal{P}_A^r\}$. Analogously, $\beta_l : A \to P(_A A)(A \in \mathcal{A})$ is an H-radical. So, from 3.5,

Proposition 3.8. *Let β be the Baer radical (or prime radical) on $\mathcal{A}$. Then $\beta = \beta_r$: $\beta_l = \beta_l : \beta_r = (\beta_r \vee \beta_l) : (\beta_r \vee \beta_l) = \overline{\beta_r \vee \beta_l}$ and β is the least H-radical that contains $\beta_r \vee \beta_l$. If A has identity then $\beta(A) = \beta_r(A) = \beta_l(A)$.*

Proof. In fact, it is well known that if β is the prime radical then $\beta(A) = \bigcap_{i \in I} \{P \trianglelefteq A : P$ is prime$\}$ and from 3.3 we have

$$\beta(A) = \bigcap \{\mathrm{An}(M) : M \in \mathcal{P}_A^r\} = \mathrm{Ker}(\mathcal{P}_A^r)$$

Analogously $\beta(A) = \mathrm{Ker}(\mathcal{P}_A^l)$. □

Remark 3.1. If $A = \bigoplus_{i \in I} A_i$ where $(A_i)_I$ is a family of ideals of A, by 3.1.c), it follows that $J_r(A) = \bigoplus_{i \in I} J_r(A_i)$, $S_r(A) = \bigoplus_{i \in I} S_r(A_i)$ and $\beta_r(A) = \bigoplus_{i \in I} \beta_r(A_i)$. Also if γ is any special radical and γ_r is defined as in 3.7, then $\gamma_r(A) = \bigoplus_{i \in I} \gamma_r(A_i)$. Analogously for J_l, S_l, β_l and γ_l.

About the preradicals J_r and J_l we can state also the following properties:

Recall that an ideal or a subalgebra B of A is *quasi-regular* (*q.r.* for short) if all its elements are quasi-regular (i.e., for every $a \in B$ there exists $a' \in A$ such that $a + a' - aa' = 0$ and $a + a' - a'a = 0$). It is also well known that $J(A)$ is the maximum quasi-regular ideal of A. Then $J(A) = A$ if and only if A is a quasi-regular algebra. We will say that A is a *semiprimitive* algebra if $J(A) = 0$.

In [6, III.6,7] it has been proved that $J(A) = A$ if and only if $J_r(A) = J_l(A) = A$. $J(A) = 0$ if and only if $J_r(A) = J_l(A) = 0$. Then $R_J = R_{J_d} = R_{J_e} = \{A \in \mathcal{A} : A$ is q.r.$\}$, $S_J = S_{J_d} = S_{J_e} = \{A \in \mathcal{A} : A$ is semiprimitive$\}$.

Proposition 3.9.

a) J_r *is an H-radical, idempotent and hereditary.*

b) J_r *is not complete, nor isotone and S_{J_r} is not hereditary.*

Proof. a) Since $J_r(A) \subseteq J(A)$, $J_r(A)$ is a quasi-regular ideal and thus $J_r(J_r(A)) = J_r(A)$. If $J_r(A) = A$, A is quasi-regular; then every $I \trianglelefteq A$ is quasi-regular and $J_r(I) = I$.

b) It is enough to notice that, since $J(A)$ is a quasi-regular ideal, $J_r(J(A)) = J(A)$, but, in general, $J(A) \not\subseteq J_r(A)$ [6, Example 2]. Therefore J_r is not complete so it is not isotone. Since J_r is a non isotone H-radical, by 1.3.b), S_{J_r} is not hereditary. □

4. A Plotkin radical dual of the Jacobson radical

It is well known that in *A-Mod*, the concepts of radical and socle are considered dual of each other. In 3.6, we proved that

1) in $(\mathcal{A}\text{-pr}, \leq)$ the Jacobson radical J is related with the preradicals J_r and J_l by the equalities $J = J_r : J_l = J_l : J_r = (J_r \vee J_l) : (J_r \vee J_l) = \overline{J_r \vee J_l}$, and
2) J is the least H-radical that contains $J_r \vee J_l$.

In this section we are concerned to define in $(\mathcal{A}\text{-pr}, \leq)$ a preradical which may be consider dual of J. For that, we define a preradical of algebras, denoted *Soc* which is related with the preradicals S_r and S_l (which may be consider dual of J_r and J_l respectively) by dual equalities of 1) and satisfies a property dual of 2). This preradical is a torsion Plotkin radical, but not a KA-radical.

Supposing $A \in \mathcal{A}$ let $\mathcal{S}$ be a complete set of representatives of isomorphism classes of simple A-modules. Recall that, if $M \in A\text{-Mod}$, the *homogeneous components* of M are the submodules $H_T = \sum \{V : V \leq M, V \simeq T\}$ $(T \in \mathcal{S})$, and

1. For every $T \in \mathcal{S}$, H_T is a complete invariant, i.e., for every endomorphism $\varphi \in \text{End}_A(M)$, $\varphi(H_T) \subseteq H_T$ [4, 11-1.9 and 11-1.10].
2. $Soc(M) = \bigoplus_{T \in \mathcal{S}} H_T$ [4, 11-1.9 and 11-1.10].
3. Every complete invariant contained in $Soc(M)$ is a direct sum of some of its homogeneous components. Therefore, the homogeneous components are the minimal complete invariants of $Soc(M)$. [Proof similar in [3], §3, n°4, Prop. 11.]
4. M is a *semisimple* (*or completely reducible*) *module* if $M = Soc(M)$ or, equivalently, if $MA = M$ and for every $N \leq M$, there is $T \leq M$ such that $M = N \oplus T$. Then, if M is semisimple, $N \leq M$ is a minimal submodule if and only if it is indecomposable. The equality $MA = M$ is valid and every submodule N of M is also semisimple. [4, 11-1.6].

Since, for every $a \in A$, the map $\varepsilon_a : A_A \to A_A$ defined by $\varepsilon_a : b \to ab$, $(b \in A)$ is an A-endomorphism, every complete invariant of A_A is an ideal of A. Conversely, if $I \trianglelefteq A$ is such that $AI = I$ then I is a complete invariant of A_A. In fact, supposing $\varphi \in \text{End}_A(A_A)$ we have $\varphi(I) = \varphi(A)I \subseteq I$. In particular if, for every $I \trianglelefteq A$, $AI = IA = I$, the following are equivalent:

1) $I \trianglelefteq A$,
2) I is a complete invariant of $_AA$,
3) I is a complete invariant of A_A.

A right ideal of A is called a *simple right ideal* if it is a simple submodule of A_A. Supposing $\Sigma_r(A)$ the set of all simple right ideals of A, the homogeneous components H_D of A_A for some $D \in \Sigma_r(A)$, will be called the *right homogeneous components* of A. Since H_D is a complete invariant of A_A, H_D is an ideal of A. Thus, if $\{D_k : k \in K\}$ is a complete set of representatives of the isomorphism classes of $\Sigma_r(A)$ and $H_k = H_{D_k}$, $(k \in K)$ we have $S_r(A) = Soc(A_A) = \bigoplus_{k \in K} H_k$ (direct sum of algebras). Analogously one defines *simple left ideal* and *left homogeneous components*.

A is a *right semisimple algebra*, (resp. *left semisimple algebra*) if A_A (resp. $_AA$) is a semisimple module. A is a *semisimple algebra* if A is a right and left semisimple algebra. Analogously, an ideal I of A is a *semisimple ideal* if I is a semisimple algebra. It is clear that if A is semisimple, then, for every $I \trianglelefteq A$, $IA = AI = I$. In particular, $A^2 = A$.

Remark 4.1. If $T \trianglelefteq_r A$ is such that $T^2 = 0$ then for every simple right ideal D, $DT = 0$. In fact if $DT \neq 0$, then $DT = D$ and therefore $DT^2 = DT = 0$.

Lemma 4.1. *Let $I \trianglelefteq A$.*

a) $\Sigma_r(I) = \Sigma_r(I^n) = \{D \in \Sigma_r(A) : DI \neq 0\} \subseteq \Sigma_r(A)$, $(n \in \mathbb{N})$.
b) $S_r(I) = S_r(I^n) = S_r(A)I = (I \cap S_r(A))I \subseteq S_r(A)$, $(n \in \mathbb{N})$.
c) *If $I \subseteq S_r(A)$, then $S_r(I) = I^2 = I^n$, for every $n \geq 2$.*
d) *I is right semisimple if and only if $I \subseteq S_r(A)$ and $I^2 = I$. $S_r(A)^2$ is the largest semisimple right ideal of A. I is semisimple if and only if $I \subseteq S_r(A) \cap S_l(A)$ and $I^2 = I$. $(S_r(A) \cap S_l(A))^2$ is the largest semisimple ideal of A.*

Proof. a) Let $D \in \Sigma_r(I)$. Since $DI = D \neq 0$, it follows that $D \trianglelefteq_r A$ and $DA \neq 0$. Supposing $0 \neq D' \trianglelefteq_r A$ such that $D' \subseteq D$, then $D' \trianglelefteq_r I$ and therefore $D' = D$. Conversely let $D \in \Sigma_r(A)$ be with $DI \neq 0$. Necessarily $DI = D \subseteq I$ and $D \trianglelefteq_r I$. As, for every $0 \neq D' \trianglelefteq_r I$ such that $D' \subseteq D$, we have $D'A = D$ then $D = DI = D'AI \subseteq D'I \subseteq D' \subseteq D$. So, $D' = D$.

By the above, it is clear that $\Sigma_r(I^n) \subseteq \Sigma_r(I)$. Conversely, supposing $D \in \Sigma_r(A)$ such that $DI \neq 0$, then $DI = D$ and it follows that $DI^n = D \neq 0$, $(n \in \mathbb{N})$.

b) The first equality is obvious. By a), $S_r(I) = S_r(A)I \subseteq S_r(A)$ and since $S_r(I)$ is a semisimple right I-module then $S_r(I) = S_r(I)I \subseteq (I \cap S_r(A))I \subseteq S_r(A)I = S_r(I)$.

c) Consequence of b).

d) By b) the first statement is clear. On the other hand, as $S_r(S_r(A)^2) = S_r(A)^3 = S_r(A)^2$, $S_r(A)^2$ is right semisimple and, if I is a semisimple right ideal of A, then $I \subseteq S_r(A)$ and $I = I^2 \subseteq S_r(A)^2$. Analogously we prove the last statement. $\square$

Remark 4.2.

I) By 4.2.b), S_r is an isotone, complete and ADS preradical. Nevertheless S_r is not idempotent, neither hereditary. In fact, considering the commutative $\mathbb{Z}$-algebra $\mathbb{Z}_{p^2} = \mathbb{Z}/p^2\mathbb{Z}$, where p is a prime number, $\mathbb{Z}_{p^2}$ has the only ideals $A \supseteq p\mathbb{Z}/p^2\mathbb{Z} \supseteq 0$. Thus $S_r(A) = p\mathbb{Z}/p^2\mathbb{Z}$ and $(S_r \circ S_r)(A) = S_r(A)^2 = (p\mathbb{Z}/p^2\mathbb{Z})^2 = 0 \neq S_r(A)$. On the other hand, let K be a field and $A = Kx+Ky$ a two-dimensional algebra over K whose basis elements $\{x, y\}$ satisfy the following product

·	x	y
x	x	0
y	y	0

[5, Example 5]. As $xA = Kx$ and $yA = Ky$ are simple right ideals, $S_r(A) = xA+yA = A$. However for the ideal $I = (y) = Ky$ we have $S_r(I) = S_r(A)I = AI = 0 \neq I$. So, S_r is not hereditary.

II) S_r commutes with sum and direct sum. Supposing $B = \sum_{i \in I} A_i$ where $A_i \trianglelefteq A$, $(i \in I)$. By 4.1.b) $S_r(B) = S_r(A) \sum_I A_i = \sum_I S_r(A) A_i = \sum_K S_r(A_i)$. The last statement is clear.

III) $S_r \circ S_l = S_l \circ S_r$ is an isotone and ADS preradical. In fact, as S_r and S_l are isotone and ADS preradicals, $S_r \circ S_l$ is a preradical (cf. beginning of Sec. 2) and it is easy to see that it is also isotone and ADS.

By 4.1.b) and its left version, $(S_r \circ S_l)(A) = S_r(A) S_l(A) = (S_l \circ S_r)(A)$.

In [13] a class $\mathcal{B} \subseteq \mathcal{A}$ is called a *Plotkin class* if it is homomorphically closed and every $A \in \mathcal{A}$ contains a largest $\mathcal{B}$-ideal denoted $\mathcal{B}(A)$, i.e., if $I \trianglelefteq A$ where $I \in \mathcal{B}$ then $I \subseteq \mathcal{B}(A)$. Also it is proved that if $\mathcal{B}$ is a Plotkin class then the summand preradical $s_{\mathcal{B}}$ is a Plotkin radical such that $R_{s_{\mathcal{B}}} = \mathcal{B}$ and, for every $A \in \mathcal{A}$, $s_{\mathcal{B}}(A) = \mathcal{B}(A)$. [13, Prop. 1.14.]

In particular the classes R_{S_r} and R_{S_l} are Plotkin classes. In fact, R_{S_r} is homomorphically closed and, from 4.1.d), for every $A \in \mathcal{A}$, there exist a largest right semisimple ideal, the ideal $S_r(A)^2$. Analogously for R_{S_l}. Supposing $\mathcal{CR}$ the class of all semisimple algebras, as $\mathcal{CR} = R_{S_r} \cap R_{S_l}$, also $\mathcal{CR}$ is a Plotkin class and it is hereditary. Indeed, for every $I \trianglelefteq A \in \mathcal{CR}$, we have $I = AI = IA$ thus $S_r(I) = S_r(A)I = AI = I$ and $S_l(I) = IS_l(A) = IA = I$. Then the summand preradical $s_{\mathcal{CR}}$ is a hereditary Plotkin radical such that $R_{s_{\mathcal{CR}}} = \mathcal{CR}$. So, we introduce the following definition:

Definition 4.2. *For every $A \in \mathcal{D}$, the socle of A, denoted $Soc(A)$, is the sum of all its semisimple ideals, if they exist. If they don't, then $Soc(A) = 0$.*

Remark 4.3.

I) From the definition it follows that $Soc = s_{\mathcal{CR}}$ is an hereditary Plotkin radical, such that $R_{Soc} = \mathcal{CR}$ and, for every A, $Soc(A)$ is the largest semisimple ideal of A. Then, supposing $S = S_r \wedge S_l$, by 4.1.d), $Soc(A) = S(A)^2$.

II) By 2.5.1), Soc is the largest idempotent preradical contained in $S = S_r \wedge S_l$. In fact as $R_S = R_{S_r} \cap R_{S_l} = \mathcal{CR}$, then $Soc = s_{\mathcal{CR}} = s_{R_S} = \hat{S}$ and, since S_r and S_l are complete, S is complete.

Lemma 4.3. *Let $I \trianglelefteq A$ be such that $I \subseteq S_r(A)$. Then*

a) $J(I) = \sum \{T \in \Sigma_r(A) : T \subseteq I, T^2 = 0\}$ *and* $IJ(I) = 0$. *Therefore* $J(I)^2 = 0$. *If I is semisimple then $J(I) = 0$.*

b) $I = S_r(I) + J(I) = I^2 + J(I)$. *In particular,* $S_r(A) = S_r(A)^2 + J(S_r(A))$.

c) *If $I \subseteq (S_r \wedge S_l)(A)$ then I^2 is a semisimple ideal and $I = I^2 \oplus J(I)$.*

d) *A is semisimple if and only if A is right (resp. left) semisimple and semiprimitive.*

Proof. a) Since $S_r(A)_A$ is a semisimple A-module, we have $I = \sum \{D \in \Sigma_r(A) : D \subseteq I\}$ and $J(I) = \sum \{T \in \Sigma_r(A) : T \subseteq J(I)\}$. Let $T \in \Sigma_r(A)$ be such that $T \subseteq J(I)$. If $T^2 \neq 0$, by the Brauer Lemma, [9, 10.22.], $T = eA$ for some idempotent $e \neq 0$; thus $e \in J(I)$ which is impossible. Therefore $T^2 = 0$. The converse is clear. On the other hand, suppose $D, T \in \Sigma_r(A)$ such that $D \subseteq I$ and $T \subseteq J(I)$. As

$T^2 = 0$, by Remark 4.1, $DT = 0$, so $IJ(I) = 0$. If I is semisimple, as $J(I)$ is a left submodule of $_I I$, then $J(I) = IJ(I) = 0$.

b) The following relations hold:

$$I = \sum \{D \in \Sigma_r(A) : D \subseteq I, D^2 \neq 0\} + \{D \in \Sigma_r(A) : D \subseteq I, D^2 = 0\}$$
$$\subseteq \sum \{D \in \Sigma_r(A) : DI \neq 0\} + J(I)$$
$$= S_r(I) + J(I) = I^2 + J(I) \subseteq I.$$

c) Let $I \subseteq (S_r \wedge S_l)(A)$. By 4.1.c) and its left version, $S_r(I) = S_l(I) = I^2 = S_r(I^2) = S_l(I^2)$; so I^2 is a semisimple ideal. By b), $I = I^2 + J(I)$ and as $J(I^2) = 0 = I^2 \cap J(I)$, we have $I = I^2 \oplus J(I)$.

d) If A is right semisimple and $J(A) = 0$ there are no nilpotent right ideals. Thus $S_r(A) = S_l(A)$ by [8, IV, 3. Theo. 1] and $A = S_r(A) = S_l(A)$. The converse is clear. $\qquad\square$

Remark 4.4. If $A \neq 0$ is a semisimple algebra, the right and left homogeneous components coincide and these are simple algebras (H is a simple algebra if $H^2 \neq 0$ and $\{0\}$ and H are its only ideals). They also coincide with its minimal ideals. Indeed, the first statement is a consequence of 4.3.d) and [8, IV, 3. Theo. 1]. On the other hand as, for every $I \triangleleft A$, $AI = IA = I$, its ideals are the complete invariants of A_A and $_A A$; so, the minimal ideals are the minimal complete invariants of A_A, i.e., its homogeneous components. Since every ideal I of A is a complete invariant of $_A A$, I is a direct sum of some of its homogeneous components.

Theorem 4.4.

  a) $Soc = S_r \circ S_l = S_l \circ S_r = (S_r \wedge S_l) \circ (S_r \wedge S_l)$ *is a torsion Plotkin radical. For every* $A \in \mathcal{A}$, $Soc(A) = S_r(A)S_l(A) = S_r(A)^2 S_l(A)^2$.
  b) Soc *commutes with sum and direct sum.*
  c) *Supposing* $S = S_r \wedge S_l$, *for every* $A \in \mathcal{A}$, $S(A) = Soc(A) \oplus J(S(A))$.

Proof. a) By the left version of 4.3.b), $S_l(A) = S_l(A)^2 + J(S_l(A))$, where $J(S_l(A))^2 = 0$. Then, by Remark 4.1., $S_r(J(S_l(A))) = S_r(A)J(S_l(A)) = 0$. Thus,

$$(S_r \circ S_l)(A) = S_r(S_l(A)^2 + J(S_l(A))) = S_r(A)S_l(A)^2 = S_l[(S_r \circ S_l)(A)].$$

Analogously, $(S_r \circ S_l)(A) = S_r[(S_r \circ S_l)(A)]$ and $(S_r \circ S_l)(A)$ is a semisimple ideal. Then we have $(S_r \circ S_l)(A) \subseteq Soc(A) = S(A)^2 \subseteq S_r(A)S_l(A) = (S_r \circ S_l)(A)$. So, $Soc = S_r \circ S_l$ and, by Remark 4.2.III), Soc is isotone. Then, by 1.1, Soc is a torsion preradical.

As $(S_r \wedge S_l)(A) \subseteq S_r(A), S_l(A)$, by 4.1.c),

$$S_r((S_r \wedge S_l)(A)) = S_l((S_r \wedge S_l)(A)) = (S_r \wedge S_l)(A)^2 = Soc(A)$$

then $[(S_r \wedge S_l) \circ (S_r \wedge S_l)](A) = Soc(A)$ and $Soc = (S_r \wedge S_l) \circ (S_r \wedge S_l)$.
  Also we have,

$$(S_r \circ S_l)(A) = (S_r(A)^2 + J(S_r(A)))(S_l(A)^2 + J(S_l(A))) = S_r(A)^2 S_l(A)^2$$

b) Let $A = \sum_{i \in I} A_i$. Then $Soc(A) = S_r(S_l(A)) = S_r\left(\sum_{i \in I} S_l(A_i)\right) = \sum_{i \in I} Soc(A_i)$.

c) Consequence of 4.3.c) $\square$

Example 4.1. Let $G = \{g_i : 1 \leq i \leq n\}$ be a finite group of order n, K a field and $A = K[G]$ its group algebra. It is easy to check that the map $h : A \to K$ defined by: for every $z = \sum_{i=1}^{n} \alpha_i g_i \in A$, $h(z) = \sum_{i=1}^{n} \alpha_i$ is an algebra epimorphism, such that $Ker(h) = \{z = \sum_{i=1}^{n} \alpha_i g_i : \sum_{i=1}^{n} \alpha_i = 0\}$ is an ideal of dimension $n-1$, over K, and $\{1 - g : g \in G\}$ is one of its basis. In particular for the element $v = \sum_{i=1}^{n} g_i$ on has $h(v) = n$ and for every $z \in A$ the equalities $vz = zv = h(z)v$ are valid. Therefore $Av = vA = Kv$. Since $\dim_K(vA) = 1$ and A is an algebra with identity, Kv is a simple right and left ideal, thus $Kv \subseteq S_r(A) \cap S_l(A)$. Also, as $A/Ker(h) \simeq K$, then $J(A) \subseteq Ker(h)$.

Suppose $c(K) = p$ and p divides n. Then $h(v) = 0$, therefore $v^2 = h(v)v = 0$ and so $(Kv)^2 = 0$. Thus $Kv \subseteq J(A)$.

As, for every $z \in S_r(A)$, $zJ(A) = 0$ one has $zv = h(z)v = 0$ therefore $h(z) = 0$ and $S_r(A) \subseteq Ker(h)$. Analogously, $S_l(A) \subseteq Ker(h)$.

In particular, let $n = 6$, G defined by $G = <a, b : a^2 = 1, b^3 = 1, aba = b^2>$ and suppose $c(K) = 2$. $A = K[G]$ admits the basis $\{1, a, b, b^2, ab, ba\}$ and the right ideals

$$D_1 = (1 + a + b + ab)A = K(1 + a + b^2 + ba) + K(b + b^2 + ab + ba)$$

$$D_2 = (1 + b^2 + ab + ba)A = K(1 + b^2 + ab + ba) + K(a + b + b^2 + ab)$$

are simple of dimension 2. Then $D_1 + D_2 + Kv \subseteq S_r(A) \subseteq Ker(h)$. Also the sum is direct. Therefore $\dim_K(D_1 \oplus D_2 \oplus Kv) = 5 = \dim_K(Ker(h))$ and

$$S_r(A) = D_1 \oplus D_2 \oplus Kv = \left\{z = \sum_{i=1}^{n} \alpha_i g_i : \sum_{i=1}^{n} \alpha_i = 0\right\}.$$

Since $z \in J(A)$ if and only if $S_r(A)z = 0$ and $a + b, b + ab \in S_r(A)$, from the equalities $(a + b)z = (b + ab)z = 0$, one obtains $z \in Kv$ therefore $J(A) = Kv$.

On the other hand, since A is an artinian algebra with identity, by [6, IV.3.b)], the equivalences hold

$$z \in S_l(A) \Leftrightarrow J(A)z = 0 \Leftrightarrow vz = h(z)v = 0 \Leftrightarrow h(z) = 0 \Leftrightarrow z \in Ker(h).$$

Thus $S_l(A) = S_r(A) = D_1 \oplus D_2 \oplus J(A) = Ker(h)$.

Since $D_1, D_2 \not\subseteq J(A)$, necessarily $D_1 = D_1^2$ and $D_2 = D_2^2$. From $D_1 \oplus D_2 \subseteq S_r(A)$ we get $(D_1 \oplus D_2)J(A) = J(A)(D_1 \oplus D_2) = 0$. Thus

$$Soc(A) = S_r(A)S_l(A) = (D_1 \oplus D_2 \oplus J(A))^2 = (D_1 \oplus D_2)^2 = D_1 \oplus D_2.$$

Remark 4.5.

I) By the dual equalities $J = J_r : J_l = J_l : J_r = (J_r \vee J_l) : (J_r \vee J_l)$ (3.6.a)) and $Soc = S_r \circ S_l = S_l \circ S_r = (S_r \wedge S_l) \circ (S_r \wedge S_l)$ (4.4.a)) the preradicals J and Soc can be considered dual in $\mathcal{A}$-pr. Also 3.6.b) and Remark 4.3.II) are dual statements.

II) By Example 4.1 we can say that $Soc = S_r \circ S_l < S_r \wedge S_l$. As by 4.4.c), $(S_r \wedge S_l)(A) = Soc(A) \oplus J((S_r \wedge S_l)(A))$, then $(S_r \wedge S_l)(A) = SocA$ if and only if $J((S_r \wedge S_l)(A)) = 0$.

III) As for every $A \in \mathcal{A}$, $J(Soc(A)) = 0 = SocA \cap J(A)$ one has $Soc \wedge J = 0$.

Proposition 4.5. *Let $A \in \mathcal{A}$ be such that $Soc(A) \neq 0$.*

a) *$Soc(A)$ is the direct sum of all minimal semisimple ideals of A and can be written as a direct sum of simple and semisimple algebras in an unique way.*

b) *Every non zero semisimple ideal of A is a direct sum of some simple components of $Soc(A)$.*

Proof. a) Since, if $I \trianglelefteq Soc(A)$, $ISoc(A) = Soc(A)I = I$ then, $I \trianglelefteq Soc(A)$. Hence a minimal ideal of $Soc(A)$ is also a minimal ideal of A and, conversely, if H is a minimal ideal of A semisimple, H is also a minimal ideal of $Soc(A)$. So, since $Soc(A)$ is a semisimple algebra, $Soc(A)$ is a direct sum of its minimal ideals, which are the set of all minimal ideal of A semisimple.

b) From Remark 4.4 every $I \trianglelefteq Soc(A)$ is a direct sum of some of its homogeneous components. $\qquad\square$

Proposition 4.6. *The following are equivalent:*

a) *$S_r(A) = Soc(A)$.*

b) *There are no nilpotent simple right ideals.*

Proof. a) $\Rightarrow$ b) If $S_r(A) = Soc(A)$, $S_r(A)$ is semisimple and therefore $J(S_r(A)) = 0$. By 4.3.a), there are no simple right ideals nilpotents.

b) $\Rightarrow$ a) By 4.3.a),b) $J(S_r(A)) = 0$ and $S_r(A) = S_r(A)^2 = S_r(S_r(A))$. So, by 4.3.d), $S_r(A)$ is semisimple. Thus $S_r(A) \subseteq Soc(A)$ and the equality is valid. $\qquad\square$

Corollary 4.7.

1) *The following are equivalent:*
   a) *$Soc(A) = S_r(A) = S_l(A)$.*
   b) *There are no one-sided nilpotent simple ideals.*
2) *In particular, if A is a semiprime algebra then $Soc(A) = S_r(A) = S_l(A)$.*

Proof. Consequence of 4.6 and its left version. $\qquad\square$

Recall that an idempotent $e \in A$ is called primitive if eR is indecomposable in A_A. Then D is a simple right ideal such that $D^2 \neq 0$ if and only if it is generated by a primitive idempotent $e \in S_r(A)$. In fact if $D^2 \neq 0$, by the Brauer Lemma, there is a idempotent $e \in S_r(A)$ such that $D = eR$ and as D is indecomposable, e is primitive. Conversely, if $D = eR$ where $e \in S_r(A)$ is a primitive idempotent then D is an indecomposable submodule of $S_r(A)$, so a simple submodule.

Proposition 4.8. $Soc(A) = \sum_{i \in I, k \in K} e_i A f_k$ *where $\{e_i : i \in I\}$ and $\{f_k : k \in K\}$ are the set of all primitive idempotents of A belonging to $S_r(A)$ and $S_l(A)$, respectively.*

In particular if $Soc(A) = S_r(A) = S_l(A)$ then

$$Soc(A) = \sum_{e \in \mathcal{P}} eA = \sum_{e \in \mathcal{P}} Ae = \sum_{e_i, e_k \in \mathcal{P}} e_i Ae_k = \sum_{e \in \mathcal{P}} AeA$$

where $\mathcal{P} = \{e \in Soc(A) : e \text{ primitive idempotent}\}$.

Proof. Since for $D \in \Sigma_r(A)$, $D^2 \neq 0$ if and only if $D = e_i A$ for some $i \in I$, by 4.3.a) $S_r(A) = \sum_{i \in I} e_i A + J(S_r(A))$ and, analogously, $S_l(A) = \sum_{k \in K} Af_k + J(S_l(A))$ where $J(S_r(A))^2 = J(S_l(A))^2 = 0$ and $J(S_l(A)) \left(\sum_{k \in K} Af_k\right) = \left(\sum_{i \in I} e_i A\right) J(S_r(A)) = 0$. Then,

$$Soc(A) = S_r(A).S_l(A) = \sum_{i \in I, k \in K} e_i AAf_k \subseteq \sum_{i \in I, k \in K} e_i Af_k \subseteq \sum_{i \in I, k \in K} e_i Af_k f_k$$

$$\subseteq \sum_{i \in I, k \in K} e_i AAf_k = Soc(A).$$

If $Soc(A) = S_r(A) = S_l(A)$, we obtain $Soc(A) = \sum_{e \in \mathcal{P}} eA = \sum_{e \in \mathcal{P}} Ae = \sum_{e_i, e_k \in \mathcal{P}} e_i Ae_k$. On the other hand, $Soc(A) = ASoc(A) = \sum_{e \in \mathcal{P}} AeA$. $\square$

Proposition 4.9. *Let A be an artinian right algebra such that $J(A) \neq A$. Then $Soc(A) = V'V$ where $V = \{a \in A : J(A)a = 0\}$ and $V' = \{a' \in A : a'J(A) = 0\}$.*

Proof. By [6, IV.3], supposing $e^2 = e \in A$ such that $e + A = 1_{A/J}$, one has $S_l(A) = eV$ and $S_r(A) = V'e$, where $V = \{a \in A : J(A)a = 0\}$ and $V' = \{a' \in A : a'J(A) = 0\}$. Thus $Soc(A) = S_r(A)S_l(A) = V'eV$.

Since, for every $a \in A$, $a - ae \in J(A)$, if $b' \in V'$ and $b \in V$, then $b'e = b' + q'$, with $q' \in J(A)$ and therefore $b'eb = (b' + q')b = b'b + q'b = b'b \in V'V$. Thus $V'eV = V'V$. $\square$

Example 4.2. Let $S = \{a, b, c, d\}$ be the semigroup defined by the multiplication table

·	a	b	c	d
a	a	b	a	a
b	a	b	a	a
c	a	b	a	c
d	a	b	a	d

K a field and $A = K[S] = \{z = \alpha a + \beta b + \gamma c + \delta d : \alpha, \beta, \gamma, \delta \in K\}$ its semigroup algebra. It is easy to check that, for every $z = \alpha a + \beta b + \gamma c + \delta d$, $z' = \alpha' a + \beta' b + \gamma' c + \delta' d \in A$,

$$zz' = [(\alpha + \beta + \gamma + \delta)(\alpha' + \gamma') + (\alpha + \beta)\delta']a + [(\alpha + \beta + \gamma + \delta)\beta']b + (\gamma\delta')c + (\delta\delta')d.$$

In [6] it was shown that $J(A) = \{z = \alpha a + \beta b + \gamma c \in A : \alpha + \beta + \gamma = 0\}$ and $d + J(A) = 1_{A/J(A)}$ where $d^2 = d$. Let $z = \alpha a + \beta b + \gamma c + \delta d \in A$, $V = \{z \in A : J(A)z = 0\}$ and $V' = \{z \in A : zJ(A) = 0\}$. Then $z \in V$, if and only if, for every $z' = \alpha' a + \beta' b + \gamma' c \in J(A)$,

$$z'z = [(\alpha' + \beta' + \gamma')(\alpha + \gamma) + (\alpha' + \beta')\delta]a + [(\alpha' + \beta' + \gamma')\beta]b + (\gamma'\delta)c = -\gamma'\delta a + \gamma'\delta c = 0,$$

i.e., if and only if $\gamma'\delta(c-a)=0$. Since γ' is arbitrary, necessarily, $\delta=0$. Thus, $V=Ka+Kb+Kc$.

$z\in V'$, if and only if, for every $z'=\alpha'a+\beta'b+\gamma'c\in J(A)$,

$$zz'=[(\alpha+\beta+\gamma+\delta)(-\beta')]a+[(\alpha+\beta+\gamma+\delta)\beta']b=0$$

and, since β' is arbitrary, necessarily, $\alpha+\beta+\gamma+\delta=0$.

Therefore $V'=\{z=\alpha a+\beta b+\gamma c+\delta d:\alpha+\beta+\gamma+\delta=0\}$.

Then $S_r(A)=V'd$ and $S_l(A)=dV$.

Since, for every $z'=\alpha a+\beta b+\gamma c+\delta d\in V'$, $z'd=(-\gamma-\delta)a+\gamma c+\delta d$ one has

$$S_r(A)=V'd=\{\alpha a+\gamma c+\delta d\in A:\alpha+\gamma+\delta=0\}$$
$$S_l(A)=dV=d(Ka+Kb+Kc)=K(da)+K(db)+K(dc)=Ka+Kb.$$

Finally, since, for every $z'=\alpha a+\beta b+\gamma c+\delta d\in V'$, $z'a=z'b=z'c=0$ it follows that $z'V=z'(Ka+Kb+Kc)=0$. Therefore $Soc(A)=V'V=0$. In this case,

$$S(A)=S_r(A)\cap S_l(A)=\{\alpha a+\gamma c+\delta d\in A:\alpha+\gamma+\delta=0\}\cap(Ka+Kb)=0$$

and $S_r(A)\cap S_l(A)=Soc(A)$, i.e., $(S_r\wedge S_l)(A)=Soc(A)$.

It is well known that every semiprimitive right artinian algebra is semisimple, has identity, and is artinian and noetherian. Supposing $I\trianglelefteq A$ such that I and A/I are semiprimitive right artinian then A is semiprimitive and right artinian.

Proposition 4.10. *Let A be a right artinian (or right noetherian) algebra. Then*

a) *$Soc(A)$ is a semisimple artinian and noetherian algebra.*

b) *$Soc(A/Soc(A))=0$.*

Proof. a) Let A be right artinian such that $Soc(A)\neq 0$. Supposing $D\trianglelefteq_r Soc(A)$, as $Soc(A)$ is semisimple one has $DSoc(A)=D$ and therefore $D\trianglelefteq_r A$. So, if A is right artinian also $Soc(A)$ is right artinian. On the other side, $J(Soc(A))=0$. Then $Soc(A)$ is semiprimitive and right artinian; therefore it is artinian and noetherian (on right and left).

b) Since $A/Soc(A)$ is also right artinian, analogously one can deduce that $Soc(A/Soc(A))=I/Soc(A)$ (where $I\trianglelefteq A$), is a semiprimitive artinian algebra. So, as $Soc(A)$ and $I/Soc(A)$ are semiprimitive right artinian, I is a semiprimitive and artinian algebra and therefore a semisimple algebra. Thus $I\subseteq Soc(A)$ and $Soc(A/Soc(A))=0$.

Analogously if A is right noetherian also $Soc(A)$ is right noetherian and therefore right artinian. The proof follows as above.   □

Supposing $\mathcal{D}$ the class of all right (or left) artinian (or noetherian) algebras, in general $\mathcal{D}$ is not hereditary and so it is not an universal class. However, by 4.10 we can consider the ideal mapping $Soc_{|\mathcal{D}}:\mathcal{D}\to\mathcal{D}$ which is also idempotent, complete and satisfies, for every $A\in\mathcal{D}$, $Soc(A/Soc(A))=0$. Therefore $Soc_{|\mathcal{D}}$ can be consider a Kurosh-Amitzur radical on $\mathcal{D}$.

Remark 4.6. One may notice that, if A is not right artinian, in general, 4.10.b) is not valid, therefore *Soc* is not an H-radical and so it is not a Kurosh-Amitzur radical. Indeed, in [8, II.6.3] one has the following example of a primitive ring, so an $\mathbb{Z}$-algebra (Kaplansky): if V is a left vector space over the division ring Δ and $(e_i)_{\mathbb{N}}$ is one of its basis, consider the subring of linear maps given by the following matrices:

$$
S_{A,d} = \begin{bmatrix} A & 0 & 0 & \cdots & \cdots \\ 0 & d & 0 & \cdots & \cdots \\ 0 & 0 & d & \cdots & \cdots \\ & & & \cdots & \cdots \\ \cdots & & & & \cdots \end{bmatrix}
$$

where $A \in M_n(\Delta)$ is any finite matrix, and d is any element of Δ. This subring is a primitive ring and therefore it is semiprime. Thus, the set $\mathcal{A}$ of all matrices $S_{A,d}$ is also a semiprime ring; so $Soc(\mathcal{A}) = S_r(\mathcal{A}) = S_l(\mathcal{A})$. By [8, IV, 9], $Soc(\mathcal{A})$ is the set of all matrices that represent finite rank endomorphisms (i.e., whose images space has finite dimension), therefore the matrices of type $S_{A,0}$. Since, for every $n \in \mathbb{N}$, $A \in M_n(\Delta)$ and $d \in \Delta$, $S_{A,d} - dI \in Soc(A)$, where dI is a scalar matrix, one has $\mathcal{A}/Soc(A) = \{dI + Soc(A) : d \in \Delta\} \simeq \Delta$ and so

$$
Soc(\mathcal{A}/Soc(A)) \simeq Soc(\Delta) = \Delta \neq 0.
$$

Acknowledgements

The author wishes to thank M. Teresa A. Nogueira for some useful scientific remarks.

References

[1] T. Anderson, N. Divinsky and A. Sulinski, Hereditary radicals in associative and alternative rings, *Canad. J. Math.* **17** (1965) 594–603.

[2] V.A. Andrunakievch, Radical of Associative Rings I, *Mat. Sb.* **44(86)** (1958) 179–212 (in Russian), English Translation: *Amer. Math. Soc. Transl.* **2(52)** (1966) 95–128.

[3] N. Bourbaki, *Modules et anneaux semisimples*, (Éléments de mathématique Livre II, Algèbre, Cap. 8, Hermann, Paris 1958).

[4] J. Dauns, *Modules and Rings* (Cambridge University Press, 1994).

[5] N.J. Divinski, *Rings and Radicals* (Mathematical Expositions N°14, University of Toronto Press, 1965).

[6] M.L. Galvão, M.T. Nogueira and J.A. Green, Radicals and Socles of an Algebra without Identity, *Commun. Algebra* **31(6)** (2003) 2883–2907.

[7] B.J. Gardner and R. Wiegandt, *Radical Theory of Rings*, Pure and Applied Mathematics. A series of Monographs and Textbooks **261** (2004) (New York, Marcel Dekker).

[8] N. Jacoson, *Structure of Rings*, Amer. Math. Soc. Colloquium Publ. **37** (1956) (Providence).

[9] T.Y. Lam, A First Course in Noncommutative Rings, 2nd edition, Graduate Texts in Mathematics **131** (1991) (Springer-Verlag, New York).

[10] G. Michler, Radikale und Sockel, *Math. Annalen* **167** (1966) (1–48).

[11] R. Mlitz and R. Wiegant, Radicals and subdirect decompositions of Ω-groups, *J. Austral. Math. Soc.*, Ser A **48** (1990) 171–198.

[12] F. Raggi, J.R. Montes, H. Rincón, R. Fernández-Alonso and C. Signoret, The lattice structure of preradicals, *Commun. Algebra* **30(3)** (2002) 1533–1544.

[13] B. de la Rosa, S. Veldsman and R. Wiegandt, On the theory of Plotkin Radicals, *Chinese J. Math.* (Taiwan, R.O.C.) **21(1)** (1993) 33–54.

M. Luísa Galvão
Centro de Álgebra da Universidade de Lisboa
Av. Prof. Gama Pinto, 2
1649-003 Lisboa, Portugal
e-mail: `mlgalvao@ptmat.fc.ul.pt`

Modules and Comodules
Trends in Mathematics, 227–241
© 2008 Birkhäuser Verlag Basel/Switzerland

On the Construction of Separable Modules

P.A. Guil Asensio, M.C. Izurdiaga and B. Torrecillas

Dedicated to Robert Wisbauer on his 65th birthday

1. Introduction

Let R be an associative ring with identity and κ, an infinite regular cardinal. A left R-module M is said to be $\kappa^{\leq}$-*generated* (resp. $\kappa^{<}$-*generated*) if there exists a generator set $\{m_i\}_I$ of M of cardinality at most κ (resp. strictly smaller than κ). And a module M is called κ-*separable* if any subset X of M of cardinality strictly smaller than κ is contained in a $\kappa^{<}$-generated direct summand of M. Let us note that any direct sum of $\kappa^{<}$-generated modules is clearly a κ-separable module that we will call trivial. This notion of separability can be extended as follows. Given an infinite regular cardinal κ and a non-empty class of modules $\mathcal{C}$, we may say that a module M is $(\kappa, \mathcal{C})$-separable if each subset X of M of cardinality strictly smaller than κ is contained in a direct summand of M belonging to $\mathcal{C}$. Again, we will say that the $(\kappa, \mathcal{C})$-separable module M is non-trivial if it is not a direct sum of elements in $\mathcal{C}$. These modules present a pathological behavior in the sense that they have enough direct summands belonging to $\mathcal{C}$, but they are not direct sums of modules in $\mathcal{C}$.

Several authors have proved that, for some particular elections of κ and $\mathcal{C}$, the absence of non-trivial $(\kappa, \mathcal{C})$-separable modules characterizes the structure of certain rings (see, e.g., [8, 10, 12, 16, 21]). These results are mainly based on the fact that the construction of non-trivial $(\kappa, \mathcal{C})$-separable modules is closely related to the decomposition properties of modules in $\mathcal{C}$ into direct summands. For instance, if $\mathcal{C}$ is the class of all countably generated left modules, left perfect rings are characterized in [12] as those rings for which there do not exist non-trivial $(\aleph_1, \mathcal{C})$-separable and torsionless left modules (in the sense that they are submodules of direct products of copies of the ring). With the same notation, it is shown in [21] that a ring R is left pure-semisimple if and only if any $(\aleph_1, \mathcal{C})$-separable left

The first author was partially supported by the DGI and by the Fundación Séneca. Part of the sources of both institutions come from the FEDER funds of the European Union.

module is trivial. Other interesting case appears when we consider the class $\mathcal{P}$ of all projective left modules. It is essentially proved in [21, Theorem 10] that the ring is left perfect if and only if any $(\aleph_0, \mathcal{P})$-separable left module is trivial.

In this paper we review some of the methods to construct κ-separable modules, with the aim of stressing the close relation between the lack of nice decomposition properties for certain classes of modules and the existence of pathological κ-separable modules, in the sense that they are not direct sums of $\kappa^<$-generated modules (i.e., they are non-trivial κ-separable modules). First, we will study the construction of $\aleph_0$-separable modules in Section 4. We will follow the approach given in [13], which has its origin in the construction developed by Hill for Abelian groups [16] (see also [21]). Let M be a module and let us denote by $AddM$ the class of all direct summands of arbitrary direct sums of copies of M. We will see that, when M is a direct sum of finitely generated modules, the existence of a cardinal number λ such that any $(\aleph_0, AddM)$-separable module is a direct sum of λ-generated modules implies that the module M has a perfect decomposition. Where a module M has a perfect decomposition when any element of $AddM$ has a decomposition that complements direct summands (see for instance [4, 11]). Moreover, when M is a direct sum of finitely presented modules, the converse also holds.

The extension of this result to arbitrary R-modules has been made in [14] by using techniques related to λ-presentable modules. Recall that, if λ is an infinite regular cardinal number, a partially ordered set I is called λ-directed when every subset of I of cardinality strictly smaller than λ has an upper bound in I. A directed system of modules $\{M_i, f_{ij}\}_I$ is λ-directed when so is its index set I; and its direct colimit is called a λ-directed colimit. A module M is λ-presentable if the functor $Hom_R(M, -) : R-\text{Mod} \to \mathbf{Ab}$ commutes with λ-directed colimits. Let us note that, for any module M, there exists a regular cardinal $\lambda \leq |M|^+$ such that M is λ-presentable (where $|M|^+$ denotes the successor cardinal of the cardinal $|M|$ of M). We will devote Section 5 to explain one of the main results of [14]. Namely, that if M is a λ-presentable module, then $AddM = \lambda - \varinjlim AddM$ precisely when every $(\lambda, \text{Add } M)$-separable module is trivial (where $\lambda - \varinjlim AddM$ is the class of all totally ordered λ-directed colimits of modules in $AddM$). The proof of this general result requires the assumption of the Generalized Continuum Hypothesis (GCH) and is based on an extension of the construction developed by Eklof in [8]. Let us point out that, when $\lambda = \aleph_0$, the GCH is not required and, consequently, we obtain a theorem in ZFC (see also [10]).

2. Preliminaries and notation

We start by recalling some known set-theoretic facts. Along this paper, the cardinality of a set I will be denoted by $|I|$, and the successor cardinal of a cardinal number λ, by λ^+. If λ is an infinite cardinal, we shall denote by $\mathcal{P}_\lambda(I)$ the set $\{J \subseteq I : |J| < \lambda\}$. The *cofinality* of a limit ordinal α is the least cardinal number, denoted $cf(\alpha)$, such that there exists an increasing sequence of ordinals

$\{\alpha_\nu : \alpha_\nu < \alpha\}_{\nu < cf(\alpha)}$ whose limit is α. An infinite cardinal κ is *regular* when $cf(\kappa) = \kappa$. The symbol $\upharpoonright$ will mean restriction.

Let κ be an uncountable regular cardinal. A subset C of κ is said to be *closed unbounded* (club for short) if $\sup C = \kappa$ and any $C' \subseteq C$ with $\sup C' < \kappa$ verifies that $\sup C' \in C$. A subset E of κ is *stationary* if $E \cap C \neq \emptyset$ for every club C of κ.

Let E be a stationary subset of κ consisting of limit ordinals. For any $e \in E$, a *ladder* on e is a strictly increasing sequence of successor ordinals, $\{\eta_e(\alpha) : \alpha < cf(e)\}$, whose limit is e. A *ladder system* on E is a family, $\eta = \{\eta_e : e \in E\}$, where η_e is a ladder on e for each $e \in E$. The ladder system η on E is said to be *tree-like* if verifies the following compatibility condition: for any $e, f \in E$ and any $\alpha \in cf(e)$ and $\mu \in cf(f)$ such that $\eta_e(\alpha) = \eta_f(\mu)$, we have that $\alpha = \mu$ and $\eta_e \upharpoonright \alpha = \eta_f \upharpoonright \alpha$. Every stationary subset of any uncountable regular cardinal κ has a ladder system. If $\kappa = \aleph_1$, this ladder system can be constructed with the tree-like property (see [9, Exercise XII.17]). However, if $\kappa > \aleph_1$, the existence of tree-like ladder systems has been only proved using additional set-theoretic hypotheses. For instance, given a successor cardinal λ^+, the existence of a tree-like ladder system on some stationary subset of λ^{++} has been obtained in [8, Theorem 9] using the hypothesis $2^\lambda \leq \lambda^{++}$. If, in addition, we assume the GCH, that is, that $2^\lambda = \lambda^+$ for any infinite cardinal λ, a similar argument to that of [9, Exercise XII.17] gives the existence of a tree-like ladder system on any stationary subset of λ^+ consisting of limit ordinals of cofinality λ.

Proposition 2.1. (see [14, Proposition 2.1]) *Let us assume GCH and let λ be an uncountable cardinal number and E, a stationary subset of λ^+ consisting of limit ordinals of cofinality λ. Then there exists a tree-like ladder system on E.*

Recall that a ring T (possibly without identity) is said to have enough idempotents if there exists a family $\{e_i : i \in I\}$ of pairwise orthogonal idempotent elements of T such that

$$T = \bigoplus_{i \in I} T e_i = \bigoplus_{i \in I} e_i T.$$

Let us fix a ring with enough idempotents T. Along this paper, the word module will mean a unitary left T-module (in the sense that $TM = M$) and morphisms will operate on the right. We will denote by $T-\text{Mod}$ the category of all unitary left T-modules. Given a T-module M, an index set I and an element $x \in M^I$, we will denote by $x(i)$ the i-entry of x, for each $i \in I$. The same notation will be used to denote the components of a morphism $f : N \to M^I$. If N is isomorphic to a direct summand of M, we shall write $N \lesssim_\oplus M$. We will call $\mathcal{P}roj$ and $\mathcal{PP}roj$ the classes of all projective and all pure-projective unitary modules, respectively.

A non-empty class $\mathcal{C}$ of modules is said to be *abstract* when it is closed under isomorphisms; every class considered in this paper will be of this type. Given an abstract class $\mathcal{C}$, we will call

$$Add_R\mathcal{C} = \left\{ N : N \lesssim_\oplus \bigoplus_{i \in I} C_i \text{ for some family } \{C_i\}_I \text{ of objects in } \mathcal{C} \right\}$$

Let κ be an infinite cardinal number. We will denote by $\mathcal{C}_\kappa$ the subclass of $\mathcal{C}$ consisting of all $\kappa^<$-generated modules of $\mathcal{C}$. In particular, if κ is regular and all modules in $\mathcal{C}$ are $\kappa^<$-generated, then $(Add_R\mathcal{C})_\kappa$ is precisely the class

$$Add_R^\kappa\mathcal{C} = \left\{ N : N \lesssim_\oplus \bigoplus_{i\in I} C_i \text{ for some family } \{C_i\}_I \text{ of objects in } \mathcal{C} \text{ with } |I| < \kappa \right\}.$$

Recall that a decomposition of the module M into direct summands, $M = \bigoplus_{i\in I} M_i$, is said to complement direct summands when for any direct summand S of M, there exists $J \subseteq I$ such that $M = S \oplus \left(\bigoplus_{j\in J} M_j\right)$ (see, e.g., [2]). We will say that a module M has a perfect decomposition if any module in $AddM$ has a decomposition that complement direct summands (see [4, 11]).

We refer to [2, 1, 9, 15, 20] for any undefined notion used along this text.

3. Separable modules

We begin this section by stating some basic properties of separable modules.

Definition 3.1. *Let κ be an infinite regular cardinal and $\mathcal{C}$, a nonempty class of modules. We will say that a module Q is $(\kappa, \mathcal{C})$-separable if any subset of Q of cardinality strictly smaller than κ is contained in a direct summand of Q that belongs to $\mathcal{C}$.*

The next easy lemma characterizes separable modules in terms of the existence of certain special filtrations.

Lemma 3.2. [14, Lemma 3.2] *Let κ be an infinite regular cardinal, Q, a $\kappa^\le$-generated module and $\mathcal{C}$, a nonempty class of modules. Suppose that $\mathcal{C}$ is closed under direct summands and that each module in $\mathcal{C}$ is a direct sum of $\kappa^<$-generated modules. The following assertions are equivalent:*

i) *Q is $(\kappa, \mathcal{C}_\kappa)$-separable.*
ii) *Q is $(\kappa, \mathcal{C})$-separable.*
iii) *There exists a continuous chain $\{Q_\alpha : \alpha < \kappa\}$ of $\kappa^<$-generated submodules of Q such that $Q = \bigcup_{\alpha<\kappa} Q_\alpha$ and for each successor ordinal $\alpha < \kappa$, we have that Q_α is a direct summand of Q that belongs to $\mathcal{C}$.*

Let Q be a left R-module. A continuous ascending chain of $\kappa^<$-generated submodules $\{Q_\alpha : \alpha < \kappa\}$ of Q with $Q = \bigcup_{\alpha<\kappa} Q_\alpha$ will be called a κ-*filtration* of Q (see [9, Definition IV.1.3]). If, in addition, this chain verifies the equivalent conditions of the above lemma, it will be called a *strong $(\kappa, \mathcal{C})$-filtration*. As we will see along this paper, these filtrations are particularly useful for checking when a separable module is trivial.

We now define what it is known as the Γ-invariant of a module (see [9, IV]). Recall that the binary relation defined on the set $\mathcal{P}(\kappa)$ of all subsets of a regular cardinal number κ by

$$X \sim Y \Leftrightarrow \exists \text{ a club } C \subseteq \kappa \text{ such that } X \cap C = Y \cap C$$

is an equivalence relation. The equivalence class of any $X \subseteq \kappa$ will be denoted by $\widetilde{X}$ (see [9, Definition 4.4]). Let us note that a subset E of κ is stationary if and only if $\widetilde{E} \neq \widetilde{\emptyset}$.

Notation 1. *Let κ be a regular infinite cardinal, $\mathcal{C}$ a nonempty class of modules and Q, a $\kappa^{\leq}$-generated module which is $(\kappa, \mathcal{C})$-separable. Suppose that $\mathcal{C}$ is closed under direct summands and that each module in $\mathcal{C}$ is a direct sum of $\kappa^{<}$-generated modules. Let $\{Q_\alpha : \alpha < \kappa\}$ be a strong $(\kappa, \mathcal{C})$-filtration of Q and consider the set*

$$E = \{\beta < \kappa : Q_\beta \text{ is not a direct summand of } Q\}.$$

We will denote $\widetilde{E}$ by $\Gamma(Q)$ and we will call it the Γ-invariant *of the filtration.*

Reasoning as in [9, p. 85], it is easy to check that $\Gamma(Q)$ does not depend on the selected strong filtration $\{Q_\alpha : \alpha < \kappa\}$ of Q. Our next proposition shows that the Γ-invariant of the filtration characterizes when a separable module is trivial.

Proposition 3.3. [14, Proposition 3.5] *Let κ be a regular infinite cardinal, $\mathcal{C}$ a nonempty class of modules and Q, a $\kappa^{\leq}$-generated module which is $(\kappa, \mathcal{C})$-separable. Suppose that $\mathcal{C}$ is closed under direct summands and arbitrary direct sums, and that each module in $\mathcal{C}$ is a direct sum of $\kappa^{<}$-generated modules. Then the following assertions are equivalent:*

i) $Q \in \mathcal{C}$.
ii) Q *is a direct sum of $\kappa^{<}$-generated modules.*
iii) $\Gamma(Q) = \widetilde{\emptyset}$.

We are now going to focus our attention on the study of separable modules over functor rings. Let R be a ring with identity and M, a left R-module. Let us assume that $M = \bigoplus_{i \in I} M_i$ for a family $\{M_i : i \in I\}$ of finitely generated modules. For any module L, let us denote by $\widehat{Hom}_R(M, L)$ the following subgroup of $\mathrm{Hom}_R(M, L)$:

$$\widehat{Hom}_R(M, L) = \{f \in \mathrm{Hom}_R(M, L) : \exists I' \subseteq I \text{ finite with } (M_i)f = 0 \ \forall i \in I \setminus I'\}.$$

It is well known that $\widehat{End}_R(M)$ is a (non necessarily unitary) ring with enough idempotents, which we will denote by $\widehat{S}$ (see, e.g., [20, Section 51]). And $\widehat{Hom}_R(M, -)$ is a functor from $R-\mathrm{Mod}$ to $\widehat{S}-\mathrm{Mod}$ which has as left adjoint the tensor product $M \otimes_{\widehat{S}} -$. The following easy lemma characterizes $(\aleph_1, \mathcal{P}roj)$-separable left $\widehat{S}$-modules.

Proposition 3.4. *Let R be a unitary ring and $M = \bigoplus_{i \in I} M_i$, a module which is a direct sum of finitely presented modules. A unitary left $\widehat{S}$-module X is $(\aleph_1, \mathcal{P}roj)$-separable if and only if there exists an $(\aleph_0, AddM)$-separable module Q such that $X \cong \widehat{Hom}_R(M, Q)$.*

Idea of the proof. Suppose that Q is an $(\aleph_1, AddM)$-separable left R-module and choose $g_1, \ldots, g_n \in \widehat{Hom}_R(M, Q)$. Then $\sum_{i=1}^{n} Img_i$ is finitely generated and there exists a decomposition $Q = S \oplus S'$ with $S \in AddM$ and $\sum_{i=1}^{n} Im f_i \subseteq S$. As $\widehat{Hom}_R(M, -)$ preserves direct sums, this decomposition induces the decomposition $Hom_R(M, Q) = T \oplus T'$ with $\{f_1, \ldots, f_n\} \subseteq T$ and $T \cong \widehat{Hom}_R(M, Q)$. By [20, 51.6.(4)]), T is projective.

In order to see the other implication, assume that X is $(\aleph_1, \mathcal{P}roj)$-separable in $\widehat{S} - Mod$. Then X is flat since it is the direct limit of its projective direct summands. By [20, 51.10], there exists a $Q \in R - Mod$ with $X \cong \widehat{Hom}(M, Q)$. But, by [20, 51.10(2)], S is a direct summand of Q which belongs to $AddM$ if and only if there exists a projective direct summand T of X such that $S \cong M \otimes_{\widehat{S}}$. As the tensor functor $M \otimes_{\widehat{S}}$ commutes with direct limits, $Q \cong M \otimes_{\widehat{S}}$ is the direct limit of its direct summands that belong to $AddM$. That is, Q is $(\aleph_1, AddM)$-separable. $\qquad\square$

4. Separable modules associated to descending chains of cyclic ideals

Let R be a unitary ring. It was proved in [2, Theorem 29.5] that a finitely generated module M has a perfect decomposition (i.e., any object in $AddM$ has a decomposition that complements direct summands) if and only if its endomorphism ring is left perfect. A similar result can be shown when the module M is a direct sum of finitely generated modules by replacing the endomorphism ring of M by $\widehat{End}_R(M)$. The proof can be obtained in a similar way as in [2, Theorem 29.5] but using [20, 51.6.(4)] and [5, Theorem 4.12].

Proposition 4.1. [13, Proposition 1.7] *Let M be a module which is a direct sum of finitely generated modules. The following assertions are equivalent:*

  i) *M has a perfect decomposition.*
  ii) *$\widehat{End}_R(M)$ is left perfect.*

On the other hand, Bass [7] constructed in his characterization of left perfect rings, a flat module F (usually called the *Bass factor module*) associated to any descending chain of cyclic right ideals of R. This flat module F is projective if and only if the descending chain is stationary. Again, this construction can be easily extended to rings with enough idempotents.

We are going to devote this section to show that we can associate to any Bass factor module F, an $\aleph_0$-separable module Q satisfying that it is trivial if and only if the descending chain of cyclic right ideals in the construction of F is stationary. More generally, if M is a direct sum of finitely generated modules, we are going to associate to any descending chain of cyclic right ideals of $\widehat{End}_R(M)$ and to any infinite cardinal λ, a left $(\aleph_0, AddM)$-separable R-module Q. And we will see that Q cannot be a direct sum of $\lambda^{\leq}$-generated if the descending chain of cyclic ideals is not stationary. Moreover, the converse is true when M is a direct sum of finitely presented modules. The origin of this construction can be traced back to the work of Griffith and Hill on Abelian groups (see [12, 16]) and has lately been used, for instance, in [13, 21] for constructing $\aleph_0$-separable modules over non-perfect (or non pure-semisimple) rings. We are going to follow the construction given in [13].

For this purpose, let us fix a module M which is a direct sum of finitely generated modules, say $M = \bigoplus_{i \in I} M_i$, and denote by $\widehat{S}$ the ring $\widehat{\mathrm{End}_R(M)}$. Let $\mathbb{N}^* = \mathbb{N} \setminus \{0\}$ and fix a subset $\{f_n : n \in \mathbb{N}^*\}$ of $\widehat{S}$. Consider the descending chain

$$f_1 \widehat{S} \geq f_1 f_2 \widehat{S} \geq f_1 f_2 f_3 \widehat{S} \geq \cdots . \tag{1}$$

of cyclic right ideals of S. Choose a finite subset $I_0 \subseteq I$ such that

$$(M_i) f_1 = 0 \quad \forall i \in I \setminus I_0,$$

and, for any $k \in \mathbb{N}^*$, let $I_k \subseteq I$ be a finite subset containing I_{k-1} such that

$$Im f_k \leq \bigoplus_{i \in I_k} M_i \quad \text{and} \quad (M_i) f_{k+1} = 0 \quad \forall i \in I \setminus I_k.$$

Let $N_k = \bigoplus_{i \in I_k} M_i$, for any $k \in \mathbb{N}$, $N = \bigoplus_{k \in \mathbb{N}^*} N_k$ and $P = \prod_{k \in \mathbb{N}^*} N_k$. Let us choose, for any $k \in \mathbb{N}$, an idempotent morphism $e_k \in \widehat{S}$ with $Im e_k = N_k$. Define, for any $n \in \mathbb{N}^*$, the morphism $F^n : M \to P$ whose coordinates are

$$F^n(k) = \begin{cases} 0 & \text{if } k < n-1 \\ e_{n-1} & \text{if } k = n-1 \\ f_n \cdots f_k & \text{if } k > n-1 \end{cases}$$

when $n > 1$, and

$$F^1(k) = f_1 \cdots f_k \quad \forall k \in \mathbb{N}^*.$$

Let $G : M^{(\mathbb{N}^*)} \to P$ be the morphism induced in the direct sum by the morphisms $\{F^n : n \in \mathbb{N}^*\}$ and denote by L its image. Note that $N \subseteq L$ since for any $k \in \mathbb{N}^*$ and any $x_k \in N_k$ we have that

$$x_k = (x_k) F^{k+1} - ((x_k) f_k) F^{k+2}.$$

Let λ be an infinite cardinal and denote by $\kappa = \lambda^+$ the successor cardinal of λ. Consider the subset E of κ consisting of all limit ordinals of cofinality ω. For any $\beta \in E$ let

$$\langle \sigma_\beta(k) : k \in \mathbb{N}^* \rangle$$

be an increasing sequence of ordinals converging to β, and denote by J_β the set

$$\{\sigma_\beta(k) : k \in \mathbb{N}^*\}.$$

Define, for any $\beta \in E$ and any $n \in \mathbb{N}^*$, the morphism $F^{(\beta,n)} : M \to N^\kappa$ whose coordinates are

$$F^{(\beta,n)}(\alpha) = \begin{cases} F^n(k) & \text{if } \alpha = \sigma_\beta(k), \\ 0 & \text{if } \alpha \notin J_\beta. \end{cases}$$

Also, let $F : M^{(E \times \mathbb{N}^*)} \to N^\kappa$ be the morphisms induced by the $F^{(\beta,n)}$. For any $\delta < \kappa$, denote by N_δ the isomorphic copy of N in the δ-coordinate of N^κ.

Definition 4.2. *The module*

$$Q = \bigoplus_{\alpha < \kappa} N_\alpha + ImF$$

will be called the $((\aleph_0, AddM)$*-separable) module associated to the chain* (1).

It has been proved in [13, Section 2] that Q is the non-trivial separable module that we can associate to the descending chain of right ideals of $\hat{S}$. In other words, it is possible to prove the following theorem.

Theorem 4.3. [13, Theorem 3.4] *Let R be a unitary ring and $M = \oplus_I M_i$, a direct sum of finitely generated left R-modules. Suppose that there exists an infinite cardinal number λ such that any $(\aleph_0, AddM)$-separable module is a direct sum of λ-generated modules. Then M has a perfect decomposition.*

We are going to close this section by giving some applications of this construction. Our first corollary gives new conditions equivalent to those of [3, Theorem 4.4], under the assumption that the module M is a direct sum of finitely presented modules.

Corollary 4.4. (see [13, Corollary 3.5]) *Let R be a unitary ring and M, a direct sum of finitely presented left R-modules. The following conditions are equivalent:*

   i) *M has a perfect decomposition.*
   ii) *$\widehat{End_R}(M)$ is left perfect.*
  iii) *Every $(\aleph_0, AddM)$-separable module belongs to $AddM$.*
   iv) *Every $(\aleph_0, AddM)$-separable module is a direct sum of countably generated modules.*
   v) *There exists an infinite cardinal number λ such that each $(\aleph_0, AddM)$-separable module is a direct sum of λ-generated modules.*
  vi) *$\lim\limits_{\rightarrow} AddM = AddM$.*
 vii) *Each countably direct colimit of modules in $AddM$ belongs to $AddM$.*

Idea of the proof. The equivalence of i) and ii) was established in Proposition 4.1.

ii) $\Rightarrow$ iii) Let Q be a $(\aleph_0, AddM)$-separable module. By Lemma 3.4, $\widehat{Hom}_R(M,Q)$ is $(\aleph_0, \mathcal{P}roj)$-separable in $\hat{S}$-Mod and in particular flat, since it is the direct limit of its projective direct summands. Then ii) says that $\widehat{Hom}_R(M,Q)$ is projective by [5, Theorem 41], and Q belongs, in fact, to $AddM$, since it is isomorphic to $M \otimes_{\hat{S}} \widehat{Hom}_R(M,Q)$ by [20, 51.10].

iii) $\Rightarrow$ iv) Note that every module in $AddM$ is a direct sum of countably generated modules by Kaplansky's Theorem (see [2, Theorem 26.1]).

iv) $\Rightarrow$ v) Trivial.

v) $\Rightarrow$ ii) This is Theorem 4.3.

vi) $\Leftrightarrow$ vii) $\Leftrightarrow$ ii). Functors $\widehat{Hom}_R(M, \_)$ and $M \otimes_{\hat{S}} \_$ define an equivalence between $\lim\limits_{\rightarrow} AddM$ and the flat $\hat{S}$-modules. Then each flat $\hat{S}$-module is projective if and only if $\lim_{\rightarrow} AddM = AddM$, which proves vi) $\Leftrightarrow$ ii) (this has been proven in [3, Theorem 4.4] too). Moreover, it is very easy to check that $\lim_{\rightarrow} AddM = \lim_{\rightarrow} Add^{\aleph_0} M$, and each countable direct colimit of modules in $Add^{\aleph_0} M$ corresponds, under the mentioned equivalence, with a countably generated flat $\widehat{S}$-module. But the proof of [5, Theorem 4.1, 9 $\Rightarrow$ 2] implies that $\widehat{S}$ is left perfect if and only if each countably generated flat module is projective, which proves vii) $\Rightarrow$ ii). $\qquad\square$

If we set $M = R$ in this corollary, for some unitary ring R, we get new characterizations of left perfect rings in terms of $(\aleph_0, \mathcal{P}roj)$-separable modules. Indeed, we may assume that R is a (non-unitary) ring with enough idempotents, since in this case the category T-Mod coincides with $\sigma[_{T^*}T]$. Where T^* is the *Dorroh* overing of T (see [20, p. 465]). Moreover, the classes of all $(\aleph_0, \mathcal{P}roj)$-separable and of all projective modules in T-Mod coincide with the class of all $(\aleph_0, AddT)$-separable and the class $AddT$ constructed in T^*-Mod, respectively. As the rings T and $\widehat{End}_R(T)$ are isomorphic, we deduce from the above corollary:

Corollary 4.5. *Let T be a ring with enough idempotents. The following assertions are equivalent:*

i) *T is left perfect;*
ii) *Every $(\aleph_0, \mathcal{P}roj)$-separable unitary left T-module is projective.*
iii) *Every $(\aleph_0, \mathcal{P}roj)$-separable unitary left T-module is a direct sum of countably generated modules.*
iv) *There exists an infinite cardinal number λ such that every $(\aleph_0, \mathcal{P}roj)$-separable unitary left T-module is a direct sum of λ-generated modules.*

Our last corollary characterizes left pure semisimple rings in terms of $(\aleph_0, \mathcal{PP}roj)$-separable modules.

Corollary 4.6. [13, Corollary 3.7] *Let R be a ring with unit. The following conditions are equivalent:*

i) *R is left pure semisimple.*
ii) *Every $(\aleph_0, \mathcal{PP}roj)$-separable left module is pure projective.*
iii) *Every $(\aleph_0, \mathcal{PP}roj)$-separable left module is a direct sum of countably generated modules.*
iv) *There exists an infinite cardinal number λ such that every $(\aleph_0, \mathcal{PP}roj)$-separable module is a direct sum of λ-generated modules.*

Idea of the proof. Note that, if $\{M_i : i \in I\}$ is a set of representatives of the finitely presented modules and $M = \bigoplus_{i \in I} M_i$, $AddM$ and the class of all $(\aleph_0, AddM)$-separable modules are precisely the classes of all pure projective and all $(\aleph_0, \mathcal{PP}roj)$-separable modules respectively. Then the result follows from Corollary 4.4 and the fact that $\widehat{End}_R(M)$ is left perfect if and only if R is left pure semisimple (see, e.g., [20, 53.6]). $\qquad\square$

5. Construction of non-trivial κ-separable modules

We have developed in Section 4 a method for associating a non-trivial separable module to any direct sum $M = \oplus_I M_i$ of finitely generated modules which does not have a perfect decomposition. Unfortunately, the assumption that M is a direct sum of finitely generated modules is essential in this construction and therefore, the above method cannot be extended to general modules. We will show in this section a new way for constructing non-trivial separable modules that can be applied to any module M. This method has been developed in [14] and it extends a transfinite construction made by Eklof in [8] (see also [10]). This new construction has, however, some disadvantages. It requires the assumption of the GCH and, on the other hand, it only partially reflects the decomposition properties of the module M into direct summands. Namely, those of them which can be expressed in terms of totally ordered λ-directed colimits of objects in $AddM$, where λ is an infinite regular cardinal such that M is a direct sum of λ-presentable modules. When $\lambda = \aleph_0$, we may drop these set-theoretic assumptions in our construction. This will allows us to improve in Corollary 5.7 the results obtained in the previous section.

An important tool in our construction will be the behavior of the so-called λ-presentable modules. Recall from Section 1 that a module M is λ-presentable, for a certain infinite regular cardinal λ, if the functor $\mathrm{Hom}_R(M, -) : R-\mathrm{Mod} \to \mathbf{Ab}$ commutes with λ-directed colimits. Note that any module M is trivially λ-presentable for a sufficiently large regular cardinal λ. We summarize in the next proposition some basic properties of these modules that will be needed in the sequel.

Proposition 5.1. [14, Proposition 5.1] *Let λ be an infinite regular cardinal.*

i) *A module M is λ-presentable if and only if it satisfies the following two conditions:*
   a) *M is $\lambda^<$-generated;*
   b) *For any exact sequence in $R - Mod$*

   $$0 \to K \to L \to M \to 0$$

   with L $\lambda^<$-generated, the module K is also $\lambda^<$-generated.

ii) *Let*

   $$0 \to N_1 \to N_2 \to N_3 \to 0$$

   be a short exact sequence in R-Mod.
   a) *If N_2 is λ-presentable and N_1 is $\lambda^<$-generated, then N_3 is also λ-presentable.*
   b) *If N_1 and N_3 are λ-presentable then so is N_2.*

iii) *If I is a set with $|I| < \lambda$ and $\{M_i : i \in I\}$ is a family of λ-presentable modules then $\bigoplus_{i \in I} M_i$ is also λ-presentable.*

iv) *Any λ-presentable module is projective respect to λ-directed limits of splitting epimorphisms in R-Mod.*

v) *If* $\{f_{ij} : M_i \to M_j\}_I$ *is a* λ-*directed system of morphisms in* R-Mod, *then the canonical epimorphism*

$$\bigoplus_I M_i \longrightarrow \varinjlim M_i \to 0$$

is a λ-*directed limit of splitting epimorphisms.*

We remark that, as a consequence of i), for any infinite regular cardinal λ, there exists a set $\mathcal{S}_\lambda$ of λ-presentable objects such that any other object in $R-$Mod is a λ-direct limit of objects in this set (it suffices to choose $\mathcal{S}_\lambda$ to be the set of all isomorphism classes of modules of cardinality strictly smaller than λ). In other words, $R-$Mod is a λ-presentable category in the sense of [15, Definition 1.17]. For any non-empty class of modules $\mathcal{S}$, we denote by $\lambda - \varinjlim \mathcal{S}$ the class consisting of all totally ordered λ-directed colimits of modules in $\mathcal{S}$. If $\lambda = \aleph_0$ we will simply write $\varinjlim \mathcal{S}$.

Let us now develop our method for producing non-trivial $(\kappa, \mathcal{C})$-separable modules. For this purpose we will need the following generalization of the concept of λ-template (see [8, p. 81]).

Definition 5.2. *Let* λ *be an infinite cardinal number. A* λ-*template is a pair of modules* $N \subseteq L$ *such that there exists a continuous chain* $\{N_\nu : \nu < \lambda\}$ *of direct summands of* L *with* $\bigcup_{\nu<\lambda} N_\nu = N$.

Let λ be an infinite regular cardinal, fix a λ-template $N = \bigcup_{\nu<\lambda} N_\nu \subseteq L$ with $N_0 = 0$, and write, for each $\nu < \lambda$, $N_{\nu+1} = N_\nu \oplus N'_\nu$ and $L = N_\nu \oplus M_\nu$ for some modules N'_ν and M_ν. Let us denote by $\kappa = \lambda^+$ the successor cardinal of λ. Along this section, we will make the following assumption.

Assumption 1. *There exists a tree-like ladder system* η *on a stationary subset* E *of* κ *consisting of limit ordinals of cofinality* λ.

Recall that this assumption is satisfied when $\lambda = \aleph^+$ is a successor cardinal number such that $2^\aleph \leq \aleph^{++}$ [8, Theorem 9] or when we assume the GCH, by Proposition 2.1.

Our goal is to construct a new module Q, which will be called *the module associated to the subset* E *of* κ *and the template* $N \subseteq L$. Q will be obtained as the union of a continuous chain of modules $\{Q_\alpha : \alpha < \kappa\}$ that will be defined recursively on all the ordinals $\alpha < \kappa$. We shall use the following notation: for each successor ordinal $\gamma < \kappa$ and each $\nu < \lambda$, let $N_{\gamma,\nu}$ be an isomorphic copy of N_ν and write $N_{\gamma,\nu+1} = N^*_{\gamma,\nu} \oplus N'_{\gamma,\nu}$ with $N^*_{\gamma,\nu} \cong N_\nu$ and $N'_{\gamma,\nu} \cong N'_\nu$; moreover, denote by $s^\nu_\gamma : N'_\nu \to N'_{\gamma,\nu}$ an isomorphism.

If $\alpha = 0$, define $Q_0 = 0$. Assume now that $\alpha > 0$ and that Q_γ has been defined for any $\gamma < \alpha$. If α is a limit ordinal, set $Q_\alpha = \bigcup_{\gamma<\alpha} Q_\gamma$. If $\alpha = \gamma+1$ is a successor with $\gamma \notin E$, let us denote by $Q'_\gamma = \bigoplus_{\nu<\lambda} N_{\gamma,\nu}$, and define $Q_{\gamma+1} = Q_\gamma \oplus Q'_\gamma$.

It remains to define Q_α when $\alpha = e+1$ with $e \in E$. Let us fix an ordinal α in this situation. We begin by constructing a homomorphism from N to Q_e, for which it is sufficient to define a direct system $\{\iota^\nu_e : N_\nu \to Q_e\}_{\nu<\lambda}$ of morphisms. We do this recursively on ν. For $\nu = 0$ take $\iota_e^0 = 0$. Let now $\nu < \lambda$ be nonzero, and suppose that ι_e^ν is defined. Then note that $N_{\nu+1} = N_\nu \oplus N_\nu'$ and take $\iota_e^{\nu+1} = \iota_e^\nu \oplus s_{\eta_e(\nu)}^\nu$. Finally, if ν is a limit ordinal, set $\iota_e^\nu = \varinjlim_{\sigma < \nu} \iota_e^\sigma$.

Let now $\iota_e : \bigcup_{\nu < \lambda} N_\nu \to Q_e$ be the colimit of this direct system. The map ι_e is in fact a monomorphism. Then take the push-out of the morphisms ι_e and the inclusion $i : N \to L$ to get the commutative diagram

$$
\begin{array}{ccccc}
0 & \longrightarrow & N & \stackrel{i}{\longrightarrow} & L \\
& & \downarrow{\scriptstyle \iota_e} & & \downarrow{\scriptstyle \theta_e} \\
0 & \longrightarrow & Q_e & \stackrel{i_e}{\longrightarrow} & P
\end{array}
\qquad (2)
$$

We set $Q_{e+1} = P$.

Definition 5.3. *The module*

$$
Q = \bigcup_{\alpha < \kappa} Q_\alpha
$$

will be called the module associated to the subset E of κ and the template $N \subseteq L$.

The reason of our interest on this construction is that we have shown in [14] that, when $\mathcal{C}$ is a non-empty class of modules that satisfies certain properties, and the template $N \subseteq L$ verifies that $L \in \mathcal{C}$, the module Q is $(\kappa, \mathcal{C})$-separable. This is proved in the following theorem.

Theorem 5.4. [14, Theorem 4.6] *Let λ be an infinite regular cardinal, $\kappa = \lambda^+$, E, an stationary subset of κ which has a tree-like ladder system and $N \subseteq L$, a λ-template. Let $Q = \bigcup_{\gamma < \kappa} Q_\gamma$ be the module associated to E and $N \subseteq L$. Suppose that $\mathcal{C}$ is a nonempty class of modules closed under direct summands and direct sums such that $L \in \mathcal{C}_\kappa$. Then Q is $(\kappa, \mathcal{C})$-separable.*

We are going to close this survey by showing how to apply the above construction to the study of the decomposition properties of modules into direct summands. Let us fix a module M and consider an infinite regular cardinal λ such that M is a direct sum of λ-presentable modules. We are going to prove, under the assumption of the GCH, an analogue to Theorem 4.4: there exist non-trivial $(\lambda, AddM)$-separable modules precisely when $AddM \neq \lambda - \varinjlim AddM$. In order to prove this theorem we will need the following technical result which states that, under the assumption of GCH, any totally ordered λ-directed colimit is actually, a well-ordered λ-directed colimit.

Lemma 5.5. [14, Lemma 5.14] *Assume (GCH). Let λ be an infinite regular cardinal and I, a totally ordered λ-directed set with $|I| \geq \lambda$. Then there exists a regular cardinal τ with $\lambda \leq \tau \leq |I|$, and an ascending chain $\{I_\alpha : \alpha < \tau\}$ of λ-directed subsets of I with $|I_\alpha| < |I|$ for any $\alpha < \tau$, such that $I = \bigcup_{\alpha < \tau} I_\alpha$.*

We can now state the announced theorem.

Theorem 5.6. (see [14, Corollary 5.17]) *Assume (GCH). Let λ be an infinite regular cardinal and M, a direct sum of λ-presentable modules.*

The following conditions are equivalent.

i) $\lambda - \lim_{\to} AddM = Add\, M$.

ii) *For each regular cardinal number $\kappa \geq \lambda$, any $\kappa^{\leq}$-generated and $(\kappa, Add\, M)$-separable module belongs to $AddM$.*

iii) *For each regular cardinal number $\kappa \geq \lambda$, any $\kappa^{\leq}$-generated and $(\kappa, Add\, M)$-separable module is a direct sum of $\kappa^{<}$-generated modules.*

Idea of the proof. ii) $\Leftrightarrow$ iii) and i) $\Rightarrow$ iii) are straightforward consequences of Proposition 5.1 and Kaplansky's Theorem [2, Theorem 26.1]. Let us give an idea of the proof of ii) $\Rightarrow$ i). Suppose that ii) is true but $\lambda - \lim_{\to} AddM \neq AddM$. Let $(X_i, f_{ij})_I$ be a λ-directed system of objects in $Add\, M$ whose direct colimit does not belong to $AddM$. We may choose this directed colimit with $|I|$ minimal, in the sense that any λ-directed colimit of modules in $AddM$ indexed in a set I' with $|I'| < |I|$ belongs to $AddM$. We can use now the preceding lemma for showing that it is possible to assume that the set I is well ordered. We can prove now that the minimality of I implies that we may identify I with a regular cardinal $\tau \geq \lambda$.

Consider the canonical short exact sequence associated in R-Mod to this directed colimit

$$0 \to K \longrightarrow \bigoplus_I X_i \longrightarrow \lim_{\to} X_i \to 0$$

This short exact sequence cannot split since we are assuming that $\lim_{\to} X_i \notin AddM$. Write, using [11, Lemma 2.1], $K = \bigcup_I K_i$ for a convenient ascending chain $\{K_i\}_I$ of direct summands of $\bigoplus_I X_i$. And define the family $\{N_i\}_I$ of submodules of $\bigoplus_I X_i$ by $N_i = K_i$ if i is a successor in the linear order of I and $N_i = \bigcup_{j<i} N_j$ otherwise. Now note that $\{N_i\}_I$ is a strong $(I, AddM)$-filtration of K and that, by Lemma 3.2, this module is $(I, AddM)$-separable. If $K \notin AddM$ then K is an $(I, AddM)$-separable that does not belong to $AddM$ and we are done since this contradicts ii). So let us suppose that K belongs to $AddM$. Then $\Gamma(K) = \widetilde{\emptyset}$ by Proposition 3.3, and this means that there is a club $C \subseteq I$ with the property that $K = \bigcup_{i \in C} N_i$ with each N_i being a direct summand of $\bigoplus_I X_i$. Then, setting $N = \bigcup_{i \in C} N_i$ and $L = \bigoplus_{i \in I} X_i$, we get an $|I|$-template $N \subseteq L$. The construction made in the beginning of this section gives us an $(|I|^+, AddM)$-separable module Q for some stationary subset E of $|I|^+$ consisting of limit ordinals of cofinality $|I|$. Now we can prove that Q does not belong to $AddM$ and get a contradiction. $\square$

Our next corollary shows that, when $\lambda = \aleph_0$, we may drop some hypothesis in the above theorem (see also [10]).

Corollary 5.7. *Let M be a direct sum of finitely generated modules. If every $(\aleph_1, AddM)$-separable module is a direct sum of countably generated modules, then M has a perfect decomposition. If, moreover, M is a direct sum of finitely presented modules, then the converse is also true.*

Idea of the proof. The proof is similar to i) ⇒ ii) in the above by noting the following facts:

- Assume that $M = \oplus_T M_t$ is a direct sum of finitely generated modules. Then M does not have a perfect decomposition if and only if its functor ring $S = \widehat{End_R(M)}$ is not left perfect by Proposition 4.1. This means that if M does not have a perfect decomposition, then there exists a countable directed system of morphisms among finitely generated unitary free S-modules, say $\{g_n : F_n \to F_{n+1}\}_{n\in\mathbb{N}}$, whose direct limit is not projective. Consider the canonical exact sequence associated in S-Mod to this direct limit (see, e.g., [20, 43.3])

$$0 \to \bigoplus_I F_n \xrightarrow{\quad g \quad} \bigoplus_I F_n \longrightarrow \varinjlim F_n \to 0$$

As M is a direct sum of finitely generated modules, the functor $\widehat{Hom_R}(M, -)$ commutes with direct limits and therefore, it is easy to check that the induced monomorphism $1_M \otimes_S g : \bigoplus_{\mathbb{N}} (M \otimes_S F_n) \to \bigoplus_{\mathbb{N}} (M \otimes_S F_n)$ cannot split. But $1_M \otimes_S g$ is a directed union of splitting monomorphisms by [20, 43.3]. This shows that, in this situation, we may assume that the set I appearing in the proof of the above theorem is countable.
- When $|I| = \aleph_0$, we do not need to assume the GCH in the construction of Q.
- Finally, it is easy to check that the proof of $i) \Rightarrow ii)$ in Theorem 5.6 also works if we assume that M is a direct sum of $\lambda^<$-generated modules, instead of a direct sum of λ-presentable modules. □

Let us finally apply our results to the whole category of modules. Recall that the category R-Mod is pure-semisimple if and only is there exists a set $\mathcal{S}$ of modules such that R-Mod $= Add_R\mathcal{S}$ (see, e.g., [18, 20]). Therefore, we get the following corollary (compare with [10, 13, 17, 19, 21]).

Corollary 5.8. [14, Corollary 5.18] *Assume (GCH). Let R be a ring and λ, an infinite regular cardinal. The following assertions are equivalent:*

i) *R-Mod is pure-semisimple.*
ii) *For each infinite regular cardinal $\kappa \geq \lambda$, every κ-separable module is a direct summand of a direct sum of λ-presentable modules.*

In particular, if R-Mod is not pure-semisimple, there exist non-trivial λ-separable modules for arbitrarily large regular cardinals λ.

References

[1] J. Adámek and J. Rosický, Locally presentable and accessible categories. London Mathematical Society Lecture Note Series, 189. Cambridge University Press, Cambridge, 1994
[2] F.W. Anderson and K.R. Fuller, Rings and Categories of Modules. Second edition. Graduate Texts in Mathematics, 13. Springer-Verlag, New York, 1992.

[3] L. Angeleri Hügel, Covers and envelopes via endoproperties of modules. Proc. London Math. Soc. 86 (3) (2003), 649–665.

[4] L. Angeleri Hügel and M. Saorín, Modules with perfect decompositions. Math. Scand. 98 (1) (2006), 19–43.

[5] T.N. Ánh, N.V. Loi and D.V. Than, Perfect rings without identity. Comm. Alg. 19 (4) (1991), 1069–1082.

[6] G. Azumaya, Finite splitness and finite projectivity. J. Algebra 106 (1987), 114–134.

[7] H. Bass, Finitistic dimension and a homological generalization of semi-primary rings. Trans. Amer. Math. Soc. 95 (1960), 466–488.

[8] P.C. Eklof, Modules with strange decomposition properties. Infinite length modules (Bielefeld, 1998), 75–87, Trends Math., Birkhäuser, Basel, 2000.

[9] P.C. Eklof and A.H. Mekler, Almost free modules. Set-theoretic methods. North-Holland Mathematical Library, 46. North-Holland Publishing Co., Amsterdam, 1990.

[10] P.C. Eklof and S. Shelah, The Kaplansky test problems for $\aleph_1$-separable groups. Proc. Amer. Math. Soc. 126 (1998), no. 7, 1901–1907.

[11] J.L. Gómez Pardo and P.A. Guil Asensio, Big direct sums of copies of a module have well behaved indecomposable decompositions. J. Algebra 232 (2000), 86–93.

[12] P. Griffith, A note on a theorem of Hill. Pacific J. Math. 29 (1969) 279–284.

[13] P.A. Guil Asensio, M.C. Izurdiaga and B. Torrecillas, Decomposition properties of strict Mittag-Leffler modules, J. Algebra 310 (2007), 290–302.

[14] P.A. Guil Asensio, M.C. Izurdiaga and B. Torrecillas, Accessible subcategories of modules and pathological objects, preprint.

[15] K. Hrbacek and T. Jech, Introduction to set theory. Third edition. Monographs and Textbooks in Pure and Applied Mathematics, 220. Marcel Dekker, Inc., New York, 1999.

[16] P. Hill, On the decomposition of groups. Canad. J. Math. 21 (1969), 762–768.

[17] S. Shelah, Kaplansky test problem for R-modules. Israel J. Math. 74 (1991), no. 1, 91–127.

[18] D. Simson, On Corner type endo-wild algebras. J. Pure Appl. Algebra 202 (2005), no. 1-3, 118–132.

[19] D. Simson, On large indecomposable modules, endo-wild representation type and right pure semisimple rings. Algebra Discrete Math. 2003, no. 2, 93–118.

[20] R. Wisbauer, Foundations of module and ring theory. A handbook for study and research. Algebra, Logic and Applications, 3. Gordon and Breach Science Publishers, Philadelphia, PA, 1991.

[21] B. Zimmermann-Huisgen, On the abundance of $\aleph_1$-separable modules. Abelian groups and noncommutative rings. Contemp. Math. 130 (1992), 167–180.

P.A. Guil Asensio
Departamento de Matemáticas, Universidad de Murcia
30100 Espinardo, Murcia, Spain

M.C. Izurdiaga and B. Torrecillas
Departamento de Álgebra y Análisis Matemático, Universidad de Almería
04071 Almería, Spain

Modules and Comodules

Trends in Mathematics, 243–246

Essential Extensions of a Direct Sum of Simple Modules-II

S.K. Jain and Ashish K. Srivastava

Dedicated to Robert Wisbauer on his 65th Birthday

Abstract. It is shown that a semi-regular ring R with the property that each essential extension of a direct sum of simple right R-modules is a direct sum of quasi-injective right R-modules is right noetherian.

Mathematics Subject Classification (2000). 16P40, 16D60, 16D50, 16D80.

Keywords. Right noetherian, essential extension, quasi-injective module, simple module, von Neumann regular ring.

1. Introduction

The question whether a von Neumann regular ring R with the property that every essential extension of a direct sum of simple right R-modules is a direct sum of quasi-injective right R-modules is noetherian was considered in [2]. This question was answered in the affirmative under a stronger hypothesis. The purpose of this note is to answer the question in the affirmative even for a more general class of rings, namely, semi-regular, semi-perfect.

2. Definitions and notations

All rings considered in this paper have unity and all modules are right unital. Let M be an R-module. We denote by $\operatorname{Soc}(M)$, the socle of M. We shall write $N \subseteq_e M$ whenever N is an essential submodule of M. A module M is called N-injective, if every R-homomorphism from a submodule L of N to M can be lifted to an R-homomorphism from N to M. A module M is said to be quasi-injective if it is M-injective. A ring R is said to be right $q.f.d.$ if every cyclic right R-module has finite uniform (Goldie) dimension, that is, every direct sum of submodules of a cyclic module has finite number of terms. We shall say that Goldie dimension of

N with respect to U, $G \dim_U(N)$, is less than or equal to n, if for any independent family $\{V_j : j \in \mathcal{J}\}$ of nonzero submodules of N such that each V_j is isomorphic to a submodule of U, we have that $|\mathcal{J}| \leq n$. Next, $G \dim_U(N) < \infty$ means that $G \dim_U(N) \leq n$ for some positive integer n. The module N is said to be q.f.d. relative to U if for any factor module $\bar{N}$ of N, $G \dim_U(\bar{N}) < \infty$. A ring R is called von Neumann regular if every principal right (left) ideal of R is generated by an idempotent. A regular ring is called abelian if all its idempotents are central. A ring R is called semiregular if $R/J(R)$ is von Neumann regular. A ring R is said to be a semilocal ring if $R/J(R)$ is semisimple artinian. In a semilocal ring R every set of orthogonal idempotents is finite, and R has only finitely many simple modules up to isomorphism. A ring R is called a q-ring if every right ideal of R is quasi-injective [4]. For any term not defined here, we refer the reader to Wisbauer [6].

3. Main results

Throughout, we will refer to the condition 'every essential extension of a direct sum of simple R-modules is a direct sum of quasi-injective R-modules' as property $(*)$.

We note that the property $(*)$ is preserved under ring homomorphic images.

We first state some of the results that are used throughout the paper.

Lemma 3.1. [2] *Let R be a ring which satisfies the property $(*)$ and let N be a finitely generated R-module. Then there exists a positive integer n such that for any simple R-module S, we have*

$$G \dim_S(N) \leq n.$$

Lemma 3.2. [2] *Let R be a right nonsingular ring which satisfies the property $(*)$. Then R has a bounded index of nilpotence.*

The following lemma is a key to the proof of main results.

Lemma 3.3. *Let R be an abelian regular ring with the property $(*)$. Then R is right noetherian.*

Proof. Recall that an abelian regular ring is duo. Assume R is not right noetherian. Then there exists an infinite family $\{e_i R : e_i = e_i^2, i \in I\}$ of independent ideals in R. Now for each $i \in I$, there exists a maximal right ideal M_i such that each $e_i R \not\subseteq M_i$, for otherwise $e_i R \subseteq J(R)$ which is not possible. Set $A = \oplus_{i \in I} e_i R$ and $M = \oplus_{i \in I} e_i M_i$. Note $M \neq R$ and $A/M = (\oplus_{i \in I} e_i R)/(\oplus_{i \in I} e_i M_i)$. So, R/M is a ring with nonzero socle of infinite Goldie dimension. Choose $K/M \subset R/M$ such that $\mathrm{Soc}(R/M) \oplus K/M \subseteq_e R/M$. This implies that $\mathrm{Soc}(R/M)$ is essentially embeddable in R/K and so $\mathrm{Soc}(R/M) \cong \mathrm{Soc}(R/K)$. Obviously, $\mathrm{Soc}(R/K) \subseteq_e R/K$. Set $\bar{R} = R/K$. By $(*)$, $\bar{R} = \bar{e_1} \bar{R} \oplus \cdots \oplus \bar{e_k} \bar{R}$, where each $\bar{e_i} \bar{R}$ is quasi-injective. Since each $\bar{e_i}$ is a central idempotent, $\bar{e_i} \bar{R}$ is $\bar{e_j} \bar{R}$-injective. Hence, $\bar{e_1} \bar{R} \oplus \cdots \oplus \bar{e_k} \bar{R}$ is quasi-injective. Thus, $\bar{R} = R/K$ is a right self-injective duo ring and hence R/K is a q-ring. Since R/K is a regular q-ring, $R/K = S \oplus T$, where S is semisimple artinian and T has zero socle (see Theorem 2.18, [4]). But R/K has essential socle. So $T = 0$ and hence R/K is semisimple artinian, which gives a contradiction to the fact that R/K contains an infinite independent family of right ideals. Therefore, R must be right noetherian. This completes the proof.   □

Next we claim that if the matrix ring has the property (∗) then the base ring also has the property (∗). This can be deduced from the following two lemmas whose proofs are standard and given here only for the sake of completeness.

Lemma 3.4. *If R is a ring with the property (∗) and $ReR = R$, then eRe is also a ring with the property (∗).*

Proof. We know that if $ReR = R$, then mod-R and mod-eRe are Morita equivalent under the functors given by $\mathcal{F} : \text{mod-}R \longrightarrow \text{mod-}eRe$, $\mathcal{G} : \text{mod-}eRe \longrightarrow \text{mod-}R$ such that for any M_R, $\mathcal{F}(M) = Me$ and for any module T over eRe, $\mathcal{G}(T) = T \otimes_{eRe} eR$.

Suppose R is a ring with the property (∗). Let N be an essential extension of a direct sum of simple eRe-modules $\{S_i : i \in I\}$. This gives, $\oplus_i S_i \otimes_{eRe} eR \subseteq_e N \otimes_{eRe} eR$. By Morita equivalence each $S_i \otimes_{eRe} eR$ is a simple R-module. Thus, $N \otimes_{eRe} eR$ is an essential extension of a direct sum of simple R-modules. But since R is a ring with the property (∗), we have $N \otimes_{eRe} eR = \oplus_i A_i$, where A_i's are quasi-injective R-modules. By Morita equivalence we get that each $A_i e$ is quasi-injective as an eRe-module. Then $N = NeRe = A_1 e \oplus \cdots \oplus A_n e$ is a direct sum of quasi-injective eRe-modules. Hence eRe is a ring with the property (∗).   □

As a consequence of the above lemma, we have the following:

Lemma 3.5. *If $M_n(R)$ is a ring with the property (∗), then R is also a ring with the property (∗).*

Proof. We have $R \cong e_{11} M_n(R) e_{11}$ and $M_n(R) e_{11} M_n(R) = M_n(R)$, where e_{11} is a matrix unit. Therefore, the result follows from the Lemma 3.4.   □

Now we are ready to prove the result that answers the question raised in [2].

Theorem 3.6. *Let R be a regular ring with the property (∗). Then R is right noetherian.*

Proof. By Lemma 3.2, R has bounded index of nilpotence. Hence each primitive factor ring of R is artinian. Therefore, $R \cong M_n(S)$ for some abelian regular ring S (see Theorem 7.14, [3]). By Lemma 3.5, S has the property (∗). Therefore, by Lemma 3.3, S is right noetherian. Hence, R is right noetherian.   □

We next proceed to generalize the above theorem to semiregular rings. First we prove the following:

Lemma 3.7. *Let R be a semilocal ring with the property (∗). Then R is right noetherian.*

Proof. We claim that R_R is right *q.f.d.* Consider any cyclic module R/I. Suppose there exists an infinite direct sum $A_1/I \oplus A_2/I \oplus \cdots \subset R/I$, where $\frac{A_i}{I} = \frac{a_i R + I}{I}$. Let M_i/I be a maximal submodule of A_i/I for each i, and set $M/I = \oplus M_i/I$. Then $\frac{A_1}{M_1} \times \frac{A_2}{M_2} \times \cdots \cong \frac{A_1 \oplus A_2 \oplus \cdots}{M_1 \oplus M_2 \oplus \cdots} \subset R/M$. Each A_i/M_i is a simple module. Set $S_i = A_i/M_i$. Since the semilocal ring R/M has only finitely many simple modules up to isomorphism, copies of at least one of the A_i/M_i must appear infinitely many times, and so $G \dim_{S_i}(R/M) = \infty$, for some i. This gives a contradiction to Lemma 3.1. Therefore, R is right *q.f.d.* Hence, by (Theorem 2.2, [1]), R is right noetherian. $\qquad\square$

Corollary 3.8. *A right self-injective ring with the property* ($*$) *is Quasi-Frobenius.*

Theorem 3.9. *Let R be a semiregular ring with the property* ($*$). *Then R is right noetherian.*

Proof. $R/J(R)$ is a von Neumann regular ring with the property ($*$). Therefore, by Theorem 3.6, $R/J(R)$ is a right noetherian and hence a semisimple artinian ring. So, R is a semilocal ring. Finally, by Lemma 3.7, R is right noetherian. $\qquad\square$

References

[1] K.I. Beidar and S.K. Jain, When Is Every Module with Essential Socle a Direct Sums of Quasi-Injectives?, Communications in Algebra, 33, 11 (2005), 4251–4258.

[2] K.I. Beidar, S.K. Jain and Ashish K. Srivastava, Essential Extensions of a Direct Sum of Simple Modules, Groups, Rings and Algebras, Contemporary Mathematics, AMS, 420 (2006), 15–23.

[3] K.R. Goodearl, von Neumann Regular Rings, Krieger Publishing Company, Malabar, Florida, 1991.

[4] S.K. Jain, S.H. Mohamed, Surjeet Singh, Rings In Which Every Right Ideal Is Quasi-Injective, Pacific Journal of Mathematics, Vol. 31, No. 1 (1969), 73–79.

[5] S.H. Mohamed and B.J. Muller, Continuous and Discrete Modules, Cambridge University Press, 1990.

[6] R. Wisbauer, Foundations of Module and Ring Theory, Gordon and Breach, 1991.

S.K. Jain
Department of Mathematics
Ohio University
Athens, Ohio-45701, USA
e-mail: jain@math.ohiou.edu

Ashish K. Srivastava
Assistant Professor
Department of Mathematics & Computer Science
St. Louis University
St. Louis, MO-63103, USA
e-mail: asrivas3@slu.edu
URL: http://euler.slu.edu/~srivastava

Modules and Comodules
Trends in Mathematics, 247–264
© 2008 Birkhäuser Verlag Basel/Switzerland

Pseudo-Galois Extensions and Hopf Algebroids

Lars Kadison

Abstract. A pseudo-Galois extension is shown to be a depth two extension. Studying its left bialgebroid, we construct an enveloping Hopf algebroid for the semi-direct product of groups, or more generally involutive Hopf algebras, and their module algebras. It is a type of cofibered sum of two inclusions of the Hopf algebra into the semi-direct product and its derived right crossed product. Van Oystaeyen and Panaite observe that this Hopf algebroid is non-trivially isomorphic to a Connes-Moscovici Hopf algebroid.

Mathematics Subject Classification (2000). Primary 13B05, 16W30, 20L05, 81R50.

1. Introduction

The analytic notion of finite depth for subfactors was widened to the algebraic setting of Frobenius extensions in [11], the main theorem of which states that a certain depth two Frobenius extension $A \mid B$ with trivial centralizer is a Hopf-Galois extension. The theorem and its proof is essentially a reconstruction theorem, which uses an Ocneanu-Szymanski pairing of two centralizers on the tower of algebras above $A \mid B$, isomorphic to the two main players in this paper $\operatorname{End}{}_B A_B$ and $\operatorname{End}{}_A A \otimes_B A_A$; then shows that the resulting algebra-coalgebra is a Hopf algebra. In the paper [13] the notion of depth two was widened to arbitrary algebra extensions whereby it was shown that the bimodule endomorphism ring $\operatorname{End}{}_B A_B$ of a depth two extension $A \mid B$ has a bialgebroid structure as in Lu [15]. Interesting classes of examples were noted such as finite-dimensional algebras, weak Hopf-Galois extensions, H-separable extensions and various normal subobjects in quantum algebra. Later in [10], the underlying fact emerged that any (including infinite index) algebra extension $A \mid B$ is right Galois w.r.t. a bialgebroid coaction if and only if it is right depth two and the natural module A_B is balanced. This is

The author thanks Caenepeel, Van Oystaeyen and Torrecillas for a pleasant stay in Belgium in September 2005.

in several respects analogous to the characterization of a Galois field extension as normal and separable (the splitting field of separable polynomials).

A bialgebroid is in simplest terms a bialgebra over a noncommutative base, which is of interest from the point of view of tensor categories [7] and mathematical physics [2]. A depth two extension $A \mid B$ has a Galois action machinery consisting of a bialgebroid with base algebra the centralizer $C_A(B)$ of the extension. For example, if A is a commutative ring, then this Galois bialgebroid specializes to the Galois biring in Winter [25]. If $A \mid B$ is Frobenius and $C_A(B)$ is semisimple in all base field extensions, as with reducible subfactors, the extension is a weak Hopf-Galois extension, where the Galois bialgebroid is a weak Hopf algebra [11, 5]. Bialgebroids equipped with antipodes are Hopf algebroids, a recent object of study with competing definitions of what constitutes an antipode [3, 15, 14]. In Section 3, we investigate the Galois bialgebroid of a new type of depth two extension called a pseudo-Galois extension, which is a notion generalizing the notions of H-separable extension (e.g., an Azumaya algebra) and group-Galois extension (e.g., a Galois algebra) [18]. Its Galois bialgebroid is shown in Theorem 2.3 to be closely related to a certain Hopf algebroid (in the sense of Böhm and Szlachányi [3]) we obtain from a cofibered sum of a semi-direct product of an algebra with a group of automorphisms and its opposite right crossed product. This Hopf algebroid extends Lu's basic Hopf algebroid on the enveloping algebra over an algebra [15], to the group action setting, and in Theorem 3.1 to the involutive Hopf algebra action setting. In the last section of this paper we discuss Van Oystaeyen and Panaite's isomorphism of this "enveloping" Hopf algebroid given in the preprint `QA/0508411` to this paper with a Hopf algebroid in Connes-Moscovici [6], which is known as a para-Hopf algebroid in [14]. The isomorphism may be derived from a universal condition in Proposition 3.4 or from the unit representation of a bialgebroid on its base algebra. The isomorphism of bialgebroid invariants naturally raises the possibility of a relation between certain pseudo-Galois extensions, or a generalization we propose, with Rankin-Cohen brackets [6].

2. Depth two and pseudo-Galois extensions

All algebras in this paper are unital associative algebras over a commutative ground ring K. An algebra extension $A \mid B$ is a unit-preserving algebra homomorphism $B \to A$, a proper extension if this mapping is monic. The induced bimodule ${}_B A_B$ plays the main role below. Unadorned tensors, hom-groups and endomorphism-groups (in a category of modules) between algebras are over the ground ring unless otherwise stated. For example, End A denotes the linear endomorphisms of an algebra A, but not the algebra endomorphisms of A; and Hom (A_B, B_B) denotes the B-A-bimodule of right B-bimodule homomorphisms from A into B. The default setting is the natural module structure unless otherwise specified.

An algebra extension $A \,|\, B$ is *left depth two* (D2) if its tensor-square $A \otimes_B A$ as a natural B-A-bimodule is isomorphic to a direct summand of a finite direct sum of the natural B-A-bimodule A: for some positive integer N, we have

$$A \otimes_B A \oplus * \cong A^N \tag{1}$$

An extension $A \,|\, B$ is *right* D2 if Eq. (1) holds instead as natural A-B-bimodules.

Since condition (1) implies maps in two hom-groups satisfying $\sum_{i=1}^{N} g_i \circ f_i = \mathrm{id}_{A \otimes_B A}$, where $g_i \in \mathrm{Hom}\,({}_B A_A, {}_B A \otimes_B A_A) \cong (A \otimes_B A)^B$ (via $g \mapsto g(1)$) and

$$f_i \in \mathrm{Hom}\,({}_B A \otimes_B A_A, {}_B A_A) \cong \mathrm{End}\,{}_B A_B := \mathcal{S}$$

via $f \mapsto (a \mapsto f(a \otimes_B 1))$, we obtain an equivalent condition for extension $A \,|\, B$ to be left D2: there is a positive integer N, $\beta_1, \ldots, \beta_N \in \mathcal{S}$ and $t_1, \ldots, t_N \in (A \otimes_B A)^B$ (i.e., satisfying for each $i = 1, \ldots, N$, $bt_i = t_i b$ for every $b \in B$) such that

$$\sum_{i=1}^{N} t_i \beta_i(x) y = x \otimes_B y \tag{2}$$

for all $x, y \in A$.

Like dual bases for projective modules, this equation is useful. For example, to show $\mathcal{S}$ finite projective as a left $C_A(B)$-module (module action given by $r \cdot \alpha = \lambda_r \circ \alpha$), apply $\alpha \in \mathcal{S}$ to the first tensorands of the equation, set $y = 1$ and apply the multiplication mapping $\mu : A \otimes_B A \to A$ to obtain

$$\alpha(x) = \sum_i \alpha(t_i^1) t_i^2 \beta_i(x), \tag{3}$$

where we suppress a possible summation in $t_i \in A \otimes_B A$ using a Sweedler notation, $t_i = t_i^1 \otimes_B t_i^2$. But for each $i = 1, \ldots, N$, we note that

$$T_i(\alpha) := \alpha(t_i^1) t_i^2 \in C_A(B) := R$$

defines a homomorphism $T_i \in \mathrm{Hom}\,({}_R \mathcal{S}, {}_R R)$, so that Eq. (3) shows that $\{T_i\}$, $\{\beta_i\}$ are finite dual bases for ${}_R \mathcal{S}$.

As another example, Eq. (2) is used in the explicit formula for coproduct in Eq. (5), which in the case A is a commutative ring and B a subfield would give an explicit formula for "preservations" in [25] in terms of dual bases and implying a simpler proof for the Galois biring correspondence theorem [25, Theorem 6.1]. (We note that the generalized Jacobson-Bourbaki correspondence theorem [25, Theorem 2.1] may be extended with almost the same argument to any noncommutative algebra A possessing algebra homomorphism into a division algebra.)

Similarly, an algebra extension $A \,|\, B$ is right D2 if there is a positive integer N, elements $\gamma_j \in \mathrm{End}\,{}_B A_B$ and $u_j \in (A \otimes_B A)^B$ such that

$$x \otimes_B y = \sum_{j=1}^{N} x \gamma_j(y) u_j \tag{4}$$

for all $x, y \in A$. We call the elements $\gamma_j \in \mathcal{S}$ and $u_j \in (A \otimes_B A)^B$ right D2 quasibases for the extension $A \,|\, B$. Fix this notation and the corresponding notation

$\beta_i \in \mathcal{S}$ and $t_i \in (A \otimes_B A)^B$ for left D2 quasibases throughout this paper. An algebra extension is of course D2 if it is both left and right D2.

Recall from [13] that a depth two extension $A \mid B$ has left bialgebroid structure on the algebra $\mathrm{End}\,_B A_B$ with noncommutative base algebra $C_A(B)$. Again let R denote $C_A(B)$ and $\mathcal{S}$ denote the bimodule endomorphism algebra $\mathrm{End}\,_B A_B$ (under composition). We sketch the left R-bialgebroid structure on $\mathcal{S}$ since we will need it in the case of pseudo-Galois extensions.

Its R-R-bimodule structure $_R \mathcal{S}_R$ is generated by the algebra homomorphism $\lambda : R \to \mathcal{S}$ given by $r \mapsto \lambda_r$, left multiplication by r, and the algebra anti-homomorphism $\rho : R \to \mathcal{S}$ given by $r \mapsto \rho_r$, right multiplication by r. These are sometimes called the source map λ and the target map ρ of the bialgebroid. These two mappings commute within $\mathcal{S}$ at all values:

$$\rho_r \circ \lambda_s = \lambda_s \circ \rho_r, \quad r, s \in R := C_A(B),$$

whence we may define a bimodule by composing strictly from the left:

$$_R \mathcal{S}_R : \quad r \cdot \alpha \cdot s = \lambda_r \circ \rho_s \circ \alpha = r\alpha(?)s, \quad \alpha \in \mathcal{S}, \ r, s \in R.$$

Next we equip $\mathcal{S}$ with an R-coring structure $(\mathcal{S}, R, \Delta, \varepsilon)$ as follows. The comultiplication $\Delta : \mathcal{S} \to \mathcal{S} \otimes_R \mathcal{S}$ is an R-R-homomorphism given by

$$\Delta(\alpha) := \sum_i \alpha(?t_i^1) t_i^2 \otimes_R \beta_i = \sum_j \gamma_j \otimes_R u_j^1 \alpha(u_j^2?) \tag{5}$$

in terms of left D2 quasibases in the first equation or right D2 quasibases in the second equation. There is a simplification that shows this comultiplication is a generalization of the one in [15] (for the linear endomorphisms of an algebra):

$$\phi : \mathcal{S} \otimes_R \mathcal{S} \xrightarrow{\cong} \mathrm{Hom}\,(_B A \otimes_B A_B, \,_B A_B) \quad \phi(\alpha \otimes_R \beta)(x \otimes_B y) = \alpha(x)\beta(y) \tag{6}$$

for all $\alpha, \beta \in \mathcal{S}$ and $x, y \in A$. From a variant of Eq. (3) we obtain:

$$\phi(\Delta(\alpha))(x \otimes y) = \alpha(xy), \quad (x, y \in A, \ \alpha \in \mathrm{End}\,_B A_B). \tag{7}$$

The counit $\varepsilon : \mathcal{S} \to R$ is given by evaluation at the unity element, $\varepsilon(\alpha) = \alpha(1_A)$, again an R-R-homomorphism. It is then apparent from Eq. (7) that

$$(\varepsilon \otimes_R \mathrm{id}_\mathcal{S})\Delta(\alpha) = \varepsilon(\alpha_{(1)}) \cdot \alpha_{(2)} = \lambda_{\alpha_{(1)}(1)}\alpha_{(2)} = \alpha, \quad (\alpha \in \mathcal{S})$$

using a reduced Sweedler notation for the coproduct of an element, and a similar equation corresponding to $(\mathrm{id}_\mathcal{S} \otimes_R \varepsilon)\Delta = \mathrm{id}_\mathcal{S}$.

Finally, the comultiplication and counit satisfy additional bialgebra-like axioms that make $(\mathcal{S}, R, \lambda, \rho, \Delta, \varepsilon)$ a left R-bialgebroid [13, p. 80]. These are:

$$\varepsilon(1_\mathcal{S}) = 1_R, \tag{8}$$

which is obvious,

$$\Delta(1_\mathcal{S}) = 1_\mathcal{S} \otimes_R 1_\mathcal{S}, \tag{9}$$

which follows from Eq. (7),

$$\forall \alpha \in \mathcal{S}, \ \alpha_{(1)} \circ \rho_r \otimes_R \alpha_{(2)} = \alpha_{(1)} \otimes_R \alpha_{(2)} \circ \lambda_r \tag{10}$$

which follows from the equation defining ϕ (since both sides yield $\alpha(xry)$),

$$\Delta(\alpha \circ \beta) = \Delta(\alpha)\Delta(\beta) = \alpha_{(1)} \circ \beta_{(1)} \otimes_R \alpha_{(2)} \circ \beta_{(2)} \tag{11}$$

where Eq. (10) justifies the use of a tensor algebra product in $\operatorname{Im}\Delta \subseteq \mathcal{S} \otimes_R \mathcal{S}$ and the equation follows again from equation defining ϕ (as both sides equal $\alpha(\beta(xy))$), and at last the easy

$$\varepsilon(\alpha \circ \beta) = \varepsilon(\alpha \circ \lambda_{\varepsilon(\beta)}) = \varepsilon(\alpha \circ \rho_{\varepsilon(\beta)}). \tag{12}$$

On occasion the bialgebroid $\mathcal{S}$ is a *Hopf algebroid* [3, Def. 4.1], i.e., possesses an *antipode* $\tau : \mathcal{S} \to \mathcal{S}$. This is *anti-automorphism* of the algebra $\mathcal{S}$ which satisfies:

1. $\tau \circ \rho = \lambda$,
2. $\tau^{-1}(\alpha_{(2)})_{(1)} \otimes_R \tau^{-1}(\alpha_{(2)})_{(2)}\alpha_{(1)} = \tau^{-1}(\alpha) \otimes 1_{\mathcal{S}}$ $(\forall \alpha \in \mathcal{S})$;
3. $\tau(\alpha_{(1)})_{(1)}\alpha_{(2)} \otimes_R \tau(\alpha_{(1)})_{(2)} = 1_{\mathcal{S}} \otimes_R \tau(\alpha)$ $(\forall \alpha \in \mathcal{S})$.

Examples of Hopf algebroids are weak Hopf algebras [7] and Hopf algebras, including group algebras and enveloping algebras of Lie algebras [21]. Lu [15] provides the example $A \otimes A^{\mathrm{op}}$ of a Hopf algebroid over any algebra A with twist being the antipode, and another bialgebroid the linear endomorphisms $\operatorname{End}A$, which is a particular case of the construction $\mathcal{S}$.

A homomorphism of R-bialgebroids $\mathcal{S}' \to \mathcal{S}$ is an algebra homomorphism which commutes with the source and target mappings, which additionally is an R-coring homomorphism (so there are three commutative triangles and a commutative square for such an algebra homomorphism to satisfy) [4, 9]. If $\mathcal{S}'$ and $\mathcal{S}$ additionally come equipped with antipodes τ' and τ, respectively, then the homomorphism $\mathcal{S}' \to \mathcal{S}$ is additionally a homomorphism of Hopf algebroids if it commutes with the antipodes in an obvious square diagram.

2.1. Pseudo-Galois extensions

If σ is automorphism of the algebra A, we let A_σ denote the bimodule A twisted on the right by σ, with module actions defined by $x \cdot a \cdot y = xa\sigma(y)$ for $x, y, a \in A$. Two such bimodules A_σ and A_τ twisted by automorphisms $\sigma, \tau : A \to A$ are A-A-bimodule isomorphic if and only if there is an invertible element $u \in A$ such that $\tau \circ \sigma^{-1}$ is the inner automorphism by u (for send $1 \mapsto u$).

Let B be a subalgebra of A. Recall the characterization of a group-Galois extension $A \,|\, B$, where only two conditions need be met. First, there is a finite group G of automorphisms of A such that $B = A^G$, i.e., the elements of B are fixed under each automorphism of G and each element of A in the complement of B is moved by some automorphism of G. Second, there are elements $a_i, b_i \in A$, $i = 1, \ldots, n$ such that $\sum_i a_i b_i = 1$ and $\sum_i a_i \sigma(b_i) = 0$ if $\sigma \in G$ and $\sigma \neq \mathrm{id}_A$.

Since $E : A \to B$ defined by $E(a) = \sum_{\sigma \in G} \sigma(a)$ is a Frobenius homomorphism with dual bases a_i, b_i, it follows that there is an A-A-bimodule isomorphism between the tensor-square and the semi-direct product of A and G:

$$h : A \otimes_B A \longrightarrow A \rtimes G, \quad h(x \otimes y) := \sum_{\sigma \in G} x\sigma y = \sum_{\sigma \in G} x\sigma(y)\sigma \tag{13}$$

(For the inverse is given by $h^{-1}(a\tau) = \sum_i a\tau(a_i) \otimes_B b_i$.) Thus $A \otimes_B A$ is isomorphic to $\oplus_{\sigma \in G} A_\sigma$ as A-A-bimodules. Mewborn and McMahon [18] relax this condition as follows:

Definition 2.1. *The algebra A is a pseudo-Galois extension of a subalgebra B if there is a finite set G of B-automorphisms (i.e., fixing elements of B) and a positive integer N such that $A \otimes_B A$ is isomorphic to a direct summand of $\oplus_{\sigma \in G} A_\sigma^N$: in symbols this becomes*

$$A \otimes_B A \oplus * \cong \oplus_{\sigma \in G} A_\sigma^N, \tag{14}$$

in terms of A-A-bimodules. Assume with no loss of generality that G is minimal in the group of B-automorphisms $\mathrm{Aut}_B(A)$ with respect to this property and $A_\sigma \not\cong A_\tau$ if $\sigma \neq \tau$ in G.

It is clear that Galois extensions are pseudo-Galois. For example, let A be a simple ring with finite group G of outer automorphisms of A, then A is Galois over its fixed subring $B = A^G$, cf. [20, 2.4]. Another example of a pseudo-Galois extension is an *H-separable extension* $A \mid B$, which by definition satisfies $A \otimes_B A \oplus * \cong A^N$ as A-A-bimodules, so we let $G = \{\mathrm{id}_A\}$ in the definition above [8, 9, 18]. For example, if A is a simple ring, G a finite group of outer automorphisms of A such that each nonidentity automorphism moves an element of the center of A, then the skew group ring $A \rtimes G$ is H-separable over A [19].

Note that the definition of pseudo-Galois extension $A \mid B$ leaves open the possibility that B is a proper subset of the invariant subalgebra A^G: if $B \subset C \subset A^G$ and C is a separable extension of B, then $A \mid C$ is also a pseudo-Galois extension, since one may show that $A \otimes_C A \oplus * \cong A \otimes_B A$ as natural A-A-bimodules via the separability element. Conversely, if $A \mid C$ is a pseudo-Galois extension and $C \supset B$ is H-separable, then $A \mid B$ is pseudo-Galois by noting that $A \otimes_B A \cong A \otimes_C C \otimes_B C \otimes_C A$. For example, if $E \mid F$ is a finite Galois extension of fields where F is the quotient field of a domain R, then the ring extension $E \mid R$ is pseudo-Galois.

In the next proposition, we note that pseudo-Galois extensions are depth two by means of a characterization of pseudo-Galois extensions using *pseudo-Galois elements*.

Proposition 2.2. *An algebra extension $A \mid B$ is pseudo-Galois iff there is a finite set G of B-automorphisms and N elements $r_{i,\sigma} \in C_A(B)$ and N elements $e_{i,\sigma} \in (_\sigma A \otimes_B A)^A$ for each element $\sigma \in G$ satisfying*

$$1 \otimes_B 1 = \sum_{\sigma \in G} \sum_{i=1}^N r_{i,\sigma} e_{i,\sigma} \tag{15}$$

As a consequence, $A \mid B$ is left and right D2 with left and right D2 quasibases derived from the elements $r_{i,\sigma}$ and $e_{i,\sigma} \in (A \otimes_B A)^B$.

Proof. ($\Rightarrow$) We note that the condition (14) implies the existence of N pairs of mappings $f_{i,\sigma}$ and $g_{i,\sigma}$ for each B-automorphism $\sigma \in \mathcal{G}$ satisfying

$$\sum_{i=1}^{N} \sum_{\sigma \in \mathcal{G}} g_{i,\sigma} \circ f_{i,\sigma} = \mathrm{id}_{A \otimes_B A}. \tag{16}$$

The mappings simplify as

$$f_{i,\sigma} \in \mathrm{Hom}\left(_A A \otimes_B A_A, {}_A(A_\sigma)_A\right) \cong C_A(B),$$

via $F \mapsto F(1 \otimes 1)$ with inverse $r \mapsto (a \otimes a' \mapsto ar\sigma(a'))$, as well as mappings

$$g_{i,\sigma} \in \mathrm{Hom}\left(_A(A_\sigma)_A, {}_A A \otimes_B A_A\right) \cong ({}_\sigma A \otimes_B A)^A,$$

via $f \mapsto f(1)$ with inverse $e \mapsto (a \mapsto ae)$ where $e \in ({}_\sigma A \otimes_B A)^A$ iff $ea = \sigma(a)e$ for each $a \in A$.

If $r_{i,\sigma}$ corresponds via the isomorphism above with $f_{i,\sigma}$, then $f_{i,\sigma}(x \otimes_B y) = xr_{i,\sigma}\sigma(y)$ for each $x, y \in A$. If $e_{i,\sigma}$ corresponds via the other isomorphism above with $g_{i,\sigma}$, then $g_{i,\sigma}(a) = ae_{i,\sigma}$. We compute:

$$x \otimes_B y = \sum_{i,\sigma \in \mathcal{G}} (g_{i,\sigma} \circ f_{i,\sigma})(x \otimes y) = \sum_{\sigma \in \mathcal{G}} \sum_{i=1}^{N} xr_{i,\sigma}\sigma(y)e_{i,\sigma}, \tag{17}$$

which shows that

$$\lambda_{r_{i,\sigma}} \circ \sigma \in \mathrm{End}\,_B A_B, \quad e_{i,\sigma} \in (A \otimes_B A)^B \tag{18}$$

are right D2 quasibases. Setting $x = y = 1$ we obtain the Eq. (15). Finally use the twisted centralizer property $ea = \sigma(a)e$ for $a \in A$ and $e \in ({}_\sigma A \otimes_B A)^A$ to obtain

$$x \otimes_B y = \sum_{\sigma \in \mathcal{G}} \sum_{i=1}^{N} e_{i,\sigma}\sigma^{-1}(x)\sigma^{-1}(r_{i,\sigma})y.$$

Hence, the following are left D2 quasibases for $A \,|\, B$:

$$e_{i,\sigma} \in (A \otimes_B A)^B, \quad \sigma^{-1} \circ \rho_{r_{i,\sigma}} \in \mathrm{End}\,_B A_B. \tag{19}$$

($\Leftarrow$) Conversely, suppose we are given a finite set $\mathcal{G}$ of B-automorphisms, and for each $\sigma \in \mathcal{G}$, N centralizer elements $r_{i,\sigma} \in C_A(B)$, and N twisted A-central elements $e_{i,\sigma} \in ({}_\sigma A \otimes_B A)^A$ for $i = 1, \ldots, N$ such that Eq. (15) holds. By multiplying the equation from the left by $x \in A$ and from the right by $y \in A$, we obtain Eq. (17) and then Eq. (16) by defining $f_{i,\sigma}$ and $g_{i,\sigma}$ as before, which is of course equivalent to the condition (14) for pseudo-Galois extension. $\qquad \square$

We note that pseudo-Galois elements in Eq. (15) specialize to H-separability elements in case $\mathcal{G} = \{\mathrm{id}_A\}$ [8, 2.5].

Recall that in homological algebra the enveloping algebra of an algebra A is denoted by $A^e := A \otimes A^{op}$.

Theorem 2.3. *Suppose $A \mid B$ is pseudo-Galois extension satisfying condition (14) with G the subgroup generated by $\mathcal{G}$ within $\mathrm{Aut}_B A$ and R the centralizer $C_A(B)$. Then there is a Hopf algebroid denoted by $R^e \bowtie G$ which maps epimorphically as R-bialgebroids onto the left bialgebroid $\mathcal{S} = \mathrm{End}\,{}_B A_B$.*

Proof. Denote the identity in G by e and the canonical anti-isomorphism $R \to R^{op}$ by $r \mapsto \bar{r}$ satisfying $\bar{r}\,\bar{s} = \overline{sr}$ for every $r, s \in R$. Note that the B-automorphisms of G restrict to automorphisms of the centralizer R. The notation C for $R^e \bowtie G$ is adopted at times. The Hopf algebroid structure of $R^e \bowtie G$ is given by

1. as a K-module (over the ground ring K) $R^e \bowtie G = R \otimes R^{op} \otimes K[G]$ where $K[G]$ is the group K-algebra of G;
2. multiplication given by

$$(r \otimes \bar{s} \bowtie \sigma)(u \otimes \bar{v} \bowtie \tau) = r\sigma(u) \otimes \overline{v\tau^{-1}(s)} \bowtie \sigma\tau \qquad (20)$$

   with unity element $1_C = 1_R \otimes \overline{1_R} \bowtie e$,
3. source map $s_L : R \to R^e \bowtie G$ given by $s_L(r) = r \otimes \bar{1} \bowtie e$,
4. target map $t_L : R^{op} \to R^e \bowtie G$ given by $t_L(r) = 1 \otimes \bar{r} \bowtie e$,
5. counit $\varepsilon_C : R^e \bowtie G \to R$ given by $\varepsilon_C(r \otimes \bar{s} \bowtie \sigma) = r\sigma(s)$,
6. comultiplication $\Delta_C : C \to C \otimes_R C$ is given by

$$\Delta(r \otimes \bar{s} \bowtie \sigma) = (r \otimes \bar{1} \bowtie \sigma) \otimes_R (1 \otimes \bar{s} \bowtie \sigma), \qquad (21)$$

7. and the antipode $\tau : C \to C$ by $\tau(r \otimes \bar{s} \bowtie \sigma) = s \otimes \bar{r} \bowtie \sigma^{-1}$

We will postpone the proof that this defines a Hopf algebroid over R until the next section where it is shown more generally for an involutive Hopf algebra H and its H-module algebras.

The epimorphism of left R-bialgebroids $\Psi : R^e \bowtie G \to \mathcal{S}$ is given by

$$\Psi(r \otimes \bar{s} \bowtie \sigma) = \lambda_r \circ \sigma \circ \rho_s \qquad (22)$$

We note that Ψ is an algebra homomorphism by comparing Eq. (20) with

$$\lambda_r \circ \sigma \circ \rho_s \circ \lambda_u \circ \tau \circ \rho_v = \lambda_{r\sigma(u)} \circ \sigma \circ \tau \circ \rho_{v\tau^{-1}(s)},$$

and $\Psi(1_C) = 1_{\mathcal{S}}$. The mapping Ψ is epimorphic since each $\beta \in \mathcal{S}$ may be expressed as a sum of mappings of the form $\lambda_r \circ \sigma \circ \rho_s$ where $\sigma \in G$ and $r, s \in R$. To see this, apply $\mu(\mathrm{id}_A \otimes \beta)$ to Eq. (17) with $x = 1$, which yields

$$\beta(y) = \sum_{i,\sigma \in G} r_{i,\sigma}\sigma(y)e_{i,\sigma}^1 \beta(e_{i,\sigma}^2)$$

where $e_{i,\sigma}^1 \beta(e_{i,\sigma}^2) \in R$ for each i and σ.

Note next that Ψ commutes with source, target and counit maps. For $\Psi(s_L(r)) = \Psi(r \otimes \bar{1} \bowtie e) = \lambda_r$ and $\Psi(t_L(r)) = \Psi(1 \otimes \bar{r} \bowtie e) = \rho_r$ for $r \in R$ (so $\Psi : C \to \mathcal{S}$ is an R-R-bimodule map). The map Ψ is counital since

$$\varepsilon(\Psi(r \otimes \bar{s} \bowtie \sigma)) = \lambda_r(\sigma(\rho_s(1))) = r\sigma(s) = \varepsilon_C(r \otimes \bar{s} \bowtie \sigma).$$

Using the isomorphism $\phi : \mathcal{S} \otimes_R \mathcal{S} \to \mathrm{Hom}\,(_B A \otimes_B A_B, _B A_B)$ for a depth two extension $A\,|\,B$ defined as above by $\phi(\alpha \otimes_R \beta)(x \otimes_B y) = \alpha(x)\beta(y)$, note from Eq. (21) that Ψ is comultiplicative:

$$\phi((\Psi \otimes_R \Psi)(\Delta_C(r \otimes \bar{s} \bowtie \sigma)))(x \otimes_B y) = \lambda_r(\sigma(x))\sigma(\rho_s(y))$$

$$= \phi(\Delta(\lambda_r \circ \sigma \circ \rho_s))(x \otimes_B y) = \phi(\Delta(\Psi(r \otimes \bar{s} \bowtie \sigma)))(x \otimes_B y)$$

since $\sigma \in G$ is a group-like element satisfying $\sigma_{(1)} \otimes_R \sigma_{(2)} = \sigma \otimes_R \sigma$ (corresponding to the automorphism condition).   □

Corollary 2.4. *If the algebra extension $A\,|\,B$ is H-separable, then $G = \{\mathrm{id}_A\}$ and $\Psi :$ $R^e \to \mathcal{S}$ is an isomorphism of bialgebroids, whence $\mathcal{S}$ has the antipode $\Psi \circ \tau \circ \Psi^{-1}$. If $A\,|\,B$ is G-Galois, then $\Psi : R^e \bowtie G \to \mathcal{S}$ is a split epimorphism of bialgebroids.*

Proof. Note that R^e is isomorphic as algebras to the subalgebra $R^e \bowtie \{e\}$. The first statement follows from [9], since $\Psi(r \otimes \bar{s}) = \lambda_r \circ \rho_s$ is shown there to be an isomorphism of bialgebroids.

If $A\,|\,B$ is G-Galois, then $A \otimes_B A \cong A \rtimes G$ via h above. Since each $\sigma \in G$ fixes elements of B, it follows that $(A \otimes_B A)^B \cong R \rtimes G$. Since $A\,|\,B$ is a Frobenius extension, $\mathrm{End}\,A_B \cong A \otimes_B A$ via $f \mapsto \sum_i f(a_i) \otimes b_i$ with inverse $x \otimes_B y \mapsto \lambda_x \circ E \circ \lambda_y$. This restricts to $\mathrm{End}\,_B A_B \cong (A \otimes_B A)^B$. Putting the two together yields

$$\Phi : \mathcal{S} \xrightarrow{\cong} R \rtimes G, \quad \Phi(\alpha) := \sum_{\sigma \in G} \sum_{i=1}^{n} \alpha(a_i)\sigma(b_i)\sigma$$

with inverse $r \rtimes \tau \mapsto \lambda_r \circ \tau$. Then the algebra epimorphism $\Phi \circ \Psi : R^e \bowtie G \to R \rtimes G$ simplifies to

$$(\Phi \circ \Psi)(r \otimes \bar{s} \bowtie \sigma) = \Phi(\lambda_r \circ \sigma \circ \rho_s) = \sum_{\tau \in G} \sum_{i=1}^{n} \lambda_{r\sigma(a_i s \tau(b_i))} \circ \sigma \circ \tau.$$

which is split by the monomorphism $R \rtimes G \to R^e \bowtie G$ given by $r \rtimes \sigma \mapsto r \otimes \bar{1} \bowtie \sigma$, an algebra homomorphism by an application of Eq. (20).   □

3. An enveloping Hopf algebroid over algebras in certain tensor categories

Let H be a Hopf algebra with bijective antipode S and A a left H-module algebra, i.e., an algebra in the tensor category of H-modules. Motivated by the left bialgebroid of a pseudo-Galois extension as studied in Section 3, we define a type of enveloping algebra $A^e \bowtie H$ for the smash product algebra $A \rtimes H$. It is a left bialgebroid over A, and a Hopf algebroid in case H is involutive such as a group algebra or the enveloping algebra of Lie algebra. In terms of noncommutative algebra, it is the minimal algebra which contains subalgebras isomorphic to the Hopf algebra H, the standard enveloping algebra A^e of an algebra A, and the semi-direct or crossproduct algebra $A \rtimes H$ as well as its derived right crossproduct algebra $H \ltimes A^{op}$. In terms of category theory, it is derived from the push-out construction [16] of the inclusion $H \hookrightarrow A \rtimes H$ and its opposite via the isomorphism $S : H \to H^{cop,\,op}$.

Theorem 3.1. *Suppose $B := A^e \bowtie H$ is the vector space $A \otimes A^{op} \otimes H$ with multiplication*

$$(a \otimes \bar{b} \bowtie h)(c \otimes \bar{d} \bowtie k) := a(h_{(1)} \cdot c) \otimes \overline{d(S(k_{(2)}) \cdot b)} \bowtie h_{(2)}k_{(1)}. \qquad (23)$$

Then B is a left bialgebroid over A with structure given in Eqs. (24) through (28). If $S^2 = \mathrm{id}_H$, then B is a Hopf algebroid with antipode Eq. (29).

Proof. Clearly the unity element $1_B = 1_A \otimes \overline{1_A} \bowtie 1_H$. The multiplication is associative, since

$$[(a \otimes \bar{b} \bowtie h)(c \otimes \bar{d} \bowtie k)](e \otimes \bar{f} \bowtie j)$$
$$= (a(h_{(1)} \cdot c) \otimes \overline{d(S(k_{(2)}) \cdot b)} \bowtie h_{(2)}k_{(1)})(e \otimes \bar{f} \bowtie j)$$
$$= a(h_{(1)} \cdot c)(h_{(2)}k_{(1)} \cdot e) \otimes \overline{f(S(j_{(3)}) \cdot d)(S(k_{(3)}j_{(2)}) \cdot b)} \bowtie h_{(3)}k_{(2)}j_{(1)}$$
$$= (a \otimes \bar{b} \bowtie h)(c(k_{(1)} \cdot e) \otimes \overline{f(S(j_{(2)}) \cdot d)} \bowtie k_{(2)}j_{(1)})$$
$$= (a \otimes \bar{b} \bowtie h)[(c \otimes \bar{d} \bowtie k)(e \otimes \bar{f} \bowtie j)].$$

It follows that B is an algebra.

Define a source map $s_L : A \to B$ and target map $t_L : A \to B$ by

$$s_L(a) = a \otimes \overline{1_A} \bowtie 1_H \qquad (24)$$
$$t_L(a) = 1_A \otimes \bar{a} \bowtie 1_H, \qquad (25)$$

an algebra homomorphism and anti-homomorphism, respectively. It is evident that $t_L(x)s_L(y) = s_L(y)t_L(x)$ for all $x, y \in A$. The A-A-bimodule structure induced from $x \cdot b \cdot y = s_L(x)t_L(y)b$ for $b \in B$ is then given by $(a, c \in A,\ h \in H)$

$$x \cdot (a \otimes \bar{c} \bowtie h) \cdot y = xa \otimes \overline{c(S(h_{(2)}) \cdot y)} \bowtie h_{(1)}. \qquad (26)$$

The counit $\varepsilon : B \to A$ is defined by

$$\varepsilon(a \otimes \bar{c} \bowtie h) := a(h \cdot c). \qquad (27)$$

Note that ε is an A-A-bimodule homomorphism via its application to the RHS of Eq. (26):

$$\varepsilon(xa \otimes \overline{c(S(h_{(2)}) \cdot y)} \bowtie h_{(1)}) = xah_{(1)} \cdot (c(S(h_{(2)}) \cdot y)) = xa(h \cdot c)y = x\varepsilon(a \otimes \bar{c} \bowtie h)y,$$

since $h \cdot (xy) = (h_{(1)} \cdot x)(h_{(2)} \cdot y)$, the measuring axiom on A.

The comultiplication $\Delta : B \to B \otimes_A B$ is defined by

$$\Delta(a \otimes \bar{c} \bowtie h) := (a \otimes \overline{1_A} \bowtie h_{(1)}) \otimes_A (1_A \otimes \bar{c} \bowtie h_{(2)}). \qquad (28)$$

It is an A-A-homomorphism:

$$\Delta(xa \otimes \overline{c(S(h_{(2)}) \cdot y)} \bowtie h_{(1)}) = (xa \otimes \overline{1_A} \bowtie h_{(1)}) \otimes_A (1_A \otimes \overline{c(S(h_{(3)}) \cdot y)} \bowtie h_{(2)}$$
$$= x \cdot \Delta(a \otimes \bar{c} \bowtie h) \cdot y$$

by Eq. (26). The left counit equation $(\varepsilon \otimes_A \mathrm{id}_B)\Delta = \mathrm{id}_B$ follows from

$$\varepsilon(a \otimes \bar{1} \bowtie h_{(1)}) \cdot (1 \otimes \bar{c} \bowtie h_{(2)}) = a(h_{(1)} \cdot 1) \cdot (1 \otimes \bar{c} \bowtie h_{(2)})$$
$$= a \otimes \bar{c} \bowtie h,$$

since $h \cdot 1_A = \varepsilon(h)1_A$ in the H-module algebra A. The right counit equation $(\mathrm{id}_B \otimes \varepsilon)\Delta = \mathrm{id}_B$ follows from

$$(a \otimes \bar{1} \bowtie h_{(1)}) \cdot \varepsilon(1 \otimes \bar{c} \bowtie h_{(2)}) = (a \otimes \bar{1} \bowtie h_{(1)}) \cdot (h_{(2)} \cdot c)$$
$$= a \otimes \overline{(S(h_{(2)})h_{(3)} \cdot c)} \bowtie h_{(1)} = a \otimes \bar{c} \bowtie h.$$

Hence $(B, A, \Delta, \varepsilon)$ is an A-coring.

We check the remaining bialgebroid axioms:

$$\Delta(1_B) = 1_B \otimes 1_B, \quad \varepsilon(1_B) = 1_A$$

are apparent from Eqs. (28) and (27). The axiom corresponding to Eq. (10) computes as:

$$\Delta(a \otimes b \bowtie h)(t_L(c) \otimes_A 1_B) = (a \otimes \bar{1} \bowtie h_{(1)})(1 \otimes \bar{c} \bowtie 1_H) \otimes_A (1 \otimes \bar{b} \bowtie h_{(2)})$$
$$= (a \otimes \bar{c} \bowtie h_{(1)}) \otimes_A (1 \otimes \bar{b} \bowtie h_{(2)}),$$

and on the other hand

$$\Delta(a \otimes \bar{b} \bowtie h)(1_B \otimes_A s_L(c)) = (a \otimes \bar{1} \bowtie h_{(1)}) \otimes_A (1 \otimes \bar{b} \bowtie h_{(2)})(c \otimes \bar{1} \bowtie 1_H)$$
$$= (a \otimes \bar{1} \bowtie h_{(1)}) \otimes_A (h_{(2)} \cdot c \otimes \bar{b} \bowtie h_{(3)})$$
$$= (a \otimes \overline{S(h_{(2)})h_{(3)} \cdot c} \bowtie h_{(1)}) \otimes_A (1 \otimes \bar{b} \bowtie h_{(4)})$$

which equals the RHS expression for $\Delta(a \otimes b \bowtie h)(t_L(c) \otimes_A 1_B)$.

Next, the comultiplication is multiplicative:

$$\Delta((a \otimes \bar{b} \bowtie h)(c \otimes \bar{d} \bowtie k)) = \Delta(a(h_{(1)} \cdot c) \otimes \overline{d(S(k_{(2)}) \cdot b)} \bowtie h_{(2)}k_{(1)})$$
$$= (a(h_{(1)} \cdot c) \otimes \bar{1} \bowtie h_{(2)}k_{(1)}) \otimes_A (1_A \otimes \overline{d(S(k_{(3)}) \cdot b)} \bowtie h_{(3)}k_{(2)})$$
$$= (a \otimes \bar{1} \bowtie h_{(1)})(c \otimes \bar{1} \bowtie k_{(1)}) \otimes_A (1 \otimes \bar{b} \bowtie h_{(2)})(1 \otimes \bar{d} \bowtie k_{(2)})$$
$$= \Delta(a \otimes \bar{b} \bowtie h)\Delta(c \otimes \bar{d} \bowtie k).$$

The counit satisfies

$$\varepsilon((a \otimes \bar{b} \bowtie h)(c \otimes \bar{d} \bowtie k)) = \varepsilon(a(h_{(1)} \cdot c) \otimes \overline{d(S(k_{(2)}) \cdot b)} \bowtie h_{(2)}k_{(1)})$$
$$= a(h_{(1)} \cdot c)(h_{(2)}k_{(1)} \cdot (d(S(k_{(2)}) \cdot b))) = a(h_{(1)} \cdot c)(h_{(2)}k \cdot d)(h_{(3)} \cdot b)$$
$$= \varepsilon((a \otimes \bar{b} \bowtie h)(c(k \cdot d) \otimes \bar{1} \bowtie 1_H)) = \varepsilon((a \otimes \bar{b} \bowtie h)s_L(\varepsilon(c \otimes \bar{d} \bowtie k))).$$

Similarly, $\varepsilon((a \otimes \bar{b} \bowtie h)(c \otimes \bar{d} \bowtie k)) = \varepsilon((a \otimes \bar{b} \bowtie h)t_L(\varepsilon(c \otimes \bar{d} \bowtie k)))$ for all $a, b, c, d \in A$, $h, k \in H$. Thus B is a bialgebroid over A.

Suppose the antipode on H is bijective and satisfies $S^2 = \mathrm{id}_H$. Define an antipode on B by $(a, b \in A, h \in H)$

$$\tau(a \otimes \bar{b} \bowtie h) = b \otimes \bar{a} \bowtie S(h) \tag{29}$$

Denote the compositional inverse of S by $\overline{S}$. Then τ has inverse,

$$\tau^{-1}(a \otimes \overline{b} \bowtie h) = b \otimes \overline{a} \bowtie \overline{S}(h).$$

Note that τ is an anti-automorphism of B:

$$\tau(c \otimes \overline{d} \bowtie k)\tau(a \otimes \overline{b} \bowtie h) = (d \otimes \overline{c} \bowtie S(k))(b \otimes \overline{a} \bowtie S(h))$$

$$= d(S(k_{(2)}) \cdot b) \otimes \overline{a(S^2(h_{(1)}) \cdot c)} \bowtie S(k_{(1)})S(h_{(2)})$$

$$= \tau(a(h_{(1)} \cdot c) \otimes \overline{d(S(k_{(2)}) \cdot b)} \bowtie h_{(2)}k_{(1)}) = \tau((a \otimes \overline{b} \bowtie h)(c \otimes \overline{d} \bowtie k)).$$

The antipode satisfies the three axioms (1)–(3):

$$\tau(t_L(a)) = \tau(1 \otimes \overline{a} \bowtie 1_H) = a \otimes \overline{1} \bowtie 1_H = s_L(a),$$

for all $a \in A$. Next, for $b := a \otimes \overline{c} \bowtie h \in B$,

$$\tau^{-1}(b_{(2)})_{(1)} \otimes_A \tau^{-1}(b_{(2)})_{(2)} b_{(1)}$$

$$= \tau^{-1}(1 \otimes \overline{c} \bowtie h_{(2)})_{(1)} \otimes_A \tau^{-1}(1 \otimes \overline{c} \bowtie h_{(2)})_{(2)}(a \otimes \overline{1} \bowtie h_{(1)})$$

$$= (c \otimes \overline{1} \bowtie \overline{S}(h_{(4)})) \otimes_A (\overline{S}(h_{(3)}) \cdot a \otimes \overline{1} \bowtie \overline{S}(h_{(2)})h_{(1)}$$

$$= c \otimes \overline{S(\overline{S}(h_{(2)})\overline{S}(h_{(1)}) \cdot a} \bowtie \overline{S}(h_{(3)}) \otimes_A 1_B$$

$$= (c \otimes \overline{a} \bowtie \overline{S}(h)) \otimes_A 1_B = \tau^{-1}(b) \otimes_A 1_B.$$

Continuing our notation $b = a \otimes \overline{c} \bowtie h \in B$, note too that

$$\tau(b_{(1)})_{(1)} b_{(2)} \otimes_A \tau(b_{(1)})_{(2)} = (1 \otimes \overline{1} \bowtie S(h_{(2)}))(1 \otimes \overline{c} \bowtie h_{(3)}) \otimes_A (1 \otimes \overline{a} \bowtie S(h_{(1)}))$$

$$= (1 \otimes \overline{c} \bowtie 1_H) \otimes_A (1 \otimes \overline{a} \bowtie S(h))$$

$$= 1_B \otimes_A (c \otimes \overline{a} \bowtie S(h)) = 1_B \otimes_A \tau(b).$$

Hence, B is a Hopf algebroid. □

Given a group G, its group algebra $K[G]$ over a commutative ring K is an involutive Hopf algebra [21]. Moreover, if G acts by automorphisms on a K-algebra A, then A is a left $K[G]$-module algebra and $A \rtimes G$ is identical with the semidirect product [21]. Thus the construction $R^e \bowtie G$ (covering the left bialgebroid of a pseudo-Galois extension in Section 3) is a Hopf algebroid, and we record the following.

Corollary 3.2. *Given a K-algebra A and a group G of algebra automorphisms of A, the algebra $A^e \bowtie K[G]$ is a Hopf algebroid over A.*

Recall that Lu [15] defines over an algebra A a Hopf algebroid A^e. This is a Hopf subalgebroid of the construction in the theorem above.

Corollary 3.3. *Let H be an involutive Hopf algebra and A a left H-module algebra. Then the Hopf algebroid $A^e \bowtie H$ contains subalgebras isomorphic to*

1. *Lu's Hopf algebroid $A \otimes A^{op}$*
2. *the semidirect product $A \rtimes H$*
3. *its derived right crossproduct $H \ltimes A^{op}$*
4. *the Hopf algebra H*

Proof. It is easy to check from Eq. (23) that the following mappings

1. $A^e \hookrightarrow A^e \bowtie H$ given by $a \otimes \bar{b} \mapsto a \otimes \bar{b} \bowtie 1_H$ is an algebra monomorphism as well as a homomorphism of Hopf algebroids over A (i.e., it commutes with the source, target, counit, comultiplication and antipode maps above and those given in [15]);

2. $j_1 : A \rtimes H \hookrightarrow A^e \bowtie H$ given by $j_1(a\#h) := a \otimes \overline{1_A} \bowtie h$ is an algebra monomorphism, where we recall that the multiplication in $A \rtimes H$ is given by

$$(a\#h)(b\#k) = a(h_{(1)} \cdot b)\#h_{(2)}k$$

3. $j_2 : H \ltimes A^{op} \hookrightarrow A^e \bowtie H$ given by $j_2(h\#\bar{a}) := 1 \otimes \bar{a} \bowtie h$ is an algebra monomorphism, where $\bar{a} \cdot h := \overline{S(h) \cdot a}$ defines the derived right action of H on A^{op} and the multiplication in $H \ltimes A^{op}$ (cf. [17, p. 22]) is given by

$$(h\#\bar{a})(k\#\bar{b}) = hk_{(1)}\#(\bar{a} \cdot k_{(2)})\bar{b}.$$

4. $H \hookrightarrow A^e \bowtie H$ given by $h \mapsto 1 \otimes \bar{1} \bowtie h$ is an algebra monomorphism as well as a Hopf algebra homomorphism (for it commutes with the counit, comultiplication and antipode mappings of H and $A^e \bowtie H$ if A is a faithful K-algebra and K is identified with $K1_A$.).   □

The construction $A^e \bowtie H$ for a Hopf algebra and a left H-module algebra is a type of cofibered sum [23, p. 99] of the algebra monomorphisms $\iota_1 : H \hookrightarrow A \rtimes H$ and $\iota_2 : H \hookrightarrow H \ltimes A^{op}$ defined by $\iota_1(h) := 1_A\#h$ and $\iota_2(h) := h\#\overline{1_A}$ for each $h \in H$. We note that $j_1 \circ \iota_1 = j_2 \circ \iota_2$, both sending $h \mapsto 1_A \otimes \overline{1_A} \bowtie h$. Also define the algebra monomorphism $k_1 : A \hookrightarrow A \rtimes H$ by $k_1(a) := a\#1_H$ and anti-monomorphism $k_2 : A \hookrightarrow H \ltimes A^{op}$ by $k_2(a) = 1_H\#\bar{a}$. Note that

$$j_2(k_2(a))j_1(k_1(b)) = (1_A \otimes \bar{a} \bowtie 1_H)(b \otimes \overline{1_A} \bowtie 1_H)$$
$$= (b \otimes \overline{1_A} \bowtie 1_H)(1_A \otimes \bar{a} \bowtie 1_H)$$
$$= j_1(k_1(b))j_2(k_2(a)),$$

for all $a, b \in A$.

Proposition 3.4. *Suppose B is an algebra with monomorphisms $f_1 : A \rtimes H \hookrightarrow B$ and $f_2 : H \ltimes A^{op} \hookrightarrow B$ such that $f_1 \circ \iota_1 = f_2 \circ \iota_2$ (i.e., satisfying the commutative square in Figure 1) and $f_1(k_1(a))f_2(k_2(b)) = f_2(k_2(b))f_1(k_1(a))$ for all $a, b \in A$. Then there is a uniquely defined algebra homomorphism $F : A^e \bowtie H \to B$ such that $F \circ j_i = f_i$ for $i = 1, 2$.*

Proof. Define $F : A^e \bowtie H \to B$ by

$$F(a \otimes \bar{b} \bowtie h) := f_1(a\#h)f_2(1_H\#\bar{b}) = f_1(a\#1_H)f_2(h\#\bar{b}). \quad (30)$$

The second equality follows from $a\#h = (a\#1_H)(1_A\#h)$ and $f_1 \circ \iota_1 = f_2 \circ \iota_2$. It follows that $F \circ j_i = f_i$ for $i = 1, 2$ since $f_i(1) = 1_B$. Then the uniqueness of F

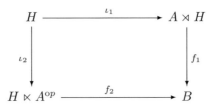

FIGURE 1. $A^e \bowtie H$ is the cofibered sum of $\iota_{1,2} : H \hookrightarrow A \rtimes H$, $H \ltimes A^{op}$ such that A^e embeds homomorphically.

follows from noting $a \otimes \bar{b} \bowtie h = (a \otimes \overline{1_A} \bowtie h)(1 \otimes \bar{b} \bowtie 1_H)$ and the homomorphic property of F. We compute that F is an algebra homomorphism by using $f_1(k_1(c))f_2(k_2(b)) = f_2(k_2(b))f_1(k_1(c))$ in the third equality:

$$F(a \otimes \bar{b} \bowtie h)F(c \otimes \bar{d} \bowtie k) = f_1(a\#h)f_2(1_H\#\bar{b})f_1(c\#1_H)f_2(k\#\bar{d})$$
$$= f_1(a\#h)f_1(c\#1_H)f_2(1_H\#\bar{b})f_2(k\#\bar{d})$$
$$= f_1(a(h_{(1)} \cdot c)\#h_{(2)})f_2(k_{(1)}\#\overline{d(S(k_{(2)}) \cdot b)}) = F((a \otimes \bar{b} \bowtie h)(c \otimes \bar{d} \bowtie k)),$$

by comparing the last equation with Eq. (23).   □

The homomorphism F may fail to be monic as for example when A is a commutative algebra, H acts trivially on A and $B = A \otimes H$.

Example 3.5. Enveloping algebras of Lie algebras are involutory Hopf algebras with comultiplication defined via primitive elements and the antipode via sign change. The Weyl algebra $K[X, Z \,|\, XZ+1 = ZX]$ is isomorphic to the semi-direct product $A \ltimes K[Z]$ of the one-dimensional Lie algebra $K[Z]$ acting by Leibniz derivation on the one-variable polynomial algebra $A = K[X]$ [21]. The enveloping Hopf algebroid is then the push-out of the inclusion $K[Z] \hookrightarrow K[X, Z \,|\, XZ + 1 = ZX]$ with itself:

$$K[X,Y] \bowtie K[Z] \cong K[X,Y,Z \,|\, XZ + 1 = ZX, \; XY = YX, \; YZ + 1 = ZY] \quad (31)$$

with Hopf algebroid comultiplication given on monomials by (integers $k \geq p, q \geq 0$)

$$\Delta(X^n Y^m Z^k) = \sum_{p+q=k} \binom{k}{p} X^n Z^p \otimes_A Y^m Z^q \quad (32)$$

counit by

$$\varepsilon(X^n Y^m Z^k) = \begin{cases} \frac{m!}{(m-k)!} X^{n+m-k} & \text{if } m \geq k \\ 0 & \text{if } k > m \end{cases} \quad (33)$$

and antipode by

$$\tau(X^n Y^m Z^k) = (-1)^k X^m Y^n Z^k. \quad (34)$$

4. Discussion

Böhm and Brzeziński [2, A.1] generalize the construction $A^e \bowtie H$ in the previous section to a certain module algebra A w.r.t. the action of a Hopf algebroid H which is twisted by an A-valued cocycle on H.

Panaite and Van Oystaeyen [22] observe that the Hopf algebroid $A^e \bowtie H$ constructed in the last section is isomorphic to the Hopf algebroid $A \odot H \odot A$ in Connes-Moscovici [6] with antipode given in [14], which arises in a quite different context. The algebra $A \odot H \odot A$ formed from a Hopf algebra H and a left H-module algebra A is linearly just $A \otimes H \otimes A$ with multiplication given by

$$(a \odot h \odot b)(c \odot k \odot d) = a(h_{(1)} \cdot c) \odot h_{(2)} k \odot (h_{(3)} \cdot d)b. \tag{35}$$

Note the algebra homomorphism $f_1 : A \ltimes H \to A \odot H \odot A$ given by $f_1(a \# h) := a \odot h \odot 1_A$. Note that $f_2 : H \rtimes A^{op} \to A \odot H \odot A$ given by

$$f_2(k \# \bar{b}) := 1_A \odot k_{(1)} \odot k_{(2)} \cdot b \tag{36}$$

is an algebra homomorphism satisfying with f_1 the hypotheses of Prop. 3.4. This leads to a mapping $F : A^e \bowtie H \to A \odot H \odot A$ given by

$$a \otimes \bar{b} \bowtie h \longmapsto a \odot h_{(1)} \odot h_{(2)} \cdot b, \tag{37}$$

which is the isomorphism in [22, 2.4].

Comparing the two isomorphic Hopf algebroids (see [22] for details) we note that the antipode in $A^e \bowtie H$ is given by a simpler formula, while the A-A-bimodule structure in $A \odot H \odot A$ is simpler. The multiplication in $A^e \bowtie H$ is closer to the smash product of a Hopf algebra with a bimodule algebra, which is the method of proof in [22].

It should also be noted that [22, 3.1, 3.2] provides an equivalent condition to that in proposition 3.4 which shows $A^e \bowtie H$ is a certain universal *bialgebroid*.

The picture of universals for bialgebroids over a fixed base ring A is the following. As observed in [15], for any (finite projective) algebra A there is a homomorphism of bialgebroids $A^e \to \text{End}\, A$, where $x \otimes \bar{y} \mapsto \lambda_x \circ \rho_y$, since $\text{End}\, A$ is a terminal object in a category of A-bialgebroids (existence in [15, Prop. 3.7], uniqueness: an easy argument). For similar reasons, A^e is an initial object in this category. For A a left H-module algebra, this homomorphism factors through the bialgebroids $A^e \bowtie H$, $A \odot H \odot A$, or any bialgebroid over A as follows.

Let $\mathcal{S}$ be a bialgebroid over A with source, target mappings $s_L, t_L : A \to \mathcal{S}$ and counit $\varepsilon : \mathcal{S} \to A$. In addition to Lu's mapping above, define bialgebroid arrows $A^e \to \mathcal{S}$, $a \otimes \bar{b} \mapsto s_L(a)t_L(b)$ and the Xu anchor mapping $\mathcal{S} \to \text{End}\, A$ given by $x \mapsto \varepsilon(? s_L(x))$. P. Xu's anchor map [24] corresponds to the action of $\mathcal{S}$ on A via source and counit [15, 3.7], for which A becomes the unit module in the tensor category of $\mathcal{S}$-modules.

Proposition 4.1. *The natural arrows defined above form a commutative triangle of bialgebroid homomorphisms.*

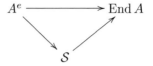

Proof. This follows readily from bialgebroid identities such as $\varepsilon \circ s_L = \mathrm{id}_A = \varepsilon \circ t_L$. $\square$

The anchor mapping $A^e \bowtie H \to \mathrm{End}\, A$ is given by $(a, b \in A,\, h \in H)$

$$a \otimes \bar{b} \bowtie h \longmapsto \lambda_a \circ \lambda_{h \triangleright} \circ \rho_b \tag{38}$$

where $\lambda_{h \triangleright}$ denotes the endomorphism given by left action by h, $x \mapsto h \triangleright x$. (Note that this is an algebra homomorphism since for $a, b, c, d \in A,\, h, k \in H$,

$$\lambda_a \circ \lambda_{h \triangleright} \circ \rho_b \circ \lambda_c \circ \lambda_{k \triangleright} \circ \rho_d = \lambda_{a(h_{(1)} \triangleright c)} \circ \lambda_{h_{(2)} k_{(1)} \triangleright} \circ \rho_{d(S(k_{(2)}) \triangleright b)},$$

which is the Eq. (23) up to a simple re-writing.)

Xu's anchor mapping for the Connes-Moscovici bialgebroid $A \odot H \odot A$ is the mapping $A \odot H \odot A \to \mathrm{End}\, A$ given by sending $a \odot h \odot b$ into the endomorphism

$$\varepsilon((a \odot h \odot b)(x \odot 1_H \odot 1_A)) = \varepsilon(a(h_{(1)} \triangleright x) \odot h_{(2)} \odot b) = a(h_{(1)} \triangleright x)\varepsilon(h_{(2)})b$$
$$= \lambda_a \circ \rho_b \circ \lambda_{h \triangleright}(x).$$

Note that in $\mathrm{End}\, A$ we have

$$\lambda_a \circ \lambda_{h \triangleright} \circ \rho_b = \lambda_a \circ \rho_{h_{(2)} \triangleright b} \circ \lambda_{h_{(1)} \triangleright}, \tag{39}$$

which lifts to the isomorphism (37) $A^e \bowtie H \to A \odot H \odot A$.

We propose a generalization of pseudo-Galois extension to *pseudo-Hopf-Galois extension* as follows. Let H be a finite dimensional (or finite projective) Hopf algebra acting from the left on an H-module algebra A, and B be a subalgebra contained in the subalgebra of invariants

$$A^H = \{b \in A : \forall h \in H,\, h \triangleright b = \varepsilon(h)b\}.$$

With $R := C_A(B)$ denoting the centralizer as usual, note that H restricts to an action on R. To be a pseudo-Hopf-Galois extension, we require the algebra extension $A \,|\, B$ be D2, and we require the bialgebroid homomorphism

$$R^e \bowtie H \to \mathrm{End}\,_B A_B, \quad r \otimes \bar{s} \bowtie h \mapsto \lambda_r \circ \lambda_{h \triangleright} \circ \rho_s \tag{40}$$

to be surjective. For example, if $A \,|\, B$ is a Hopf-Galois extension (technically, right H^*-Galois), it is pseudo-Hopf-Galois since it is D2 and by [13]

$$\Psi : R \ltimes H \xrightarrow{\cong} \mathrm{End}\,_B A_B, \quad \Psi(r \# h) := \lambda_r \circ \lambda_{h \triangleright}. \tag{41}$$

In addition, if $A \,|\, B$ is H-separable, it is pseudo-Hopf-Galois since it is D2 [13] and $R^e \cong \mathrm{End}\,_B A_B$ via $r \otimes \bar{s} \mapsto \lambda_r \circ \rho_s$. These are the two examples we wish to generalize at once.

The following is a third class of example of a pseudo-Hopf-Galois extension. Let $A \mid B$ have a split injective Galois mapping $\beta : A \otimes_B A \to A \otimes H^*$ as A-B-bimodules and let its trace function $A \to B$ be (a non-surjective) Frobenius homomorphism [21, chs. 4, 8]. Then $A \mid B$ is D2 and the mapping in Eq. (41) is a split epimorphism via the commutative square below.

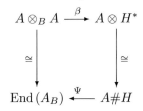

References

[1] F.W. Anderson and K.R. Fuller, *Rings and Categories of Modules*, GTM **13**, Springer, New York, 2nd edition, 1992.

[2] G. Böhm and T. Brzeziński, Pre-torsors and equivalences, *J. Algebra* **317** (2007) no. 2, 544–580.

[3] G. Böhm and K. Szlachányi, Hopf algebroids with bijective antipodes: axioms, integrals and duals, *J. Algebra* **274** (2004), 708–750.

[4] T. Brzeziński and R. Wisbauer, *Corings and Comodules*, LMS **309**, Cambridge University Press, 2003.

[5] S. Caenepeel and E. De Groot, Galois theory for weak Hopf algebras, *Rev. Roumaine Math. Pures Appl.*, to appear. RA/0406186.

[6] A. Connes and H. Moscovici, Rankin-Cohen brackets and the Hopf algebra of transverse geometry, *Moscow Math. J.* **4** (2004), 111–130.

[7] P. Etingof, D. Nikshych, and V. Ostrik, On fusion categories, *Annals of Math.* **162** (2005), 1–62.

[8] L. Kadison, *New Examples of Frobenius Extensions*, University Lecture Series **14**, AMS, Providence, 1999.

[9] L. Kadison, Hopf algebroids and H-separable extensions, *Proc. A.M.S.* **131** (2003), 2993–3002.

[10] L. Kadison, Infinite index subalgebras of depth two, *Proc. Amer. Math. Soc.* **136** (2008), 1523–1532.

[11] L. Kadison and D. Nikshych, Hopf algebra actions on strongly separable extensions of depth two, *Adv. Math.* **163** (2001), 258–286.

[12] L. Kadison and D. Nikshych, Frobenius extensions and weak Hopf algebras, *J. Algebra* **244** (2001), 312–342.

[13] L. Kadison and K. Szlachányi, Bialgebroid actions on depth two extensions and duality, *Adv. Math.* **179** (2003), 75–121.

[14] M. Khalkhali and B. Rangipour, Para-Hopf algebroids and their cyclic cohomology, *Lett. Math. Phys.* **70** (2004), 259–272.

[15] J.-H. Lu, Hopf algebroids and quantum groupoids, *Int. J. Math.* **7** (1996), 47–70.

[16] S. Mac Lane, *Categories for the Working Mathematician*, GTM **5**, Springer, New York, 2nd edition, 1998.

[17] S. Majid, *Foundations of Quantum Group Theory*, Cambridge Univ. Press, 1995.

[18] E. McMahon and A.C. Mewborn, Separable extensions of noncommutative rings, *Hokkaido Math. J.* **13** (1984), 74–88.

[19] E. McMahon and J. Shapiro, On strong and H-separability in ordinary and skew group rings, *Houston Math. J.* **15** (1989), 395–408.

[20] S. Montgomery, *Fixed Subrings of Finite Automorphism Groups of Associative Rings*, Springer Lecture Notes **818**, New York, 1980.

[21] S. Montgomery, *Hopf Algebras and Their Actions on Rings*, CBMS Regional Conf. Series in Math. Vol. 82, AMS, Providence, 1993.

[22] F. Van Oystaeyen and F. Panaite, Some bialgebroids constructed by Kadison and Connes-Moscovici are isomorphic, *Appl. Categ. Struct.* **14** (2006), 627–632.

[23] E.H. Spanier, *Algebraic Topology*, Springer, New York, 1966.

[24] P. Xu, Quantum groupoids, *Commun. Math. Physics* **216** (2001), 539–581.

[25] D.J. Winter, A Galois theory of commutative rings, *J. Algebra* **289** (2005), 380–411.

Lars Kadison
Department of Mathematics
University of Pennsylvania
David Rittenhouse Laboratory
209 South 33rd Street
Philadelphia, PA 19104-6395, USA
e-mail: `lkadison@c2i.net`

Modules and Comodules
Trends in Mathematics, 265–279
© 2008 Birkhäuser Verlag Basel/Switzerland

Cohereditary Modules in $\sigma[M]$

Derya Keskin Tütüncü, Nil Orhan Ertaş and Rachid Tribak

This paper is dedicated to Professor Robert Wisbauer on his 65th birthday

Abstract. A module $N \in \sigma[M]$ is called cohereditary in $\sigma[M]$ if every factor module of N is injective in $\sigma[M]$. This paper explores the properties and the structure of some classes of cohereditary modules. Among others, we prove that any cohereditary lifting semi-artinian module in $\sigma[M]$ is a direct sum of Artinian uniserial modules. We show that over a commutative ring a lifting module N with small radical is cohereditary in $\sigma[M]$ if and only if N is semisimple M-injective. It is also shown that if E is an indecomposable injective module over a commutative Noetherian ring R with associated prime ideal p, then E is cohereditary lifting if and only if there is only one maximal ideal m over p and the ring R_m is a discrete valuation ring.

Mathematics Subject Classification (2000). 16D10; 16D50; 16D70; 16D99.

Keywords. Cohereditary module; Lifting module; Injective module; Semisimple module.

1. Introduction

Throughout this article, R denotes an associative ring with identity and all modules will be unitary right R-modules. Mod-R denotes the category of all right R-modules. Let M be an R-module and $A \leq M$. The notation $A \ll M$ means that A is a small submodule of M. Let $End_R(M)$ denotes the endomorphism ring of M. We will denote by $\sigma[M]$ the full subcategory of Mod-R whose objects are isomorphic to a submodule of an M-generated module. A module $N \in \sigma[M]$ is said to be M-small if N is small in its injective hull in $\sigma[M]$. It is easy to see that N is M-small if and only if there exists a module $L \in \sigma[M]$ such that $N \ll L$. A module $N \in \sigma[M]$ is called *cohereditary* in $\sigma[M]$ if every factor module of N is M-injective. The module N is called *cohereditary* if N is cohereditary in Mod-R ([5],[25]). Recall that a module M is called *lifting* if for every submodule A of M there exists a direct summand B of M such that $B \leq A$ and $A/B \ll M/B$.

In Section 2 some examples of cohereditary modules are given. In particular, we show that a direct sum of cohereditary modules in $\sigma[M]$ need not be in general cohereditary in $\sigma[M]$. We also prove that if M is a generator in $\sigma[M]$, then M is cohereditary in $\sigma[M]$ if and only if M is semisimple (Proposition 2.4).

Section 3 is devoted to the study of cohereditary lifting modules in $\sigma[M]$. We prove that any cohereditary lifting semi-artinian module in $\sigma[M]$ is a direct sum of Artinian uniserial modules (Corollary 3.8). It is also shown that over a commutative ring a lifting module $N \in \sigma[M]$ with small radical is cohereditary in $\sigma[M]$ if and only if N is semisimple M-injective (Corollary 3.12). We give an example of a finitely generated cohereditary lifting right R-module M over a noncommutative ring R such that M is not semisimple.

Section 4 deals with the cohereditary lifting modules over commutative Noetherian rings. Our main result shows that if E is an indecomposable injective module over a commutative Noetherian ring R with associated prime ideal p, then E is cohereditary lifting if and only if there is only one maximal ideal m over p and the ring R_m is a discrete valuation ring (Corollary 4.13).

2. Cohereditary modules in $\sigma[M]$

Recall that a module $N \in \sigma[M]$ is called *cohereditary* in $\sigma[M]$ if every factor module of N is M-injective. The module N is called *cohereditary* if N is cohereditary in Mod-R. Obviously factor modules of cohereditary modules in $\sigma[M]$ are again cohereditary in $\sigma[M]$. Of course, this property of N depend on the surrounding category $\sigma[M]$. For example, if M is a semisimple R-module, then for every module $N \in \sigma[M]$, N is cohereditary in $\sigma[M]$ by [24, 20.3] but N need not be cohereditary in Mod-R. The ring R is called right *cohereditary* if the module R_R is cohereditary. In [19], Osofsky proved that the ring R is semisimple if and only if every cyclic right R-module is injective. Therefore right cohereditary rings are precisely the semisimple rings. On the other hand, it is clear from the next example that a cohereditary module need not be semisimple and a semisimple module need not be cohereditary.

Examples 2.1. (1) Let R be a Dedekind domain which is not a field. The quotient field of R will be denoted by K. Let $\mathbb{P}$ denote the set of all non-zero prime ideals of R. If $P \in \mathbb{P}$, let $R(P^\infty)$ denote the set of all $x \in \frac{K}{R}$ such that $P^n x = 0$ for some integer $n \geq 0$. By [24, 39.16], a ring R is hereditary if and only if every injective R-module is cohereditary. In particular, over a Dedekind domain R, it is clear that a module M is cohereditary if and only if M is injective if and only if M is radical if and only if M is a direct sum of copies of K and $R(P^\infty)$, for various $P \in \mathbb{P}$ (see [8, Lemma 2.1]). So a $\mathbb{Z}$-module M is cohereditary if and only if M is a direct sum of copies of $\mathbb{Q}$ and $\mathbb{Z}(p^\infty)$, for various primes p.

(2) It is clear that a semisimple module is cohereditary if and only if it is injective.

Proposition 2.1. *Let M be an R-module and N a cohereditary module in $\sigma[M]$. Then every submodule of an N-projective module in $\sigma[M]$ is N-projective.*

Proof. Let T be an N-projective module in $\sigma[M]$ and $A \leq T$. Let $f : A \to K$ be a homomorphism and $\pi : N \to K$ be an epimorphism, where K is any module. Since K is injective in $\sigma[M]$, there exists a homomorphism $g : T \to K$ such that $gi = f$, where $i : A \to T$ is the inclusion map. Since T is N-projective, there exists a homomorphism $h : T \to N$ such that $\pi h = g$. hi is the desired homomorphism and $\pi hi = f$. Hence A is N-projective. $\square$

Theorem 2.2. *Let M be a projective module in $\sigma[M]$. Then the following are equivalent for a module N in $\sigma[M]$:*

(1) *N is cohereditary in $\sigma[M]$.*
(2) *Every submodule of an N-projective module in $\sigma[M]$ is N-projective and N is injective in $\sigma[M]$.*

Proof. $(1) \Rightarrow (2)$. By Proposition 2.1.

$(2) \Rightarrow (1)$. Since M is N-projective, every submodule of M is N-projective. Now the result follows by [24, 39.2 (2)]. $\square$

The following lemma is [25, Proposition 6.2(1)]. We give its proof for completeness.

Lemma 2.3. *Let M be an R-module and $\{N_i\}_{i=1}^n$ be a family of modules in $\sigma[M]$. Then $N = \oplus_{i=1}^n N_i$ is cohereditary in $\sigma[M]$ if and only if all N_i are cohereditary in $\sigma[M]$.*

Proof. It is sufficient to show that the direct sum of two cohereditary modules N_1, N_2 in $\sigma[M]$ is again cohereditary. Let $N = N_1 \oplus N_2$ and $K \leq N$. We will show that N/K is M-injective. Since N_1 is cohereditary in $\sigma[M]$, $(N_1 + K)/K$ is M-injective. By [24, 16.3], $(N_1 + K)/K$ is a direct summand of N/K. There exists a submodule T/K of N/K such that $N/K = (N_1 + K)/K \oplus T/K$. Then $T/K \cong N_2/[(N_1 + K) \cap N_2]$. Since N_2 is cohereditary in $\sigma[M]$, T/K is M-injective. Then N/K is M-injective. $\square$

A module $P \in \sigma[M]$ is called *hereditary* in $\sigma[M]$ if every submodule of P is projective in $\sigma[M]$. A ring R is right *hereditary* if the module R_R is hereditary in Mod-R. It is well known that a ring R is right hereditary if and only if every injective module is cohereditary [24, 39.16]. The following Remark shows that a direct sum of cohereditary modules in $\sigma[M]$ need not be in general cohereditary in $\sigma[M]$.

Remark 2.1. Let M be a projective hereditary module in $\sigma[M]$ which is not a locally Noetherian R-module. By [24, 39.8], every injective module in $\sigma[M]$ is cohereditary in $\sigma[M]$. By [24, 27.3], there exists a family of M-injective modules $(N_i)_{i \in I}$ such that $N = \oplus_{i \in I} N_i$ is not M-injective. This proves that a direct sum of cohereditary modules in $\sigma[M]$ need not be in general cohereditary in $\sigma[M]$. In particular, if R is a hereditary ring which is not Noetherian, then there exists a family of cohereditary R-modules $(N_i)_{i \in I}$ such that $N = \oplus_{i \in I} N_i$ is not cohereditary. As an example of hereditary non-Noetherian ring we can take the ring $\begin{bmatrix} \mathbb{Q} & \mathbb{R} \\ 0 & \mathbb{Q} \end{bmatrix}$.

In fact, by [2, Example 28.12], R is a hereditary semiprimary ring that is neither left nor right Artinian. Therefore R is neither left nor right Noetherian by [2, Theorem 15.20].

Proposition 2.4. *Let M be an R-module such that M is a generator in $\sigma[M]$. Then M is cohereditary in $\sigma[M]$ if and only if M is semisimple.*

Proof. Assume M is semisimple. Then M is cohereditary in $\sigma[M]$ by [24, 20.3]. Conversely, assume M is cohereditary in $\sigma[M]$. Let X be a cyclic module in $\sigma[M]$. Since M is a generator in $\sigma[M]$, there exist a finite index set I and an epimorphism $f : M^{(I)} \to X$. By Lemma 2.3, $M^{(I)}$ is cohereditary in $\sigma[M]$. Therefore X is injective in $\sigma[M]$. Since every cyclic module in $\sigma[M]$ is injective in $\sigma[M]$, M is semisimple by [6, Corollary 7.14]. $\square$

Corollary 2.5. *Let M be a right R-module such that M is a generator in $Mod - R$. Then M is cohereditary in Mod-R if and only if M is semisimple if and only if R is semisimple.*

Example 2.2. Consider the $\mathbb{Z}$- module $\mathbb{Q}$. It is well known that $\mathbb{Q}$ is an injective module. Since $\mathbb{Z}$ is a Dedekind ring, $\mathbb{Q}$ is a cohereditary $\mathbb{Z}$-module by [24, 40.5]. In particular, $\mathbb{Q}$ is cohereditary in $\sigma[\mathbb{Q}]$. On the other hand, it is clear that $\mathbb{Q}$ is not semisimple. Thus by [24, 20.3], $\sigma[\mathbb{Q}]$ contains a module which is not cohereditary in $\sigma[\mathbb{Q}]$. Moreover, this example shows that the condition "M is a generator in $\sigma[M]$" is not superfluous in Proposition 2.4.

A module M has the *summand sum property (SSP)* (resp. *summand intersection property (SIP)*) if the sum (resp. intersection) of two direct summands of M is a direct summand of M.

Theorem 2.6. (See [1, Theorem 8]). *M has the SSP iff for every decomposition $M = A \oplus B$ and every homomorphism $f : A \longrightarrow B$, Imf is a direct summand of B.*

Recall that a module M is said to be a *(D_3)-module* if for every two direct summands U, V of M with $U + V = M$, the submodule $U \cap V$ is also a direct summand of M. Note that every quasi-projective module is (D_3).

Lemma 2.7. (See [1, Lemma 19]). *If M has the SSP and M is a (D_3)-module, then M has the SIP.*

Proposition 2.8.

(1) *Let N be a cohereditary module in $\sigma[M]$. Then N has the SSP.*
(2) *Let N be a quasi-projective cohereditary module in $\sigma[M]$. Then N has the SIP.*

Proof. (1) Let $N = A \oplus B$ and let $f : A \longrightarrow B$ be any homomorphism. Clearly, $N/\mathrm{Ker}f = (A/\mathrm{Ker}f) \oplus ((B + \mathrm{Ker}f)/\mathrm{Ker}f)$. Therefore $A/\mathrm{Ker}f \cong \mathrm{Im}f$ is injective in $\sigma[M]$. Thus $\mathrm{Im}f$ is a direct summand of B. Hence by Theorem 2.6, M has the SSP.

(2) Clear by Lemma 2.7 and (1). $\square$

3. Cohereditary lifting modules in $\sigma[M]$

Let M and N be right R-modules. In [22], Talebi and Vanaja define

$$\bar{Z}_M(N) = \cap\{Ker g \mid g : N \to L, L \in \mathcal{S}\}$$

where $\mathcal{S}$ denotes the class of all M-small modules. They call N an *M-cosingular (non-M-cosingular) module* if $\bar{Z}_M(N) = 0$ ($\bar{Z}_M(N) = N$). It is easy to see that a module $N \in \sigma[M]$ is non-M-cosingular if and only if every nonzero factor module of N is not M-small. Note that if $M = R$, then we say that "M is non-cosingular" instead of "M is non-R-cosingular".

Lemma 3.1. *Let $N \in \sigma[M]$. If N is cohereditary in $\sigma[M]$, then N is non-M-cosingular.*

Proof. Straightforward. $\square$

A module M is called *discrete* if M is a lifting module such that every submodule N of M with M/N isomorphic to a direct summand of M is itself a direct summand of M. Note that every quasi-projective lifting module is discrete. A nonzero module M is called *hollow* if every proper submodule of M is small in M. The module M is called *local* if it is hollow and cyclic.

If every injective module in $\sigma[M]$ is lifting, then M is called a *Harada module*. Note that if M is a non-M-cosingular Harada module in $\sigma[M]$, then M is cohereditary in $\sigma[M]$ by [20, Theorem 2.2].

The following example shows that a non-M-cosingular module need not be cohereditary in $\sigma[M]$.

Example 3.1. Let K be a field and let $R = \prod_{n \geq 1} K_n$ with $K_n = K$ for all $n \geq 1$. Then the ring R is a von Neumann regular ring which is not semisimple (see [13, p. 264]). Hence the R-module R is not cohereditary in $Mod - R$ by Corollary 2.5. On the other hand, by [22, Corollary 2.6], the R-module R is non-cosingular.

Proposition 3.2. *Let M be an R-module. The following are equivalent for a module N in $\sigma[M]$:*

(1) *N is cohereditary in $\sigma[M]$.*
(2) *N is non-M-cosingular and every non-M-cosingular finitely N-generated module in $\sigma[M]$ is injective in $\sigma[M]$.*

Proof. (1) $\Rightarrow$ (2). N is non-M-cosingular by Lemma 3.1. Let T be a non-M-cosingular finitely N-generated module in $\sigma[M]$. Then there exist a finite index set I and an epimorphism $f : N^{(I)} \to T$. By Lemma 2.3, T is injective in $\sigma[M]$.

(2) $\Rightarrow$ (1). Let $K \leq N$. By [22, Proposition 2.4], N/K is non-M-cosingular. Since N/K is a finitely N-generated non-M-cosingular module, N/K is injective in $\sigma[M]$. Therefore N is cohereditary in $\sigma[M]$. $\square$

Lemma 3.3. *Let $N \in \sigma[M]$. If N is non-M-cosingular discrete, then $End_R(N)$ is von Neumann regular.*

Proof. By [17, Theorem 5.4], $\mathrm{End}_R(N)/J(\mathrm{End}_R(N))$ is von Neumann regular and $J(\mathrm{End}_R(N)) = \{f : N \to N \mid \mathrm{Im}f \ll N\}$. Let $f \in J(\mathrm{End}_R(N))$. Then $\mathrm{Im}f$ is M-small. On the other hand, $N/\mathrm{Ker}f \cong \mathrm{Im}f$ implies that $\mathrm{Im}f = 0$ since N is non-M-cosingular. Therefore $J(\mathrm{End}_R(N)) = 0$ and hence $\mathrm{End}_R(N)$ is von Neumann regular. $\qquad\square$

Corollary 3.4. *Let N be a cohereditary discrete module in $\sigma[M]$. Then $\mathrm{End}_R(N)$ is von Neumann regular.*

Proof. By Lemma 3.1 and Lemma 3.3. $\qquad\square$

Proposition 3.5.

(1) *Let N be a self-injective lifting R-module. Then N has a decomposition $N = \oplus_{i \in I} N_i$ such that each N_i is hollow with local endomorphism ring.*

(2) *Let H be a hollow non-M-cosingular module in $\sigma[M]$. Then H is discrete if and only if $\mathrm{End}_R(H)$ is a division ring.*

Proof. (1) By [5, 22.20], [24, 19.9] and [17, Corollary 4.9].

(2) Let $0 \neq f : H \to H$ be any homomorphism. Assume $f(H) \neq H$. Then $f(H) \ll H$. Therefore $f = 0$, a contradiction. Therefore f is epic. Now $H/\mathrm{Ker}f \cong H$ implies that $\mathrm{Ker}f$ is a direct summand of H since H is discrete. Therefore $\mathrm{Ker}f = 0$. Thus $\mathrm{End}_R(H)$ is a division ring. The converse is a consequence of [17, Lemma 5.1]. $\quad\square$

We say that a module M has the *strong summand sum property*, denoted by *SSSP*, if the sum of any family of direct summands of M is a direct summand of M.

Proposition 3.6. *Let N be a cohereditary module in $\sigma[M]$. If N is lifting, then N has SSSP.*

Proof. By Proposition 2.8, N has the SSP. Therefore N has the SSSP by [7, Proposition 4.9]. $\qquad\square$

Theorem 3.7. *Let N be a cohereditary module in $\sigma[M]$. If N is lifting, then N is a direct sum of uniserial modules.*

Proof. By Proposition 3.5, $N = \oplus_{i \in I} N_i$, where each N_i is hollow. Now we show that each N_i is uniserial. Let $i \in I$. Assume that T is any submodule of N_i. Then N_i/T is indecomposable injective in $\sigma[M]$. Therefore every factor module of N_i is uniform by [24, 19.9]. Thus for every factor module L of N_i, $Soc(L)$ is simple or zero. By [24, 55.1], each N_i is uniserial. $\qquad\square$

Corollary 3.8. *Let N be a cohereditary module in $\sigma[M]$. If N is lifting semi-artinian, then N is a direct sum of Artinian uniserial modules.*

Proof. Let U be a uniserial semi-artinian module. By [24, 55.1], every factor module of U has a simple essential socle. Therefore U is Artinian. The result follows from Theorem 3.7. $\qquad\square$

Example 3.2. Let R be as in Example 2.1. Let M be an R-module. By [17, Proposition A.7 and Proposition A.8], M is injective lifting if and only if $M \cong \oplus_{P \in \mathbb{P}}[R(P^\infty)]^{n_P}$ with for each $P \in \mathbb{P}$, n_P is a natural number which vary with P. In particular, a $\mathbb{Z}$-module N is injective lifting if and only if $N \cong \oplus_p[\mathbb{Z}(p^\infty)]^{m_p}$ where all p are primes and for each prime p, m_p is a natural number which vary with p. It is clear that every p-primary component of N is Artinian and N is cohereditary semi-artinian. In fact, let X be any proper submodule of N. Since X is torsion, it is a direct sum of primary groups X_p (see, e.g., [12, Theorem 1]). It is clear that $X_p \leq N_p$ where N_p is the p-component of N. Hence $\frac{N}{X} \cong \oplus_p(\frac{N_p}{X_p})$. Let q be a prime such that $\frac{N_q}{X_q} \neq 0$. Since N_q is Artinian as a finite direct sum of $\mathbb{Z}(q^\infty)$, we have $Soc(\frac{N_q}{X_q}) \neq 0$. Therefore N is semi-artinian.

Lemma 3.9. *Any local module over a commutative ring is discrete.*

Proof. By [17, Lemma 5.1] and [5, 4.27]. $\square$

Lemma 3.10. *Suppose that the ring R is commutative and let M be an R-module. A local module L in $\sigma[M]$ is non-M-cosingular if and only if L is a simple M-injective module.*

Proof. Suppose that R is a commutative ring and let L be a local R-module. Let I be an ideal of R such that $L \cong \frac{R}{I}$. By Lemma 3.9, L is a discrete module. Since L is non-M-cosingular, $End_R(L)$ is a division ring by Proposition 3.5. But it is not hard to see that the ring $\frac{R}{I}$ is isomorphic to $End_R(\frac{R}{I})$. Therefore the ring $\frac{R}{I}$ is a division ring. Thus L is a simple R-module. By [16, Proposition 5.1.4], L is M-injective. The converse is clear. $\square$

Theorem 3.11. *Suppose that the ring R is commutative and let M be an R-module. The following are equivalent for a module $N = \oplus_{i \in I} N_i$ in $\sigma[M]$ which is a direct sum of local submodules N_i:*

(1) *N is non-M-cosingular.*
(2) *For every $i \in I$, N_i is a simple M-injective module.*

Proof. (1) $\Rightarrow$ (2). By Lemma 3.10, each N_i is a simple M-injective module.

(2) $\Rightarrow$ (1). Since N is semisimple, N is quasi-injective lifting. By Lemma 3.10, every N_i is non-M-cosingular. Then N is non-M-cosingular by [22, Proposition 2.4]. $\square$

Corollary 3.12. *Suppose that the ring R is commutative and let M be an R-module. The following are equivalent for a lifting R-module N in $\sigma[M]$ with $Rad(N) \ll N$:*

(1) *N is non-M-cosingular M-injective.*
(2) *N is cohereditary in $\sigma[M]$.*
(3) *N is semisimple M-injective.*

Proof. If N satisfies any of these three conditions, then N is a direct sum of hollow submodules by Proposition 3.5. But $\mathrm{Rad}(N) \ll N$. Then M is a direct sum of local submodules. The equivalence of the three statements follows from Theorem 3.11. □

Corollary 3.13. *Suppose that the ring R is commutative and let M be an R-module. The following are equivalent for a finitely generated lifting module N in $\sigma[M]$:*

(1) N *is non-M-cosingular.*
(2) N *is cohereditary in $\sigma[M]$.*
(3) N *is semisimple M-injective.*

Proof. It is a consequence of Theorem 3.11 and the fact that every finitely generated lifting module is a direct sum of local submodules. □

The following example shows that, when the ring R is not commutative, there may exist cohereditary and lifting cyclic modules which are not semisimple. Thus the commutativity of the ring R is necessary in Corollary 3.12.

Example 3.3. Let F be any field. Let R denote the ring of all upper triangular 2×2 matrices with entries in F. We know that R is a left and right hereditary Artinian ring (see [6, Example 13.6]). Consider the right R-module $M = \left[\begin{smallmatrix} F & F \\ 0 & 0 \end{smallmatrix} \right]$. It is easy to see that M is cyclic Artinian. Clearly M is a direct summand of $E(R) = \left[\begin{smallmatrix} F & F \\ F & F \end{smallmatrix} \right]$. Therefore M is an injective right R-module. Since R is right hereditary, M is cohereditary by [24, 39.16]. Since $R = \left[\begin{smallmatrix} F & F \\ 0 & 0 \end{smallmatrix} \right] \oplus \left[\begin{smallmatrix} 0 & 0 \\ 0 & F \end{smallmatrix} \right]$, M is projective. Since R is a right perfect ring, M is a lifting right R-module by [17, Theorem 4.41]. It is not hard to see that $\mathrm{Soc}(M) = \left[\begin{smallmatrix} 0 & F \\ 0 & 0 \end{smallmatrix} \right]$. Therefore M is not semisimple.

Let M and N be two modules. If for every module F, any epimorphism $f : M \to F$ and any homomorphism $h : N \to F$ either there exists a homomorphism $\varphi : N \to M$ with $f\varphi = h$ or there exist a nonzero direct summand M_1 of M and $\varphi : M_1 \to N$ with $h\varphi = f \mid_{M_1}$, then N is called *almost M-projective* (see [3]).

Lemma 3.14. *Let U be a simple module and N an indecomposable module. If U is almost N-projective, then U is almost N/X-projective for every submodule X of N.*

Proof. Let U be almost N-projective and $X \le N$. Let $f : U \to (N/X)/(Y/X)$ be any nonzero homomorphism and $\pi : N/X \to (N/X)/(Y/X)$ be the natural epimorphism, where $Y/X \le N/X$. Consider the isomorphism $\alpha : (N/X)/(Y/X) \to N/Y$ and the natural epimorphism $\nu : N \to N/X$. Now, we have the homomorphism $\alpha f : U \to N/Y$ and the epimorphism $\alpha\pi\nu : N \to N/Y$. Assume that there exists a homomorphism $g : U \to N$ such that $\alpha\pi\nu g = \alpha f$. Therefore νg lifts f. Now, assume that there exists a nonzero direct summand N_1 of N and a homomorphism $h : N_1 \to U$ such that $\alpha f h = \alpha\pi\nu \mid_{N_1}$. Since N is indecomposable, $N_1 = N$ and hence $h : N \to U$ and $\alpha f h = \alpha\pi\nu$. Since U is simple, h is epic and f is monic. Therefore $\mathrm{Ker} h = Y$. Define the homomorphism $\bar{h} : N/X \to U$ with $\bar{h}(n + X) = h(n)$ where $n \in N$. It is easy to check that $f\bar{h} = \pi$. Thus U is almost N/X-projective. □

Theorem 3.15. *Let N be a cohereditary lifting module in $\sigma[M]$ with the decomposition $N = \oplus_{i \in I} N_i$, where each N_i is hollow. Assume that every simple subfactor S of N_i is almost N_i-projective $(i \in I)$. Then N is a direct sum of Artinian hollow modules with local endomorphism rings.*

Proof. We want to show that every N_i is Artinian. As we saw in the proof of Theorem 3.7, every factor module of each N_i $(i \in I)$ is a uniform module with zero or simple socle. Let $i \in I$. Let N_i/T be a nonzero factor of N_i. We will prove that N_i/T has nonzero socle. Suppose that N_i/T is not simple. Then there exists a nonzero element $x + T \in N_i/T$ with $\bar{x}R = (x + T)R \neq N_i/T$. Let P/T be a maximal submodule of $\bar{x}R$. Now, $U = (xR + T)/P$ is simple. Therefore by hypothesis, Lemma 3.14 and [3, Theorem 1], $U \oplus N_i/T$ is lifting. Consider the inclusion map $f : U \to N_i/P$ and the epimorphism $\alpha : N_i/T \to N_i/P$. By [14, Lemma 1], there exists a homomorphism $g : U \to N_i/T$ with $\alpha g = f$. Therefore $0 \neq g(U)$ is the only simple submodule in N_i/T. Hence every N_i is Artinian. □

4. Cohereditary lifting modules over commutative Noetherian rings

Throughout this section, R will be a commutative and Noetherian ring. The full ring of quotients of R will be denoted by $Q(R)$. If p is a prime ideal of R, we will denote by R_p the localization of R at p. If M is an R-module, we will mean by M *is cohereditary* that M is cohereditary in $Mod - R$.

Proposition 4.1. *Let M be a cohereditary lifting R-module. Then $M = \oplus_{i \in I} H_i$ is a direct sum of uniserial modules H_i in which each $H_i (i \in I)$ is either radical or simple and isomorphic to $E(\frac{R}{p_i})$, for some prime ideal p_i of R.*

Proof. By [21, Corollary p. 53], Theorem 3.7 and Lemma 3.10. □

Lemma 4.2. (See [27, Folgerung 2.7]) *Let I be an ideal of R. Then the following are equivalent:*

(1) $\frac{R}{I}$ *is non-cosingular.*

(2) $\frac{R}{I}$ *is semisimple and I is a direct summand of R.*

Remarks 4.1. (1) If R is a local ring which is not a field or R is a ring with $Soc(R) = 0$ then, by Lemma 4.2, for every proper ideal I of R, the R-module $\frac{R}{I}$ is non-cosingular and hence is not cohereditary. Therefore every cohereditary lifting R-module is a direct sum of radical uniserial modules such that each of them is isomorphic to $E(\frac{R}{p_i})$ for some prime ideal p_i of R.

(2) If R is a local ring which is not a field, then every cohereditary lifting R-module is a finite direct sum of radical uniserial modules (see [11, Proposition 2.2]).

(3) If R is a local Artinian ring with maximal ideal m, then $E(\frac{R}{m})$ is cohereditary hollow if and only if R is a field. In this case, we have $E(\frac{R}{m}) \cong R$ (see [9, Remark 2] and Corollary 2.5).

Now our purpose is to investigate the structure of hollow cohereditary R-modules.

Lemma 4.3. (See [15, Theorem 3.72]) *Let m be a maximal ideal in a commutative ring R (not necessarily Noetherian). Then $\frac{R}{m}$ is an injective R-module if and only if R_m is a field.*

Proposition 4.4. *Suppose that R is a commutative ring (not necessarily Noetherian). Let p be a prime ideal of R. Then the following are equivalent:*

(1) $E(\frac{R}{p})$ *is cohereditary local.*

(2) p *is maximal and R_p is a field.*

In this case, we have $E(\frac{R}{p}) \cong \frac{R}{p}$.

Proof. (1) $\Rightarrow$ (2). By Lemma 3.10, $E(\frac{R}{p})$ is simple. Then $E(\frac{R}{p}) \cong \frac{R}{p}$ and p is a maximal ideal of R. Therefore R_p is a field by Lemma 4.3.

(2) $\Rightarrow$ (1). By Lemma 4.3, $E(\frac{R}{p}) \cong \frac{R}{p}$ is simple injective. This completes the proof. $\qquad\square$

Corollary 4.5. *Suppose that R is a commutative local ring (not necessarily Noetherian) with maximal ideal m and let p be a prime ideal of R. Then the following are equivalent:*

(1) $E(\frac{R}{p})$ *is cohereditary local.*

(2) R *is a field.*

Proof. (1) $\Rightarrow$ (2). By Proposition 4.4, $p = m$ and R_m is a field. But $R_m \cong R$. Then R is a field.

(2) $\Rightarrow$ (1). Clear. $\qquad\square$

Corollary 4.6. *Suppose that R is a commutative ring (not necessarily Noetherian). Then the following statements are equivalent:*

(1) R *is von Neumann regular.*

(2) *For every maximal ideal m of R, $E(\frac{R}{m})$ is cohereditary local.*

Proof. (1) $\Rightarrow$ (2). By [15, Theorem 3.71] and Proposition 4.4.

(2) $\Rightarrow$ (1). From Proposition 4.4 we conclude that for every maximal ideal m of R, R_m is a field. Therefore R is von Neumann regular by [15, Theorem 3.71]. $\quad\square$

As in [23], we say that an R-module M is *almost finitely generated* (a.f.g.) if M is not finitely generated but every proper R-submodule of M is finitely generated.

A ring R is *almost DVR* if R is a local Noetherian domain of Krull dimension 1 and the integral closure R' of R in $Q(R)$ is a discrete valuation ring which is finitely generated as R-module [23, page 194].

Proposition 4.7. *Let E be a cohereditary hollow radical R-module. Then E is almost finitely generated.*

Proof. Let F be any proper submodule of E. Then there exists an element x of E such that $x \notin F$. But E is uniserial (see Proposition 4.1). Thus F is contained in Rx. Since Rx is a Noetherian R-module, F is a finitely generated R-module. □

Definitions 4.8. An element of R that is not a zero divisor in R will be called a *regular element* of R. We shall say that an ideal of R is *regular* if it contains a regular element of R.

We will say that a ring R is a 1-*dimensional Cohen-Macaulay ring* if it is a commutative Noetherian ring of Krull dimension 1 such that every maximal ideal of R contains a regular element. A Noetherian domain of Krull dimension 1 is a 1-dimensional Cohen-Macaulay ring.

Let R be a local, 1-dimensional Cohen-Macaulay ring. Let I be a regular ideal of R. Then I is said to be a *canonical ideal* for R if $\frac{Q(R)}{I}$ is an injective R-module.

Theorem 4.9. *Suppose that R is local with maximal ideal m. Then the following are equivalent:*

(1) $E = E(\frac{R}{m})$ *is cohereditary hollow radical.*
(2) R *is a discrete valuation ring.*

If these conditions hold, then $E(\frac{R}{m}) \cong \frac{Q(R)}{R}$.

Proof. $(1) \Rightarrow (2)$. Following [9, Proposition 4] and Proposition 4.7, R is an almost DVR. By [9, Proposition 3], there exists a nonzero ideal I of R such that $E \cong \frac{Q(R)}{I}$. So $\frac{Q(R)}{R}$ is cohereditary hollow. Let X be any proper nonzero R-submodule of $Q(R)$ and let x be a nonzero element in X. Then we have the exact sequence $\frac{Q(R)}{Rx} \to \frac{Q(R)}{X} \to 0$. Since $\frac{Q(R)}{Rx} \cong \frac{Q(R)}{R}$, $\frac{Q(R)}{X}$ is indecomposable injective Artinian (see [23, Proposition 1.4]). Therefore $\frac{Q(R)}{X} \cong E$. Thus for every nonzero ideal I of R, we have $\frac{Q(R)}{I} \cong E$, and hence I is a canonical ideal. By [18, Theorem 15.8], for every nonzero ideal I of R, we have $I \cong R$. So R is a principal ideal ring. The result follows.

$(2) \Rightarrow (1)$. If R is a discrete valuation ring, it is well known that $E(\frac{R}{m}) \cong \frac{Q(R)}{R}$ and so $E(\frac{R}{m})$ is cohereditary hollow. □

Let Ω be the set of all maximal ideals of R. As in [28, page 53], given an $m \in \Omega$ and an R-module M, we will denote the m-local component of M by $K_m(M) = \{x \in M \mid x = 0$ or the only maximal ideal over $\operatorname{Ann}_R(x)$ is $m\}$.

We call M m-local if $K_m(M) = M$, or equivalently if m is the only maximal ideal over each $p \in \operatorname{Ass}(M)$. In this case M is an R_m-module with the following operation: $(\frac{r}{s})x := rx'$ with $x = sx'$ ($r \in R$, $s \in R\backslash m$). The submodules of M over R and over R_m are identical.

For $K(M) = \{x \in M \mid \frac{R}{\operatorname{Ann}_R(x)}$ is semiperfect$\}$ we always have the decomposition $K(M) = \oplus_{m \in \Omega} K_m(M)$ by [28, Satz 2.3].

Lemma 4.10. ([28, Lemma 1.5(b)]) *The following are equivalent for an R-module M:*

(1) $M = K(M)$.

(2) $\frac{R}{p}$ *is a local ring for all* $p \in \mathrm{Ass}(M)$.

Lemma 4.11. *Let p be a prime ideal of R such that* $\frac{R}{p}$ *is a local ring and let m be the only maximal ideal over p. Then:*

(1) $E(\frac{R}{p})$ *has the structure of an* R_m-*module.*

(2) *The submodules of* $E(\frac{R}{p})$ *over R and over* R_m *are identical.*

Moreover, as R_m-*module,* $E(\frac{R}{p})$ *is isomorphic to an injective envelope of* $\frac{R_m}{S^{-1}p}$ *where* $S = R \backslash m$.

Proof. By Lemma 4.10, $E(\frac{R}{p})$ is m-local. The rest of the proof runs as the proof of [10, Proposition 5.9]. $\square$

Remarks 4.2. By Theorem 4.9, Lemma 4.11 and [21, Proposition 5.5], we can easily get the following results:

(1) Let p be a prime ideal of R such that $\frac{R}{p}$ is a local ring and let m be the only maximal ideal over p. Then $E(\frac{R}{p})$ has the structure of an R_m-module such that the submodules of $E(\frac{R}{p})$ over R and over R_m are identical, and the following are equivalent:

(i) $E(\frac{R}{p})$ is cohereditary hollow as an R-module.

(ii) $E(\frac{R}{p})$ is cohereditary hollow as an R_m-module.

(2) Let m be a maximal ideal of R. The following are equivalent:

(i) $E(\frac{R}{m})$ is cohereditary hollow radical.

(ii) The ring R_m is a discrete valuation ring.

A principal ideal ring is called *special* if it has only one prime ideal $p \neq R$ and p is nilpotent [26, page 245].

Theorem 4.12. *Suppose that R is local with maximal ideal m and let p be a prime ideal of R different from m. The following are equivalent:*

(1) $E(\frac{R}{p})$ *is cohereditary hollow radical.*

(2) *R is a discrete valuation ring.*

In this case we have $p = 0$ *and* $E(R) \cong Q(R)$.

Proof. (1) $\Rightarrow$ (2). Since $E(\frac{R}{p})$ is hollow, $E(\frac{R}{p}) \cong Q(\frac{R}{I})$ with $I = \mathrm{Ann}_R(E(\frac{R}{p}))$, by [9, Proposition 5]. By Proposition 4.1, $Q(\frac{R}{I})$ is uniserial. Thus $\frac{R}{I}$ is a uniserial R-module. Therefore $\frac{R}{I}$ is a uniserial ring. Since $\frac{R}{I}$ is Noetherian, $\frac{R}{I}$ is a principal ideal ring. By [26, Chapter IV, Section 15, Theorem 33], $\frac{R}{I}$ is a discrete valuation ring or a special principal ideal ring. But $E(\frac{R}{p})$ is radical. Then $\frac{R}{I}$ is a discrete valuation ring (see [2, Corollary 15.21]). Since $\frac{p}{I}$ is the only minimal prime ideal of $\frac{R}{I}$, by [9, Proposition 5], $p = I$ and $E(\frac{R}{p}) \cong Q(\frac{R}{p})$. By [23, Proposition 1.4],

$Q(\frac{R}{p})/\frac{R}{p}$ is Artinian, as an $(\frac{R}{p})$-module. So $Q(\frac{R}{p})/\frac{R}{p}$ is Artinian, as an R-module. But $Q(\frac{R}{p})$ is a cohereditary R-module. Then $Q(\frac{R}{p})/\frac{R}{p}$ is an Artinian injective R-module. By [21, Corollary p. 53], $Q(\frac{R}{p})/\frac{R}{p} \cong E(\frac{R}{m})$, as R-modules. It follows that $E(\frac{R}{m})$ is a cohereditary hollow radical R-module. By Theorem 4.9, R is a discrete valuation ring. It is clear that $p = 0$.

$(2) \Rightarrow (1)$. This is obvious. $\qquad\square$

Corollary 4.13. *Let p be a prime ideal of R. The following are equivalent:*

(1) *$E(\frac{R}{p})$ is cohereditary hollow radical.*

(2) *There is only one maximal ideal m over p and the ring R_m is a discrete valuation ring.*

Proof. If p is maximal, then $(1) \Leftrightarrow (2)$ by Remarks 4.2(2). Suppose that p is not maximal.

$(1) \Rightarrow (2)$. By [28, Satz 2.3 and Satz 2.5], $E(\frac{R}{p}) = K(E(\frac{R}{p}))$. By Lemma 4.10, $\frac{R}{p}$ is a local ring. Therefore there is only one maximal ideal m over p. The result follows from Lemma 4.11, Remarks 4.2(1) and Theorem 4.12.

$(2) \Rightarrow (1)$. By Lemma 4.11, Remarks 4.2(1) and Theorem 4.12. $\qquad\square$

Proposition 4.14. *Let R be a commutative Noetherian domain which is not a field. Then the following statements are equivalent:*

(1) *R is a Dedekind domain.*

(2) *For every maximal ideal m of R, $E(\frac{R}{m})$ is cohereditary hollow radical.*

Proof. By Remarks 4.2(2) and [4, Théorème 1 p. 217]. $\qquad\square$

References

[1] M. Alkan and A. Harmanci, *On summand sum and summand intersection property of modules*, Turkish J. Math. **26** (2002), 131–147.

[2] F.W. Anderson and K.R. Fuller, *Rings and Categories of Modules*, Springer-Verlag, Berlin, Heidelberg, New York, 1974.

[3] Y. Baba and M. Harada, *On almost M-projectives and almost M-injectives*, Tsukuba J. Math. **14 (1)** (1990), 53–69.

[4] N. Bourbaki, *Algèbre commutative*, Chapitres 5 à 7. Masson, Paris, 1985.

[5] J. Clark, C. Lomp, N. Vanaja and R. Wisbauer, *Lifting Modules*, Frontiers in Mathematics, Birkhäuser, 2006.

[6] N.V. Dung, D.V. Huynh, P.F. Smith and R. Wisbauer, *Extending Modules*, Pitman Research Notes in Mathematics series, Longman, Harlow, 1994.

[7] L. Ganesan and N. Vanaja, *Modules for which every submodule has a unique coclosure*, Comm. Algebra **30(5)** (2002), 2355–2377.

[8] J. Hausen, *Supplemented modules over Dedekind domains*, Pacific J. Math. **100 (2)** (1982), 387–402.

[9] A. Idelhadj and R. Tribak, *On injective ⊕-supplemented modules*, Proceedings of the first Moroccan-Andalusian meeting on algebras and their applications, Tétouan, Morocco, September 2001. Morocco: Université Abdelmalek Essaadi, Faculté des Sciences de Tétouan, Dépt. de Mathématiques et Informatique, UFR-Algèbre et Géométrie Différentielle. (2003), 166–180.

[10] A. Idelhadj and R. Tribak, *On some properties of ⊕-supplemented modules*, Int. J. Math. Math. Sci. **69** (2003), 4373–4387.

[11] A. Idelhadj and R. Tribak, *Modules for which every submodule has a supplement that is a direct summand*, Arab. J. Sci. Eng. Sect. C Theme Issues **25 (2)**(2000), 179–189.

[12] Irving Kaplansky, *Infinite Abelian Groups*, University of Michigan, 1969.

[13] F. Kasch, *Modules and Rings*, New York: Academic Press, 1982.

[14] D. Keskin, *Finite direct sum of (D₁) modules*, Turkish J. Math. **22** (1998), 85–91.

[15] T.Y. Lam, Lectures on Modules and Rings, vol. 189 of Graduate Texts in Mathematics, New York: Springer-Verlag, 1998.

[16] C. Lomp, *On Dual Goldie Dimension*, Diplomarbeit (M. Sc. Thesis), University of Düsseldorf, Germany, 1996.

[17] S.H. Mohamed and B.J. Müller, *Continuous and Discrete Modules*, London Math. Soc. Lecture Notes Series, **147**, Cambridge, 1990.

[18] E. Matlis, *1-Dimensional Cohen-Macaulay Rings*, Lecture Notes in Mathematics, **327**, Springer-Verlag, 1973.

[19] B.L. Osofsky, *Rings all of whose finitely generated modules are injective*, Pacific J. Math. **14** (1964), 645–650.

[20] K. Oshiro and R. Wisbauer, *Modules with every subgenerated module lifting*, Osaka J. Math. **32** (1995), 513–519.

[21] D.W. Sharpe and P. Vamos, *Injective Modules*, Lecture in Pure Mathematics, University of Sheffield, 1972.

[22] Y. Talebi and N. Vanaja, *The torsion theory cogenerated by M-small modules*, Comm. Algebra **30 (3)** (2002), 1449–1460.

[23] William D. Weakley, *Modules whose proper submodules are finitely generated*, J. Algebra **84** (1983), 189–219 .

[24] R. Wisbauer, *Foundations of Module and Ring Theory*, Gordon and Breach, Philadelphia, 1991.

[25] R. Wisbauer, *Tilting in Module Categories*, Abelian groups, module theory, and topology (Padua, 1997), Lect. Notes Pure Appl. Math., **201**, Marcel Dekker, New York, (1998), 421–444.

[26] O. Zarisky and P. Samuel. *Commutative Algebra*, **1**. Springer-Verlag, New York, Heidelberg, Berlin, 1979.

[27] H. Zöschinger, *Schwach-Injektive Moduln*, Period. Math. Hungar. **52 (2)** (2006), 105–128.

[28] H. Zöschinger, *Gelfandringe und Koabgeschlossene Untermoduln*, Bayer. Akad. Wiss. Math.-Natur. Kl., Sitzungsber., **3** (1982), 43–70.

Derya Keskin Tütüncü
Hacettepe University
Department of Mathematics
06800 Beytepe Ankara, Turkey
e-mail: keskin@hacettepe.edu.tr

Nil Orhan Ertaş
Süleyman Demirel University
Department of Mathematics
32260 Çünür Isparta, Turkey
e-mail: orhannil@yahoo.com

Rachid Tribak
Département de mathématiques
Faculté des sciences de Tétouan
B.P 21.21
Tétouan, Morocco
e-mail: tribak12@yahoo.com

Modules and Comodules
Trends in Mathematics, 281–293
© 2008 Birkhäuser Verlag Basel/Switzerland

When Maximal Linearly Independent Subsets of a Free Module Have the Same Cardinality?

Farid Kourki

Dedicated to Robert Wisbauer

Abstract. We call a ring R right Lazarus if any two maximal linearly independent subsets of a free right R-module have the same cardinality. We study these rings via weakly right semi-Steinitz rings. As an application, several classes of right Lazarus rings are given.

1. Introduction

Let R be a ring. Consider the two following properties on R:

(1) Any two maximal linearly independent subsets of a free right R-module have the same cardinality.

(2) Any finite linearly independent subset of a free right R-module F can be extended to a basis of F.

Commutative rings with property (1) were introduced and studied by Lazarus [15] and a characterization is given by Bouanane and Kourki [3]. Rings with property (2) are actually called *weakly right semi-Steinitz*. These rings were introduced by Nashier and Nichols [17] and characterized as rings R that are right Hermite (i.e., every stably free right R-module is free) and left self-associated (i.e., every finitely generated proper left ideal of R has nonzero right annihilator). Although properties (1) and (2) are obvious generalizations of properties satisfied by vector spaces, the literature on this subject is quite meager. In this work we continue our contribution to this subject. First we study weakly right semi-Steinitz endomorphism rings. When doing so, a condition related to continuous modules, called C2-*condition*, arises. A right module M is called a C2-*module* if N is a summand of M whenever N is a submodule of M isomorphic to a summand of M, and it is called a *strongly* C2-*module* if every finite direct sum of copies of M is a C2-module. A ring R is called a right C2-*ring* (*strongly right* C2-*ring*) [18] if R_R is a C2-module (strongly C2-module). We see that, when M is a generator, $S = \text{End}\,(M)$ is left self-associated if and only if M is a strongly C2-module. Since there exist commutative C2-rings that are not self-associated, the condition being "right C2-ring" is not a Morita invariant and this answers in the negative a question of Nicholson and Yousif [18]. Right continuous rings are shown to be left self-associated, and we characterize when they are Hermite. As an application we see that every right and left continuous ring is weakly right and left semi-Steinitz. We also show that if R is a ring having finite right Goldie dimension or if R satisfies the ACC on right annihilators, then R is weakly right semi-Steinitz if and only if it is left self-associated. Thus, every left self-associated right noetherian ring is weakly right semi-Steinitz. Let us call a ring R right *Lazarus* if it satisfies property (1), and let us call it right *strong stably finite* if for every finitely generated free right R-module F having a basis of n elements, any linearly independent subset of F having n elements is a basis of F. We show that these two conditions are not left-right symmetric. We also prove that, when R is right strong stably finite, R is weakly right semi-Steinitz if and only if R is right Lazarus. In view of this last result and the study made on weakly right semi-Steinitz rings, large classes of right Lazarus rings are deduced: directly finite right continuous rings, right or left pseudo-Frobenius rings, unit regular rings, left self-associated rings having finite right Goldie dimension and left self-associated rings satisfying the ACC on right annihilators.

Throughout this paper R is an associative ring with identity and modules are right modules, on which homomorphisms act on the left. We write $\text{Rad}R$ for the Jacobson radical of R and $M_n(R)$ for the ring of $n \times n$ matrices over R. The right (left) annihilator of a subset X of R is denoted by $r(X)$ $(l(X))$.

The ring R is called right *Hermite* if $P \oplus R_R^m \cong R_R^n$ ($m \in \mathbb{N}$ and $n \in \mathbb{N}$) implies $P \cong R_R^r$ for some $r \in \mathbb{N}$. It is called right *self-associated* if $l(I) \neq 0$ for every finitely generated proper right ideal I of R. This is equivalent to saying that any monomorphism $R_R^m \to R_R^k$ ($m \in \mathbb{N}^*$ and $k \in \mathbb{N}^*$) splits [2, Theorem 5.4]. Recall that R is said to be *weakly right semi-Steinitz* if any finite linearly independent subset of a free right R-module F can be extended to a basis of F, and it is said to be right *Lazarus* if any two maximal linearly independent subsets of a free right R-module have the same cardinality.

A module M is called *directly finite* if for every module N, the condition $N \oplus M \cong M$ implies $N = 0$. A ring R is called *directly finite* if R_R is directly finite. R is called *IBN* if the condition $R_R^m \cong R_R^n$ implies $m = n$. We say that R satisfies the *rank condition* if, for any positive integer n, any generating set of R_R^n has cardinality$\geq n$. Equivalently, if there is an epimorphism of right free modules $R^k \to R^n$, then $k \geq n$. R is said to satisfy the right *strong rank condition* if, for any positive integer n, any linearly independent subset of R_R^n has cardinality$\leq n$. Equivalently, if there is a monomorphism of right free modules $R^m \to R^n$, then $m \leq n$. Call a ring R *stably finite* if for every free right R-module F having a basis of n elements, any generating set for F of n elements is a basis of F. Equivalently, any epimorphism $R_R^n \to R_R^n$ is an isomorphism (i.e., every f.g. free right R-module is hopfian). Examples of stably finite rings include commutative rings, semilocal rings and rings having finite right (or left) Goldie dimension. If R is stably finite then R is IBN; and we have the equivalence when R is Hermite. Unlike the directly finite, IBN, rank condition and stably finite cases, the strong rank condition is not left-right symmetric.

Consider the following conditions on a module M:

C1 Every submodule of M is essential in a summand of M.

C2 If a submodule N of M is isomorphic to a summand of M, then N is a summand of M.

C3 If N and L are summands of M such that $N \cap L = 0$, then $N \oplus L$ is a summand of M.

A module M is called CS if it has C1, and M is called a C2-*module* if it has C2. If M has both C1 and C2 then M is called *continuous* while if it has C1 and C3 then it is called *quasi-continuous*. We have the following implications:

Injective $\Rightarrow$ Quasi-injective $\Rightarrow$ Continuous $\Rightarrow$ Quasi-continuous $\Rightarrow$ CS

2. Weakly right semi-Steinitz endomorphism rings

We begin by taking up the question of when an endomorphism ring is weakly right semi-Steinitz. The following theorem is the key.

Theorem 2.1. [17, Theorem 2.2] *A ring R is weakly right semi-Steinitz if and only if R is right Hermite and left self-associated.*

In a previous paper [12] we introduced the notion of Hermite modules: a nonzero module M is said to be *Hermite* if $N \oplus M^m \cong M^n$ ($m \in \mathbb{N}$ and $n \in \mathbb{N}$) implies $N \cong M^r$ for some $r \in \mathbb{N}$. Hence, R is right Hermite means that R_R is Hermite. In the same paper we showed that being right Hermite is a left-right symmetric condition, so there is no need to use the prefixes "right" or "left". Any semilocal ring is Hermite [12, Proposition 2.5]. The following proposition characterizes Hermite endomorphism rings.

Proposition 2.2. [12, Theorem 2.1] *Let M be a nonzero module and let $S = \mathrm{End}\,(M)$. Then M is Hermite if and only if S is an Hermite ring.*

Proposition 2.3. *For a module M, the following conditions are equivalent:*

(1) *For every n in $\mathbb{N}^*$, M^n is a C2-module.*

(2) *For every m and k in $\mathbb{N}^*$, any monomorphism $\alpha : M^m \to M^k$ splits.*

Proof. By definition, a module M is a C2-module if and only if every monomorphism $P \to M$, where P is a summand of M, splits.

(1) $\Rightarrow$ (2). Let t in $\mathbb{N}$ such that $t > m$ and $t > k$ and let i be the natural monomorphism $M^k \to M^t$. $i \circ \alpha : M^m \to M^t$ is a monomorphism and M^m is a summand of M^t, so $i \circ \alpha$ splits. Therefore α splits.

(2) $\Rightarrow$ (1). Let $\beta : P \to M^n$ be a monomorphism where P is a summand of M^n. There exists Q such that $P \oplus Q = M^n$. To show that β splits it is enough to

show that $\gamma = \beta \oplus 1_Q : M^n \to M^n \oplus Q$ splits. Since there exists a monomorphism $i : M^n \oplus Q \to M^n \oplus Q \oplus P = M^{2n}$, then $i \circ \gamma : M^n \to M^{2n}$ is a monomorphism, which splits by hypothesis. So γ splits. $\square$

Call a module M a *strongly* C2-*module* if it satisfies the equivalent conditions in Proposition 2.3. A ring R is called a right C2-*ring* (*strongly right* C2-*ring*) [18] if R_R is a C2-module (strongly C2-module). The equivalence (2) $\Leftrightarrow$ (4) in the following corollary was also obtained by Jianlong and Wenxi [10, Lemma 2.3].

Corollary 2.4. *Let R be a ring. The following are equivalent:*

(1) *Every finitely generated free right R-module is a C2-module.*
(2) *R is a right strongly C2-ring.*
(3) *$M_n(R)$ is a right C2-ring for every $n \in \mathbb{N}^*$.*
(4) *R is left self-associated.*

Proof. (1) $\Leftrightarrow$ (2). By definition.
(2) $\Leftrightarrow$ (3). [18, Corollary 3.10].
(2) $\Leftrightarrow$ (4). Proposition 2.3. $\square$

By the preceding corollary, if M is a module and $S = \operatorname{End}(M)$, then S is left self-associated if and only if $\operatorname{End}_R(M^n)$ is a right C2-ring for every $n \in \mathbb{N}^*$. An R-module is called a *generator* if it generates the category of right R-modules.

Proposition 2.5. *Let M be a module and let $S = \operatorname{End}(M)$.*

(1) *If S is left self-associated then M is a strongly C2-module.*
(2) *If M is a generator then: S is left self-associated $\Leftrightarrow$ M is a strongly C2-module.*

Proof. Use the above remark and [18, Proposition 3.9]. $\square$

What happens if we remove the condition "finitely generated" from (1) of Corollary 2.4? The following proposition gives an answer. Let $rFPD(R) = \{Pd_R(M) \mid M$ is a right R-module and $Pd_R(M) < \infty\}$ where $Pd_R(M)$ is the projective dimension of M.

Proposition 2.6. *Let R be a ring. The following are equivalent:*

(1) *Every free right R-module is a C2-module.*
(2) *For every free right R-module F, $S = \operatorname{End}(F)$ is left self-associated.*
(3) *R is right perfect and left self-associated.*
(4) *$rFPD(R) = 0$.*

Proof. (1) $\Rightarrow$ (2). Let F be a free right R-module. For every positive integer n, F^n is a free module and so it is a C2-module. Since every free module is a generator, $S = \operatorname{End}(F)$ is left self-associated (Proposition 2.5).

(2) $\Rightarrow$ (1). Let F be a free right R-module. Since $S = \operatorname{End}(F)$ is left self-associated, F is a strongly C2-module (Proposition 2.5) and therefore it is a C2-module.

(2) $\Leftrightarrow$ (3). [4, Theorem 2]
(3) $\Leftrightarrow$ (4). [2, Theorem 6.3] $\square$

Question [18]: Is "right C2-ring" a Morita invariant?

The answer is "no" as shown in the following example:

Let A be a commutative local UFD, but not a principal ideal domain (for instance R can be the ring of power series in two variables over a field). Let M be the direct sum of all A/pA, p ranging over the primes of A. Let $R = A \alpha M$, the trivial extension of A by M. Then R is not self-associated [11, Exercise 7, p. 63] and so R is not a strongly C2-ring (Corollary 2.4). Since every regular element of R is invertible and R is local, then R is a C2-ring [18, Corollary 3.5].

Proposition 2.7. *Consider the following conditions on a module $M \neq 0$:*

(1) *M is Hermite and a strongly C2-module.*
(2) *For all positive integers m and k, every exact sequence of the form $0 \to M^m \to M^k \to P \to 0$ splits and there exists t in $\mathbb{N}$ such that $P \cong M^t$.*

(3) *$S = \mathrm{End}\,(M)$ is weakly right semi-Steinitz.*

Then $(1) \Leftrightarrow (2) \Rightarrow (3)$. If M is a generator then these three conditions are equivalent.

Proof. $(1) \Rightarrow (2)$. The exact sequence in (2) splits since M is a strongly C2-module (Proposition 2.3). But M is Hermite, so $P \cong M^t$ for some t.

$(2) \Rightarrow (1)$. By Proposition 2.3 M is a strongly C2-module. An isomorphism of the form $M^m \oplus P \cong M^k$ leads to an exact sequence:

$$0 \to M^m \to M^k \to P \to 0$$

By hypothesis $P \cong M^t$, hence M is Hermite.

Propositions 2.2 and 2.5 give either the implication $(3) \Rightarrow (1)$ or the equivalence when M is a generator. $\qquad\square$

3. Classes of weakly right semi-Steinitz rings

A ring R is called right *weakly continuous* [18] if R is a right C2-ring and for every element a of R the right annihilator of a is essential in a summand of R_R. Clearly, every right continuous ring is weakly right continuous. Every (von Neumann) regular ring is right and left weakly continuous. Being right weakly continuous is a Morita invariant property ([18], Theorem 2.60), thus:

Proposition 3.1. *If R is right weakly continuous then R is a strongly right C2-ring and hence left self-associated. In particular every right continuous ring is left self-associated.*

Lemma 3.2. *Let R be a right continuous ring. Then R is IBN if and only if $R_R^2 \not\cong R_R$.*

Proof. $R = P \oplus D$, where P and D are right R-modules such that D is directly finite and $P^2 \cong P$ and P and D are orthogonal [16, Theorem 2.29]. Let $S = \mathrm{End}(P)$ and $T = \mathrm{End}(D)$. Then $R/\mathrm{Rad}R \cong S/\mathrm{Rad}S \times T/\mathrm{Rad}T$ and $R/\mathrm{Rad}R$, $S/\mathrm{Rad}S$ and $T/\mathrm{Rad}T$ are right continuous and regular [16, Lemma 3.3, Proposition 3.5 and Theorem 3.11]. The rings S and T also satisfy $(S_S/\mathrm{Rad}S_S)^2 \cong S_S/\mathrm{Rad}S_S$ and $T/\mathrm{Rad}T$ is directly finite. If R is IBN then obviously $R_R^2 \not\cong R_R$. Suppose now that $R_R^2 \not\cong R_R$. Then $(R_R/\mathrm{Rad}R_R)^2 \not\cong R_R/\mathrm{Rad}R_R$. If $T/\mathrm{Rad}T = 0$ then $R/\mathrm{Rad}R \cong S/\mathrm{Rad}S$ and so $(R_R/\mathrm{Rad}R_R)^2 \cong R_R/\mathrm{Rad}R_R$, a contradiction. It follows that $T/\mathrm{Rad}T$ is a nonzero right continuous directly finite ring, so it has the cancellation property [16, Corollary 3.25]. In particular, it is an IBN ring and so is $R/\mathrm{Rad}R$ [9, Theorem 2.1]. Therefore R is IBN. □

Theorem 3.3. *Let R be a right continuous ring. The following conditions are equivalent:*

(1) *R is weakly right semi-Steinitz.*
(2) *If $P_R \oplus R_R \cong R_R$ then $P = 0$ or $P_R \cong R_R$.*

Proof. Suppose that $R_R^2 \not\cong R_R$.

$(1) \Rightarrow (2)$. Since R is Hermite the condition $P_R \oplus R_R \cong R_R$ implies $P_R \cong R_R^n$ for some n, and hence $R_R^{n+1} \cong R_R$. But R is IBN (Lemma 3.2), so $n = 0$ and then $P = 0$.

$(2) \Rightarrow (1)$. By Proposition 3.1 R is left self-associated. If $P_R \oplus R_R \cong R_R$, then $P = 0$ or $P_R \cong R_R$. We cannot have $P_R \cong R_R$ because $R_R^2 \not\cong R_R$, so $P = 0$ and hence R is directly finite. It follows from [16, Corollary 3.25] that R has the cancellation property, so R is Hermite. Theorem 2.1 shows that R is weakly right semi-Steinitz.

Now suppose that $R_R^2 \cong R_R$.

$(1) \Rightarrow (2)$. If $P_R \oplus R_R \cong R_R$, then $P \cong R_R^n$ for some n, since R is Hermite. If $n = 0$ then $P = 0$. If $n \neq 0$ then $P \cong R_R$ since $R_R^n \cong R_R$.

$(2) \Rightarrow (1)$. By Proposition 3.1 R is left self-associated. If $P_R \oplus R_R \cong R_R^m$ then $P_R \oplus R_R \cong R_R$ since $R_R^m \cong R_R$. Hence $P \cong 0$ or $P_R \cong R_R$. This means that P is free, so R is Hermite. Now use Theorem 2.1. □

Corollary 3.4. *Let R be a right quasi-continuous ring such that $R_R^2 \not\cong R_R$. The following are equivalent:*

(1) *R is weakly right semi-Steinitz.*
(2) *Every right regular element of R is invertible.*
(3) *R is directly finite and right continuous.*

Proof. $(1) \Rightarrow (3)$. R is left self-associated, so it is a right C2-ring and hence right continuous. Since $R_R^2 \not\cong R_R$, by the proof of Theorem 3.3, R is directly finite.

$(3) \Rightarrow (2)$. R is left associated by Proposition 3.1, so every right regular element of R is left invertible and hence invertible since R is directly finite.

$(2) \Rightarrow (3)$. Let f be a monomorphism from R_R to R_R. There exists a right regular element a such that $f(x) = ax$ for every $x \in R$. But a is invertible, hence f is an isomorphism. This last condition plus the fact that R is right quasi-continuous imply that R is right continuous [16, Lemma 3.14]. If $xy = 1$ then y is right regular so invertible by hypothesis. Therefore R is directly finite.

$(3) \Rightarrow (1)$. Theorem 3.3.   □

Remark. If R is right and left continuous then R is directly finite [19, Lemma 5.3]. By Corollary 3.4 R is weakly right and left semi-Steinitz.

Recall that a module M is said to have *finite Goldie dimension* if there are no infinite direct sums of non-zero submodules in M. If M is noetherian or artinian then M has finite Goldie dimension.

Proposition 3.5. *If M is a generator and a strongly C2-module having finite Goldie dimension then* $\mathrm{End}(M)$ *is weakly right semi-Steinitz.*

Proof. By [18, Proposition 3.7] $\mathrm{End}(M)$ is semilocal and hence Hermite. By Proposition 2.5 $\mathrm{End}(M)$ is left self-associated and hence weakly right semi-Steinitz.   □

Corollary 3.6. *If R is a left self-associated ring having finite right Goldie dimension then R is weakly right semi-Steinitz.*

In particular, every right noetherian left self-associated ring is weakly right semi-Steinitz.

Proposition 3.7. *If R is a right self-associated ring satisfying the maximum condition on left annihilators of elements of R then R is weakly left semi-Steinitz.*

Proof. Since R satisfies the maximum condition on left annihilators of elements of R, it is not hard to see that R does not contain an infinite set of pairwise orthogonal idempotents. So R is directly finite [6, p. 85]. Let x be a non-invertible element of R. x is neither right nor left invertible, hence $xR \neq R$ and so $l(x) \neq 0$ since R is right self-associated. Therefore x is a right zerodivisor. By [5, Theorem 3] R is semilocal and hence Hermite. So R is weakly right semi-Steinitz.   □

Corollary 3.8. *If R is a right self-associated ring satisfying the ACC on left annihilators then R is weakly left semi-Steinitz.*

Remark. A ring R is said to be *its own classical quotient* ring if every non-unit of R is a zerodivisor, that is, either a right zerodivisor or a left zerodivisor. If R is a commutative ring which is its own classical ring of quotients having finite Goldie dimension or satisfying the ACC on annihilators then R is weakly semi-Steinitz [13, Corollary 2] and [3, Example (5), p. 135]. So Corollaries 3.6 and 3.8 generalize this last result to the noncommutative case. However, we cannot replace the condition "R is left self-associated" or the condition "R is right self-associated" by the condition "R is its own classical ring of quotients" as it is shown in the following example:

Let K be a field and $R = \left(\begin{smallmatrix} K & 0 \\ K & K \end{smallmatrix}\right)$, then R is right and left artinian so it is its own classical ring of quotients and R has finite right and left Goldie dimension and R satisfies the ACC on right and left annihilators. R is also left and right hereditary. Suppose that R is left (right) self-associated, then every right (left) finitely generated ideal is a summand of R and hence R is regular. But this cannot happen since $\mathrm{Rad}R = \left(\begin{smallmatrix} 0 & 0 \\ K & 0 \end{smallmatrix}\right) \neq 0$. So R is neither weakly left nor weakly right semi-Steinitz. Remark that $\dim R_R = \dim{}_R R = 2$ (the Goldie dimension). For the Goldie dimension 1 we have:

Proposition 3.9. *Let R be a ring. If $\dim R_R = 1$ and every non-unit element of R is a left zerodivisor, then R is weakly right semi-Steinitz.*

Proof. Let $a_1, a_2, \ldots, a_n$ be elements of R such that $Ra_1 + \cdots + Ra_n \neq R$. We have $Ra_i \neq R$ and so a_i is not invertible. Each a_i is then a left zerodivisor, that is, $r(a_i) \neq 0$. Since $\dim R_R = 1$, $r(a_1) \cap \cdots \cap r(a_n) \neq 0$. Hence $r(Ra_1 + \cdots + Ra_n) \neq 0$, implying that R is left self-associated. By Corollary 3.6 R is weakly right semi-Steinitz.   □

4. Rings over which every free right module has property (P)

We say that a free right R-module F satisfies *property* (P) if any two maximal linearly independent subsets of F have the same cardinality. Lazarus [15] showed that, for a commutative ring R, if R is noetherian or if R is an integral domain then every free R-module satisfies (P). Recall that a ring R is right *Lazarus* if every free right R-module satisfies property (P).

Lemma 4.1. *Let R be a ring.*

(1) *If R satisfies the right strong rank condition, then it satisfies the rank condition. If R is left self-associated then we have the equivalence.*

(2) *If every finitely generated free right R-module satisfies property* (P) *then R satisfies the right strong rank condition.*

Proof. (1) The first part of the assertion comes from the fact that every epimorphism $R_R^k \to R_R^n$ splits [14, Proposition 1.21], and the equivalence comes from the fact that, for a left self-associated ring R, every monomorphism $R_R^m \to R_R^n$ splits (Proposition 2.3).

(2) Let G be a finitely generated free submodule of a finitely generated free right R-module F. Let B' be a basis of G of m elements and let B be a basis of F of n elements. B' can be completed to a maximal linearly independent subset of F, say S. But F satisfies (P), so $cardS = n$. Hence $m \leq n$.   □

Proposition 4.2. *Let R be a ring. The following conditions are equivalent:*

(1) *R is right Lazarus.*

(2) *every finitely generated free right R-module satisfies property* (P).

Proof. (1) ⇒ (2) Clear.

(2) ⇒ (1) By Lemma 4.1, R satisfies the right strong rank condition. In [15, 2.1], Lazarus showed that if R is a commutative ring and if F is a free R-module having an infinite basis, then F satisfies property (P). The proof of this result shows that the only property of commutative rings used is the right strong rank condition. □

Being right Lazarus is not a right-left symmetric property. To show this we need the following theorem:

Theorem 4.3. *Let R be a domain and $Q = Q^r_{\max}(R)$, the maximal right ring of quotients of R. The following are equivalent:*

(1) R *is right Lazarus.*
(2) R *has the right strong rank condition.*
(3) R *is right Ore.*
(4) $\dim(R_R) = 1$ *(Goldie dimension).*
(5) $\dim(R_R) < \infty$ *(Goldie dimension).*
(6) Q *is right Lazarus.*
(7) Q *is directly finite.*
(8) Q *is IBN.*
(9) $Q^2_Q \not\cong Q_Q.$
(10) Q *is a division ring.*

In this case $Q = Q^r_{cl}(R)$, the classical right ring of quotients of R.

Proof. (3) ⇔ (4) ⇔ (5). Goldie (see [14, Theorem 10.22]).

(2) ⇔ (3). [14, Exercise 21, p. 319].

(1) ⇒ (2). Lemma 4.1.

(5) ⇒ (1). By Proposition 4.2, it is enough to show that every finitely generated free right R-module satisfies property (P). Let F be a finitely generated free right R-module and let S be a maximal linearly independent subset of F. Let G be the free right R-module spanned by S. Since R satisfies the right strong rank condition then S is finite. Let $0 \neq x \in F$. By the maximality of S there exists $0 \neq \lambda \in R$ such that $x\lambda \in G$. Let $\{e_1, \ldots, e_n\}$ be a basis of F and let $x = e_1 a_1 + \cdots + e_n a_n$ for some $a_1, \ldots, a_n$ in R. Since $x \neq 0$, there exists i such that $a_i \neq 0$. But R is a domain, so $a_i \lambda \neq 0$. Therefore $0 \neq x\lambda \in G$, this means that G is essential in F and hence $m \dim R_R = \dim G = \dim F = n \dim R_R$, where m is the cardinality of S. So $n = m$. Consequently (1) ⇔ (2) ⇔ (3) ⇔ (4) ⇔ (5).

[14, Exercise 20, p. 382] gives (7) ⇒ (3) and [14, Corollary 13.43] gives (3) ⇔ (10). The implication (10) ⇒ (7) is clear. So (3) ⇔ (10) ⇔ (7).

We have (10) ⇒ (6) ⇒ (8) ⇒ (9).

(9) ⇒ (7). Since R is right nonsingular, Q is a right selfinjective regular ring [14, Theorem 13.36]. Suppose that Q is not directly finite. Then Q has a nonzero right ideal I which is a summand such that $I^2 \cong I$ [8, Proposition 6.10]. But $I \cong Q_Q$

[14, Exercise 20, p. 382], so $Q_Q^2 \cong Q_Q$, a contradiction. Hence $(10) \Leftrightarrow (6) \Leftrightarrow (8)$ $\Leftrightarrow (9) \Leftrightarrow (7)$. $\qquad\qquad\qquad\qquad\qquad\qquad\qquad\qquad\qquad\qquad\qquad\qquad\qquad\qquad\Box$

Remarks. (1) Let R be a right Ore domain which is not left Ore. By the preceding theorem R is right Lazarus but it is not left Lazarus. For the existence of such a domain, see [8, Exercise 1, p. 101].

(2) Let R be a domain and let $Q = Q_{\max}^r(R)$. Q is a right selfinjective regular ring. Let P be a nonzero finitely generated projective right Q-module. Since Q is regular then P is a direct sum of submodules each of which is isomorphic to nonzero principal right ideals of Q. These latter ideals are isomorphic to Q [14, Exercise 20, p. 382]. Hence P is free and so R is Hermite. Q is regular, so Q is right and left self-associated and hence Q is weakly right and left semi-Steinitz.

(3) In [3, Corollary 2] is it shown that if R is any commutative ring, then the polynomial ring $R[X]$ is Lazarus. This result is no longer true in the noncommutative case. By Shock's Theorem [14, Theorem 6.65] $\dim R_R = \dim R[X]_{R[X]}$. Therefore, by Theorem 4.3, if R is a domain then R is right Lazarus if and only if $R[X]$ is right Lazarus.

Let call a ring R right *strong stably finite* if for every finitely generated free right R-module F having a basis of n elements, any linearly independent set in F of n elements is a basis of F. Equivalently, any monomorphism $R_R^n \to R_R^n$ is an isomorphism (i.e., every f.g. free right R-module is cohopfian). If R is right or left perfect then every injective endomorphism of a finitely generated right (or left) R-module is bijective [1]. So R is right and left strong stably finite. Every right strong stably finite ring is stably finite; and we have the equivalence when the ring is left self-associated. A ring R is right *Johns* if R is right noetherian and every right ideal is an annihilator. Faith and Menal [7] gave an example of a right Johns ring which is not right artinian. Let R be such a ring. Since every right ideal is an annihilator then R is right self-associated. R is right noetherian so it is stably finite and therefore left strong stably finite. Suppose that R is also right strong stably finite. R_R is cohopfian and therefore for every element a of R, $r(a) = 0$ implies a is invertible, in particular $aR = R$. By [7, Proposition 3.3] R is right artinian, a contradiction. So being right strong stably finite is not a left-right symmetric property.

Theorem 4.4. *Let R be a right strong stably finite ring. The following conditions are equivalent:*

(1) *R is weakly right semi-Steinitz.*
(2) *R is right Lazarus.*

Proof. $(1) \Rightarrow (2)$. Let F be a free right R-module and let S be a maximal linearly independent subset of F. By Proposition 4.2, we can assume that F is finitely generated. R is stably finite and hence IBN. But R is weakly right semi-Steinitz, hence R satisfies the right strong rank condition and so S is necessarily finite. S can be completed to a basis of F and therefore S is itself a basis of F since it is maximal. R is IBN so all bases of F and hence all maximal linearly independent subsets of F have the same cardinality.

$(2) \Rightarrow (1)$. Let S be a finite linearly independent subset of a finitely generated free right R-module F. By Zorn's Lemma, S can be completed to a maximal linearly independent subset S' of F. Let n be the cardinality of a basis of F. Since R is right Lazarus we have $card S' = n$. But R is right strong stably finite, so S' is a basis of F. $\qquad \square$

By [3, Corollary 2.1], it is not difficult to see that a commutative ring R is strong stably finite if and only if every regular element of R is invertible.

Corollary 4.5. [3, Proposition 3.1] *Let R be a commutative ring where every regular element is invertible. Then R is weakly semi-Steinitz if and only if R is Lazarus.*

Corollary 4.6. *Let R be a stably finite ring. The following are equivalent:*

(1) *R is weakly right semi-Steinitz.*
(2) *R is right Lazarus left self-associated.*

Proof. Use Theorem 4.4 since a left self-associated stably finite ring is right strong stably finite. $\qquad \square$

Corollary 4.7. *Let R be a right (or left) perfect ring. The following are equivalent:*

(1) *R is weakly right semi-Steinitz.*
(2) *R is right Lazarus.*
(3) *R is left self-associated.*

Proof. R is right strong stably finite, by Theorem 4.4 we have the equivalence $(1) \Leftrightarrow (2)$. R is semilocal so R is Hermite. Hence Theorem 2.1 gives the equivalence $(1) \Leftrightarrow (3)$. $\qquad \square$

Remarks. (1) Let $R = \left(\begin{smallmatrix} K & 0 \\ K & K \end{smallmatrix} \right)$. Then R is left and right artinian which is neither weakly right nor weakly left semi-Steinitz. By Corollary 4.7 R is neither left nor right Lazarus.

(2) Let R be a domain which is not right Ore. $Q = Q^r_{\max}(R)$ is weakly left and right semi-Steinitz. Since Q is not IBN (Theorem 4.3) then Q is neither left nor right Lazarus.

(3) If R is a commutative domain which is not a field. Then R is Lazarus but it is not weakly semi-Steinitz.

The study made on weakly right semi-Steinitz rings and their connection with right Lazarus rings enables us to provide examples of these latter rings. Some of our examples are generalizations to the noncommutative case of the results of Lazarus [15] and Bouanane and Kourki [3].

Examples. (1) If R is a directly finite right continuous ring, then R cancels from direct sums and hence is stably finite. Since R is weakly right semi-Steinitz (Corollary 3.4) then R is right Lazarus (Corollary 4.6). In particular any right and left continuous ring is left and right Lazarus. If R is a commutative CS ring, then it is not difficult to see that $T(R)$, the total ring of quotients of R, is continuous, and hence $T(R)$ is weakly semi-Steinitz. By [3, Corollary 3.2] R is Lazarus.

(2) If R is a right or left PF (pseudo-Frobenius) ring then R is semilocal and hence stably finite. R is also weakly left and right semi-Steinitz [12, Proposition 4.9]. By Corollary 3.4 R is left and right Lazarus.

(3) If R is a unit-regular ring then R is a stably finite weakly left and right semi-Steinitz ring [12, Proposition 4.10]. By Corollary 3.4 R is left and right Lazarus.

(4) If R is a left self-associated ring having finite right Goldie dimension, then R is stably finite and weakly right semi-Steinitz (Corollary 3.6). By Corollary 3.4 R is right Lazarus. In particular any left self-associated right noetherian ring is right Lazarus.

(5) If R is a left self-associated ring satisfying the ACC on right annihilators, then R is semilocal and so stably finite. It is also weakly right semi-Steinitz (Corollary 3.6). By Corollary 3.4 R is right Lazarus.

References

[1] E.P. Armendariz, J.W. Fisher and R.L. Snider, *On injective and surjective endomorphisms of finitely generated modules*, Comm. Algebra **6** (2001), 659–672.

[2] H. Bass, *Finitistic dimension and a homological characterization of semi-primary rings*, Trans. Amer. Math. Soc. **95** (1960), 466–485.

[3] A. Bouanane and F. Kourki, *On weakly semi-Steinitz rings*, Lecture Notes Pure Appl. Math. **185** (1997), 131–139.

[4] G.M. Brodskiĭ, *Endomorphism rings of free modules over perfect rings*, Math. USSR Sb. **17** (1) (1972), 138–147.

[5] R. Camps and W. Dicks, *On semi-local rings*, Israel J. Math. **81** (1993), 203–211.

[6] C. Faith, *Algebra II*, Springer-Verlag, Berlin-Heidelberg, 1976.

[7] C. Faith and P. Menal, *A counterexample to a conjecture of Johns*, Proc. Amer. Math. Soc. **116** (1992), 21–26.

[8] K.R. Goodearl, *Ring Theory: Nonsingular rings and modules*, Marcel Dekker, New York, 1976.

[9] A. Haghany and K. Varadarajan, *IBN and related properties for rings*, Acta Math. Hungar. **93** (3) (2002), 251–261.

[10] C. Jianlong and L. Wenxi, *On the artiness of right CF rings*, Comm. Algebra **32** (11) (2004), 4485–4494.

[11] I. Kaplansky, *Commutative Rings*, Chicago (1974).

[12] F. Kourki, *Hermite and weakly semi-Steinitz rings*, Comptes Rendus de la Première Rencontre Maroco-Andalouse sur les Algèbres et leurs Applications 26-28 septembre 2001,(2003), 200–210.

[13] F. Kourki, *On some annihilator conditions over commutative rings*, Lecture Notes Pure Appl. Math. **231** (2003), 323–334.

[14] T.Y. Lam, *Lectures on Modules and Rings*, Graduate Texts in Math. **189**, Springer-Verlag, 1999.

[15] M. Lazarus, *Les familles libres maximales d'un module ont-elles le même cardinal?* Pub. Sém. Math. Rennes **4** (1973), 1–12.

[16] S.H. Mohamed and B.J. Müller, *Continuous and Discrete Modules*, Cambridge univ. Press, Cambridge, 1990.

[17] B. Nashier and W. Nichols, *On Steinitz properties*, Arch. Math. **57** (1991), 247–253.

[18] W.K. Nicholson and M.F. Yousif, *Weakly continuous and C2-rings*, Comm. Algebra **29** (6) (2001), 2429–2446.

[19] Y. Utumi, *On continuous rings and self-injective rings*, Trans. Amer. Math. Soc. **118** (3)(1965), 158–173.

Farid Kourki
Département de Mathématiques et Informatique
Université Abdelmalek Essaâdi
Faculté des Sciences
B.P. 2121
Tétouan, Morocco

Modules and Comodules
Trends in Mathematics, 295–300
© 2008 Birkhäuser Verlag Basel/Switzerland

Embedding Group Algebras into Finite von Neumann Regular Rings

Peter A. Linnell

Abstract. Let G be a group and let K be a field of characteristic zero. We shall prove that KG can be embedded into a von Neumann unit-regular ring. In the course of the proof, we shall obtain a result relevant to the Atiyah conjecture.

Mathematics Subject Classification (2000). Primary: 16S34; Secondary: 16E50, 20C07.

Keywords. Group von Neumann algebra, ultrafilter.

The main result of this paper is

Theorem 1. *Let G be a group and let K be a field of characteristic zero. Then KG can be embedded into a finite self-injective von Neumann $*$-regular ring.*

This result has been stated in [8, p. 217]. However the proof there depends on a result of [5]. The proof presented here will not use that result, and will also give a result related to the Atiyah conjecture [9, §10]. We shall apply the techniques of ultrafilters and ultralimits, as used in [4].

Let G be a group and let $\mathcal{U}(G)$ denote the algebra of unbounded operators on $\ell^2(G)$ affiliated to the group von Neumann algebra $\mathcal{N}(G)$ of G [9, §8.1]. Then $\mathcal{U}(G)$ is a finite von Neumann regular $*$-ring that is left and right self-injective, and also unit-regular [1, §2,3]. For $\alpha, \beta \in \mathcal{U}(G)$, we have $\alpha^*\alpha + \beta^*\beta = 0$ if and only if $\alpha = \beta = 0$ [1, p. 151]. Furthermore there is a unique projection $e \in \mathcal{N}(G)$ (so $e = e^2 = e^*$) such that $\alpha\mathcal{U}(G) = e\mathcal{U}(G)$ and we have the following useful result.

Lemma 2. *Let G be a group and let $\alpha, \beta \in \mathcal{U}(G)$. Then $(\alpha\alpha^* + \beta\beta^*)\mathcal{U}(G) \supseteq \alpha\mathcal{U}(G)$.*

Proof. Write $U = \mathcal{U}(G)$ and let $e \in U$ be the unique projection such that $\alpha U = eU$. Then $(1-e)U = \{u \in U \mid \alpha^*u = 0\}$. Let $f \in U$ be the unique projection such that $(\alpha\alpha^* + \beta\beta^*)U = fU$. Then $(1-f)U = \{u \in U \mid (\alpha\alpha^* + \beta\beta^*)u = 0\}$. Therefore if $u \in (1-f)U$, we have $(\alpha\alpha^* + \beta\beta^*)u = 0$, hence

$$0 = u^*\alpha\alpha^*u + u^*\beta\beta^*u = (\alpha^*u)^*(\alpha^*u) + (\beta^*u)^*(\beta^*u)$$

and we deduce that $\alpha^* u = 0$. Thus $u \in (1-e)U$ and we conclude that $(1-f)U \subseteq (1-e)U$. Therefore we may write $1 - f = (1-e)v$ for some $v \in U$. Then if $w \in eU$, we find that $(1-f)w = v^*(1-e)w = 0$, consequently $w = fw$ and hence $w \in fU$. Thus $eU \subseteq fU$ and the result follows. $\square$

Now given $\alpha \in \mathcal{N}(G)$, we may write $\alpha = \sum_{g \in G} \alpha_g g$ with $\alpha_g \in \mathbb{C}$, and then $\operatorname{tr}\alpha = \alpha_1$ is the trace of α. Moreover for $\alpha \in \mathcal{U}(G)$, there is a unique projection $e \in \mathcal{N}(G)$ such that $\alpha\mathcal{U}(G) = e\mathcal{U}(G)$, and then we set $\operatorname{rk}\alpha = \operatorname{tr} e$, the rank of α. Then $\operatorname{rk}\colon \mathcal{U}(G) \to [0,1]$ is a rank function [6, Definition, p. 226, §16]. This means that rk satisfies (a)–(d) of Property 3 below; it also satisfies Properties 3(e),(f).

Property 3.

(a) $\operatorname{rk}(1) = 1$.
(b) $\operatorname{rk}(\alpha\beta) \leq \operatorname{rk}(\alpha), \operatorname{rk}(\beta)$ *for all* $\alpha, \beta \in \mathcal{U}(G)$.
(c) $\operatorname{rk}(e + f) = \operatorname{rk}(e) + \operatorname{rk}(f)$ *for all orthogonal idempotents* $e, f \in \mathcal{U}(G)$.
(d) $\operatorname{rk}(\alpha) > 0$ *for all nonzero* $\alpha \in \mathcal{U}(G)$.
(e) $\operatorname{rk}(\alpha^*\alpha) = \operatorname{rk}(\alpha\alpha^*) = \operatorname{rk}(\alpha)$ *for all* $\alpha \in \mathcal{U}(G)$.
(f) *If* $G \leq H$ *and* $\alpha \in \mathcal{U}(G)$, *then* $\operatorname{rk}(\alpha)$ *is the same whether we view* $\alpha \in \mathcal{U}(G)$ *or* $\alpha \in \mathcal{U}(H)$.

There is also a well-defined dimension $\dim_{\mathcal{U}(G)}$ for $\mathcal{U}(G)$-modules [9, Theorem 8.29] which satisfies $\dim_{\mathcal{U}(G)} \alpha\mathcal{U}(G) = \operatorname{rk}(\alpha)$.

We need the following result for generating units in $\mathcal{U}(G)$. Recall that an ICC group is a group in which all conjugacy classes except the identity are infinite, and that a unitary element is an element u of $\mathcal{U}(G)$ such that $u^*u = 1$ (equivalently $uu^* = 1$, because $\mathcal{U}(G)$ is a finite von Neumann algebra).

Lemma 4. *Let* G *be an ICC group, let* $\alpha \in \mathcal{U}(G)$, *and let* n *be a positive integer. Suppose* $\operatorname{rk}(\alpha) \geq 1/n$. *Then there exist unitary elements* $\tau_1, \ldots, \tau_n \in \mathcal{U}(G)$ *such that* $\sum_{i=1}^n \tau_i \alpha \alpha^* \tau_i^{-1}$ *is a unit in* $\mathcal{U}(G)$.

Proof. Suppose e and f are projections in $\mathcal{U}(G)$ such that $e\mathcal{U}(G) \cong f\mathcal{U}(G)$. Since $\mathcal{U}(G)$ is a unit-regular ring, we see that $(1-e)\mathcal{U}(G) \cong (1-f)\mathcal{U}(G)$ by [6, Theorem 4.5]. Therefore e and f, and also $1 - e$ and $1 - f$, are algebraically equivalent projections and hence equivalent [1, §5]. We deduce that e and f are unitarily equivalent [2, p. 69 and Exercise 17.12]; this means that there is a unitary element $u \in \mathcal{U}(G)$ such that $ueu^{-1} = f$.

Suppose now that e, f are projections in $\mathcal{U}(G)$ with $\operatorname{tr}(e) \leq \operatorname{tr}(f)$. Since G is an ICC group, the center of $\mathcal{N}(G)$ is $\mathbb{C}$. Therefore by [9, Theorem 8.22 and Theorem 9.13(1)], two finitely generated projective $\mathcal{U}(G)$-modules P, Q are isomorphic if and only if $\dim_{\mathcal{U}(G)}(P) = \dim_{\mathcal{U}(G)}(Q)$. Using [7, Theorem 8.4.4(ii)], we see that there is a finitely generated projective $\mathcal{U}(G)$-module P such that $\dim_{\mathcal{U}(G)} P = \operatorname{tr}(f) - \operatorname{tr}(e)$ and then $e\mathcal{U}(G) \oplus P \cong f\mathcal{U}(G)$. From the previous paragraph, we deduce that there is a unitary element $u \in \mathcal{U}(G)$ such that $ueu^{-1}\mathcal{U}(G) \subseteq f\mathcal{U}(G)$.

Set $\beta = \alpha\alpha^*$. Then $\mathrm{rk}\beta = \mathrm{rk}\alpha\alpha^* \geq 1/n$ by Property 3(e). Suppose $0 \leq r \leq 1/\mathrm{rk}\beta - 1$ and we have chosen unitary elements $\tau_1, \ldots, \tau_r \in \mathcal{U}(G)$ such that

$$\mathrm{rk}(\tau_1\beta\tau_1^{-1} + \cdots + \tau_r\beta\tau_r^{-1}) = r\,\mathrm{rk}(\beta);$$

certainly we can do this for $r = 0$. Since $\mathcal{U}(G)$ is a regular von Neumann $*$-algebra, there is a unique projection $f \in \mathcal{U}(G)$ such that $(\tau_1\beta\tau_1^{-1} + \cdots + \tau_r\beta\tau_r^{-1})\mathcal{U}(G) = f\mathcal{U}(G)$. Also $\mathrm{rk}(1 - f) \geq \mathrm{rk}\beta$, hence there is a unitary element $\tau_{r+1} \in \mathcal{U}(G)$ such that $\tau_{r+1}\beta\tau_{r+1}^{-1} \in (1 - f)\mathcal{U}(G)$, and then

$$\mathrm{rk}(\tau_1\beta\tau_1^{-1} + \cdots + \tau_{r+1}\beta\tau_{r+1}^{-1}) = (r + 1)\mathrm{rk}(\beta)$$

by Lemma 2.

Now suppose $r > 1/\mathrm{rk}(\beta) - 1$ and we have chosen unitary elements $\tau_1, \ldots, \tau_r \in \mathcal{U}(G)$ such that

$$\mathrm{rk}(\tau_1\beta\tau_1^{-1} + \cdots + \tau_r\beta\tau_r^{-1}) \geq 1 - \mathrm{rk}(\beta);$$

by the previous paragraph, we can certainly do this if $r \leq 1/\mathrm{rk}(\beta)$. Again, let $f \in \mathcal{U}(G)$ be the unique projection such that $(\tau_1\beta\tau_1^{-1} + \cdots + \tau_r\beta\tau_r^{-1})\mathcal{U}(G) = f\mathcal{U}(G)$. Then $\mathrm{rk}(\beta) \geq \mathrm{tr}(1 - f)$ and therefore there is a projection $e \in \mathcal{U}(G)$ such that $1 - f \in e\mathcal{U}(G)$ and $\mathrm{tr}(e) = \mathrm{rk}(\beta)$. Then we may choose a unitary element $\tau_{r+1} \in \mathcal{U}(G)$ such that $\tau_{r+1}\beta\tau_{r+1}^{-1}\mathcal{U}(G) = e\mathcal{U}(G)$. Applying Lemma 2, we see that

$$\mathrm{rk}(\tau_1\beta\tau_1^{-1} + \cdots + \tau_{r+1}\beta\tau_{r+1}^{-1}) = 1,$$

which tells us that $\tau_1\beta\tau_1^{-1} + \cdots + \tau_{r+1}\beta\tau_{r+1}^{-1}$ is a unit in $\mathcal{U}(G)$. We deduce that $\tau_1\beta\tau_1^{-1} + \cdots + \tau_n\beta\tau_n^{-1}$ is a unit in $\mathcal{U}(G)$ as required. $\qquad\square$

Proof of Theorem 1. We shall use the techniques of [4, §2]. By [10, Theorem 1], we may embed G in a group which is algebraically closed, so we may assume that G is algebraically closed. It now follows from [3, Corollary 1] that the augmentation ideal $\omega(KG)$ is the only proper two-sided ideal of KG. Let $\mathcal{F}$ denote the set of all finitely generated subfields of K. For each $F \in \mathcal{F}$, let $\mathcal{C}(F)$ denote the set of all finitely generated subfields containing F: this is a subset of $\mathcal{F}$. Now let

$$\mathcal{C} = \{X \subseteq \mathcal{F} \mid X \supseteq \mathcal{C}(F) \text{ for some } F \in \mathcal{F}\}.$$

The following are clear:

- $\mathcal{C} \neq \emptyset$.
- If $A \subseteq B \subseteq \mathcal{F}$ and $A \in \mathcal{C}$, then $B \in \mathcal{C}$ (if $\mathcal{C}(F) \subseteq A$, then $\mathcal{C}(F) \subseteq B$).
- If $A, B \in \mathcal{C}$, then $A \cap B \in \mathcal{C}$ (if $\mathcal{C}(E) \subseteq A$ and $\mathcal{C}(F) \subseteq B$, then $\mathcal{C}(\langle E, F\rangle) \subseteq A \cap B$).

This means that the sets $\mathcal{C}$ form a filter, and hence they are contained in an ultrafilter $\mathcal{D}$, that is a maximal filter. Thus $\mathcal{D}$ has the properties of $\mathcal{C}$ listed above, and the additional property

- If $X \subseteq \mathcal{F}$, then either X or $\mathcal{F} \setminus X$ is in $\mathcal{D}$.

Now set $R = \prod_{f \in \mathcal{F}} \mathcal{U}(G)$, the Cartesian product of the $\mathcal{U}(G)$ (so infinitely many coordinates of an element of R may be nonzero). Since the Cartesian product of self-injective von Neumann unit-regular rings is also self-injective von Neumann unit-regular, we see that R is also a self-injective von Neumann unit-regular ring. The general element of $\alpha \in R$ has coordinates α_f for $f \in \mathcal{F}$, and we define $\rho(\alpha) = \lim_{\mathcal{D}} \rho(\alpha_f)$, where lim indicates the limit of $\rho(\alpha_f)$ associated to the ultrafilter $\mathcal{D}$. Thus $\rho(\alpha)$ has the property that it is the unique number which is in the closure of $\{\mathrm{rk}(\alpha_d) \mid d \in X\}$ for all $X \in \mathcal{D}$. It is easy to check that ρ is a pseudo-rank function [6, Definition, p. 226, §16]; thus ρ satisfies Properties 3(a)–(c), but not Property 3(d), because we can have $\rho(\alpha) = 0$ with $\alpha \neq 0$.

Let $I = \{r \in R \mid \rho(r) = 0\}$. Using [6, Proposition 16.7], we see that I is a two-sided ideal of R and ρ induces a rank function on R/I. Of course, R/I will also be a von Neumann unit-regular *-ring. Next we show that R/I is self injective. In view of [6, Theorem 9.32], we need to prove that I is a maximal ideal of R; of course this will also show that R/I is a simple ring. Suppose $\alpha \in R \setminus I$. Then we may choose a real number ϵ such that $0 < \epsilon < \rho(\alpha)$. Set $S = \{s \in \mathcal{F} \mid \mathrm{rk}(\alpha_s) < \epsilon\}$. Then $\rho(\alpha)$ is not in the closure of $\{\mathrm{rk}(\alpha_s) \mid s \in S\}$ and therefore $S \notin \mathcal{D}$. Let $T = \mathcal{F} \setminus S$, so $T \in \mathcal{D}$ and choose a positive integer n such that $n > 1/\epsilon$. Since G is an ICC group, for each $t \in T$ there exist by Lemma 4 units $\tau(t)_1, \dots, \tau(t)_n \in \mathcal{U}(G)$ such that

$$\tau(t)_1 \alpha_t \tau(t)_1^{-1} + \cdots + \tau(t)_n \alpha_t \tau(t)_n^{-1}$$

is a unit in $\mathcal{U}(G)$ (the important thing here is that n is independent of t). Now for $r = 1, \dots, n$, define $\tau_r \in R$ by $(\tau_r)_t = \tau(t)_r$ for $t \in T$ and $(\tau_r)_s = 1$ for $s \in S$. Then

$$(\tau_1 \alpha \tau_1^{-1} + \cdots + \tau_n \alpha \tau_n^{-1})_t$$

is a unit for all $t \in T$ and it follows that its image in R/I is a unit. This proves that R/I is a simple ring.

Now we want to embed KG into R/I. For each $f \in \mathcal{F}$, choose an embedding of f into $\mathbb{C}$. This will in turn induce an embedding θ_f of fG into $\mathcal{U}(G)$.

For $\alpha \in KG$ and $f \in \mathcal{F}$, we define $\alpha_f = \theta_f(\alpha)$ if $\alpha \in fG$, and $\alpha_f = 0$ otherwise. This yields a well-defined map (not a homomorphism) $\phi \colon KG \to R$ which in turn induces a map $\psi \colon KG \to R/I$. Let $\alpha, \beta \in KG$ and let F denote the subfield of K generated by the supports of α and β, so $F \in \mathcal{F}$. Since

$$\alpha_f + \beta_f - (\alpha + \beta)_f = 0 = \alpha_f \beta_f - (\alpha\beta)_f$$

for all $f \in \mathcal{C}(F)$, we see that

$$\mathrm{rk}(\alpha_f + \beta_f - (\alpha + \beta)_f) = 0 = \mathrm{rk}(\alpha_f \beta_f - (\alpha\beta)_f)$$

for all $f \in \mathcal{C}(F)$. Since $\mathcal{C}(F) \in \mathcal{D}$, we deduce that $\psi(\alpha) + \psi(\beta) - \psi(\alpha + \beta) = 0 = \psi(\alpha)\psi(\beta) - \psi(\alpha\beta)$ and hence ψ is a ring homomorphism.

Finally we show that $\ker \psi = 0$. Since $\omega(KG)$ is the only proper ideal of KG, we see that if $\ker \psi \neq 0$, then $g - 1 \in \ker \psi$ for $1 \neq g \in G$. But $\mathrm{rk}((g-1)_f)$ is a constant positive number for $f \in \mathcal{F}$. Therefore $\rho(g - 1) \neq 0$ and hence $g - 1 \notin \ker \psi$. We conclude that $\ker \psi = 0$ and the proof is complete. $\qquad\square$

If K is a field, G is a group and θ is an automorphism of K, then θ induces an automorphism θ_* of KG by setting $\theta_*(\sum_g a_g g) = \sum_g \theta(a_g)g$. It is not difficult to deduce from the proof of Theorem 1 the following result related to the Atiyah conjecture [9, §10].

Proposition 5. *Let G be a group, let K be a subfield of $\mathbb{C}$ and let $0 \neq \alpha \in KG$. Then there exists $\epsilon > 0$ such that $\mathrm{rk}(\theta_*\alpha) > \epsilon$ for every automorphism θ of K.*

However we shall give an independent proof. It ought to be true that $\mathrm{rk}(\theta_*\alpha) = \mathrm{rk}(\alpha)$ for every automorphism θ of K.

Proof of Proposition 5. Using [10, Theorem 1], we may embed G in a group which is algebraically closed and has an element of infinite order. Thus by Property 3(f), we may assume that G is algebraically closed and contains an element x of infinite order. Consider the two-sided ideal generated by $\alpha(x - 1)$. Since $0 \neq \alpha(x - 1) \in \omega(KG)$ and $\omega(KG)$ is the only proper two-sided ideal of KG by [3, Corollary 1], we see that there exists a positive integer n and $\beta_i, \gamma_i \in KG$ such that $\sum_{i=1}^n \beta_i \alpha \gamma_i = x - 1$. Then

$$\sum_{i=1}^{n} \theta_*(\beta_i)\theta_*(\alpha)\theta_*(\gamma_i) = x - 1.$$

Since $\mathrm{rk}(x - 1) = 1$ and $\mathrm{rk}(\theta_*(\beta_i)\theta_*(\alpha)\theta_*(\gamma_i)) \leq \mathrm{rk}(\theta_*\alpha)$ for all i, we see that $\mathrm{rk}(\theta_*\alpha) \geq 1/n$ and the result follows. $\qquad\square$

References

[1] S.K. Berberian. The maximal ring of quotients of a finite von Neumann algebra. *Rocky Mountain J. Math.*, 12(1):149–164, 1982.

[2] Sterling K. Berberian. *Baer ∗-rings.* Springer-Verlag, New York, 1972. Die Grundlehren der mathematischen Wissenschaften, Band 195.

[3] K. Bonvallet, B. Hartley, D.S. Passman, and M.K. Smith. Group rings with simple augmentation ideals. *Proc. Amer. Math. Soc.*, 56:79–82, 1976.

[4] Gábor Elek and Endre Szabó. Sofic groups and direct finiteness. *J. Algebra*, 280(2):426–434, 2004.

[5] Carl Faith. Dedekind finite rings and a theorem of Kaplansky. *Comm. Algebra*, 31(9):4175–4178, 2003.

[6] K.R. Goodearl. *von Neumann regular rings.* Robert E. Krieger Publishing Co. Inc., Malabar, FL, second edition, 1991.

[7] Richard V. Kadison and John R. Ringrose. *Fundamentals of the theory of operator algebras. Vol. II*, volume 16 of *Graduate Studies in Mathematics.* American Mathematical Society, Providence, RI, 1997. Advanced theory, Corrected reprint of the 1986 original.

[8] Dinesh Khurana and T.Y. Lam. Rings with internal cancellation. *J. Algebra*, 284(1):203–235, 2005.

[9] Wolfgang Lück. L^2-invariants: theory and applications to geometry and K-theory, volume 44 of Ergebnisse der Mathematik und ihrer Grenzgebiete. 3. Folge. A Series of Modern Surveys in Mathematics [Results in Mathematics and Related Areas. 3rd Series. A Series of Modern Surveys in Mathematics]. Springer-Verlag, Berlin, 2002.

[10] W.R. Scott. Algebraically closed groups. Proc. Amer. Math. Soc., 2:118–121, 1951.

Peter A. Linnell
Department of Mathematics, Virginia Tech
Blacksburg, VA 24061-0123, USA
e-mail: linnell@math.vt.edu
URL: http://www.math.vt.edu/people/plinnell/

Modules and Comodules

Trends in Mathematics, 301–312

© 2008 Birkhäuser Verlag Basel/Switzerland

The Local Multiplier Algebra: Blending Noncommutative Ring Theory and Functional Analysis

Martin Mathieu

Dedicated to Robert Wisbauer on the occasion of his 65th birthday.

Abstract. We discuss some basic features of the local multiplier algebra of a *C**-algebra, the analytic analogue of the well-known Kharchenko–Martindale symmetric ring of quotients, and also the more recent maximal *C**-algebra of quotients, which is the analytic companion to the Utumi–Lanning maximal symmetric ring of quotients, together with some of the applications to operator theory on *C**-algebras. The emphasis lies in illustrating the interrelations between noncommutative ring theory and functional analysis.

Mathematics Subject Classification (2000). 46L05; 16-06, 16A08, 16S90, 46-02, 46L07, 46L57.

Keywords. Noncommutative rings of quotients, *C**-algebras, *AW**-algebras, local multiplier algebra, injective envelope, operator modules, derivations.

1. Introduction

Rings of quotients are a widely used concept in noncommutative ring theory and intimately connected with the important technique of localisation; see, e.g., [9]. Various 'breeds' have been developed depending on the kind of application one had in mind: for instance, the Kharchenko–Martindale symmetric ring of quotients of a semiprime ring serves particularly well in Galois theory [10] and in extending Herstein's programme on nonassociative derivations and isomorphisms from the simple case [6]. To a lesser extent analogous constructs have appeared in analysis, mainly due to the additional difficulties that arise from the need of complete spaces (*closed* ideals) and continuous mappings (*bounded* homomorphisms). In the early 1990's, together with Pere Ara, we started a systematic study of an

This paper is part of a research project supported by the Royal Society.

analogue of the symmetric ring of quotients in the context of C^*-algebras, the local multiplier algebra, and more recently extended this to a more comprehensive treatment of C^*-algebras of quotients. Important precursors were papers by Elliott [7] and Pedersen [18] aiming at the structure of automorphisms and derivations of C^*-algebras; and, in fact, many of the uses of the local multiplier algebra are found in operator theory on C^*-algebras.

While this is documented in detail in our monograph [3], the purpose of the present article is to underline the similarities between the purely algebraic theory and its C^*-algebraic companion and to point out where modifications must be made. We shall also discuss briefly some of the applications of local multipliers to the structure theory of various classes of operators between C^*-algebras.

2. C*-algebras of quotients

For basic terminology and facts in C^*-algebra theory, we refer the reader to Section 1.2 of [3], where many further references can be found.

We begin by recalling the definition of a two-sided ring of quotients in a form which is especially well suited for our setting.

Definition 2.1. Let R be a semiprime ring with involution $*$. A unital ring S is called a *two-sided ring of quotients of R* if

(i) $R \subseteq S$;
(ii) $\forall\, b \in S$: $bJ + J^*b \subseteq R$ for some $J \in \mathfrak{I}$;
(iii) $\forall\, b \in S, J \in \mathfrak{I}$: $bJ = 0 \implies b = 0$;

where $\mathfrak{I}$ is a 'good' set of (right) ideals in R.

We shall not pause to spell out in detail the requirements on the 'good' set of ideals (since we will mainly be interested in two examples) nor discuss the properties the involution is supposed to have (such as positive-definiteness, e.g.), since in a moment all our rings will be C^*-algebras anyway.

Here are our main examples.

Examples 2.2. Let R be a semiprime ring with involution $*$.

1. If we choose $\mathfrak{I} = \{R\}$ then we obtain $S = M(R)$, the multiplier ring of R.
2. If we choose $\mathfrak{I} = \mathfrak{I}_e$, the set of all essential two-sided ideals of R, then we obtain $S = Q_s(R)$, the Kharchenko–Martindale symmetric ring of quotients of R.
3. If we choose $\mathfrak{I} = \mathfrak{I}_{er}$, the set of all essential right ideals of R, then we obtain $S = Q^s_{\max}(R)$, the Utumi–Lanning maximal symmetric ring of quotients of R.

Each of the above enjoys a universal property with respect to the prescribed set of ideals. Example 3 was already implicitly contained in Utumi's work in the 1950's but it was Lanning who started a systematic study [11]. Some explicit computations of $Q^s_{\max}(R)$ are carried out, e.g., in [16].

Following the above pattern we now introduce the concept of a C^*-algebra of quotients.

Definition 2.3. Let A be a C^*-algebra. A unital C^*-algebra B is called a C^*-*algebra of quotients of* A if

(i) $A \subseteq B$;
(ii) $\{b \in B \mid bJ + J^*b \subseteq A \text{ for some } J \in \mathfrak{J}\}$ is dense in B;
(iii) $\forall\, b \in B,\ J \in \mathfrak{J}:\ bJ = 0 \implies b = 0$;

where $\mathfrak{J}$ is a 'good' set of closed (right) ideals in A.

Evidently the main difference in the two concepts lies in condition (ii); the necessity of completeness forces us to make the adjustment in Definition 2.3.

As before we have three main examples in mind.

Examples 2.4. Let A be a C^*-algebra.

1. If we choose $\mathfrak{J} = \{A\}$ then we obtain $B = M(A)$, the multiplier algebra of A. This is a well-studied C^*-algebra and the maximal *unitization* of A. It has found manifold applications in various areas of C^*-algebra theory.

2. If we choose $\mathfrak{J} = \mathfrak{J}_{ce}$, the set of all closed essential two-sided ideals of A, then we obtain $B = M_{\text{loc}}(A)$, the local multiplier algebra of A. This C^*-algebra first came up in work by Elliott [7] and Pedersen [18] in the mid 1970s but seems to have lain dormant until the author and Pere Ara started a systematic investigation from 1990 onwards. Nowadays, its structure is fairly well understood and many uses have been found, see [3] and Section 4 below.

3. If we choose $\mathfrak{J} = \mathfrak{J}_{cer}$, the set of all closed essential right ideals of A, then we obtain $B = Q_{\max}(A)$, the maximal C^*-algebra of quotients of A. This algebra was first introduced in [1] and is now the topic of an ongoing research project by Ara and the author, see [5].

At this point we want to stress a subtle difference between Examples 2 and 3 in (2.4). For a two-sided ideal in a C^*-algebra, the concepts "closed essential" and "essential closed" coincide; that is to say if $I \subseteq A$ is a closed two-sided ideal which is essential as a closed ideal – $I \cap J \neq 0$ for every non-zero closed two-sided ideal $J \subseteq A$ – then it is *algebraically* essential, that is, essential in A-$\mathcal{M}od$-A. The reason is that, in this case, "essential" can be expressed by an annihilator condition. This is not the case for one-sided, say right, ideals in general. So, a priori, a closed right ideal in a C^*-algebra which is essential in $\mathcal{B}an$-A, the category of Banach A-right modules, need not be essential in $\mathcal{M}od$-A. But, fortunately, it turns out that these concepts agree nevertheless. For this, and a comprehensive discussion of various notions of essentiality for one-sided ideals, see [5].

We use the chart below to discuss the interrelations between the purely algebraic and the analytic constructs.

$$
\begin{array}{cc}
M_{\text{loc}}(A) & Q_{\max}(A) \\
\hline
Q_s(A) & Q_{\max}^s(A) \\
Q_b(A) & Q_{\max}^s(A)_b
\end{array}
$$

The symmetric algebra of quotients $Q_s(A)$ of A of Definition 2.1 is endowed with a positive-definite involution, inherited from A, and a good order structure, thus allowing us to define *bounded elements* in the sense of Handelman–Vidav:

$$q \in Q_s(A) \text{ is bounded if } q^*q \leq \lambda 1 \text{ for some } \lambda \in \mathbb{R}_+; \tag{1}$$

in this case, the *norm* $\|q\|$ of q is defined to be $\sqrt{\lambda}$ for the least λ in (1) above. With this norm, the set of all bounded elements in $Q_s(A)$ – the *bounded part* $Q_b(A)$ of $Q_s(A)$ – becomes a pre-C^*-algebra, and its completion is $M_{\mathrm{loc}}(A)$. Consequently, $Q_s(A)$ and $M_{\mathrm{loc}}(A)$ contain a common *-subalgebra $Q_b(A)$; the latter is dense in $M_{\mathrm{loc}}(A)$, and $Q_s(A)$ can be reconstructed from $Q_b(A)$ by central localisation. For more details on this, see [3, Section 2.2].

We call $Q_b(A)$ the *bounded symmetric algebra of quotients of A*.

A similar, if slightly more complicated mechanism creates a bounded part $Q_{\mathrm{max}}^s(A)_b$ in $Q_{\mathrm{max}}^s(A)$ and the completion of this pre-C^*-algebra is the *maximal C^*-algebra of quotients* $Q_{\mathrm{max}}(A)$ of A, see [1], [5]. Since we shall devote most of our attention to the local multiplier algebra, we will not discuss any further details at this point.

One way to construct the symmetric algebra of quotients $Q_s(A)$ is to employ (equivalence classes of) essentially defined double centralisers; that is, pairs of left and right module homomorphisms defined on essential ideals. These may not be continuous and hence may not be defined on closed ideals. The elements in $Q_b(A)$ correspond precisely to those which are defined via continuous (i.e., bounded) module homomorphisms, and, in general, $Q_b(A)$ is strictly smaller than $Q_s(A)$; see [3, Proposition 2.2.13].

The reader will also note that the two examples 2.2 (1) and 2.4 (1) agree with each other. This is a consequence of the fact that a right A-module homomorphism from A into A (or, more generally, from $I \in \mathfrak{I}_{ce}$ into A) is automatically continuous.

Since closed two-sided ideals in C^*-algebras are particularly well behaved, the above constructions can be performed in alternative ways; the one discussed in the next section is of fundamental importance for the applications.

3. The local multiplier algebra

Suppose I and J are closed essential two-sided ideals in a C^*-algebra A. Then $I \cap J$ also belongs to $\mathfrak{I}_{ce}$ and, moreover, $I \cap J$ is an essential ideal in both $M(I)$ and $M(J)$. (This uses the property of closed ideals in a C^*-algebra to be idempotent.) By the universal property of the multiplier algebra, we obtain injective *-homomorphisms $M(I) \to M(I \cap J)$ and $M(J) \to M(I \cap J)$. These are given by "restricting the multipliers" to the smaller ideal. Since injective *-homomorphisms between C^*-algebras are isometries, we thus obtain a directed family of C^*-algebras $\{M(I) \mid I \in (\mathfrak{I}_{ce}, \supseteq)\}$ and *-monomorphisms $\{\rho_{JI} \mid I, J \in (\mathfrak{I}_{ce}, \supseteq)\}$, where $\rho_{JI} \colon M(I) \to M(J)$ is the above restriction homomorphism, if $J \subseteq I$. Taking the direct limit in the category of C^*-algebras yields the local multiplier algebra.

Definition 3.1. For a C^*-algebra A, $M_{\mathrm{loc}}(A) = \varinjlim_{\mathfrak{J}_{ce}} M(I)$ is the *local multiplier algebra* of A.

One of the basic achievements of our work with Ara was to show that this definition of the local multiplier algebra – which was first used in [7] and [18], but under a different name – agrees with the one presented in Section 2, thus providing the link between the algebraic and the analytic theory. See [3, Section 2.3].

In the case of C^*-algebras we can moreover describe the symmetric algebra of quotients in an alternative way.

Proposition 3.2. *For every C^*-algebra A, we have $Q_s(A) = \mathrm{alg}\varinjlim_{\mathfrak{J}_{ce}} M(K_I)$, where K_I denotes the Pedersen ideal of an ideal $I \in \mathfrak{J}_{ce}$.*

The *Pedersen ideal* of a closed two-sided ideal I in a C^*-algebra is the smallest two-sided ideal which is dense in I. If $I, J \in \mathfrak{J}_{ce}$ and $J \subseteq I$ then $K_J \subseteq K_I$ and $K_I = K_I^2$ is essential too. This enables us to prove the above result in [3, Proposition 2.2.4].

Examples 3.3. Let us look at some examples of local multiplier algebras.

(1) Let A be a commutative unital C^*-algebra; then $A = C(X)$, the complex-valued continuous functions on a compact Hausdorff space X. In this case, $M_{\mathrm{loc}}(A) = B(X)$, the algebra of all bounded Borel functions modulo the ideal of those functions that vanish off a rare subset of X. This algebra is sometimes called the *Dixmier algebra*, since in Dixmier's work it provided the first example of an AW^*-algebra which is not a von Neumann algebra (for $X = [0,1]$). See [3, Proposition 3.4.5].

(2) A C^*-algebra A is called simple if it does not contain any closed two-sided ideal other than 0 and A. Evidently, for a simple C^*-algebra A, we have $M_{\mathrm{loc}}(A) = M(A)$. In [2] we gave examples of unital non-simple C^*-algebras A such that $M_{\mathrm{loc}}(A)$ is simple; hence $A \neq M_{\mathrm{loc}}(A) = M_{\mathrm{loc}}(M_{\mathrm{loc}}(A))$ in this case.

(3) A C^*-algebra A is said to be an AW^*-*algebra* if the left annihilator of every subset of A is principal, that is, of the form Ap for a projection $p \in A$. If A is an AW^*-algebra then $M_{\mathrm{loc}}(A) = A$; see, e.g., [3, Theorem 2.3.8]. Note that this in particular applies to every von Neumann algebra, that is, weakly closed unital C^*-subalgebra of the algebra $B(H)$ of all bounded linear operators on a Hilbert space H.

(4) Let $A = C(X) \otimes B(H)$ be the C^*-tensor product of $C(X)$, for a compact Hausdorff space X, and $B(H)$. (In this case, there is only one C^*-tensor norm on the algebraic tensor product.) Then

$$M_{\mathrm{loc}}(A) = \varinjlim_{U \in \mathfrak{D}} C_b(U, B(H)_s),$$

where $\mathfrak{D}$ is the filter of dense open subsets of X and $C_b(U, B(H)_s)$ denotes the C^*-algebra of all bounded continuous functions from U into $B(H)$ endowed with the strict topology. This result is proved in [5, Corollary 5.3].

The theory of the local multiplier algebra bears some similarity with the one of the symmetric ring of quotients but generally the additional analytic structure leads to complications. For instance, it is well known that Q_s is not a closure operation. For some time, it was an open problem, first raised in [18], whether M_{loc} is a closure operation or not, that is, whether there can be a C^*-algebra A such that $M_{loc}(A)$ is different from $M_{loc}(M_{loc}(A))$. This question was recently settled in [4].

Theorem 3.4. *There is a unital separable primitive approximately finite-dimensional C^*-algebra A such that $M_{loc}(M_{loc}(A)) \neq M_{loc}(A)$.*

The proof of this result uses non-stable K-theory, Elliott's classification of AF-algebras and a detailed study of strict limits of sequences of projections in the local multiplier algebra, among others.

4. Applications of local multipliers

In this section we shall discuss some typical applications of local multipliers of C^*-algebras. The symmetric ring of quotients has been put to good use in the study of a number of classes of additive mappings on semiprime rings, notably automorphisms and derivations; see [6] and [10], for example. It is thus no surprise that the local multiplier algebra has too been exploited for this purpose.

One of Pedersen's original motivations to investigate multipliers of closed essential ideals of a C^*-algebra in [18] was to find a bigger C^*-algebra in which every derivation of the original C^*-algebra becomes inner, that is, is implemented as a commutator. A derivation d on a C^*-algebra A is a linear mapping $d: A \to A$ satisfying the usual Leibniz product rule; such a mapping is automatically bounded, as was first shown by Sakai. If there is an element a such that $dx = xa - ax$ for all $x \in A$, the derivation is called *inner*. In most cases such an element does not exist within A; therefore one tries to extend the derivation d to a bigger C^*-algebra which may contain an implementing element. Whether the local multiplier algebra has this property for every C^*-algebra is still unknown, though there have been some advances in this direction; see, e.g., [13] and [20]. Pedersen proved in [18] that the answer is positive if A is separable.

Once one knows that d is inner, one has a better chance to estimate its norm, which, of course, is important from the analytic point of view. It is easy to see that, if $dx = xa - ax$ for all $x \in A$, then $||d|| \leq 2 \operatorname{dist}(a, Z(A))$, where $Z(A)$ stands for the centre of A. In general, this estimate is strict and, in fact, a lower estimate is related to cohomological properties of A. Various kinds of C^*-algebras are known to have the property that the above inequality is indeed an equality for every inner derivation, such as von Neumann algebras, e.g., (a result by Zsido). We were able to show that this is true for local multiplier algebras and, more generally, for boundedly centrally closed C^*-algebras (see below for the definition). Moreover, Somerset proved that the distance $\operatorname{dist}(a, Z(A))$ from any element a to the centre is always attained, regardless of the nature of the C^*-algebra A [19].

Putting all this together one obtains full information on derivations on separable C^*-algebras with the aid of local multipliers; the details of the argument take up the major part of Sections 4.1 and 4.2 in [3].

Theorem 4.1 (Theorem 4.2.20 in [3]). *For every derivation d on a separable C^*-algebra A there exists $a \in M_{\mathrm{loc}}(A)$ such that $dx = xa - ax$ for all $x \in A$ and $\|d\| = 2\,\|a\|$.*

In the background of the above arguments to calculate the norm of an inner derivation works another category, the category of operator spaces together with completely bounded mappings (we shall briefly review this category in the next section), and the fact that the norm and the completely bounded norm of an inner derivation agree. The interplay with local multipliers becomes even more apparent when we now turn our attention to elementary operators.

Already in the mid 1950's Grothendieck proposed to use tensor products of Banach spaces to study operators defined between them. In the case of a C^*-algebra A, a very natural class arising in this way is the one consisting of elementary operators. Define

$$\theta \colon M(A) \otimes M(A) \longrightarrow B(A), \quad a \otimes b \longmapsto M_{a,b},$$

where $M_{a,b}x = axb$ for $x \in A$, $a, b \in M(A)$ and $B(A)$ denotes the Banach algebra of all bounded linear operators on the C^*-algebra A with the operator norm. Elements in the image of θ are operators of the form $S \colon x \mapsto \sum_{j=1}^{n} a_j x b_j$, $a_j, b_j \in M(A)$ and are called *elementary operators* on A.

Once again it is easy to give upper estimates for the norm of an elementary operator S in terms of the norms of a_j and b_j but hard to give a precise description. To this end, Haagerup introduced a new norm on the tensor product of two C^*-algebras which is no longer a C^*-tensor norm but a good norm in the category of operator spaces. This so-called Haagerup norm is defined as follows. For $u \in M(A) \otimes M(A)$ put

$$\|u\|_h = \inf_{u = \sum_j a_j \otimes b_j} \left\{ \left\| \sum_{j=1}^{n} a_j a_j^* \right\|^{1/2} \left\| \sum_{j=1}^{n} b_j^* b_j \right\|^{1/2} \right\},$$

the *Haagerup norm* of u. Then the following result holds.

Theorem 4.2. *For every infinite-dimensional simple unital C^*-algebra A, the mapping θ is an isometry on $A \otimes_h A$, the completion of the algebraic tensor product with respect to the Haagerup norm.*

This theorem is a special case of [3, Corollary 5.4.35] and rests heavily on results by Haagerup and Magajna. It is a prime example of a result in operator theory on C^*-algebras in the formulation of which local multipliers are absent – but in the proof of which they are essential. First of all we note that the assumption that A is unital is vital; otherwise the compact operators provide a counterexample. Under the hypothesis of Theorem 4.2, $M_{\mathrm{loc}}(A) = A$ and $Z(M_{\mathrm{loc}}(A)) = \mathbb{C}$;

this is already important for the injectivity of θ! (As was first noticed in [12].) Furthermore the assumption entails that A is antiliminal and hence the operator norm of S and the completely bounded norm coincide ([3, Corollary 5.4.36]).

Once we move beyond a trivial ideal space, local multipliers become fully visible.

For a general C^*-algebra A with multiplier algebra $M(A)$ let $^cA = \overline{AZ}$ and $^cM(A) = \overline{M(A)Z}$ denote the *bounded central closure* of A and of $M(A)$, respectively, where $Z = Z(M_{\mathrm{loc}}(A))$ is the centre of the local multiplier algebra. (See also Section 5 below.) Both ^cA and $^cM(A)$ are modules over Z, and θ induces a mapping θ_Z on the module tensor product over Z. This module tensor product can be endowed with a central version of the Haagerup norm and will thus be denoted by $^cM(A) \otimes_{Z,h} {}^cM(A)$. Instead of $B(A)$ we now have to use $CB(^cA)$ on the right-hand side, the Banach algebra of all completely bounded operators on ^cA with the completely bounded norm (which, in general, is bigger than the operator norm). With this notation, the general result reads as follows.

Theorem 4.3 (Theorem 5.4.30 in [3]). *For every C^*-algebra A, the mapping θ_Z is an isometry from $^cM(A) \otimes_{Z,h} {}^cM(A)$ into $CB(^cA)$.*

Other classes of operators that were studied in [3] with the help of local multipliers include generalised derivations, automorphisms, Jordan and Lie isomorphisms and centralising mappings, and results that use, but do not show, local multiplier theory (in the same vein as Theorem 4.2 above) are found, for instance, in the structure theory of Lie derivations on C^*-algebras; see [14].

5. Multipliers and the injective envelope

Lately, categories of operator spaces and completely bounded mappings have gained in importance in the study of C^*-algebras but also many other branches of infinite-dimensional analysis. One reason for this is the existence of injective hulls (which is not guaranteed in the category of C^*-algebras), which will lead us to a third picture of the local multiplier algebra in this section.

Just as the most general Banach space is $C(X)$, where X is a compact Hausdorff space, – since, by the Banach–Alaoglu theorem, every Banach space is isometrically isomorphic to a closed subspace of some space $C(X)$ – the most general operator space is $B(H)$, the space of all bounded linear operators from H into itself, where H is a complex Hilbert space. Already in the early work by von Neumann in the 1920s it emerged that it is vital to consider matrices of operators as well; that is, matrices whose entries are from $B(H)$. Since such a matrix space $M_n(B(H))$ is isomorphic to $B(H^n)$, where H^n is the n-fold direct sum of H, $n \in \mathbb{N}$, there is a natural choice of a norm on $M_n(B(H))$; incidentally, the only one turning it into a C^*-algebra.

An abstract definition of operator spaces is provided by Ruan's axioms; see [17, Chapter 13]. However, as every C^*-algebra can be regarded as a concrete algebra of bounded linear operators on Hilbert space by the Gelfand–Naimark–Segal theorem, we can content ourselves with a concrete picture here.

Definition 5.1. An *operator space* E is a linear subspace of $B(H)$, for some Hilbert space H, such that, for each $n \in \mathbb{N}$, the space of matrices $M_n(E)$ is complete with respect to the canonical norm on $M_n(B(H)) = B(H^n)$.

The morphisms are given as follows.

Definition 5.2. For a linear mapping $T \colon E \to F$ between operator spaces E and F we denote by $T_n \colon M_n(E) \to M_n(F)$ the n-fold ampliation given by

$$T_n\big((x_{ij})_{1 \le i,j \le n}\big) = (Tx_{ij})_{1 \le i,j \le n}.$$

The operator T is called *completely bounded* if $\|T\|_{cb} = \sup_{n \in \mathbb{N}} \|T_n\| < \infty$, where $\|T_n\|$ denotes the operator norm of each T_n. For a completely bounded operator the quantity $\|T\|_{cb}$ is called the *cb-norm* of T.

In the case where each T_n is an isometry, the operator T is called a *complete isometry*. If $\|T_n\| \le 1$ for all $n \in \mathbb{N}$ then T is said to be a *complete contraction*.

The two categories thus arising in a natural way are $\mathcal{O}_\infty$, consisting of operator spaces as the objects and completely bounded operators as the morphisms ("isomorphism" in this case means the existence of a bijective completely bounded linear mapping), and $\mathcal{O}_1$, consisting of operator spaces together with complete contractions; in this case, "isomorphism" is completely isometric linear isomorphism. We will work in the latter category, since two unital C*-algebras are completely isometric if and only if they are isomorphic as C*-algebras.

The notion of 'injective envelope' now is the usual one in a category.

Definition 5.3. An operator space I is called *injective* if, whenever $h \colon E \to F$ is a complete isometry between operator spaces E and F and $f_0 \colon E \to I$ is a complete contraction, there exists $f \colon F \to I$, a complete contraction, such that $f \circ h = f_0$.

Definition 5.4. Let E be an operator space.
  (i) A complete isometry $h \colon E \to F$ into another operator space F is said to be *essential* if, whenever $g \colon F \to G$ is a complete contraction into an operator space G such that $g \circ h$ is a complete isometry, then g is a complete isometry.
  (ii) An injective operator space $I(E)$ together with an essential complete isometry $\iota \colon E \to I(E)$ is called an *injective envelope of* E.

Since, as usual, injective envelopes are unique up to isomorphism in $\mathcal{O}_1$, we will speak of *the* injective envelope of an operator space E and denote it by $I(E)$.

Wittstock was the first to show that $B(H)$ is injective in the above sense, and Hamana established the existence of the injective envelope and studied injective envelopes of C*-algebras in great detail. For a nice exposition, see [17].

Theorem 5.5 (Hamana). *For every C*-algebra A, there is an injective envelope $I(A)$ which is a unital C*-algebra containing A as a C*-subalgebra. Moreover,*
  (i) *$I(A)$ is monotone complete, hence an AW*-algebra;*
  (ii) *if A is prime then $I(A)$ is prime, hence an AW*-factor;*
  (iii) *if A is unital and simple then $I(A)$ is simple.*

The injective envelope of a C^*-algebra is, in general, a fairly big C^*-algebra, big enough to contain both the local multiplier algebra and the maximal C^*-algebra of quotients. For a C^*-algebra A, the $*$-subalgebra of $I(A)$

$$\{x \in I(A) \mid xI + Ix \subseteq A \ \text{ for some ideal } I \in \mathfrak{I}_{ce}\}$$

is isomorphic to $Q_b(A)$. This was first observed by Frank and Paulsen [8]; see also [5, Section 3]. Under this isomorphism we can, and will, identify $M_{\mathrm{loc}}(A)$ with a C^*-subalgebra of $I(A)$.

Additionally, the same kind of identification establishes an isomorphism between

$$\{y \in I(A) \mid yJ + J^*y \subseteq A \ \text{ for some right ideal } J \in \mathfrak{I}_{cer}\}$$

and $Q_{\max}^s(A)_b$. As $Q_{\max}(A) = \overline{Q_{\max}^s(A)_b}$, we can also consider $Q_{\max}(A)$ as a C^*-subalgebra of $I(A)$.

This leads to the following result; see Theorems 3.8 and 3.12 of [5].

Theorem 5.6. *For every C^*-algebra A, we have*

$$A \subseteq M(A) \subseteq M_{\mathrm{loc}}(A) \subseteq Q_{\max}(A) \subseteq \bar{A} \subseteq I(A)$$

and

$$Z(M_{\mathrm{loc}}(A)) = Z(Q_{\max}(A)) = Z(\bar{A}) = Z(I(A)).$$

The C^*-algebra $\bar{A}$ appearing in the above theorem is the *regular monotone completion* of A, defined as the smallest monotone complete C^*-subalgebra of $I(A)$ which contains A as an order-dense C^*-subalgebra. An immediate consequence of Theorem 5.6 is that $Q_{\max}(A) = A$ for every monotone complete C^*-algebra A, in particular, every von Neumann algebra.

The equality of all the centres of the C^*-algebras from $M_{\mathrm{loc}}(A)$ upward in Theorem 5.6 is very interesting in itself. The local Dauns–Hofmann theorem [3, Theorem 3.1.1] describes the centre $Z = Z(M_{\mathrm{loc}}(A))$ as a direct limit:

$$Z(M_{\mathrm{loc}}(A)) = \varinjlim_{\mathfrak{I}_{ce}} Z(M(I))$$

and combining this with Theorem 5.6, we obtain a new concrete description of the centre of the injective envelope.

A C^*-algebra A is called *boundedly centrally closed* if $Z(M(A)) = Z$; equivalently, $^cA = A$ and $^cM(A) = M(A)$ [3, Proposition 3.2.3]. As a consequence of the local Dauns–Hofmann theorem, $M_{\mathrm{loc}}(A)$ and, more generally, every C^*-subalgebra of $M_{\mathrm{loc}}(A)$ containing both A and Z is boundedly centrally closed [3, Theorem 3.2.8]. Boundedly centrally closed C^*-algebras behave analogously to centrally closed semiprime rings (where the extended centroid agrees with the centroid of the ring R, that is, $Z(Q_s(R)) = Z(M(R))$. As a result many statements in operator theory on C^*-algebras become much simpler for this class of C^*-algebras.

It follows from the above that, for a unital boundedly centrally closed C^*-algebra A, $Z(A) = Z(I(A))$ although the injective envelope itself might be much bigger than A.

6. The maximal C*-algebra of quotients

Despite the fact that the maximal C^*-algebra of quotients $Q_{\max}(A)$ of a C^*-algebra A contains all the multiplier algebras of essential hereditary C^*-subalgebras of A, there does not seem to be a way to describe $Q_{\max}(A)$ as a direct limit of C^*-algebras, in contrast to the direct limit description of the local multiplier algebra; see Section 3. This makes the study of its properties much harder.

Fortunately there is at least a direct limit construction in the category of operator modules, which we will briefly indicate; for more details, see [5].

Given a unital C^*-algebra A we denote by $CB_A(E, F)$ the space of all completely bounded right A-module maps between the operator right A-modules E and F. If F is even an operator A-bimodule, this is an operator left A-module, where the operator space structure is given by $M_n(CB_A(E, F)) = CB_A(E, M_n(F))$, $n \in \mathbb{N}$ and the A-module structure on $CB_A(E, F)$ is defined by $(ag)(x) = ag(x)$, $a \in A$, $x \in E$ and $g \in CB_A(E, F)$.

For $I, J \in \mathfrak{I}_{cer}$, $J \subseteq I$, denote by $\rho_{JI} \colon CB_A(I, A) \to CB_A(J, A)$ the restriction map, which turns out to be a complete isometry. Letting

$$E_b(A) = \mathrm{alg}\lim_{\longrightarrow \mathfrak{I}_{cer}} CB_A(I, A)$$

we obtain an (uncompleted) direct limit in the category of operator left A-modules, which in fact is an (incomplete) unital operator algebra. The relation with the maximal C^*-algebra of quotients is given by the following result.

Theorem 6.1 ([5]). *For every C^*-algebra A, we can consider $E_b(A)$ as an operator subalgebra of $I(A)$. Under this identification, $E_b(A) \cap E_b(A)^* = Q_{\max}^s(A)_b$.*

At this stage, there are several basic questions still open for the maximal C^*-algebra of quotients.

Open Questions. 1. Is it true that $Q_{\max}(A) = A$ whenever A is an AW*-algebra? We know that the answer is positive if A is finite; if A is σ-finite (countably decomposable); or if A is monotone complete [5], but not yet in general.

2. Is $Q_{\max}(A)$ always an AW*-algebra? We expect the answer to be "no" but do not have a counterexample yet.

3. We know of examples such that $M_{\mathrm{loc}}(M_{\mathrm{loc}}(A)) \neq M_{\mathrm{loc}}(A)$, but is it possible that $Q_{\max}(Q_{\max}(A)) = Q_{\max}(A)$ for every C^*-algebra A?

New ways to obtain the maximal symmetric ring of quotients, as outlined in [15], e.g., may be useful to tackle these problems.

Acknowledgement

This paper is a slightly expanded version of a talk entitled *What can we say about local multipliers?* delivered on 8 September 2006 as part of the International Conference on Modules and Comodules dedicated to Robert Wisbauer and held at the Universidade do Porto. The support received from the *Deutsche Akademische Austauschdienst (DAAD)* to attend this meeting is gratefully acknowledged.

References

[1] P. ARA, Some interfaces between noncommutative ring theory and operator algebras, Irish Math. Soc. Bull. **50** (2003), 7–26.

[2] P. ARA AND M. MATHIEU, A simple local multiplier algebra, Math. Proc. Cambridge Phil. Soc. **126** (1999), 555–564.

[3] P. ARA AND M. MATHIEU, Local multipliers of C^*-algebras, Springer-Verlag, London, 2003.

[4] P. ARA AND M. MATHIEU, A not so simple local multiplier algebra, J. Funct. Anal. **237** (2006), 721–737.

[5] P. ARA AND M. MATHIEU, Maximal C^*-algebras of quotients and injective envelopes of C^*-algebras, preprint.

[6] K.I. BEIDAR, W.S. MARTINDALE AND A.V. MIKHALEV, Rings with generalized identities, Marcel Dekker, New York, 1996.

[7] G.A. ELLIOTT, Automorphisms determined by multipliers on ideals of a C^*-algebra, J. Funct. Anal. **23** (1976), 1–10.

[8] M. FRANK AND V.I. PAULSEN, Injective envelopes of C^*-algebras as operator modules, Pacific J. Math. **212** (2003), 57–69.

[9] K.R. GOODEARL, Ring theory: Nonsingular rings and modules, Marcel Dekker, New York, 1976.

[10] V.K. KHARCHENKO, Automorphisms and derivations of associative rings, Kluwer Acad. Publ., Dordrecht, 1991.

[11] S. LANNING, The maximal symmetric ring of quotients, J. Algebra **179** (1996), 47–91.

[12] M. MATHIEU, Elementary operators on prime C^*-algebras, I, Math. Ann. **284** (1989), 223–244.

[13] M. MATHIEU, Derivations implemented by local multipliers, Proc. Amer. Math. Soc. **126** (1998), 1133–1138.

[14] M. MATHIEU AND A.R. VILLENA, The structure of Lie derivations on C^*-algebras, J. Funct. Anal. **202** (2003), 504–525.

[15] E. ORTEGA, The maximal symmetric ring of quotients: Path algebras, incidence algebras and bicategories, PhD Thesis, Universitat Autònoma de Barcelona, Barcelona, 2006.

[16] E. ORTEGA, Rings of quotients of incidence algebras and path algebras, J. Algebra **303** (2006), 225–243.

[17] V. PAULSEN, Completely bounded maps and operator algebras, Cambridge Studies in Adv. Math. **78**, Cambridge Univ. Press, Cambridge, 2002.

[18] G.K. PEDERSEN, Approximating derivations on ideals of C^*-algebras, Invent. math. **45** (1978), 299–305.

[19] D.W.B. SOMERSET, The proximinality of the centre of a C^*-algebra, J. Approx. Theory **89** (1997), 114–117.

[20] D.W.B. SOMERSET, The local multiplier algebra of a C^*-algebra, II, J. Funct. Anal. **171** (2000), 308–330.

Martin Mathieu
Department of Pure Mathematics, Queen's University Belfast
Belfast BT7 1NN, Northern Ireland
e-mail: m.m@qub.ac.uk

Modules and Comodules
Trends in Mathematics, 313–329
© 2008 Birkhäuser Verlag Basel/Switzerland

On Some Injective Modules In $\sigma[M]$

A.Ç. Özcan, D.K. Tütüncü and M.F. Yousif

Dedicated to Professor Robert Wisbauer on his 65th birthday

Abstract. In this paper, we study the notions (strongly) soc-injective, (strongly) simple-injective and (strongly) mininjective modules in $\sigma[M]$. For any module N in $\sigma[M]$, N is strongly mininjective in $\sigma[M]$ if and only if it is strongly simple-injective in $\sigma[M]$. A module M is locally Noetherian if and only if every strongly simple-injective module in $\sigma[M]$ is strongly soc-injective. We also characterize Noetherian QF-modules.

1. Introduction

Let M be any R-module. Any R-module N is *generated by M* or *M-generated* if there exists an epimorphism $M^{(\Lambda)} \longrightarrow N$ for some index set Λ. An R-module N is said to be *subgenerated* by M if N is isomorphic to a submodule of an M-generated module. We denote by $\sigma[M]$ the full subcategory of the right R-modules whose objects are all right R-modules subgenerated by M.

Let M be a module and let N and T be in $\sigma[M]$. N is called *soc-T-injective* if any R-homomorphism $f : \text{Soc}(T) \to N$ extends to T. Equivalently, for any semisimple submodule K of T, any homomorphism $f : K \to N$ extends to T. A module $N \in \sigma[M]$ is called *soc-quasi-injective in $\sigma[M]$* if N is soc-N-injective. N is called *soc-injective in $\sigma[M]$* if N is soc-M-injective. N is called *strongly soc-injective in $\sigma[M]$* if N is soc-T-injective for all $T \in \sigma[M]$.

According to Harada [7], if M and N are modules, M is called *simple-N-injective* if, for every submodule L of N, every homomorphism $\gamma : L \longrightarrow M$ with $\gamma(L)$ simple extends to N. If $N = R$, M is called *simple-injective*, and if $M = N$, M is called *simple-quasi-injective*. Dually, M is called *min-N-injective* if, for every simple submodule L of N, every homomorphism $\gamma : L \to M$ extends to N. If $N = R$, M is called *mininjective*, and if $M = N$, M is called *min-quasi-injective*. Let $T \in \sigma[M]$. T is called *strongly simple-injective* in $\sigma[M]$ if T is simple-N-injective for all $N \in \sigma[M]$, and T is called *strongly mininjective* in $\sigma[M]$, if T is min-N-injective for all $N \in \sigma[M]$ (see [2]).

Throughout this article, all rings are associative and have an identity, and all modules are unitary right R-modules. Let M be an R-module. For a direct summand N of M we write $N \leq_d M$ and for an essential submodule N of M, $N \leq_e M$. Let $\widehat{N}$ be the M-injective hull of N in $\sigma[M]$. A module N in $\sigma[M]$ is called M-singular (or singular in $\sigma[M]$) if $N \cong L/K$ for an $L \in \sigma[M]$ and $K \leq_e L$ (see [6]). Every module $N \in \sigma[M]$ contains a largest M-singular submodule which is denoted by $Z_M(N)$. If $Z_M(N) = 0$, then N is called non-M-singular. We will use $\mathrm{Soc}(K)$ to indicate the socle of any module K.

In Section 2, we prove that, for any finitely generated module T in $\sigma[M]$, direct sums of soc-T-injective modules in $\sigma[M]$ is soc-T-injective if and only if $\mathrm{Soc}(T)$ is finitely generated. Also it is proven that if $N \in \sigma[M]$ is soc-(N)-lifting, then any module K in $\sigma[M]$ is soc-N-injective if and only if K is N-injective.

In Section 3, we consider the strongly soc-injective modules in $\sigma[M]$. Semi-artinian and Noetherian QF-modules are characterized in terms of strongly soc-injective modules in $\sigma[M]$. For example, any module M is semiartinian if and only if every strongly soc-injective module in $\sigma[M]$ is injective in $\sigma[M]$ (quasi-continuous). Let M be a finitely generated self-projective module. Then M is a Noetherian QF-module if and only if every strongly soc-injective module in $\sigma[M]$ is projective in $\sigma[M]$ if and only if M is a self-generator, $\mathrm{Soc}(M) \leq_e M$ and every projective module in $\sigma[M]$ is strongly soc-injective in $\sigma[M]$ if and only if $M/\mathrm{Soc}(M)$ has finite length and M is a self-generator strongly soc-injective in $\sigma[M]$. In this section we also characterize GCO-modules and cosemisimple modules in terms of strongly soc-injective modules in $\sigma[M]$.

In Section 4, we consider soc-injective modules. Let S and R be any rings and let M be a left S-, a right R-bimodule. We prove that if M_R is soc-injective, then $l_S(T_1 \cap T_2) = l_S(T_1) + l_S(T_2)$ for all semisimple submodules T_1 and T_2 of M_R while $l_S(A \cap B) = l_S(A) + l_S(B)$ for all semisimple submodules A and all submodules B of M_R in the case where $S = End_R(M)$.

In the last section, it is shown that the notions of strongly mininjective and strongly simple-injective coincide. We also prove that any module M is locally Noetherian if and only if every strongly simple-injective module in $\sigma[M]$ is strongly soc-injective, and that if M is finitely generated self-projective, then M is a Noetherian QF-module if and only if every strongly simple-injective module in $\sigma[M]$ is projective in $\sigma[M]$.

2. Soc-Injective Modules in $\sigma[M]$

Theorem 2.1. *Let M be a module.*

(1) *Let $N \in \sigma[M]$ and $\{M_i : i \in I\}$ a family of right R-modules in $\sigma[M]$. Then the direct product $\prod_{i \in I} M_i$ is soc-N-injective if and only if M_i is soc-N-injective for all $i \in I$.*

(2) *Let T, N and $K \in \sigma[M]$ with $K \leq N$. If T is soc-N-injective, then T is soc-K-injective.*

(3) *Let T, N and $K \in \sigma[M]$ with $T \cong N$. If T is soc-K-injective, then N is soc-K-injective.*

(4) *Let $N \in \sigma[M]$ and $\{A_i : i \in I\}$ a family of right R-modules in $\sigma[M]$. Then N is soc-$\oplus_{i \in I} A_i$-injective if and only if N is soc-A_i-injective for all $i \in I$.*

(5) *Let M be a projective module in $\sigma[M]$. Any module $N \in \sigma[M]$ is soc-injective if and only if N is soc-P-injective for every M-generated projective module P in $\sigma[M]$.*

(6) *Let T, N and $K \in \sigma[M]$ with $N \leq_d T$. If T is soc-K-injective, then N is soc-K-injective.*

(7) *If A, B and $N \in \sigma[M]$, $A \cong B$ and N is soc-A-injective, then N is soc-B-injective.*

Proof. Clear. □

The next corollary is an immediate consequence of Theorem 2.1.

Corollary 2.2.

(1) *If $N \in \sigma[M]$, then a finite direct sum of soc-N-injective modules in $\sigma[M]$ is again soc-N-injective. In particular, a finite direct sum of soc-injective (strongly soc-injective) modules in $\sigma[M]$ is again soc-injective (strongly soc-injective).*

(2) *A direct summand of soc-quasi-injective (soc-injective, strongly soc-injective) module in $\sigma[M]$ is again soc-quasi-injective (soc-injective, strongly soc-injective).*

Proposition 2.3. *Suppose $N \in \sigma[M]$ is a soc-quasi-injective module.*

(1) *(Soc-C_2) If K and L are semisimple submodules of N, $K \cong L$ and $K \leq_d N$, then $L \leq_d N$.*

(2) *(Soc-C_3) Let K and L be semisimple submodules of N with $K \cap L = 0$. If $K \leq_d N$ and $L \leq_d N$, then $K \oplus L \leq_d N$.*

Proof. (1) Since $K \cong L$, and K is soc-N-injective, being a direct summand of the soc-quasi-injective module N, L is soc-N-injective . If $i : L \to N$ is the inclusion map, the identity map $id_L : L \to L$ has an extension $\eta : N \to L$ such that $\eta i = id_L$, and so $L \leq_d N$.

(2) Then both K and L are soc-N-injective. Thus the semisimple module $K \oplus L$ is soc-N-injective, and so a direct summand of N. □

Proposition 2.4. *For $N \in \sigma[M]$, the following are equivalent:*

(1) *Every module in $\sigma[M]$ is soc-N-injective.*

(2) *Every semisimple module in $\sigma[M]$ is soc-N-injective.*

(3) *$Soc(N) \leq_d N$.*

Proof. Straightforward. □

Theorem 2.5. *For a projective module $N \in \sigma[M]$, the following are equivalent:*

(1) *Every quotient of a soc-N-injective module in $\sigma[M]$ is soc-N-injective.*

(2) *Every quotient of an injective module in $\sigma[M]$ is soc-N-injective.*

(3) *$Soc(N)$ is projective in $\sigma[M]$.*

Proof. (1) $\Rightarrow$ (2) Clear.

(2) $\Rightarrow$ (3) Consider the following diagram

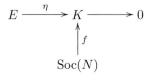

where E and K are in $\sigma[M]$, η is an epimorphism and f any homomorphism. By Cartan and Eilenberg [4], we may assume that E is injective in $\sigma[M]$. Since K is soc-N-injective, f can be extended to $g : N \to K$. Since N is projective in $\sigma[M]$, g can be lifted to $\tilde{g} : N \to E$ such that $\eta\tilde{g} = g$. Now define $\tilde{f} : Soc(N) \to E$ by $\tilde{f} = \tilde{g}|_{Soc(N)}$. Clearly, $\eta\tilde{f} = f$. Hence $Soc(N)$ is projective in $\sigma[M]$.

(3) $\Rightarrow$ (1) Let $K \in \sigma[M]$ be soc-N-injective. Assume $\eta : K \to L$ is an epimorphism. We want to show that L is soc-N-injective. Consider the following diagram

$$0 \longrightarrow Soc(N) \xrightarrow{inc.} N$$
$$\downarrow f$$
$$K \xrightarrow{\eta} L \longrightarrow 0$$

Since $Soc(N)$ is projective, f can be lifted to $g : Soc(N) \to K$. Since K is soc-injective, g can be extended to $\tilde{g} : N \to K$. Clearly $\eta\tilde{g} : N \to L$ extends f. □

Corollary 2.6. *The following are equivalent for a projective module M in $\sigma[M]$:*

(1) *Every quotient of a soc-injective module in $\sigma[M]$ is soc-injective in $\sigma[M]$.*

(2) *Every quotient of an injective module in $\sigma[M]$ is soc-injective in $\sigma[M]$.*

(3) *Every semisimple submodule of a projective module in $\sigma[M]$ is projective in $\sigma[M]$.*

(4) *$Soc(M)$ is projective in $\sigma[M]$.*

Proof. (1) $\Leftrightarrow$ (2) $\Leftrightarrow$ (4) By Theorem 2.5.

(3) $\Rightarrow$ (4) Since M is projective in $\sigma[M]$, $Soc(M)$ is projective in $\sigma[M]$.

(4) $\Rightarrow$ (3) If P is a projective module in $\sigma[M]$, then it is a direct summand of a direct sum of finitely generated submodules of $M^{(\mathbb{N})}$ by [9, 18.4]. Then $Soc(P)$ is a direct summand of a direct sum of socles of finitely generated submodules of $M^{(\mathbb{N})}$. Since $Soc(M)$ is projective in $\sigma[M]$, then $Soc(P)$ is projective in $\sigma[M]$. Hence (3) follows. □

Theorem 2.7. *Let $T \in \sigma[M]$ be finitely generated. Then the following are equivalent:*

(1) *Direct sums of soc-T-injective modules in $\sigma[M]$ is soc-T-injective.*
(2) *$\mathrm{Soc}(T)$ is finitely generated.*

Proof. (1) $\Rightarrow$ (2) Let $\mathrm{Soc}(T) = \oplus_{i \in I} S_i$ where each S_i is a simple submodule of T. Let $\widehat{S}_i$ be the injective hull of S_i in $\sigma[M]$, $i \in I$, and $\iota : \oplus_{i \in I} S_i \to \oplus_{i \in I} \widehat{S}_i$ be the inclusion map. Since $\oplus_{i \in I} \widehat{S}_i$ is soc-T-injective, ι can be extended to an R-homomorphism $\hat{\iota} : T \to \oplus_{i \in I} \widehat{S}_i$. Since T is finitely generated, $\hat{\iota}(T) \leq \oplus_{i=1}^{n} \widehat{S}_i$, for some positive integer n. Therefore $\mathrm{Soc}(T) \leq \oplus_{i=1}^{n} \widehat{S}_i$ implies that $\mathrm{Soc}(T)$ is finitely generated.

(2) $\Rightarrow$ (1) Let $E = \oplus_{i \in I} E_i$ be a direct sum of soc-T-injective modules in $\sigma[M]$ and $f : \mathrm{Soc}(T) \to E$ be an R-homomorphism. Since $\mathrm{Soc}(T)$ is finitely generated, $f(\mathrm{Soc}(T)) \leq \oplus_{i=1}^{n} E_i$, for some positive integer n. Since $\oplus_{i=1}^{n} E_i$ is soc-T-injective, f can be extended to an R-homomorphism $\hat{f} : T \to T$. $\square$

Corollary 2.8. *Let M be finitely generated. Then the following are equivalent:*

(1) *Direct sums of soc-injective modules in $\sigma[M]$ are soc-injective.*
(2) *$\mathrm{Soc}(M)$ is finitely generated.*

Corollary 2.9. *The following are equivalent:*

(1) *Direct sums of soc-T-injective modules in $\sigma[M]$ are soc-T-injective for every cyclic R-module T in $\sigma[M]$.*
(2) *Finitely generated R-modules in $\sigma[M]$ are finite dimensional.*

Definition 2.10. Let X be a submodule of a module M. We say that $Soc(M)$ *respects* X if there exists a direct summand A of M contained in X such that $X = A \oplus B$ and $B \leq \mathrm{Soc}(M)$. M is called $Soc(M)$-*lifting* if $\mathrm{Soc}(M)$ respects every submodule of M.

Proposition 2.11. *Let $N \in \sigma[M]$. If N is $Soc(N)$-lifting, then any module K in $\sigma[M]$ is soc-N-injective if and only if K is N-injective.*

Proof. Assume that a module $K \in \sigma[M]$ is soc-N-injective. Let L be any submodule of N, $i_2 : L \to N$ the inclusion map and $f : L \to K$ any homomorphism. By hypothesis, L has a decomposition $L = A \oplus B$ such that A is a direct summand of N and $B \leq \mathrm{Soc}(N)$. $N = A \oplus A'$ for some submodule A' of N. Then $L = A \oplus (L \cap A')$ and $L \cap A'$ is semisimple. Let $i_1 : L \cap A' \to L$ be the inclusion map and $f|_{L \cap A'} : L \cap A' \to K$. Since K is soc-N-injective, there exists a homomorphism $g : N \to K$ such that $g i_2 i_1 = f|_{L \cap A'}$. Now define $h : N \to K$ by $h(a + a') = f(a) + g(a')$ $(a \in A, a' \in A')$. Then $h i_2 = f$. $\square$

Corollary 2.12. [11, Lemma 2.14] *If $R/\mathrm{Soc}(R_R)$ is semisimple, then a right R-module M is soc-injective in Mod-R if and only if M is injective.*

Proof. $R/\mathrm{Soc}(R_R)$ is semisimple if and only if $\mathrm{Soc}(R_R)$ respects every right ideal of R [11, Theorem 2.3]. Hence by Proposition 2.11, the result holds. $\square$

Clearly if $\mathrm{Soc}(M)$ respects every submodule of M, then $M/\mathrm{Soc}(M)$ is semi-simple. We don't know if the converse is true or not.

3. Strongly soc-injective modules in $\sigma[M]$

Theorem 3.1. *Let $N \in \sigma[M]$. The following are equivalent:*
(1) *N is strongly soc-injective in $\sigma[M]$.*
(2) *N is soc-$\widehat{N}$-injective.*
(3) *$N = E \oplus T$, where E is injective in $\sigma[M]$ and T has zero socle.*

Moreover, if N has a nonzero socle, then E can be taken to have essential socle.

Proof. $(1) \Rightarrow (2)$ Clear.

$(2) \Rightarrow (3)$ If $\mathrm{Soc}(N) = 0$, we are done. Assume that $\mathrm{Soc}(N) \neq 0$, and consider the following diagram

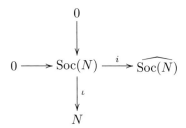

where ι and i are inclusion maps. Since N is soc-$\widehat{N}$-injective, N is soc-$\widehat{\mathrm{Soc}(N)}$-injective. So, there exists an R-homomorphism $\sigma : \widehat{\mathrm{Soc}(N)} \to N$, which extends ι. Since $\mathrm{Soc}(N) \leq_e \widehat{\mathrm{Soc}(N)}$, σ is an embedding of $\widehat{\mathrm{Soc}(N)}$ in N. If we write $E = \sigma(\widehat{\mathrm{Soc}(N)})$, then $N = E \oplus T$ for some submodule T of N. Clearly, E is injective and T has zero socle.

$(3) \Rightarrow (1)$ This is clear, since modules with zero socle are strongly soc-injective in $\sigma[M]$ and finite direct sum of strongly soc-injective modules are strongly soc-injective in $\sigma[M]$.

For the last statement of the theorem, then $\sigma(\mathrm{Soc}(N)) \leq_e E$. On the other hand, $\mathrm{Soc}(E) = \mathrm{Soc}(N) = \sigma(\mathrm{Soc}(N)) \leq_e E$ implies that $\mathrm{Soc}(E) \leq_e E$.   □

Corollary 3.2. *Let $N \in \sigma[M]$ be a module with essential socle. Then the following are equivalent:*

(1) *N is strongly soc-injective in $\sigma[M]$.*
(2) *N is injective in $\sigma[M]$.*

A module M is called *locally Noetherian* if every finitely generated submodule of M is Noetherian. It is well known that M is locally Noetherian if and only if every direct sum of M-injective modules is M-injective [9, 27.3], if and only if every (countable) direct sum of M-injective hulls of simple modules (in $\sigma[M]$) is M-injective ([9, 27.3] and [6, 2.5]).

Theorem 3.3. *The following are equivalent for a module M:*

(1) M *is locally Noetherian.*

(2) *Every direct sum of strongly soc-injective modules in $\sigma[M]$ is strongly soc-injective in $\sigma[M]$.*

Proof. (1) $\Rightarrow$ (2) Let $\{M_i\}_{i\in I}$ be a family of strongly soc-injective modules in $\sigma[M]$. By Theorem 3.1, for each $i \in I$, write $M_i = E_i \oplus T_i$ where E_i is injective in $\sigma[M]$ and $\text{Soc}(T_i) = 0$. If $E = \oplus_{i\in I}E_i$ and $T = \oplus_{i\in I}T_i$, then $\oplus_{i\in I}M_i = E \oplus T$ with $\text{Soc}(T) = 0$. Since M is locally Noetherian, E is M-injective, that is injective in $\sigma[M]$, and by Theorem 3.1, $\oplus_{i\in I}M_i$ is strongly soc-injective in $\sigma[M]$.

(2) $\Rightarrow$ (1) In order to prove that M is locally Noetherian, we only need to show that if $K_1, K_2, \ldots$ are simple modules (in $\sigma[M]$), then $\oplus_{i=1}^{\infty}\widehat{K_i}$ is injective in $\sigma[M]$, where $\widehat{K_i}$ is the M-injective hull of K_i. Since $\oplus_{i=1}^{\infty}\widehat{K_i}$ is strongly soc-injective in $\sigma[M]$ with essential socle, by Corollary 3.2, $\oplus_{i=1}^{\infty}\widehat{K_i}$ is injective in $\sigma[M]$. $\square$

Proposition 3.4. *If $N \in \sigma[M]$ is strongly soc-injective in $\sigma[M]$, then every semisimple submodule K of N is essential in a direct summand of N.*

Proof. This is clear if $\text{Soc}(N) = 0$. If $\text{Soc}(N) \neq 0$, then by Theorem 3.1, $N = \widehat{\text{Soc}(N)} \oplus T$ with $\text{Soc}(T) = 0$. Then $K \leq_e L \leq_d \widehat{\text{Soc}(N)}$ for some submodule L of N. $\square$

M is called *CESS* if every closure of every semisimple submodule of M is a direct summand of M. By Theorem 3.1, if $N \in \sigma[M]$ is strongly soc-injective in $\sigma[M]$, then $N = E \oplus T$ with $E = \widehat{\text{Soc}(N)}$ and $\text{Soc}(T) = 0$, and by [5], if T is E-injective, then N is a CESS-module. In particular, if T is non-M-singular, then T is E-injective and so N is a CESS-module.

Proposition 3.5. *Let $N \in \sigma[M]$ be $N = E \oplus T$ with $E = \widehat{\text{Soc}(N)}$, $\text{Soc}(T) = 0$ and T is E-injective. If S is a semisimple submodule of N, then every closure in N, of S is injective in $\sigma[M]$.*

Proof. By the above remark, if K is a closure of S in N, then K is a direct summand of N, and by Corollary 2.2 (2), K is strongly soc-injective in $\sigma[M]$. Let K' be a closure of S in E. Then K' is a direct summand of E and so is injective in $\sigma[M]$. Now we consider the following diagram

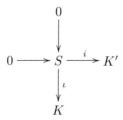

where ι and i are inclusion maps. Since K is strongly soc-injective in $\sigma[M]$, there exists a homomorphism $\sigma : K' \to K$ which extends ι. Since $S \leq_e K'$, σ is an embedding of K' in K, and so $S \leq_e \sigma(K') \leq_e K$, since $\sigma(K')$ is injective in $\sigma[M]$, it is a direct summand of K, and so $\sigma(K') = K$ is injective in $\sigma[M]$. □

A module M is called *semiartinian* if every nonzero homomorphic image of M has essential socle. Equivalently, every nonzero homomorphic image of M has nonzero socle. M is semiartinian if and only if every module in $\sigma[M]$ is semiartinian (see [6, 3.12]).

Theorem 3.6. *The following are equivalent for a module M:*

(1) M *is semiartinian.*
(2) *Every strongly soc-injective module in $\sigma[M]$ is injective in $\sigma[M]$.*
(3) *Every strongly soc-injective module in $\sigma[M]$ is quasi-continuous.*

Proof. (1) $\Rightarrow$ (2) Since M is semiartinian, $\mathrm{Soc}(N) \leq_e N$ for every module $N \in \sigma[M]$. By Corollary 3.1, (2) holds.

(2) $\Rightarrow$ (3) Clear.

(3) $\Rightarrow$ (1) Let N be a proper submodule of M. We claim that $\mathrm{Soc}(M/N) \neq 0$. If $\mathrm{Soc}(M/N) = 0$, let X/N be an arbitrary nonzero submodule of M/N. By hypothesis, $(X/N) \oplus (M/N)$ is quasi-continuous. By [8, Corollary 2.14], X/N is M/N-injective and hence $X/N \leq_d M/N$. This means that M/N is semisimple, a contradiction. Hence M is semiartinian. □

If M is a Noetherian injective cogenerator in $\sigma[M]$, then it is called a *Noetherian Quasi-Frobenius (QF)-module*. For a finitely generated quasi-projective module M, M is Noetherian QF-module if and only if every injective module in $\sigma[M]$ is projective in $\sigma[M]$ if and only if M is a self-generator and every projective module in $\sigma[M]$ is injective in $\sigma[M]$ by [9, 48.14].

Proposition 3.7. *Let M be a finitely generated self-projective module. Then the following are equivalent:*

(1) M *is a Noetherian QF-module.*
(2) *Every strongly soc-injective module in $\sigma[M]$ is projective in $\sigma[M]$.*

Proof. (1) $\Rightarrow$ (2) If M is a Noetherian QF-module, then M is Artinian by [9, 48.14]. By Theorem 3.6, every strongly soc-injective module in $\sigma[M]$ is injective in $\sigma[M]$, and hence projective in $\sigma[M]$ by [9, 48.14].

(2) $\Rightarrow$ (1) Clear. □

Observe that if $\mathrm{Soc}(M) = 0$, then every projective module in $\sigma[M]$ has zero socle by [9, 18.4(1)], and hence strongly soc-injective in $\sigma[M]$. On the other hand we have the following result by Corollary 3.2 and the above remark.

Proposition 3.8. *Let M be a finitely generated self-projective module. Then the following are equivalent:*

(1) M *is a Noetherian QF-module.*
(2) M *is a self-generator, $\mathrm{Soc}(M) \leq_e M$ and every projective module in $\sigma[M]$ is strongly soc-injective in $\sigma[M]$.*

Proof. (1) $\Rightarrow$ (2) Clear.

(2) $\Rightarrow$ (1) Let P be a nonzero projective module in $\sigma[M]$. Then P is strongly soc-injective in $\sigma[M]$. By Theorem 3.1, $P = E \oplus T$ with E injective in $\sigma[M]$ and $\mathrm{Soc}(T) = 0$. On the other hand, P is a direct summand of a direct sum of nonzero finitely generated submodules M_i of $M^{(\mathbb{N})}$. Since every M_i has essential socle, $\mathrm{Soc}(P) \leq_e P$. Therefore $P = E$, and hence P is injective in $\sigma[M]$. Since M is a self-generator, the proof is completed by [9, 48.14]. $\qquad\square$

Any module M is called $\sum$-*injective* if the direct sum of any number of copies of M is injective.

Proposition 3.9. *Let M be a projective module in $\sigma[M]$. Then the following are equivalent:*

(1) *Every projective M-generated module in $\sigma[M]$ is strongly soc-injective in $\sigma[M]$.*

(2) *$M = E \oplus T$ where E is $\sum$-injective in $\sigma[M]$ and $\mathrm{Soc}(T) = 0$.*

Proof. (1) $\Rightarrow$ (2) If $\mathrm{Soc}(M) = 0$, we are done. Assume $\mathrm{Soc}(M)$ is nonzero. Since M is projective, it follows from Theorem 3.1 that $M = E \oplus T$ where E is injective in $\sigma[M]$ with essential socle and $\mathrm{Soc}(T) = 0$. Since for any ordinal number α, $E^{(\alpha)}$ is projective in $\sigma[M]$ and M-generated, $E^{(\alpha)}$ is strongly soc-injective with essential socle. Therefore by Corollary 3.2, $E^{(\alpha)}$ is injective in $\sigma[M]$. Hence E is $\sum$-injective in $\sigma[M]$.

(2) $\Rightarrow$ (1) By (2), $M^{(\Lambda)} = E^{(\Lambda)} \oplus T^{(\Lambda)}$ for any ordinal number Λ. Since $E^{(\Lambda)}$ is injective in $\sigma[M]$, $M^{(\Lambda)}$ is strongly soc-injective in $\sigma[M]$ by Theorem 3.1. Let P be a projective M-generated module in $\sigma[M]$. Then P is isomorphic to a direct summand of $M^{(\Lambda)}$ for some Λ. Since every direct summand of strongly soc-injective module in $\sigma[M]$ is strongly soc-injective in $\sigma[M]$, P is strongly soc-injective in $\sigma[M]$. $\qquad\square$

Proposition 3.10. *Let $N \in \sigma[M]$ be a strongly soc-injective module. If $N/\mathrm{Soc}(N)$ is finite dimensional (Noetherian, Artinian, respectively), then $N = T \oplus S$, where T is finite dimensional (Noetherian, Artinian, respectively) and S is semisimple injective in $\sigma[M]$.*

Proof. By Theorem 3.1, $N = E \oplus K$ with E is injective in $\sigma[M]$ and $\mathrm{Soc}(K) = 0$. Now, $N/\mathrm{Soc}(N) \cong E/\mathrm{Soc}(E) \oplus K$. So both $E/\mathrm{Soc}(E)$ and K are finite dimensional (Noetherian, Artinian, respectively). By [3, Corollary 3], $E = L \oplus S$ with L finite dimensional and S semisimple. If $E/\mathrm{Soc}(E)$ is Noetherian (Artinian), then by [3, Lemma 4 and Proposition 5], $E = L \oplus S$ where L is Noetherian (Artinian) and S is semisimple. Consequently, $N = T \oplus S$ with S semisimple injective in $\sigma[M]$ and $T = K \oplus L$ finite dimensional (Noetherian, Artinian, respectively). $\qquad\square$

Corollary 3.11. *Let M be a finitely generated self-projective module in $\sigma[M]$. Then the following are equivalent:*

(1) *M is a Noetherian QF-module.*

(2) $M/\mathrm{Soc}(M)$ *has finite length and* M *is a self-generator strongly soc-injective in* $\sigma[M]$.

Proof. (2) $\Rightarrow$ (1) By Proposition 3.10, $\mathrm{Soc}(M) \leq_e M$. Then by Corollary 3.2, M is injective in $\sigma[M]$. Again by Proposition 3.10, M is Noetherian. By [9, 48.14], M is a Noetherian QF-module.

(1) $\Rightarrow$ (2) By [9, 48.14], $M/\mathrm{Soc}(M)$ has finite length and M is a self-generator. Since M is projective in $\sigma[M]$, by [9, 48.14], M is injective in $\sigma[M]$ and hence M is strongly soc-injective in $\sigma[M]$. $\qquad\square$

Lemma 3.12. *Let* $N \in \sigma[M]$ *be semisimple. The following are equivalent:*
(1) N *is injective in* $\sigma[M]$.
(2) N *is strongly soc-injective in* $\sigma[M]$.
(3) N *is soc-K-injective for every factor module* K *of* M.

Proof. (1) $\Leftrightarrow$ (2) By Corollary 3.2.
(1) $\Rightarrow$ (3) Clear.
(3) $\Rightarrow$ (1) Consider the following diagram

$$0 \longrightarrow L \overset{i}{\longrightarrow} M$$
$$\downarrow f$$
$$N$$

where $L \leq M$ and $f : L \longrightarrow N$ is any homomorphism. Then we have the diagram

where α is an isomorphism and ι is the inclusion map.

Since N is soc-$M/Kerf$-injective and $L/Kerf$ is semisimple, there exists a homomorphism $g : M/Kerf \longrightarrow N$ such that $g\bar{i} = \iota\alpha$. Then the homomorphism $h = g\pi$ extends f where $\pi : M \longrightarrow M/Kerf$ is the natural epimorphism. $\qquad\square$

A module M is called *cosemisimple* (*or a* V-module) if every simple module (in $\sigma[M]$) is M-injective. Clearly, M is cosemisimple if and only if every simple module is strongly soc-injective in $\sigma[M]$.

Proposition 3.13. *The following are equivalent for a module* M:
(1) *Every semisimple module in* $\sigma[M]$ *is strongly soc-injective in* $\sigma[M]$.
(2) *Every semisimple module in* $\sigma[M]$ *is soc-K-injective for every factor module* K *of* M.

(3) *Every module in $\sigma[M]$ is strongly soc-injective in $\sigma[M]$.*
(4) *Every module in $\sigma[M]$ is soc-K-injective for every factor module K of M.*
(5) *Every semisimple module in $\sigma[M]$ is injective in $\sigma[M]$.*
(6) *M is locally Noetherian and cosemisimple.*

Proof. (5) $\Leftrightarrow$ (6) by [6, 15.5].

(1) $\Leftrightarrow$ (3) By Proposition 2.4.

(1) $\Leftrightarrow$ (2) $\Leftrightarrow$ (5) By Lemma 3.12.

(4) $\Rightarrow$ (2) Clear.

(2) $\Rightarrow$ (4) By (2) $\Leftrightarrow$ (3). $\square$

A module M is called *generalized cosemisimple* (or a *GCO-module*) if every simple singular module is M-injective or M-projective. Equivalently, every M-singular simple module is M-injective by [6, 16.4].

By adopting the above proof we have the following proposition. Note that (5) $\Leftrightarrow$ (6) of Proposition 3.14 is well known from [6, 16.16].

Proposition 3.14. *The following are equivalent for a module M:*

(1) *Every semisimple M-singular module is strongly soc-injective in $\sigma[M]$.*
(2) *Every semisimple M-singular module in $\sigma[M]$ is soc-K-injective for every factor module K of M.*
(3) *Every M-singular module in $\sigma[M]$ is strongly soc-injective in $\sigma[M]$.*
(4) *Every M-singular module in $\sigma[M]$ is a direct sum of an injective module in $\sigma[M]$ and a module with zero socle.*
(5) *Every M-singular semisimple module in $\sigma[M]$ is injective in $\sigma[M]$.*

If M is self-projective, then they are equivalent to

(6) *M is a GCO-module and $M/\mathrm{Soc}(M)$ is locally Noetherian.*

4. When M is soc-injective

Proposition 4.1. *Let M be a module. The following are equivalent:*

(1) *M is soc-injective.*
(2) *If $\mathrm{Soc}(M) = X \oplus Y$ and $\gamma : X \longrightarrow M$ is an R-homomorphism, then there exists $c : M \longrightarrow M$ such that $\gamma(x) = c(x)$ for all $x \in X$ and $c(Y) = 0$.*
(3) *If $X \subseteq \mathrm{Soc}(M)$ and $\gamma : X \longrightarrow M$ is an R-homomorphism, then there exists $c : M \longrightarrow M$ such that $\gamma(x) = c(x)$ for all $x \in X$.*

If M is finitely generated and self-projective in $\sigma[M]$, then (1)–(3) are equivalent to

(4) *If K is semisimple, P is projective M-generated in $\sigma[M]$, Q is a finitely generated projective M-generated in $\sigma[M]$, $\iota : K \longrightarrow P$ is a monomorphism and $f : K \longrightarrow Q$ is an R-homomorphism, then f can be extended to an R-homomorphism $\tilde{f} : P \longrightarrow Q$.*

Proof. (1) $\Rightarrow$ (2) Let $\mathrm{Soc}(M) = X \oplus Y$ and $\gamma : X \longrightarrow M$ be an R-homomorphism. Define the homomorphism $\tilde{\gamma} : X \oplus Y \longrightarrow M$ by $x+y \mapsto \gamma(x)$ ($x \in X, y \in Y$). Since M is soc-M-injective, $\tilde{\gamma}$ can be extended to the homomorphism $c : M \longrightarrow M$. Let $x \in X$. Then $c(x) = \tilde{\gamma}(x) = \gamma(x)$. Let $y \in Y$. Then $c(y) = \tilde{\gamma}(y) = \gamma(0) = 0$. Thus $c(Y) = 0$.

(2) $\Rightarrow$ (3) $\Rightarrow$ (1) and (4) $\Rightarrow$ (1) are clear.

(1) $\Rightarrow$ (4) Since M is soc-injective, M is soc-P-injective. Clearly, Q is isomorphic to a direct summand of $M^{(n)}$, for some positive integer n. Therefore Q is soc-P-injective by Theorem 2.1. Thus f can be extended to $\tilde{f} : P \longrightarrow Q$. $\qquad \square$

Proposition 4.2. *Let M be a soc-injective module. Then the following holds.*

(1) *M satisfies $(Soc\text{-}C_2)$,*
(2) *M satisfies $(Soc\text{-}C_3)$.*

Proof. Take $N = M$ in Proposition 2.3. $\qquad \square$

Let R and S be rings with identity and M a left S-, a right R-bimodule. For any $X \subseteq M$ and any $T \subseteq S$ denote $l_S(X) = \{s \in S \mid sX = 0\}$ and $r_M(T) = \{m \in M \mid Tm = 0\}$.

Note that if M is a right R-module then M is a left $End_R(M)$-module. If $l_S(A \cap B) = l_S(A) + l_S(B)$ for all submodules A and B of M_R, where $S = End_R(M)$, M is called an *Ikeda-Nakayama* module [10]. Note that every quasi-injective module is an Ikeda-Nakayama module [10, Lemma 1]. For a soc-injective module we have the following result.

Proposition 4.3. *Let S and R be any rings and M a left S-, a right R-bimodule. If M_R is soc-injective, then*

(1) *$l_S(T_1 \cap T_2) = l_S(T_1) + l_S(T_2)$ for all semisimple submodules T_1, T_2 of M_R.*
(2) *If Sk is a simple left S-module ($k \in M$), then $\mathrm{Soc}(kR)$ is zero or simple.*
(3) *$r_M l_S(\mathrm{Soc}(M)) = \mathrm{Soc}(M) \Leftrightarrow r_M l_S(K) = K$ for all semisimple submodule K of M_R.*

Proof. (1) By [10, Lemma 1].

(2) Assume Sk ($k \in M$) is a simple left S-module and $\mathrm{Soc}(kR)$ is nonzero. Let $y_1 R$ and $y_2 R$ be simple submodules of M_R with $y_i \in kR, 1 \leq i \leq 2$. If $y_1 R \cap y_2 R = 0$, then by (1), $l_S(y_1) + l_S(y_2) = S$ and so $l_S(y_1) = l_S(y_2) = l_S(k)$, since $y_i \in kR$ and $l_S(k)$ is a maximal left ideal of S. Thus $l_S(k) = S$, a contradiction, hence $\mathrm{Soc}(kR)$ is simple.

(3) Assume that $r_M l_S(\mathrm{Soc}(M)) = \mathrm{Soc}(M)$ and let K be a semisimple submodule of M_R. We claim that K is essential in $r_M l_S(K)$. If $K \cap xR = 0$ for some $x \in r_M l_S(K)$, then by (1), $l_S(K \cap xR) = l_S(K) + l_S(xR) = S = l_S(xR)$ since $x \in r_M l_S(K) \leq r_M l_S(\mathrm{Soc}(M)) = \mathrm{Soc}(M)$ and $l_S(K) \leq l_S(xR)$. Then $x = 0$. Hence $K \leq_e r_M l_S(K) \leq r_M l_S(\mathrm{Soc}(M)) = \mathrm{Soc}(M)$. It follows that $K = r_M l_S(K)$. The converse is clear. $\qquad \square$

Proposition 4.4. *Let M be a right R-module and $S = End_R(M)$. Then the following are equivalent:*

(1) $r_M l_S(K) = K$ for all semisimple submodules K of M_R.
(2) $r_M[l_S(K) \cap Sa] = K + r_M(a)$ for all semisimple submodules K of M_R and all $a \in S$.

Proof. (1) $\Rightarrow$ (2) Clearly, $K + r_M(a) \leq r_M[l_S(K) \cap Sa]$. Let $x \in r_M[l_S(K) \cap Sa]$ and $y \in l_S(aK)$. Then $yaK = 0$ and $ya \in Sa \cap l_S(K)$, so $yax = 0$ and $y \in l_S(ax)$. Thus $l_S(aK) \leq l_S(ax)$, and so $ax \in r_M l_S(ax) \leq r_M l_S(aK)$. Since Soc$(M)$ is fully invariant, aK is a semisimple submodule of M_R. By (1), $ax \in aK$. Hence $ax = ak$ for some $k \in K$ and so $x - k \in r_M(a)$. This means that $x \in r_M(a) + K$.

(2) $\Rightarrow$ (1) The case when $a = 1_S$. $\qquad\qquad\qquad\qquad\qquad\qquad\qquad\qquad\square$

Proposition 4.5. *Let M be a right R-module and $S = End_R(M)$. If M_R is strongly soc-injective in $\sigma[M]$, then $l_S(A \cap B) = l_S(A) + l_S(B)$ for all semisimple submodules A and all submodules B of M_R.*

Proof. Let $x \in l_S(A \cap B)$ and define $\psi : A + B \longrightarrow M_R$ by $\psi(a + b) = xa$ for all $a \in A$ and $b \in B$. This induces an R-homomorphism $\tilde{\psi} : (A + B)/B \longrightarrow M_R$ in the obvious way. Since $(A+B)/B$ is semisimple and M_R is strongly soc-injective in $\sigma[M]$, $\tilde{\psi}$ can be extended to an R-homomorphism $\varphi : M/B \longrightarrow M$. Now let $\pi : M \longrightarrow M/B$ be the natural epimorphism. Let denote $s = \varphi\pi \in S$. Let $b \in B$. Then $sb = \varphi\pi(b) = \varphi(b+B) = 0$. For any $a \in A$, $(x-s)a = xa - sa = xa - \varphi\pi(a) = 0$. It follows that $x = (x - s) + s \in l_S(A) + l_S(B)$. $\qquad\square$

5. Strongly simple-injective modules in $\sigma[M]$

Theorem 5.1. *The following are equivalent for $N \in \sigma[M]$:*

(1) N is strongly mininjective in $\sigma[M]$.
(2) N is strongly simple-injective in $\sigma[M]$.
(3) Every homomorphism from a finitely generated semisimple submodule K of any module $T \in \sigma[M]$ into N extends to T.
(4) Every homomorphism γ from a submodule K of any module $T \in \sigma[M]$ into N, with $\gamma(K)$ finitely generated semisimple, extends to T.

Proof. (4) $\Rightarrow$ (3) $\Rightarrow$ (1) Clear.

(1) $\Rightarrow$ (2) Let L be a submodule of N and $\gamma : L \longrightarrow K$ a homomorphism with $\gamma(L)$ simple. If $T = Ker\gamma$, then γ induces an embedding $\tilde{\gamma} : L/T \longrightarrow K$ defined by $\tilde{\gamma}(x + T) = \gamma(x)$ for all $x \in L$. Since K is strongly mininjective and L/T is simple, $\tilde{\gamma}$ extends to a homomorphism $\overline{\gamma} : N/T \longrightarrow K$. If $\eta : N \longrightarrow N/T$ is the natural epimorphism, the homomorphism $\overline{\gamma}\eta : N \longrightarrow K$ is an extension of γ, for if $x \in L$, $(\overline{\gamma}\eta)(x) = \overline{\gamma}(x + T) = \tilde{\gamma}(x + T) = \gamma(x)$, as required.

(2) $\Rightarrow$ (4) Let T be any module in $\sigma[M]$, K a submodule of T, $\gamma : K \to N$ a homomorphism with $\gamma(K)$ finitely generated semisimple and consider the following diagram

$$0 \longrightarrow K \overset{i}{\longrightarrow} T$$
$$\gamma \downarrow$$
$$N$$

Write $\gamma(K) = \oplus_{i=1}^{n} S_i$ where each S_i is simple. Let $\pi_i \oplus_{i=1}^{n} S_i \to S_i$ be the canonical projection, $1 \le i \le n$, and consider the following diagram

$$0 \longrightarrow K \overset{i}{\longrightarrow} T$$
$$\pi_i \gamma \downarrow$$
$$N$$

Since N is strongly simple-injective in $\sigma[M]$, for each i, $1 \le i \le n$, there exists a homomorphism $\gamma_i : T \to N$ such that $\gamma_i(x) = \pi_i\gamma(x)$, for all $x \in K$. Now, define the map $\hat{\gamma} : T \to N$ by $\hat{\gamma}(x) = \sum_{i=1}^{n} \gamma_i(x)$. Then $\hat{\gamma}(x) = \gamma(x)$ for all $x \in K$.   □

Hence we have the following implications:

$$\text{soc-}N\text{-injective} \quad \Longrightarrow \quad \text{min-}N\text{-injective}$$
$$\text{simple-}N\text{-injective} \quad \Longrightarrow \quad \text{min-}N\text{-injective}$$
$$\text{strongly mininjective} \quad \Longleftrightarrow \quad \text{strongly simple-injective}$$

Min-N-injective modules need not be soc-N-injective (see [1, Example 4.5] and [1, Example 4.15]), and strongly simple-injective modules need not be strongly soc-injective (see [2, Remark 2.4] and [1]).

Proposition 5.2. (1) *Let $N \in \sigma[M]$ and $\{M_i : i \in I\}$ be a family of modules in $\sigma[M]$. Then the direct product $\prod_{i \in I} M_i$ is min-N-injective if and only if each M_i is min-N-injective, $i \in I$. In particular, $\prod_{i \in I} M_i$ is strongly simple-injective if and only if each M_i is strongly simple-injective, $i \in I$.*

(2) *If $\{M_i : i \in I\}$ is a family of modules in $\sigma[M]$, then the direct sum $\oplus_{i \in I} M_i$ is strongly simple-injective if and only if each M_i is strongly simple-injective, $i \in I$.*

(3) *A direct summand of a strongly simple-injective module is strongly simple-injective.*

(4) *Let M be projective. M is strongly simple-injective if and only if every M-generated projective module $N \in \sigma[M]$ is strongly simple-injective.*

Proof. Routine.   □

Note 5.3. As in Corollary 2.6, for a projective module M, every quotient of a simple-injective module in $\sigma[M]$ is simple-injective if and only if $Soc(M)$ is projective in $\sigma[M]$.

Corollary 5.4. *Let $N \in \sigma[M]$ such that $Soc(N)$ is finitely generated (in particular, if M is finite dimensional), then the following are equivalent:*

(1) *N is strongly mininjective in $\sigma[M]$.*
(2) *N is strongly simple-injective in $\sigma[M]$.*
(3) *N is strongly soc-injective in $\sigma[M]$.*

Moreover, if in addition $Soc(N) \leq_e N$, then each of the above conditions is equivalent to

(4) *M is injective.*

Proof. By Theorem 5.1 and Corollary 3.2. □

Theorem 5.5. *The following are equivalent for $N \in \sigma[M]$:*

(1) *N is strongly simple-injective in $\sigma[M]$.*
(2) *N is min-$\widehat{M}$-injective.*
(3) *N is min-$\widehat{S}$-injective for every simple module $S \in \sigma[M]$.*
(4) *N is min-$\widehat{S}$-injective for every simple submodule S of N.*

Proof. $(1) \Rightarrow (2) \Rightarrow (3) \Rightarrow (4)$ Clear.

$(4) \Rightarrow (1)$ Let $T \in \sigma[M]$, $\gamma : K \to N$ a non-zero homomorphism with $\gamma(K)$ simple, and consider the following diagram

$$0 \longrightarrow \gamma(K) \overset{i}{\longrightarrow} \widehat{\gamma(K)}$$
$$\downarrow{\scriptstyle i}$$
$$N$$

where i is the inclusion map. Since N is min-$\widehat{\gamma(K)}$-injective, there exists an embedding $\sigma : \widehat{\gamma(K)} \to N$ such that $\sigma\gamma(x) = \gamma(x)$ for every $x \in K$. Now, the map γ may be viewed as a map from K into an M-injective submodule of N, and hence has an extension $\widehat{\gamma} : T \to N$. □

Corollary 5.6. *If $N \in \sigma[M]$ is strongly simple-injective, then every simple submodule of N is essential in an M-injective direct summand of N.*

Proof. Let S be a simple submodule of N and consider the following diagram

$$0 \longrightarrow S \overset{i}{\longrightarrow} \widehat{S}$$
$$\downarrow{\scriptstyle i}$$
$$N$$

where i is the inclusion map. Since N is min-$\widehat{S}$-injective and $S \leq_e \widehat{S}$, there exists an embedding σ of $\widehat{S}$ in N such that $\sigma(x) = x$ for all $x \in S$. If $E = \sigma(\widehat{S}) \cong \widehat{S}$, then $S \leq_e E \leq_d N$. □

Proposition 5.7. *The following are equivalent for M:*

(1) *M is locally Noetherian.*

(2) *Every strongly simple-injective module in $\sigma[M]$ is strongly soc-injective.*

Proof. (1) $\Rightarrow$ (2) Suppose M is locally Noetherian, and N is strongly simple-injective in $\sigma[M]$. Write $Soc(N) = \oplus_{i \in I} S_i$, where each S_i is simple, $i \in I$. By Corollary 5.6, each $S_i \leq_e E_i \leq_d N$, where E_i is M-injective, $i \in I$. Since M is locally Noetherian, $E = \oplus_{i \in I} E_i$ is M-injective and hence E is a direct summand of N, and so $N = E \oplus T$, with $Soc(T) = 0$. By Theorem 3.1, N is strongly soc-injective in $\sigma[M]$.

(2) $\Rightarrow$ (1) Let $\{K_i\}_{i \in I}$ be a family of simple modules in $\sigma[M]$. Consider $\widehat{K_i}$ for each $i \in I$. Therefore every $\widehat{K_i}$ is strongly simple-injective in $\sigma[M]$. Then by Proposition 5.2(2), $E = \oplus_{i=1}^{\infty} \widehat{K_i}$ is strongly simple-injective in $\sigma[M]$, and hence strongly soc-injective in $\sigma[M]$. Since E has essential socle, by Corollary 3.2, E is injective in $\sigma[M]$. Therefore M is locally Noetherian by [9, 27.3]. $\square$

Proposition 5.8. *Let M be a finitely generated self-projective module. Then the following are equivalent:*

(1) *M is a Noetherian QF-module.*

(2) *Every strongly simple-injective module in $\sigma[M]$ is projective in $\sigma[M]$.*

Proof. (1) $\Rightarrow$ (2) By Proposition 5.7 and Proposition 3.7.

(2) $\Rightarrow$ (1) By Proposition 3.7. $\square$

Acknowledgment

The first author is supported by the project of Hacettepe University of number 05 G 602 001.

References

[1] Amin I., Yousif M.F., Zeyada N. (2005) Soc-injective rings and modules, *Comm. Alg.* 33, 4229–4250.

[2] Amin I., Fathi Y., Yousif M.F. (2008) Strongly simple-injective rings and modules, *Alg. Coll.* 15(1): 135–144.

[3] Camillo V., Yousif M.F. (1991) CS-modules with acc or dcc, *Comm. Algebra*, 19(2): 655–662.

[4] Cartan H., Eilenberg S. (1956) *Homological Algebra*, Princeton:Princeton University Press.

[5] Çelik C., Harmancı A. and Smith P.F. (1995) A generalization of CS-modules, *Comm. Alg.* 23: 5445–5460.

[6] Dung N.V., Huynh D.V., Smith P.F., Wisbauer R. (1994), *Extending Modules* Pitman RN Mathematics 313, Longman, Harlow.

[7] Harada M. (1982) On Modules with Extending Properties, *Osaka J. Math.*, 19: 203–215.

[8] Mohamed S.H., Müller B.J. (1990) *Continuous and Discrete Modules*, London Math. Soc. LNS 147 Cambridge Univ. Press, Cambridge.

[9] Wisbauer R. (1991) *Foundations of Module and Ring Theory.* Gordon and Breach, Reading.

[10] Wisbauer R., Yousif M.F.; Zhou Y. (2002) Ikeda-Nakayama Modules, *Contributions to Algebra and Geometry*, 43(1): 111–119.

[11] Yousif M.F., Zhou Y. (2002) Semiregular, semiperfect and perfect rings relative to an ideal, *Rocky Mountain J. Math.*, 32(4):1651–1671

A.Ç. Özcan and D. Keskin Tütüncü
Hacettepe University
Department of Mathematics
06800 Beytepe Ankara, Turkey
e-mail: ozcan@hacettepe.edu.tr
e-mail: keskin@hacettepe.edu.tr

M.F. Yousif
Department of Mathematics
The Ohio State University
Lima, Ohio, USA
e-mail: yousif.1@osu.edu

Modules and Comodules

Trends in Mathematics, 331–355

Biproducts and Two-cocycle Twists of Hopf Algebras

David E. Radford and Hans Jürgen Schneider

Abstract. Let H be a Hopf algebra with bijective antipode over a field k and suppose that $R\#H$ is a bi-product. Then R is a bialgebra in the Yetter-Drinfel'd category ${}^{H}_{H}\mathcal{Y}D$. We describe the bialgebras $(R\#H)^{op}$ and $(R\#H)^{o}$ explicitly as bi-products $R^{\underline{op}}\#H^{op}$ and $R^{\underline{o}}\#H^{o}$ respectively where $R^{\underline{op}}$ is a bialgebra in ${}^{H^{op}}_{H^{op}}\mathcal{Y}D$ and $R^{\underline{o}}$ is a bialgebra in ${}^{H^{o}}_{H^{o}}\mathcal{Y}D$. We use our results to describe two-cocycle twist bialgebra structures on the tensor product of bi-products.

Introduction

In [11] the irreducible representations of a certain class of pointed Hopf algebras over a field k are parameterized by pairs of characters, or by characters. These Hopf algebras are twists $H = (U \otimes A)_\sigma$ of the tensor product of pointed Hopf algebras or quotients of them. The twist structures are in one-one correspondence with bialgebra maps $U \longrightarrow A^{op\ o}$. In many cases the pointed Hopf algebras U and A are bi-products. We are thus led to consider the multiplicative opposite $(R\#H)^{op}$ and the dual $(R\#H)^{o}$ of a bi-product $R\#H$. Generally the multiplicative opposite and dual of a bi-product is a bi-product. One purpose of this paper is to characterize $(R\#H)^{op}$ and $(R\#H)^{o}$ as bi-products when H has bijective antipode. Recall that pointed Hopf algebras have bijective antipodes.

Let H be a Hopf algebra over k with bijective antipode and suppose that $R\#H$ is a bi-product. Then R is a bialgebra in the category of Yetter-Drinfel'd modules ${}^{H}_{H}\mathcal{Y}D$. We construct a bialgebra $R^{\underline{op}}$ in the Yetter-Drinfel'd category ${}^{H^{op}}_{H^{op}}\mathcal{Y}D$ such that $(R\#H)^{op} \simeq R^{\underline{op}}\#H^{op}$. Likewise we construct a bialgebra $R^{\underline{o}}$ in the Yetter-Drinfel'd category ${}^{H^{o}}_{H^{o}}\mathcal{Y}D$ such that $(R\#H)^{o} \simeq R^{\underline{o}}\#H^{o}$. These bialgebra constructions are based on more general procedures: the construction of an algebra (respectively a coalgebra) $A^{\underline{op}}$ in ${}^{H^{op}}_{H^{op}}\mathcal{Y}D$ from an algebra (respectively a

Research by the first author partially supported by NSA Grant H98230-04-1-0061.

coalgebra) A in $_H^H \mathcal{YD}$ and the construction of an algebra (respectively a coalgebra) $A^{\underline{o}}$ in $_{H^o}^{H^o} \mathcal{YD}$ from an algebra (respectively a coalgebra) A in $_H^H \mathcal{YD}$. We note that the $A^{\underline{op}}$ construction is described more generally in the context of braided categories [7].

Important to us is the case $U = \mathfrak{B}(W) \# k[\Gamma]$ and $A = \mathfrak{B}(V) \# k[\Lambda]$, where Γ and Λ are abelian groups and $\mathfrak{B}(W)$, $\mathfrak{B}(V)$ are Nichols algebras in the categories $_{k[\Gamma]}^{k[\Gamma]} \mathcal{YD}$, $_{k[\Lambda]}^{k[\Lambda]} \mathcal{YD}$ respectively. These are of primary interest in [11]. For these Hopf algebras an extensive class of bialgebra maps $U \longrightarrow A^{op\,o}$ can be given in terms of two linear forms $\tau : k[\Lambda] \otimes k[\Gamma] \longrightarrow k$ and $\beta : V \otimes W \longrightarrow k$. The forms can easily be produced and thus, in particular, our results provide a way of generating a large number of two-cocycles twist bialgebras, that is bialgebra maps $\mathfrak{B}(W) \# k[\Lambda] \longrightarrow (\mathfrak{B}(V) \# k[\Gamma])^{op\,o}$ without checking the relations of $\mathfrak{B}(W)$ or $\mathfrak{B}(V)$ which are unknown in general.

For finite abelian groups $\Gamma, V \in {}_{k[\Gamma]}^{k[\Gamma]} \mathcal{YD}$ and finite-dimensional Nichols algebras $\mathfrak{B}(V)$, the dual of $\mathfrak{B}(V) \# k[\Gamma]$ was already computed in [4, Theorem 2.2]. The two-cocycles in Corollary 9.1 were determined in [2] for finite-dimensional Nichols algebras with known relations by explicitly checking the relations. However, in [2] the more general case when U is not coradically graded, that is of the form $\mathfrak{B}(W) \# k[\Lambda]$, was considered.

This paper is organized as follows. In Section 1 we deal with the somewhat extensive prerequisites for the paper. First we discuss notations for algebras, coalgebras, and their representations, and then review algebraic objects in the Yetter-Drinfel'd category $_H^H \mathcal{YD}$ of a Hopf algebra H with bijective antipode in detail for the reader's convenience. This discussion is important for Sections 2 and 3 where we describe algebra, coalgebra, and bialgebra constructions in the Yetter-Drinfel'd categories and $_{H^{op}}^{H^{op}} \mathcal{YD}$ and $_{H^o}^{H^o} \mathcal{YD}$ based on the their counterparts in $_H^H \mathcal{YD}$. These constructions are basic ingredients in the realization of the multiplicative opposite and dual of a bi-product as a bi-product.

Certain bilinear forms on objects in Yetter-Drinfel'd categories which play a role in our construction of our two-cocycle twist Hopf algebras are introduced and studied in Section 4. In Section 5 we consider morphisms of bi-products. The multiplicative opposite of a bi-product is characterized as a bi-product in Section 6 and the dual of a bi-product is characterized as a bi-product in Section 7.

We apply the main results of Sections 6 and 7 to describe certain two-cocycle twists on the tensor product $(T \# K) \otimes (R \# H)$ of bi-products in Section 8. Here K and H are Hopf algebras with bijective antipodes over k. We focus on the basic case when $T = \mathfrak{B}(W)$ and $R = \mathfrak{B}(V)$ are Nichols algebras on finite-dimensional Yetter-Drinfel'd modules. In the last section we consider our basic case when $K = k[\Lambda]$ and $H = k[\Gamma]$ are group algebras of abelian groups. Results here fit nicely into the discussion of [11]. In Corollary 9.2 we describe the reduction to the case when $\beta : V \otimes W \to k$ is non-singular. In work in progress we show that the bilinear form in the non-singular case for braidings of finite Cartan type defines a Quasi-R-matrix.

We denote the antipode of a Hopf algebra H over k by S. Any one of [1, 6, 8, 12] will serve as a Hopf algebra reference for this paper. Throughout k is a field and all vector spaces are over k. For vector spaces U and V we will drop the subscript k from $\text{End}_k(V)$, $\text{Hom}_k(U, V)$, and $U \otimes_k V$. We denote the identity map of V by I_V. For a non-empty subset S of the dual space V^* we let $S^\perp$ denote the subspace of V consisting of the common zeros of the functionals in S. For $p \in U^*$ and $u \in U$ we denote the evaluation of p on u by $p(u)$ or $\langle p, u \rangle$.

1. Preliminaries

A good deal of prerequisite material is needed for this paper. We discuss general notation first then review specific topics in detail.

For a group G we let $\widehat{G}$ denote the group of characters of G with values in k and $k[G]$ the group algebra of G over k. For a Hopf algebra H over k we denote the group of grouplike elements of H by $G(H)$ as usual.

Let (A, m, η) be an algebra over k, which we shall usually denote by A. Generally we represent algebraic objects defined on a vector space by their underlying vector space. Observe that (A, m^{op}, η) is an algebra over k, where $m^{op} = m \circ \tau_{A,A}$. We denote A with this algebra structure by A^{op} and write $a^{op} = a$ for elements of A^{op}. Thus $a^{op} b^{op} = (ba)^{op}$ for all $a, b \in A$. This notation is very useful for computations in Yetter-Drinfel'd categories discussed below involving certain algebra constructions. We denote the category of left (respectively right) A-modules and module maps by ${}_A\mathcal{M}$ (respectively $\mathcal{M}_A$). If $\mathcal{C}$ is a category, by abuse of notation we will write $C \in \mathcal{C}$ to indicate that C is an object of $\mathcal{C}$.

Let M be a left A-module. Then M^* is a right A-module under the transpose action which is given by $(m^* \cdot a)(m) = m^*(a \cdot m)$ for all $m^* \in M^*$, $a \in A$, and $m \in M$. Likewise if M is a right A-module then M^* is a left A-module where $(a \cdot m^*)(m) = m^*(m \cdot a)$ for all $a \in A$, $m^* \in M^*$, and $m \in M$.

Suppose B is an algebra also over k, let N be a left B-module, and suppose that $\varphi : A \longrightarrow B$ is an algebra map. Then a linear map $f : M \longrightarrow N$ is φ-linear if $f(a \cdot m) = \varphi(a) \cdot f(m)$ for all $a \in A$ and $m \in M$. There is a way of expressing the last equation in terms of A-module maps. Note that N is a left A-module by pullback along φ. Thus f is φ-linear if and only if f is a map of left A-modules.

Let (C, Δ, ϵ) be a coalgebra over k, which we usually denote by C. At times it is convenient to denote the coproduct Δ by Δ_C. Generally we use a variant on the Heyneman-Sweedler notation for the coproduct and write $\Delta(c) = c_{(1)} \otimes c_{(2)}$ to denote $\Delta(c) \in C \otimes C$ for $c \in C$. Note that $(C, \Delta^{cop}, \epsilon)$ is a coalgebra over k, where $\Delta^{cop} = \tau_{C,C} \circ \Delta$. We let C^{cop} denote the vector space C with this coalgebra structure and sometimes write $c^{cop} = c$ for elements of C^{cop}. With this notation $c^{cop}{}_{(1)} \otimes c^{cop}{}_{(2)} = c_{(2)} \otimes c_{(1)}$ for all $c \in C^{cop}$.

Suppose that (M, δ) is a left C-comodule. For $m \in M$ we use the notation $\delta(m) = m_{(-1)} \otimes m_{(0)}$ to denote $\delta(m) \in C \otimes M$. If (M, δ) is a right C-comodule we denote $\delta(m) \in M \otimes C$ by $\delta(m) = m_{(0)} \otimes m_{(1)}$. Observe that our coproduct and comodule notations do not conflict.

We make an exception to our coproduct notation described above for coalgebras in Yetter-Drinfel'd categories, in which case we write $\Delta(c) = c^{(1)} \otimes c^{(2)}$ for $c \in C$. See Section 1.2.

Suppose that M is a left C-comodule, D is a coalgebra over k, N is a left D-comodule, and $\varphi : C \longrightarrow D$ is a coalgebra map. Then a linear map $f : M \longrightarrow N$ is left φ-colinear if $\varphi(m_{(-1)}) \otimes f(m_{(0)}) = f(m)_{(-1)} \otimes f(m)_{(0)}$ for all $m \in M$. There is a way of expressing the last equation in terms of D-comodule maps. Note that M is a left D-comodule by push-out along φ. Thus f is φ-colinear if and only if f is a map of left D-modules. We use the terminology φ-linear and colinear as shorthand for φ-linear and φ-colinear.

Bilinear forms play an important role in this paper. We will think of them in terms of linear forms $\beta : U \otimes V \longrightarrow k$ and will often write $\beta(u, v)$ for $\beta(u \otimes v)$. Note that β determines linear maps $\beta_\ell : U \longrightarrow V^*$ and $\beta_r : V \longrightarrow U^*$ where $\beta_\ell(u)(v) = \beta(u, v) = \beta_r(v)(u)$ for all $u \in U$ and $v \in V$. The form β is left (respectively right) non-singular if β_ℓ (respectively β_r) is one-one and β is non-singular if it is both left and right non-singular.

For subspaces $X \subseteq U$ and $Y \subseteq V$ we define subspaces $X^\perp \subseteq V$ and $Y^\perp \subseteq U$ by

$$X^\perp = \{v \in V \mid \beta(X, v) = (0)\} \quad \text{and} \quad Y^\perp = \{u \in U \mid \beta(u, Y) = (0)\}.$$

Note that there is a form $\overline{\beta} : U/V^\perp \otimes V/U^\perp \longrightarrow k$ uniquely determined by $\overline{\beta} \circ (\pi_{V^\perp} \otimes \pi_{U^\perp}) = \beta$, where $\pi_{V^\perp} : U \longrightarrow U/V^\perp$ and $\pi_{U^\perp} : U \longrightarrow V/U^\perp$ are the projections. Observe that $V^\perp = \operatorname{Ker} \beta_\ell$, $U^\perp = \operatorname{Ker} \beta_r$, and that $\overline{\beta}$ is non-singular.

1.1. Two-cocycle twist bialgebras

Let A be a bialgebra over k. A two-cocycle for A is a convolution invertible linear form $\sigma : A \otimes A \longrightarrow k$ which satisfies

$$\sigma(x_{(1)}, y_{(1)})\sigma(x_{(2)}y_{(2)}, z) = \sigma(y_{(1)}, z_{(1)})\sigma(x, y_{(2)}z_{(2)})$$

for all $x, y, z \in A$. If σ is a two-cocycle for A then A_σ is a bialgebra, where $A_\sigma = A$ as a coalgebra and multiplication $m_\sigma : A \otimes A \longrightarrow A$ is given by

$$m_\sigma(x \otimes y) = \sigma(x_{(1)}, y_{(1)})x_{(2)}y_{(2)}\sigma^{-1}(x_{(3)}, y_{(3)})$$

for all $x, y \in A$.

Let U and A be bialgebras over k and suppose that $\tau : U \otimes A \longrightarrow k$ is a linear form. Consider the axioms:

(A.1) $\tau(u, aa') = \tau(u_{(2)}, a)\tau(u_{(1)}, a')$ for all $u \in U$ and $a, a' \in A$;
(A.2) $\tau(1, a) = \epsilon(a)$ for all $a \in A$;
(A.3) $\tau(uu', a) = \tau(u, a_{(1)})\tau(u', a_{(2)})$ for all $u, u' \in U$ and $a \in A$;
(A.4) $\tau(u, 1) = \epsilon(u)$ for all $u \in U$.

Axioms (A.1)–(A.4) are equivalent to

$$\tau_\ell(U) \subseteq A^o \quad \text{and} \quad \tau_\ell : U \longrightarrow A^{o\,cop} = A^{op\,o} \text{ is a bialgebra map} \qquad (1)$$

and they are also equivalent to

$$\tau_r(A) \subseteq U^o \text{ and } \tau_r : A \longrightarrow U^{o\,op} \text{ is a bialgebra map.} \qquad (2)$$

Observe that (A.1)–(A.4) merely describe a bialgebra braiding between U and A^{op} [5], [7].

Suppose that (A.1)–(A.4) hold, τ is convolution invertible, and define a linear form $\sigma : (U \otimes A) \otimes (U \otimes A) \longrightarrow k$ by $\sigma(u \otimes a, u' \otimes a') = \epsilon(a)\tau(u', a)\epsilon(a')$ for all $u, u' \in U$ and $a, a' \in A$. Then σ is a two-cocycle. We denote the two-cocycle twist bialgebra structure on the tensor product bialgebra $U \otimes A$ by $H = (U \otimes A)_\sigma$. Observe that

$$(u \otimes a)(u' \otimes a') = u\tau(u'_{(1)}, a_{(1)})u'_{(2)} \otimes a_{(2)}\tau^{-1}(u'_{(3)}, a_{(3)})a'$$

for all $u, u' \in U$ and $a, a' \in A$.

Suppose that (A.1)–(A.4) hold for the linear form $\tau : U \otimes A \longrightarrow k$. Then τ is invertible if U has an antipode S or if A^{op} has an antipode ς. In the first case $\tau^{-1}(u, a) = \tau(S(u), a)$, and in the second $\tau^{-1}(u, a) = \tau(u, \varsigma(a))$, for all $u \in U$ and $a \in A$. See [11, Lemma 1.2].

As noted in [11, Section 1], the quantum double provides an important example of a two-cocycle twist bialgebra. This example is described in [5] where two-cocycle twist bialgebras are defined and discussed.

Let $U, \overline{U}$ and $A, \overline{A}$ be algebras over k. Suppose further that $\tau : U \otimes A \longrightarrow k$ and $\overline{\tau} : \overline{U} \otimes \overline{A} \longrightarrow k$ are convolution invertible linear forms satisfying (A.1)–(A.4). Set $H = (U \otimes A)_\sigma$ and $\overline{H} = (\overline{U} \otimes \overline{A})_{\overline{\sigma}}$. Suppose that $f : U \longrightarrow \overline{U}$ and $g : A \longrightarrow \overline{A}$ are bialgebra maps such that $\overline{\tau}(f(u), g(a)) = \tau(u, a)$ for all $u \in U$ and $a \in A$. Then $f \otimes g : H \longrightarrow \overline{H}$ is a bialgebra map.

1.2. Yetter-Drinfel'd categories and their algebras, coalgebras, and bialgebras

Here we organize well-known material for the reader's convenience and for our use in later sections. See [3] in particular.

Let H be a bialgebra over k and let $^H_H\mathcal{YD}$ be the category whose objects are triples $(M, \cdot, \delta)$, where $(M, \cdot)$ is a left H-module, (M, δ) is a left H-comodule, compatible in the sense that

$$h_{(1)}m_{(-1)} \otimes h_{(2)} \cdot m_{(0)} = (h_{(1)} \cdot m)_{(-1)}h_{(2)} \otimes (h_{(1)} \cdot m)_{(0)} \qquad (3)$$

for all $h \in H$ and $m \in M$, and whose morphisms $(M, \cdot, \delta) \longrightarrow (M', \cdot', \delta')$ are maps $f : M \longrightarrow M'$ simultaneously of left H-modules and of left H-comodules. We follow the convention of referring to an object of $^H_H\mathcal{YD}$ as a Yetter-Drinfel'd module [3]. If H has an antipode S then (3) is equivalent to

$$\delta(h \cdot m) = h_{(1)}m_{(-1)}S(h_{(3)}) \otimes h_{(2)} \cdot m_{(0)} \qquad (4)$$

for all $h \in H$ and $m \in M$, in practice a very useful formulation of the compatibility condition (3).

The category $^H_H\mathcal{YD}$ has a monoidal structure, where k is given the left H-module structure $h \cdot 1_k = \epsilon(h)$ for all $h \in H$ and left H-comodule structure determined by $\delta(1_k) = 1_H \otimes 1_k$, and the tensor product of objects $M, N \in ^H_H\mathcal{YD}$ is $M \otimes N$ as a vector space with left H-module structure given by $h \cdot (m \otimes n) = h_{(1)} \cdot m \otimes h_{(2)} \cdot n$

and left H-comodule structure given by $\delta(m \otimes n) = m_{(-1)} n_{(-1)} \otimes (m_{(0)} \otimes n_{(0)})$ for all $h \in H$, $m \in M$, and $n \in N$. When H is a Hopf algebra $^H_H \mathcal{YD}$ is a braided monoidal category with braiding $\sigma_{M,N} : M \otimes N \longrightarrow N \otimes M$ for objects $M, N \in ^H_H\mathcal{YD}$ determined by $\sigma_{M,N}(m \otimes n) = m_{(-1)} \cdot n \otimes m_{(0)}$ for all $m \in M$ and $n \in N$.

Let (A, m, η) be an algebra in $^H_H\mathcal{YD}$. Then (A, m^{op}, η) is as well, where $m^{op} = m \circ \sigma_{A,A}$. Thus $a^{op} b^{op} = (a_{(-1)} \cdot b) a_{(0)}$ for all $a, b \in A$. We denote the object A with this algebra structure by A^{op}. If B is also an algebra in $^H_H\mathcal{YD}$ then $A \otimes B$ is an algebra in $^H_H\mathcal{YD}$, where $\eta_{A \otimes B} = \eta_A \otimes \eta_B$ and $m_{A \otimes B} = (m_A \otimes m_B) \circ (I_A \otimes \sigma_{A,B} \otimes I_B)$. We write $A \underline{\otimes} B$ for $A \otimes B$ with this algebra structure and $a \underline{\otimes} b = a \otimes b$ for tensors. By definition

$$(a \underline{\otimes} b)(a' \underline{\otimes} b') = a(b_{(-1)} \cdot a') \underline{\otimes} b_{(0)} b'$$

for all $a, a' \in A$ and $b, b' \in B$. Observe that the object k with its usual k-algebra structure is an algebra in $^H_H\mathcal{YD}$.

Suppose (C, Δ, ϵ) is a coalgebra in $^H_H\mathcal{YD}$. We shall write $\Delta(c) = c^{(1)} \otimes c^{(2)}$ for $c \in C$. Observe that $(C, \Delta^{cop}, \epsilon)$ is a coalgebra in $^H_H\mathcal{YD}$, where $\Delta^{cop} = \sigma_{C,C} \circ \Delta$, or equivalently $\Delta^{cop}(c) = c^{(1)}_{(-1)} \cdot c^{(2)} \otimes c^{(1)}_{(0)}$ for all $c \in C$. We denote the object C with this coalgebra structure by C^{cop}. If D is a coalgebra in $^H_H\mathcal{YD}$ also then $C \otimes D$ is a coalgebra in $^H_H\mathcal{YD}$, where $\epsilon_{C \otimes D} = \epsilon_C \otimes \epsilon_D$ and $\Delta_{C \otimes D} = (I_C \otimes \sigma_{C,D} \otimes I_D) \circ (\Delta_C \otimes \Delta_D)$. We write $C \overline{\otimes} D$ for $C \otimes D$ with this coalgebra structure. By definition

$$\Delta(c \overline{\otimes} d) = (c^{(1)} \overline{\otimes} c^{(2)}_{(-1)} \cdot d^{(1)}) \otimes (c^{(2)}_{(0)} \overline{\otimes} d^{(2)})$$

for all $c \in C$ and $d \in D$. Observe that the object k with its usual k-coalgebra structure is a coalgebra in $^H_H\mathcal{YD}$.

Let $R \in ^H_H\mathcal{YD}$ be an algebra and a coalgebra in the category. Then $\Delta : R \longrightarrow R \underline{\otimes} R$ and $\epsilon : R \longrightarrow k$ are algebra maps if and only if $m : R \overline{\otimes} R \longrightarrow R$ and $\eta : k \longrightarrow R$ are coalgebra maps in which case we say that R with its algebra and coalgebra structure is a bialgebra in $^H_H\mathcal{YD}$. If R is a bialgebra in $^H_H\mathcal{YD}$ then R^{op}, R^{cop}, and therefore $R^{op \, cop}$, are bialgebras in $^H_H\mathcal{YD}$. Observe that the object k with its usual k-algebra and k-coalgebra structure is a bialgebra in $^H_H\mathcal{YD}$. If R and T are bialgebras in $^H_H\mathcal{YD}$ then the object $R \otimes T$ with its algebra structure $R \underline{\otimes} T$ and coalgebra structure $R \overline{\otimes} T$ is a bialgebra in $^H_H\mathcal{YD}$.

The most important bialgebra in $^H_H\mathcal{YD}$ for us is the Nichols algebra. Let $M \in ^H_H\mathcal{YD}$ and consider the tensor k-algebra $T(M) = k \oplus M \oplus (M \otimes M) \oplus \cdots = \bigoplus_{\ell=0}^{\infty} M^{\otimes \ell}$ on the vector space M. Regard $T(M)$ as an object of $^H_H\mathcal{YD}$, where

$$h \cdot (m_1 \otimes \cdots \otimes m_r) = h_{(1)} \cdot m_1 \otimes \cdots \otimes h_{(r)} \cdot m_r$$

for all $h \in H$ and $m_1, \ldots, m_r \in M$ and

$$\delta(m_1 \otimes \cdots \otimes m_r) = m_{1\,(-1)} \cdots m_{r\,(-1)} \otimes (m_{1\,(0)} \otimes \cdots \otimes m_{r\,(0)})$$

for all $m_1, \ldots, m_r \in M$, describe the left H-module and left H-comodule structures respectively. Let $i : M \longrightarrow T(M)$ be the inclusion. Then the pair $(i, T(M))$ satisfies the obvious analog in $^H_H\mathcal{YD}$ of the universal mapping property of the tensor algebra of a vector space as a k-algebra. Observe that $T(M) = \bigoplus_{\ell=0}^{\infty} M^{\otimes \ell}$ is a graded bialgebra (indeed Hopf algebra) in ${}_H^H\mathcal{YD}$.

The algebra $T(M)$ of ${}_H^H\mathcal{YD}$ is a bialgebra in the category just as the tensor algebra of a vector space is a k-bialgebra. The linear maps $d : M \longrightarrow T(M)\underline{\otimes}T(M)$ and $e : M \longrightarrow k$ defined by $d(m) = 1\underline{\otimes}m + m\underline{\otimes}1$ and $e(m) = 0$ respectively for all $m \in M$ lift to algebra morphisms $\Delta : T(M) \longrightarrow T(M)\underline{\otimes}T(M)$ and $\epsilon : T(M) \longrightarrow k$ uniquely determined by $\Delta \circ i = d$ and $\epsilon \circ i = e$. The structure $(T(M), \Delta, \epsilon)$ is a coalgebra in ${}_H^H\mathcal{YD}$ and $T(M)$ with its algebra and coalgebra structure is a bialgebra in ${}_H^H\mathcal{YD}$. The pair $(i, T(M))$ satisfies the following universal mapping property: If A is a bialgebra in ${}_H^H\mathcal{YD}$ and $f : M \longrightarrow A$ is a morphism such that $\operatorname{Im} f \subseteq P(A)$, the space of primitive elements of A, then there is a bialgebra morphism $F : T(M) \longrightarrow A$ uniquely determined by $F \circ i = f$.

The Nichols algebra $\mathfrak{B}(M)$ has a very simple theoretical description. Among the graded subobjects J of $T(M)$ which are coideals and satisfy $J \cap M = (0)$, there is a unique maximal one I. It is easy to see that I is an ideal of $T(M)$; thus the subobject I is a bi-ideal. Consequently the quotient $\mathfrak{B}(M) = T(M)/I$ is a connected graded bialgebra in the category ${}_H^H\mathcal{YD}$. Observe that $\mathfrak{B}(M)$ is generated as an algebra by $\mathfrak{B}(M)(1) = M$, which is also the space of primitive elements of $\mathfrak{B}(M)$. Since $I \cap \mathfrak{B}(M) = (0)$ we may think of M as a subspace of $\mathfrak{B}(M)$. The pair $(M, (\mathfrak{B}(M))$ satisfies the following universal mapping property:

Theorem 1.1. *Let H be a Hopf algebra and $M \in {}_H^H\mathcal{YD}$. Then:*

a) *$\mathfrak{B}(M)$ is a connected graded bialgebra in ${}_H^H\mathcal{YD}$ and $M = \mathfrak{B}(1)$ is a subobject which generates $\mathfrak{B}(M)$ as an algebra.*

b) *If A is a connected graded bialgebra in ${}_H^H\mathcal{YD}$ generated as an algebra by $A(1)$ and $f : A(1) \longrightarrow M$ is a morphism, then there is a bialgebra morphism $F : A \longrightarrow \mathfrak{B}(M)$ determined by $F|_{A(1)} = f$.*

Proof. We need only show part b). By the universal mapping property of the bialgebra $(T(A(1)), i)$ there is a bialgebra morphism $F : T(A(1)) \longrightarrow A$ determined by $F|_{A(1)} = i$. Since $T(A(1))$ is generated by $A(1)$ as an algebra, F is an onto morphism of graded bialgebras. Let $J = \operatorname{Ker} F$. Then J is a sub-object of $T(A(1))$ which is a graded bi-ideal of $T(A(1))$ satisfying $J \cap A(1) = (0)$. Using the universal mapping property again we see that the morphism $f : A(1) \longrightarrow M$ induces a bialgebra morphism $T(f) : T(A(1)) \longrightarrow T(M)$ determined by $T(f)|_{A(1)} = f|_{A(1)}$. Observe that $T(f)(J)$ is a subobject of $T(M)$ which is a graded bi-ideal of $T(M)$ whose intersection with M is (0). This means $T(f)(J) \subseteq I$, where the latter is defined above. The composite $A \simeq T(A(1))/J \longrightarrow T(M)/I = \mathfrak{B}(M)$, where the second map is defined by $x + J \mapsto T(f)(x) + I$, is our desired bialgebra morphism F. $\square$

We have noted that $\mathfrak{B}(M)(1)$ is the subspace of primitive elements of $\mathfrak{B}(M)$. Thus $\mathfrak{B}(M)$ is a connected graded primitively generated bialgebra in ${}_H^H\mathcal{YD}$ with subspace of primitive elements $\mathfrak{B}(M)(1)$. These are defining properties.

Corollary 1.2. *Let H be a Hopf algebra over the field k and suppose that A is a connected graded primitively generated bialgebra in ${}^H_H\mathcal{YD}$ with subspace of primitive elements $A(1)$. Then there is an isomorphism of bialgebras $A \simeq \mathfrak{B}(A(1))$ which extends the identity map $I_{A(1)}$.*

Proof. Let $F : A \longrightarrow \mathfrak{B}(A(1))$ be the bialgebra morphism of part b) of Theorem 1.1 which extends $I_{A(1)}$. Since $A(1)$ generates $\mathfrak{B}(A(1))$ the map F is onto. Now $\operatorname{Ker} F \cap P(A) = \operatorname{Ker} F \cap A(1) = (0)$. Generally if C is a connected coalgebra and $f : C \longrightarrow C'$ is a coalgebra map which satisfies $\operatorname{Ker} f \cap P(C) = (0)$ then f is one-one [12, Lemma 11.0.1]. Thus the onto map F is one-one. $\square$

We leave the reader with the exercise of verifying the following corollary to the theorem above.

Corollary 1.3. *Let K and H be Hopf algebras with bijective antipodes and let $\varphi : K \longrightarrow H$ be a bialgebra map. Suppose that $W \in {}^K_K\mathcal{YD}$, $V \in {}^H_H\mathcal{YD}$, and $f : W \longrightarrow V$ is φ-linear and colinear. Then:*

  a) *There is a map of algebras and coalgebras $\mathfrak{B}(f) : \mathfrak{B}(W) \longrightarrow \mathfrak{B}(V)$ determined by $\mathfrak{B}(f)|_W = f$. Furthermore $\mathfrak{B}(f)$ is φ-linear and colinear.*
  b) *If f is one-one (respectively onto) then $\mathfrak{B}(f)$ is one-one (respectively onto).*

$\square$

2. Associated constructions in ${}^{H^{op}}_{H^{op}}\mathcal{YD}$

Throughout this section H has bijective antipode S. Starting with objects, algebras, coalgebras, and bialgebras in ${}^H_H\mathcal{YD}$ we construct counterparts in ${}^{H^{op}}_{H^{op}}\mathcal{YD}$ which are important for the analysis of bi-products in Section 8. First we start with objects.

Let $(M, \cdot, \delta) \in {}^H_H\mathcal{YD}$. Then $(M, \cdot_{op}, \delta) \in {}^{H^{op}}_{H^{op}}\mathcal{YD}$, where

$$h \cdot_{op} m = S^{-1}(h) \cdot m \tag{5}$$

for all $h \in H$ and $m \in M$. We denote $(M, \cdot_{op}, \delta)$ by $M^{\underline{op}}$. If N is also an object of ${}^H_H\mathcal{YD}$ and $f : M \longrightarrow N$ is a morphism, then $f^{\underline{op}} : M^{\underline{op}} \longrightarrow N^{\underline{op}}$ is a morphism, where $f^{\underline{op}} = f$.

When M has the structure of an algebra, coalgebra, or bialgebra, then $M^{\underline{op}} \in {}^{H^{op}}_{H^{op}}\mathcal{YD}$ does as well. If (A, m, η) is an algebra in ${}^H_H\mathcal{YD}$ then $(A^{\underline{op}}, m^{\underline{op}}, \eta)$ is an algebra in ${}^{H^{op}}_{H^{op}}\mathcal{YD}$, where

$$m^{\underline{op}}(a \otimes b) = ba \tag{6}$$

for all $a, b \in A$. If $\overline{A}$ is also an algebra in ${}^H_H\mathcal{YD}$ and $f : A \longrightarrow \overline{A}$ is an algebra morphism then $f : A^{\underline{op}} \longrightarrow \overline{A}^{\underline{op}}$ is an algebra morphism. If (C, Δ, ϵ) is a coalgebra in ${}^H_H\mathcal{YD}$ then $(C^{\underline{op}}, \Delta^{\underline{op}}, \epsilon)$ is a coalgebra in ${}^{H^{op}}_{H^{op}}\mathcal{YD}$, where

$$\Delta^{\underline{op}}(c) = c^{(2)}{}_{(-1)} \cdot_{op} c^{(1)} \otimes c^{(2)}{}_{(0)} \tag{7}$$

for all $c \in C$. If C' is also a coalgebra in ${}^H_H\mathcal{YD}$ and $f : C \longrightarrow \overline{C}$ is a coalgebra morphism then $f : C^{\underline{op}} \longrightarrow \overline{C}^{\underline{op}}$ is a coalgebra morphism. If $(R, m, \eta, \Delta, \epsilon)$ is a bialgebra in $^H_H\mathcal{YD}$ then $(R^{op}, m^{op}, \eta, \Delta^{op}, \epsilon)$ is a bialgebra in $^{H^{op}}_{H^{op}}\mathcal{YD}$. If $\overline{R}$ is also a bialgebra in $^H_H\mathcal{YD}$ and $f : R \longrightarrow \overline{R}$ is a bialgebra morphism, then $f : R^{op} \longrightarrow \overline{R}^{op}$ is a bialgebra morphism.

Our assertions about A^{op}, C^{op}, and R^{op} can be shown directly with a good deal of effort. For this the a^{op} and c^{cop} notations are strongly recommended. A more illuminating approach which yields much easier proofs is to recognize that there is an isomorphism (F, ϑ) of the monoidal categories $^H_H\mathcal{YD}$ and $^{H^{op}}_{H^{op}}\mathcal{YD}$. The functor $F : {}^H_H\mathcal{YD} \longrightarrow {}^{H^{op}}_{H^{op}}\mathcal{YD}$ is defined by $F(M) = M^{op}$ for objects M and $F(f) = f$ for morphisms f. The morphism of left H^{op}-modules $\vartheta_{M,N} :$ $F(M{\otimes}N) \longrightarrow F(M){\otimes}F(N)$ is defined by $\vartheta_{M,N}(m{\otimes}n) = S^{-1}(n_{(-1)}){\cdot}m{\otimes}n_{(0)}$ for all $m \in M$ and $n \in N$. Observe that $\vartheta^{-1}_{M,N} : F(M){\otimes}F(N) \longrightarrow F(M{\otimes}N)$ is given by $\vartheta^{-1}_{M,N}(m{\otimes}n) = n_{(-1)}{\cdot}m{\otimes}n_{(0)}$ for all $m \in M$ and $n \in N$.

Let (A, m, η) be an algebra in $^H_H\mathcal{YD}$. Then $(F(A), F(m){\circ}\vartheta^{-1}_{A,A}, F(\eta))$ is an algebra of $^{H^{op}}_{H^{op}}\mathcal{YD}$. Since $m^{op} = (F(m){\circ}\vartheta^{-1}_{A,A})^{op}$ it follows that A^{op} is an algebra in $^{H^{op}}_{H^{op}}\mathcal{YD}$ as well. Let (C, Δ, ϵ) be a coalgebra of $^H_H\mathcal{YD}$. Then $(F(C), \vartheta_{C,C}{\circ}F(\Delta), F(\epsilon))$ is a coalgebra of $^{H^{op}}_{H^{op}}\mathcal{YD}$ which is C^{op}. If R is a bialgebra in $^H_H\mathcal{YD}$ then the object R^{op} with its algebra and coalgebra structures R^{op} is a bialgebra in $^{H^{op}}_{H^{op}}\mathcal{YD}$. Our assertions about A^{op}, C^{op}, and R^{op} also follow by the diagrammatic formalism of [7] as well.

Suppose that H has bijective antipode and let V be an object of $^H_H\mathcal{YD}$. Using Corollary 1.2 we are able to relate $\mathfrak{B}(V)^{op}$ to a Nichols algebra.

Observe that the grading of $\mathfrak{B}(V)$ is a bialgebra grading for $\mathfrak{B}(V)^{op}$. It is not hard to see that $\mathfrak{B}(V)(1)$ generates $\mathfrak{B}(V)^{op}$ and is also the space of primitives of $\mathfrak{B}(V)^{op}$. As a subobject of $\mathfrak{B}(V)^{op}$ note that $\mathfrak{B}(V)^{op}(1) = V^{op}$. Thus

$$\mathfrak{B}(V^{op}) = \mathfrak{B}(V)^{op} \tag{8}$$

by Corollary 1.2.

3. Associated constructions in $^{H^o}_{H^o}\mathcal{YD}$

We now turn to constructions in $^{H^o}_{H^o}\mathcal{YD}$. As in the previous section H has bijective antipode S. Starting with objects, algebras, coalgebras, and bialgebras in $^H_H\mathcal{YD}$ we construct counterparts in $^{H^o}_{H^o}\mathcal{YD}$ which are important for the analysis of biproducts in Section 8.

First we consider the objects. Let $(M, \eta, \delta) \in {}^H_H\mathcal{YD}$. We construct an object (M^r, δ^o, η^o) of $^{H^o}_{H^o}\mathcal{YD}$. Regard $H^*{\otimes}M^*$ as a subspace of $(H{\otimes}M)^*$ in the usual way. Recall that M^r, the subspace of all $m^* \in M^*$ which vanish on $I{\cdot}M$ for some cofinite ideal I of H, can be characterized as $M^r = (\eta^*)^{-1}(H^*{\otimes}M^*)$. Furthermore $\eta^*(M^r) \subseteq H^o{\otimes}M^r$ and (M^r, η^o) is a left H^o-comodule, where $\eta^o = \eta^*|_{M^r}$. Thus the comodule action $\eta^o(m^*) = m^*_{(-1)}{\otimes}m^*_{(0)}$ of η^o on $m^* \in M^r$ is determined by

$$m^*_{(-1)}(h)m^*_{(0)}(m) = m^*(h{\cdot}m) \tag{9}$$

for all $h \in H$ and $m \in M$. See [9] for example.

The left H-comodule structure (M, δ) induces a (rational) right H^*-module action in M which in turn induces a left H^*-module structure $(M^*, \cdot)$ on M^* under the transpose action. It is easy to see that $(M^*, \cdot) = (M^*, \delta^*|_{H^* \otimes M^*})$. By restriction of the H^*-action M^* is a left H^o-module. A straightforward calculation yields

$$\eta^*(h^o \cdot m^*) = h^o_{(1)} m^*_{(-1)} S(h^o_{(3)}) \otimes h^o_{(2)} \cdot m^*_{(0)}$$

for all $h^o \in H^o$ and $m^* \in M^r$, where S is the antipode of H^o. Let $\delta^o = \delta^*|_{H^o \otimes M^r}$. Thus $H^o \cdot M^r \subseteq M^r$; hence (M^r, δ^o) is an H^o-submodule of M^* which we denote $(M^r, \cdot)$. The equivalence of (3) and (4) imply that $(M^r, \cdot, \eta^o) \in {}^{H^o}_{H^o}\mathcal{YD}$. The left H^o-module action on M^r is given explicitly by

$$(h^o \cdot m^*)(m) = m^*(m \leftharpoonup h^o) = h^o(m_{(-1)}) m^*(m_{(0)}) \tag{10}$$

for all $h^o \in H^o$, $m^* \in M^r$, and $m \in M$. Note that $(M^{op})^r = M^r$ as vector spaces. If N is also an object of ${}^H_H\mathcal{YD}$ and $f : M \longrightarrow N$ is a morphism then $f^*(N^r) \subseteq M^r$ and the restriction $f^r = f^*|_{N^r}$ is a morphism $f^r : N^r \longrightarrow M^r$ since $f^*(n^*)(h \cdot m) = n^*_{(-1)}(h) f^*(n^*_{(0)})(m)$ for all $n^* \in N^r$, $h \in H$, and $m \in M$.

Suppose that (C, Δ, ϵ) is a coalgebra in ${}^H_H\mathcal{YD}$. Then the object $C^r \in {}^{H^o}_{H^o}\mathcal{YD}$ has the structure of an algebra in the category; as a k-algebra it is a subalgebra of the dual algebra C^*. We show that C^r is a subalgebra of C^* and leave the remaining details of the proof that C^r is an algebra in ${}^{H^o}_{H^o}\mathcal{YD}$ to the reader.

Since $\epsilon : C \longrightarrow k$ is a morphism $\text{Ker } \epsilon$ is a left H-submodule of C. Therefore $\epsilon \in C^r$. Suppose that $a, b \in C^r$. Then $a(I \cdot C) = (0) = b(J \cdot C)$ for some cofinite ideals I and J of H. Since Δ_H is an algebra map, $K = \Delta_H^{-1}(I \otimes H + H \otimes J)$ is an ideal of H which is cofinite since I and J are cofinite ideals of H. Since $\Delta = \Delta_C$ is a map of left H-modules it follows that $\Delta(h \cdot c) = h_{(1)} \cdot c^{(1)} \otimes h_{(2)} \cdot c^{(2)}$ for all $h \in H$ and $c \in C$. Thus

$$(ab)(K \cdot C) \subseteq (a \otimes b)(\Delta(K \cdot C)) \subseteq (a \otimes b)(I \cdot C \otimes H \cdot C + H \cdot C \otimes J \cdot C) = (0)$$

from which $ab \in C^r$ follows. Also note that if $\overline{C}$ is another coalgebra in ${}^H_H\mathcal{YD}$ and if $f : C \longrightarrow \overline{C}$ is a coalgebra morphism then $f^r : \overline{C}^r \longrightarrow C^r$ is an algebra morphism.

One can think of C^r as the counterpart in ${}^{H^o}_{H^o}\mathcal{YD}$ of the dual k-algebra C^*. Suppose A is an algebra in ${}^H_H\mathcal{YD}$. There is a counterpart $A^{\underline{o}}$ in ${}^{H^o}_{H^o}\mathcal{YD}$ to the dual k-coalgebra A^o which arises very naturally in Section 7. As a vector space $A^{\underline{o}}$ is the set of all functionals in A^* which vanish on a cofinite subspace I of A which is both an ideal of A and also a left H-submodule of A.

Now $I^\perp$ is a subcoalgebra of the dual k-coalgebra A^o and is also a left H-submodule of A^r. Since the intersection of two cofinite subspaces of A which are both ideals of A and left H-submodules of A has the same properties, it follows that $A^{\underline{o}}$ is a subcoalgebra of A^o and also a subobject of A^r. At this point is not hard to see that $A^{\underline{o}}$ is a coalgebra in ${}^{H^o}_{H^o}\mathcal{YD}$. Let $\overline{A}$ be an algebra in ${}^H_H\mathcal{YD}$ also and suppose that $f : A \longrightarrow \overline{A}$ is an algebra morphism. Then $f^r(\overline{A}^{\underline{o}}) \subseteq A^{\underline{o}}$ and the restriction $f^{\underline{o}} = f^r|_{\overline{A}^{\underline{o}}}$ is a coalgebra morphism $f^{\underline{o}} : \overline{A}^{\underline{o}} \longrightarrow A^{\underline{o}}$.

Suppose that R is a bialgebra in $^H_H \mathcal{YD}$. Then the object $R^{\underline{o}}$ of $^{H^o}_{H^o}\mathcal{YD}$ is a bialgebra in $^{H^o}_{H^o}\mathcal{YD}$ with the subalgebra structure of the k-algebra R^* and the subcoalgebra structure of the k-coalgebra R^o. Furthermore, if $\overline{R}$ is also a bialgebra in $^H_H \mathcal{YD}$ and $f : R \longrightarrow \overline{R}$ is a bialgebra morphism, then $f^{\underline{o}} : \overline{R}^{\underline{o}} \longrightarrow R^{\underline{o}}$ is a bialgebra morphism.

Let V be an object of $^H_H \mathcal{YD}$. There is a natural relationship between $\mathfrak{B}(V)^{\underline{o}}$ and a Nichols algebra. Consider the one-one map

$$i : V^r \longrightarrow \mathfrak{B}(V)^*$$

defined for $v^* \in V^{\underline{o}}$ by

$$i(v^*)(x) = \begin{cases} v^*(x) & : & x \in \mathfrak{B}(V)(1) = V \\ 0 & : & x \in \mathfrak{B}(V)(n), \ n \neq 1 \end{cases} .$$

Then $\mathrm{Im}\, i \subseteq \mathfrak{B}(V)^{\underline{o}}$ and $i : V^r \longrightarrow \mathfrak{B}(V)^{\underline{o}}$ is a one-one morphism. Let $\mathcal{I} : \mathfrak{B}(V^r) \longrightarrow \mathfrak{B}(V)^{\underline{o}}$ be the bialgebra morphism of Corollary 1.2 which extends i. Since $\mathrm{Ker}\,\mathcal{I} \cap P(\mathfrak{B}(V^r)) = \mathrm{Ker}\,\mathcal{I} \cap V^r = (0)$ it follows that $\mathcal{I}$ is one-one by [12, Lemma 11.0.1] again. We have shown

$$\mathcal{I} : \mathfrak{B}(V^r) \longrightarrow \mathfrak{B}(V)^{\underline{o}} \quad \text{is a one-one bialgebra morphism} \tag{11}$$

When V is finite-dimensional $\mathfrak{B}(V^r)$ is identified with the graded dual of $\mathfrak{B}(V)$ via the map $\mathcal{I}$. In the special case of a Yetter-Drinfel'd module over the group algebra of a finite group with finite-dimensional $\mathfrak{B}(V)$, the dual of $\mathfrak{B}(V)$ was determined in [4, Theorem 2.2].

4. Bilinear forms in the Yetter-Drinfel'd context

Let H be a Hopf algebra with bijective antipode. Let R, T be bialgebras in $^H_H \mathcal{YD}$ and suppose that $\beta : T \otimes R \longrightarrow k$ is a linear form. We will find the following analogs to (A.1)–(A.4) useful:

(B.1) $\beta(tt', r) = \beta(t, S^{-1}(r^{(2)}{}_{(-1)}) \cdot r^{(1)}) \beta(t', r^{(2)}{}_{(0)})$ for all $t, t' \in T$ and $r \in R$;

(B.2) $\beta(1, r) = \epsilon(r)$ for all $r \in R$;

(B.3) $\beta(t, rr') = \beta(t^{(2)}, r)\beta(t^{(1)}, r')$ for all $t \in T$ and $r, r' \in R$;

(B.4) $\beta(t, 1) = \epsilon(t)$ for all $t \in T$.

We leave the reader with the exercise of establishing the equivalence of (B.1)–(B.4) with analogs of (1) and (2) :

Lemma 4.1. *Let H be a Hopf algebra with bijective antipode, let T and R be bialgebras in $^H_H \mathcal{YD}$, and suppose $\beta : T \otimes R \longrightarrow k$ is a linear form. Then the following are equivalent:*

a) *(B.1)–(B.4) hold.*

b) *$\beta_\ell(T) \subseteq R^{op\,\underline{o}}$ and $\beta_\ell : T \longrightarrow R^{op\,\underline{o}}$ is a bialgebra map.*

c) *$\beta_r(R) \subseteq T^{\underline{o}\,op}$ and $\beta_r : R \longrightarrow T^{\underline{o}\,op}$ is a bialgebra map.* $\qquad\square$

Let K and H be bialgebras over k and suppose H has bijective antipode S. Let $W \in {}^K_K \mathcal{YD}$, $V \in {}^H_H \mathcal{YD}$, and let $\tau : K{\otimes}H \longrightarrow k$, $\beta : W{\otimes}V \longrightarrow k$ be linear forms. Two conditions relating τ and β will play an important part in this paper:

(C.1) $\beta(k{\cdot}w, v) = \beta(w, v{\leftharpoonup}\tau_\ell(k))$ for all $k \in K$, $w \in W$, and $v \in V$;

(C.2) $\beta(w{\leftharpoonup}\tau_r(h), v) = \beta(w, S^{-1}(h){\cdot}v)$ for all $w \in W$, $h \in H$, and $v \in V$.

The first (C.1) implies $V^\perp = \operatorname{Ker} \beta_\ell$ is a K-submodule of W and the second (C.2) implies $W^\perp = \operatorname{Ker} \beta_r$ is an H-submodule of V. These conditions have formulations in terms of linear and colinear maps.

Proposition 4.2. *Let K and H be Hopf algebras with bijective antipodes. Suppose that $\tau : K{\otimes}H \longrightarrow k$ satisfies (A.1)–(A.4), let $W \in {}^K_K \mathcal{YD}$, $V \in {}^H_H \mathcal{YD}$, and let $\beta : W{\otimes}V \longrightarrow k$ is a linear form. Then the following are equivalent:*

a) *(C.1) and (C.2) hold.*

b) *$\beta_\ell(W) \subseteq (V^{op})^r$ and $\beta_\ell : W \longrightarrow (V^{op})^r$ is τ_ℓ-linear and colinear.*

c) *$\beta_r(V) \subseteq W^r$ and $\beta_r : V \longrightarrow W^r$ is τ_r-linear and colinear.*

Proof. We show that parts a) and b) are equivalent and leave the reader with the exercise of adapting our proof to establish the equivalence of parts a) and c). In the latter the roles of (C.1) and (C.2) are reversed.

Suppose that $\beta_\ell(W) \subseteq (V^{op})^r$ and consider the linear map $\beta_\ell : W \longrightarrow (V^{op})^r$. Using (10) it follows that β_ℓ is τ_ℓ-linear if and only if (C.1) holds. Using (9), where $S^{-1}(h){\cdot}v = h{\cdot}_{op}v$ replaces $h{\cdot}v$, it follows that β_ℓ is τ_ℓ-colinear if and only if (C.2) holds.

Suppose that (C.2) holds. Now W is a right K^*-module under the rational action arising from (M, δ). Now $\tau_r : H \longrightarrow (K^\circ)^{op}$ is an algebra map by (2). Thus W is left H-module by pullback along τ_ℓ. Let $w \in W$. Then $H{\cdot}w = w{\leftharpoonup}\tau_r(H)$ is finite-dimensional, there is a cofinite ideal I of H such that $(0) = I{\cdot}w = w{\leftharpoonup}\tau_r(I)$. Thus

$$\beta_\ell(w)(I{\cdot}_{op}V) = \beta(w, S^{-1}(I){\cdot}V) = \beta(w{\leftharpoonup}\tau_r(I), V) = (0)$$

which means that $\beta_\ell(w) \in (V^{op})^r$. $\qquad\square$

Corollary 4.3. *Suppose that K and H are Hopf algebras with bijective antipodes and suppose that $\tau : K{\otimes}H \longrightarrow k$ satisfies (A.1)–(A.4). Let $W \in {}^K_K \mathcal{YD}$, $V \in {}^H_H \mathcal{YD}$, and τ and $\beta : W{\otimes}V \longrightarrow k$ satisfy (C.1) and (C.2). Then:*

a) *There is a form $\mathfrak{B}(\beta) : \mathfrak{B}(W){\otimes}\mathfrak{B}(V) \longrightarrow k$ determined by the properties that it satisfies (B.1)–(B.4) and $\mathfrak{B}(\beta)|_{W{\otimes}V} = \beta$. Furthermore τ and $\mathfrak{B}(\beta)$ satisfy (C.1) and (C.2).*

b) *Suppose that $\overline{K}$ and $\overline{H}$ are also Hopf algebras with bijective antipodes over k, $\overline{\tau} : \overline{K}{\otimes}\overline{H} \longrightarrow k$ satisfies (A.1)–(A.4), $\overline{W} \in {}^{\overline{K}}_{\overline{K}} \mathcal{YD}$, $\overline{V} \in {}^{\overline{H}}_{\overline{H}} \mathcal{YD}$, and $\overline{\tau}$ and $\overline{\beta} : \overline{W}{\otimes}\overline{V} \longrightarrow k$ satisfy (C.1) and (C.2). If $\overline{\beta}\circ(f{\otimes}g) = \beta$ then $\mathfrak{B}(\overline{\beta})\circ(\mathfrak{B}(f){\otimes}\mathfrak{B}(g)) = \mathfrak{B}(\beta)$.* $\qquad\square$

Proof. By Proposition 4.2 b), $\beta_\ell : W \longrightarrow (V^{op})^r$ is τ_ℓ-linear and colinear, and by Corollary 1.3

$$\mathfrak{B}(\beta_\ell) : \mathfrak{B}(W) \longrightarrow \mathfrak{B}((V^{op})^r)$$

is a bialgebra map. Then we define $\mathfrak{B}(\beta)_\ell$ as the composition of $\mathfrak{B}(\beta_\ell)$ with the maps

$$\mathfrak{B}((V^{\underline{op}})^r) \longrightarrow \mathfrak{B}(V^{\underline{op}})^{\underline{o}} = \mathfrak{B}(V)^{\underline{op}\,\underline{o}}$$

in (11) and (8). This proves part a), and part b) can be checked easily. $\qquad\square$

5. Bi-products revisited

Let H be a Hopf algebra with antipode S and suppose $R \in {}^H_H\mathcal{YD}$ is a bialgebra in the category. The biproduct $R\#H$ of R and H is a bialgebra over k described as follows. As a vector space $R\#H = R\otimes H$ and $r\#h$ stands for the tensor $r\otimes h$. As a bialgebra $R\#H$ has the smash product and smash coproduct structures. Thus $1_{R\#H} = 1_R\#1_H$,

$$(r\#h)(r'\#h') = r(h_{(1)}\cdot r')\#h_{(2)}h'$$

for all $r, r' \in R$ and $h, h' \in H$,

$$\Delta(r\#h) = (r^{(1)}\#r^{(2)}{}_{(-1)}h_{(1)})\otimes(r^{(2)}{}_{(0)}\#h_{(2)}), \quad \text{and} \quad \epsilon(r\#h) = \epsilon(r)\epsilon(h)$$

for all $r \in R$ and $h \in H$.

The map $j : H \longrightarrow R\#H$ defined by $j(h) = 1\#h$ for $h \in H$ is a bialgebra map and the map $\pi : R\#H \longrightarrow H$ defined by $\pi(r) = r\#1$ for $r \in R$ is an algebra map which satisfies $\pi\circ j = I_H$. Starting with the bialgebras $A = R\#H$, H and the maps j, π one can recover $R = R\#1$ as a bialgebra in ${}^H_H\mathcal{YD}$. Consider the convolution product

$$\Pi = I_A*(j\circ S\circ\pi) \tag{12}$$

which as an endomorphism of A given by $\Pi(a) = a_{(1)}j(S(\pi(a_{(2)})))$ for all $a \in A$. Observe that $\Pi(r\#h) = (r\#1)\epsilon(h)$ for all $r \in R$ and $h \in H$. In particular $R = \mathrm{Im}\,\Pi$. As a k-algebra R is merely a subalgebra of A. As a k-coalgebra

$$\Delta_R(r) = \Pi(r_{(1)})\otimes r_{(2)} \quad \text{and} \quad \epsilon_R(r) = \epsilon(r) \tag{13}$$

for all $r \in R$. As an object of ${}^H_H\mathcal{YD}$ the left H-module action on R is given by

$$h\cdot r = j(h_{(1)})rj(S(h_{(2)})) \tag{14}$$

for all $h \in H$ and $r \in R$ and as a left H-comodule action is given by

$$\delta(r) = \pi(r_{(1)})\otimes r_{(2)} \tag{15}$$

for all $r \in R$.

We are in now in a position to look at biproducts in more abstract terms. Let A be a bialgebra over k and suppose that $j : H \longrightarrow A$, $\pi : A \longrightarrow H$ are bialgebra maps which satisfy $\pi \circ j = I_H$. Let $\Pi : A \longrightarrow A$ be defined by (12) and set $R = \operatorname{Im} \Pi$. The mapping Π has many important properties which are basic for what follows and which we use without particular reference in Section 8; see [10] for example. First of all $\Pi \circ \Pi = \Pi$ and $\pi \circ \Pi = \eta_H \circ \epsilon_A$. Since $\Delta(\Pi(a)) = a_{(1)}(j \circ S \circ \pi)(a_{(3)}) \otimes \Pi(a_{(2)})$ for all $a \in A$ it now follows that $\Delta(R) \subseteq R \otimes A$ and

$$R = \operatorname{Im} \Pi = \{a \in A \,|\, a_{(1)} \otimes \pi(a_{(2)}) = a \otimes 1\} = A^{co\,\pi}.$$

The last equation is definition. In particular R is a subalgebra of A. Since $\Delta(R) \subseteq A \otimes R$ then map $\Delta_R : A \longrightarrow A \otimes A$ defined by (13) satisfies $\Delta_R(R) \subseteq R \otimes R$. Using the fact that $\Pi(aj(h)) = \Pi(a)\epsilon(h)$ for all $a \in A$ and $h \in H$ it follows by direct calculation that $(R, \Delta_R, \epsilon|_R)$ is a k-coalgebra.

Note that $(A, \cdot_j, \delta_\pi) \in {}^H_H \mathcal{Y}D$, where $h \cdot_j a = j(h_{(1)})aj(S(h_{(2)}))$ for all $h \in H$ and $a \in A$ and $\delta_\pi(a) = \pi(a_{(1)}) \otimes a_{(2)}$ for all $a \in A$. Since $\Delta(R) \subseteq A \otimes R$ it follows that R is a left H-subcomodule of (A, δ_π). Since $h \cdot_j \Pi(a) = \Pi(j(h)a)$ for all $h \in H$ and $a \in A$ we see that R is a left H-submodule of $(A, \cdot_j)$. Therefore R is a subobject of $(A, \cdot_j, \delta_\pi)$ and the actions are those described in (14) and (15). In fact R with these structures is a bialgebra in ${}^H_H \mathcal{Y}D$ and the map $R \# H \longrightarrow A$ determined by $r \# h \mapsto rj(h)$ is an isomorphism of k-bialgebras which we call the canonical isomorphism. We refer to R with these structures as the bialgebra in ${}^H_H \mathcal{Y}D$ associated to (A, H, j, π).

The preceding discussion has been based on a bialgebra A over k with bialgebra maps $j : H \longrightarrow A$ and $\pi : A \longrightarrow H$ satisfying $\pi \circ j = I_H$. We have the same context for A^{op}, A^{cop}, thus for $A^{op\,cop}$, and A^o too. For $j : H^{op} \longrightarrow A^{op}$ and $\pi : A^{op} \longrightarrow H^{op}$, as well as $j : H^{cop} \longrightarrow A^{cop}$ and $\pi : A^{cop} \longrightarrow H^{cop}$, are bialgebra maps which satisfy $\pi \circ j = I_H$, and $\pi^o : H^o \longrightarrow A^o$ and $j^o : A^o \longrightarrow H^o$ are bialgebra maps which satisfy $j^o \circ \pi^o = (\pi \circ j)^o = I_{H^o}$. It will be important to us to understand A^{op} and A^o as biproducts. The analysis is rather detailed and will be carried out in Section 8. We do not need to deal with A^{cop}.

We now turn our attention to maps of biproducts. The result we need follows directly from definitions.

Proposition 5.1. *Let H, $\overline{H}$ be Hopf algebras over the field k and let $R \in {}^H_H \mathcal{Y}D$, $\overline{R} \in {}^{\overline{H}}_{\overline{H}} \mathcal{Y}D$ be bialgebras in their respective categories. Suppose $\varphi : H \longrightarrow \overline{H}$ is a bialgebra map and $\psi : R \longrightarrow \overline{R}$ is a map of k-algebras and coalgebras which is also φ-linear and colinear. Then the linear map $\psi \# \varphi : R \# H \longrightarrow \overline{R} \# \overline{H}$ defined by $(\psi \# \varphi)(r \# h) = \psi(r) \# \varphi(h)$ for all $r \in R$ and $h \in H$ is a map of bialgebras over k.* □

Continuing with the statement of the proposition, note that

$$(\psi \# \varphi) \circ j = \overline{j} \circ (\psi \# \varphi) \quad \text{and} \quad \overline{\pi} \circ (\psi \# \varphi) = (\psi \# \varphi) \circ \overline{\pi}.$$

Suppose that A, $\overline{A}$ are bialgebras over k. Let $j : H \longrightarrow A$, $\pi : A \longrightarrow H$ and $\overline{j} : \overline{H} \longrightarrow \overline{A}$, $\overline{\pi} : \overline{A} \longrightarrow \overline{H}$ be bialgebra maps which satisfy $\pi \circ j = I_A$, $\overline{\pi} \circ \overline{j} = I_{\overline{A}}$

respectively. In light of the proposition a natural requirement for bialgebra maps $f : A \longrightarrow \overline{A}$ is $\overline{\pi} \circ f = f \circ \pi$ and $f \circ j = \overline{j} \circ f$. When this is the case $f(\text{Im}\, j) \subseteq \text{Im}\, \overline{j}$ and $\varphi : H \longrightarrow \overline{H}$ determined by $\overline{j} \circ \varphi = f \circ j$ is a bialgebra map, $f(A^{co\, \pi}) \subseteq \overline{A}^{co\, \overline{\pi}}$, and the restriction $f_r = f|_R$ is a map $f_r : R \longrightarrow \overline{R}$ of k-algebras, k-coalgebras, and is φ-linear and colinear. Furthermore the diagram

$$
\begin{array}{ccc}
R\#H & \xrightarrow{\ f_r\#\varphi\ } & \overline{R}\#\overline{H} \\
\downarrow & & \downarrow \\
A & \xrightarrow{\quad f \quad} & \overline{A}
\end{array}
$$

commutes, where the vertical maps are the k-bialgebra isomorphisms determined by $r\#h \mapsto rj(h)$ and $\overline{r}\#\overline{h} \mapsto \overline{r}\,\overline{j}(\overline{h})$ respectively.

6. $(R\#H)^{op}$ as a bi-product

Throughout this section H is a Hopf algebra with bijective antipode S, A is a bialgebra over k, and $j : H \longrightarrow A$, $\pi : A \longrightarrow H$ are bialgebra maps which satisfy $\pi \circ j = I_H$. We will use the results of Section 5 rather freely and in most cases without particular reference. As we noted in Section 1.2 the multiplicative opposite of a bi-product is a bi-product.

Let $R = A^{co\, \pi}$, let $(R, \cdot, \delta)$ be the structure of R as an object of $_H^H \mathcal{YD}$, and let $(R, m, \eta, \Delta, \epsilon)$ the bialgebra in $_H^H \mathcal{YD}$ associated to (A, H, j, π). We recall that $h \cdot r = j(h_{(1)}) r j(S(h_{(2)}))$ for all $h \in H$ and $r \in R$ by (14), and $\delta(r) = \pi(r_{(1)}) \otimes r_{(2)}$ for all $r \in R$ by (15).

As noted in Section 5 the maps $j : H^{op} \longrightarrow A^{op}$, $\pi : A^{op} \longrightarrow H^{op}$ are bialgebra maps which satisfy $\pi \circ j = I_{H^{op}}$. We first observe that $R = A^{co\, \pi} = (A^{op})^{co\, \pi}$. Let $(R, \cdot', \delta')$ be the structure of $R = (A^{op})^{co\, \pi}$ as an object in the category $_{H^{op}}^{H^{op}} \mathcal{YD}$ and let $(R, m', \eta', \Delta', \epsilon')$ be the bialgebra in the category associated with (A^{op}, H^{op}, j, π). The calculation

$$
\begin{aligned}
h^{op} \cdot' r &= j(h^{op}{}_{(1)})^{op} r^{op} j(S^{op}(h^{op}{}_{(2)}))^{op} \\
&= j(S^{-1}(h_{(2)})) r j(h_{(1)}) \\
&= j(S^{-1}(h_{(2)})) r j(S(S^{-1}(h_{(1)}))) \\
&= j(S^{-1}(h)_{(1)}) r j(S(S^{-1}(h)_{(2)})) \\
&= S^{-1}(h) \cdot r
\end{aligned}
$$

for all $h^{op} = h \in H^{op}$ and $r \in R$ shows that $\cdot' = \cdot_{op}$. Since $\delta' = \delta$ it follows that $(R, \cdot', \delta') = (R, \cdot_{op}, \delta)$. Thus $R = R^{op}$ as an object of $_{H^{op}}^{H^{op}} \mathcal{YD}$.

It is clear that $m' = m^{op}$, $\eta' = \eta$, and $\epsilon' = \epsilon$. To calculate Δ' we work from the definition $\Pi^{op} = I_{A^{op}} *^{op} (j \circ S^{op} \circ \pi)$ and compute $\Pi^{op}(a) = ((j \circ S^{-1} \circ \pi)(a_{(2)})) a_{(1)}$ for all $a = a^{op} \in A^{op}$. Now $\Pi(R) \subseteq A \otimes R$ and Π acts as the identity on R, which thus hold for Π^{op} as well. Since $\Pi(j(h)a) = h\cdot_j\Pi(a)$ for all $h \in H$ and $a \in A$, the calculation

$$\Delta'(r) = \Pi^{op}(r^{op}{}_{(1)})\otimes r^{op}{}_{(2)}$$
$$= ((j\circ S^{-1}\circ\pi)(r_{(2)}))r_{(1)}\otimes r_{(3)}$$
$$= \Pi(j(S^{-1}(\pi(r_{(2)})))r_{(1)})\otimes r_{(3)}$$
$$= S^{-1}(\pi(r_{(2)}))\cdot_j\Pi(r_{(1)})\otimes r_{(3)}$$
$$= S^{-1}(r^{(2)}{}_{(-1)})\cdot r^{(1)}\otimes r^{(2)}{}_{(0)}$$

for all $r \in R$ shows that $\Delta' = \Delta^{op}$. We have shown that the bialgebra in ${}^{H^{op}}_{H^{op}}\mathcal{YD}$ which is associated to (A^{op}, H^{op}, j, π) is $R^{\underline{op}}$.

Proposition 6.1. *Let H be a Hopf algebra with bijective antipode, let A be a bialgebra over k, and suppose that $j : H \longrightarrow A$, $\pi : A \longrightarrow H$ are bialgebra maps which satisfy $\pi\circ j = I_H$. Let $R = A^{co\,\pi}$ and let $(R, m, \eta, \Delta, \epsilon)$ be the bialgebra in ${}^{H}_{H}\mathcal{YD}$ associated to (A, H, j, π). Then:*

a) *$j : H^{op} \longrightarrow A^{op}$ and $\pi : A^{op} \longrightarrow H^{op}$ are bialgebra maps which satisfy $\pi\circ j = I_{H^{op}}$, $R^{\underline{op}} = R$ as a vector space, and the bialgebra in ${}^{H^{op}}_{H^{op}}\mathcal{YD}$ associated to (A^{op}, H^{op}, j, π) is $(R^{\underline{op}}, m^{\underline{op}}, \eta, \Delta^{\underline{op}}, \epsilon)$.*

b) *The map $\varphi : R^{\underline{op}}\#H^{op} \longrightarrow (R\#H)^{op}$ given by $\varphi(r\#h) = (1\#h)(r\#1)$ for all $r \in R$ and $h \in H$ is an isomorphism of bialgebras. Furthermore the diagram*

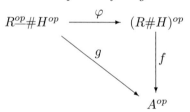

commutes, where $f : R\#H \longrightarrow A$ and $g : R^{\underline{op}}\#H^{op} \longrightarrow A^{op}$ are the canonical isomorphisms.

Proof. We have established part a). As for part b), we first note that $f(r\#h) = rj(h)$ and $g(r\#h) = r^{op}j(h)^{op} = j(h)r$ for all $r \in R$ and $h \in H$. Therefore $f\circ\varphi = g$. This means the diagram commutes and $\varphi = f^{-1}\circ g$ is an isomorphism of bialgebras. $\square$

Observe that $\varphi^{-1}(r\#h) = h_{(1)}\cdot_{op}r\#h_{(2)} = S^{-1}(h_{(1)})\cdot r\#h_{(2)}$ for all $r \in R$ and $h \in H$.

7. $(R\#H)^{\circ}$ as a bi-product

As in the preceding section, H is a Hopf algebra with bijective antipode S, A is a bialgebra over k, and $j : H \longrightarrow A$, $\pi : A \longrightarrow H$ are bialgebra maps which satisfy $\pi\circ j = I_H$. Again we will use the results of Section 5 rather freely and in most cases without particular reference. We observed in Section 1.2 that the dual of a bi-product is a bi-product.

As noted in Section 5 the maps $\pi^o : H^o \longrightarrow A^o$, $j^o : A^o \longrightarrow H^o$ are bialgebra maps which satisfy $j^o \circ \pi^o = I_{H^o}$. We will show that $(A^o)^{co\ j^o}$ can be identified with $R^{\underline{o}}$, find the structure of $R^{\underline{o}}$ as an object of ${}^{H^o}_{H^o}\mathcal{YD}$, and then find its structure as the bialgebra in ${}^{H^o}_{H^o}\mathcal{YD}$ associated to (A^o, H^o, π^o, j^o).

Let $R' = (A^o)^{co\ j^o}$ and $a^o \in A^o$. Then $a^o \in R'$ if and only if $a^o_{(1)} \otimes j^o(a^o_{(2)}) = a^o \otimes \epsilon$, or equivalently $a^o(aj(h)) = a^o(a)\epsilon(h)$ for all $a \in A$ and $h \in H$. Recall from Section 5 that the map $R\#H \longrightarrow A$ determined by $r\#h \mapsto rj(h)$ for all $r \in R$ and $h \in H$ is an isomorphism of bialgebras. Since $A = Rj(H)$ it follows that $a^o \in R'$ if and only if $a^o(rj(h)) = a^o(r)\epsilon(h)$ for all $r \in R$ and $h \in H$. The isomorphism gives rise to a linear embedding $i : R^* \longrightarrow A^*$, where $i(r^*)(rj(h)) = r^*(r)\epsilon(h)$ for all $r^* \in R^*$, $r \in R$, and $h \in H$. Observe that $R' \subseteq \text{Im}\,i$. Thus we can understand R' in terms of R^* via the embedding.

Our first claim is that $i(R^{\underline{o}}) = R'$. A consequence of the claim is that the restriction

$$i|_{R^{\underline{o}}} : R^{\underline{o}} \longrightarrow R' \tag{16}$$

is a linear isomorphism.

To prove our claim, first of all suppose that $r^{\underline{o}} \in R^{\underline{o}}$. To show that $i(r^{\underline{o}}) \in R'$ we need only show that $i(r^{\underline{o}}) \in A^o$. Since $r^{\underline{o}} \in R^{\underline{o}}$, by definition $r^{\underline{o}}(J) = (0)$ for some cofinite subspace J of R which is an ideal and a left H-submodule of R. We will use the commutation relations

$$j(h)r = (h_{(1)} \cdot_j r)j(h_{(2)}) \qquad \text{and} \qquad rj(S(h)) = j(S(h_{(1)}))(h_{(2)} \cdot_j r)$$

for all $h \in H$ and $r \in R$. Since S is onto and J is a left H-subcomodule of R it follows from the commutations relations that $j(H)J = Jj(H)$. Thus $Jj(H)$ is a left ideal of A. Now $i(r^{\underline{o}})$ vanishes on $Jj(H)$ and $Rj(H)^+ = R(j(H) \cap \text{Ker}\,\epsilon)$ as well. Since $H \cdot_j R \subseteq R$ and $j(H)^+$ is a left ideal of $j(H)$, by the first commutation relation $Rj(H)^+$ is a left ideal of A. Since $Jj(H) + Rj(H)^+$ is a cofinite left ideal of A on which $i(r^{\underline{o}})$ vanishes, it follows that $i(r^{\underline{o}})$ vanishes on a cofinite ideal of A. Thus $i(r^{\underline{o}}) \in A^o$ as required.

Now suppose that $a^o \in R'$. Since $R' \subseteq \text{Im}\,i$ it follows that $a^o = i(r^*)$ for some $r^* \in R^*$. By definition $a^o(I) = (0)$ for some cofinite ideal I of A. Since ideals of A are also left H-submodules of A, and the subalgebra R of A is also a left H-submodule, $J = R \cap I$ is a cofinite ideal of R which is also a left H-submodule of R. As $(0) = a^o(J) = i(r^*)(J) = r^*(J)$ we conclude that $r^* \in R^{\underline{o}}$. We have completed the proof of the claim.

Thus $R^{\underline{o}}$ and R' can be identified as vector spaces by the map of (16). Accordingly we will think of R' as $R^{\underline{o}}$ and show that R' is $R^{\underline{o}}$ as an object of ${}^{H^o}_{H^o}\mathcal{YD}$ and R' is $R^{\underline{o}}$ as the bialgebra in the category associated to (A^o, H^o, π^o, j^o).

Let $h^o \in H^o$ and $r^{\underline{o}} \in R^{\underline{o}}$. The left H^o-module structure on R' is given by $h^o \cdot_{\pi^o} i(r^{\underline{o}}) = \pi^o(h^o_{(1)})i(r^{\underline{o}})\pi^o(S^o(h^o_{(2)}))$. We evaluate both sides of this equation at $r \in R$. Since $r_{(1)} \otimes \pi(r_{(2)}) = r \otimes 1$ we have $\pi(r_{(1)}) \otimes r_{(2)} \otimes \pi(r_{(3)}) = \pi(r_{(1)}) \otimes r_{(2)} \otimes 1$

from which

$$\pi(r_{(1)})S(\pi(r_{(3)}))\otimes r_{(2)} = \pi(r_{(1)})\otimes r_{(2)} = r_{(-1)}\otimes r_{(0)}$$

follows. Thus

$$\begin{aligned}
(h^o \cdot_{\pi^o} i(r^o))\,(r) &= \left(\pi^o(h^o{}_{(1)})(r_{(1)})\right)\left(i(r^o)(r_{(2)})\right)\left(\pi^o(S^o(h^o{}_{(2)}))(r_{(3)})\right)\\
&= \left(h^o{}_{(1)}(\pi(r_{(1)}))\right)\left(i(r^o)(r_{(2)})\right)\left(h^o{}_{(2)}(S(\pi(r_{(3)})))\right)\\
&= h^o\left(\pi(r_{(1)})S(\pi(r_{(3)}))\right)i(r^o)(r_{(2)})\\
&= h^o(r_{(-1)})r^o(r_{(0)}).
\end{aligned}$$

We have shown that $(h^o \cdot_{\pi^o} i(r^o))\,(r) = h^o(r_{(-1)})r^o(r_{(0)})$ for all $h^o \in H^o$, $r^o \in R^o$, and $r \in R$.

The left H^o-comodule structure on R' is the subcomodule structure afforded by (A^o, δ_{j^o}). Now $\delta_{j^o}(i(r^o)) = j^o(i(r^o)_{(1)})\otimes i(r^o)_{(2)}$. Using the first commutation relation above we calculate

$$\begin{aligned}
\delta_{j^o}(i(r^o))(h\otimes r) &= i(r^o)_{(1)}(j(h))i(r^o)_{(2)}(r)\\
&= i(r^o)(j(h)r)\\
&= i(r^o)((h_{(1)}\cdot_j r)j(h_{(2)}))\\
&= r^o(h_{(1)}\cdot_j r)\epsilon(h_{(2)})\\
&= r^o(h\cdot_j r)
\end{aligned}$$

for all $h \in H$ and $r \in R$. We have shown that $R' = R^o$ as an object in ${}^{H^o}_{H^o}\mathcal{YD}$.

Next we consider the product in R'. Let r^o, $r'^o \in R^o$. Since $A = Rj(H)$ it is easy to see that $i(r^o)$ is determined on R. Thus for $r \in R$ the calculation

$$\begin{aligned}
i(r^o)(r^{(1)})i(r'^o)(r^{(2)}) &= i(r^o)(\Pi(r_{(1)}))i(r'^o)(r_{(2)})\\
&= i(r^o)\left(r_{(1)}j((S\circ\pi)(r_{(2)}))\right)i(r'^o)(r_{(3)})\\
&= i(r^o)(r_{(1)})\epsilon((S\circ\pi)(r_{(2)}))i(r'^o)(r_{(3)})\\
&= i(r^o)(r_{(1)})i(r'^o)(r_{(2)})\\
&= (i(r^o)i(r'^o))(r)
\end{aligned}$$

shows that the product $r^o r'^o$ is derived from the dual algebra of (R^o, Δ^o).

Finally we consider the coproduct of R'. First we calculate Π_{A^o} in terms of $\Pi_A = \Pi$. Let $p \in A^o$ and $a \in A$. Then

$$\begin{aligned}
(\Pi_{A^o}(p))\,(a) &= \left(p_{(1)}((\pi^o\circ S^o\circ j^o)(p_{(2)}))\right)(a)\\
&= p_{(1)}(a_{(1)})p_{(2)}((j\circ S\circ\pi)(a_{(2)}))\\
&= p(a_{(1)}((j\circ S\circ\pi)(a_{(2)})))\\
&= p(\Pi(a))
\end{aligned}$$

implies $\Pi_{A^o} = (\Pi_A)^o$. Let $r^o \in R^o$. By definition

$$\Delta_{R'}(i(r^o)) = \Pi_{A^o}(i(r^o)_{(1)})\otimes i(r^o)_{(2)}.$$

Thus for $r, r' \in R$ we compute

$$\Delta_{R'}(i(r^{\underline{o}}))(r \otimes r') = \big(\Pi_{A^{\circ}}(i(r^{\underline{o}})_{(1)})(r)\big)\,\big(i(r^{\underline{o}})_{(2)}(r')\big)$$

$$= \big(i(r^{\underline{o}})_{(1)}(\Pi(r))\big)\,\big(i(r^{\underline{o}})_{(2)}(r')\big) = i(r^{\underline{o}})(\Pi(r)r') = i(r^{\underline{o}})(rr')$$

since Π acts as the identity on R. Thus the coproduct for $R' = R^{\underline{o}}$ is that of the dual coalgebra arising from the subalgebra R of A. Therefore the bialgebra R' in the category $^{H^{\circ}}_{H^{\circ}}\mathcal{YD}$ associated to $(A^{\circ}, H^{\circ}, \pi^{\circ}, j^{\circ})$ is $R^{\underline{o}}$. We regard $R^* \otimes H^*$ as a subspace of the vector space $(R \# H)^* = (R \otimes H)^*$ in the natural way.

Proposition 7.1. *Let H be a Hopf algebra with bijective antipode, let A be a bialgebra and suppose that $j : H \longrightarrow A$, $\pi : A \longrightarrow H$ are bialgebra maps which satisfy $\pi \circ j = I_H$. Let $R = A^{co\,\pi}$ and let $(R, m, \eta, \Delta, \epsilon)$ be the bialgebra in $^H_H\mathcal{YD}$ associated to (A, H, j, π). Then:*

a) *$\pi^{\circ} : H^{\circ} \longrightarrow A^{\circ}$ and $j^{\circ} : A^{\circ} \longrightarrow H^{\circ}$ are bialgebra maps which satisfy $j^{\circ} \circ \pi^{\circ} = I_{H^{\circ}}$, $R^{\underline{o}} = A^{\circ\,co\,j^{\circ}}$ as a vector space under the identification $r^{\underline{o}}(rj(h)) = r^{\underline{o}}(r)\epsilon(h)$ for all $r \in R$ and $h \in H$, and $(R^{\underline{o}}, \Delta^{\underline{o}}, \epsilon, m^{\underline{o}}, \eta)$ is the bialgebra in $^{H^{\circ}}_{H^{\circ}}\mathcal{YD}$ associated to $(A^{\circ}, H^{\circ}, \pi^{\circ}, j^{\circ})$.*

b) *The map $\vartheta : R^{\underline{o}} \# H^{\circ} \longrightarrow (R \# H)^{\circ}$ given by $\vartheta(r^{\underline{o}} \# h^{\circ}) = r^{\underline{o}} \otimes h^{\circ}$ for all $r^{\underline{o}} \in R^{\underline{o}}$ and $h^{\circ} \in H^{\circ}$ is an isomorphism of bialgebras. Furthermore the diagram*

$$
\begin{array}{ccc}
R^{\underline{o}} \# H^{\circ} & \xrightarrow{\ \ \vartheta\ \ } & (R \# H)^{\circ} \\[2pt]
{\scriptstyle i|_{R^{\underline{o}}} \# I_{H^{\circ}}} \downarrow & & \downarrow {\scriptstyle (f^{-1})^{\circ}} \\[2pt]
R' \# H^{\circ} & \xrightarrow{\ \ g\ \ } & A^{\circ}
\end{array}
$$

commutes, where $R' = A^{\circ\,co\,j^{\circ}}$, $i|_{R^{\underline{o}}}$ is the linear isomorphism of (16), and the maps $f : R \# H \longrightarrow A$, $g : R' \# H^{\circ} \longrightarrow A^{\circ}$ are the canonical isomorphisms.

Proof. We have established part a). As for part b), we first note that the structures on $R^{\underline{o}}$ are due to the identification $i|_{R^{\underline{o}}} : R^{\underline{o}} \longrightarrow R'$. Thus $i|_{R^{\underline{o}}} \# I_{H^{\circ}}$ is an isomorphism of bialgebras by Proposition 5.1. Now g and $(f^{-1})^{\circ}$ are bialgebra isomorphisms also. Thus part b) will follow once we show that the diagram commutes.

Let $r^{\underline{o}} \in R^{\underline{o}}$, $h^{\circ} \in H^{\circ}$, $r \in R$, $h \in H$, and set $a = rj(h)$. Then

$$\big(((f^{-1})^{\circ} \circ \vartheta)\,(r^{\underline{o}} \# h^{\circ})\big)(a) = \vartheta(r^{\underline{o}} \# h^{\circ})(f^{-1}(a)) = (r^{\underline{o}} \otimes h^{\circ})(r \# h) = r^{\underline{o}}(r)h^{\circ}(h).$$

On the other hand

$$\big(\big(g \circ (i|_{R^{\underline{o}}} \# I_{H^{\circ}})\big)\,(r^{\underline{o}} \# h^{\circ})\big)(a) = \big(g(i(r^{\underline{o}}) \# h^{\circ})\big)(a) = \big(i(r^{\underline{o}})\pi^{\circ}(h^{\circ})\big)(rj(h))$$

$$= i(r^{\underline{o}})(r_{(1)}j(h_{(1)}))\pi^{\circ}(h^{\circ})(r_{(2)}j(h_{(2)})) = i(r^{\underline{o}})(r_{(1)}j(h_{(1)}))h^{\circ}(\pi(r_{(2)}j(h_{(2)})))$$

$$= r^{\underline{o}}(r_{(1)})\epsilon(h_{(1)})h^{\circ}(\pi(r_{(2)})(\pi \circ j)(h_{(2)})) = r^{\underline{o}}(r_{(1)})h^{\circ}(\pi(r_{(2)})h))$$

$$= r^{\underline{o}}(r)h^{\circ}(1h)) = r^{\underline{o}}(r)h^{\circ}(h).$$

Since $A = Rj(H)$ our calculations show that the diagram of part b) commutes. □

8. Bialgebras $\mathcal{H} = (\mathcal{U} \otimes \mathcal{A})_\sigma$ when $\mathcal{U}, \mathcal{A}$ are bi-products

Let K and H be Hopf algebras with bijective antipodes over the field k and suppose that $U = T\#K$, $A = R\#H$ are bi-products. We describe an extensive class of linear forms $(T\#K) \otimes (R\#H) \longrightarrow k$ which satisfy (A.1)–(A.4). All such forms are in bijective correspondence with the bialgebra maps $T\#K \longrightarrow (R\#H)^{o\,cop} = (R\#H)^{op\,o}$. Our forms are derived from certain bialgebra maps and determine two-cocycles σ.

8.1. The linear form $\beta\#\tau : (T\#K) \otimes (R\#H) \longrightarrow k$

We use the isomorphisms of the two preceding sections to construct bialgebra maps

$$T\#K \longrightarrow (R\#H)^{op\,o} \cong R^{\underline{op}\,o}\#H^{op\,o}$$

which determine linear forms $\beta\#\tau : (T\#K) \otimes (R\#H) \longrightarrow k$ satisfying (A.1)–(A.4). In the subsequent section we will investigate the special case when K and H are group algebras of abelian groups. These are fundamental, interesting in their own right, and arise in representations theory of pointed Hopf algebras [3, 11].

Theorem 8.1. *Let K and H be Hopf algebras with bijective antipodes. Suppose that T and R are bialgebras in the categories ${}^K_K \mathcal{YD}$ and ${}^H_H \mathcal{YD}$ respectively and that $\tau : K \otimes H \longrightarrow k$ and $\beta : T \otimes R \longrightarrow k$ are linear forms, where τ satisfies (A.1)–(A.4), β satisfies (B.1)–(B.4), and τ and β satisfy (C.1)–(C.2). Let*

$$\beta\#\tau : (T\#K) \otimes (R\#H) \longrightarrow k$$

be the linear form determined by

$$(\beta\#\tau)(t\#k, r\#h) = \beta(t, S^{-1}(h_{(1)}) \cdot r) \tau(k, h_{(2)})$$

for all $t \in T$, $k \in K$, $r \in R$, and $h \in H$. Then:

a) *$(\beta\#\tau)_\ell$ is the composite*

$$T\#K \xrightarrow{\beta_\ell \#\tau_\ell} R^{\underline{op}\,o}\#H^{op\,o} \xrightarrow{\vartheta} (R^{\underline{op}}\#H^{op})^o \xrightarrow{(\varphi^{-1})^o} (R\#H)^{op\,o},$$

where ϑ and φ are defined in part b) *of Propositions* 7.1 *and* 6.1 *respectively. Thus $\beta\#\tau$ satisfies* (A.1)–(A.4).

b) *$(\beta\#\tau)_\ell = (\varphi^{-1})^* \circ (\beta_\ell \otimes \tau_\ell)$ and $(\beta\#\tau)_r = (\beta_r \otimes \tau_r) \circ \varphi^{-1}$.*

c) *$\beta\#\tau$ is left (respectively right) non-singular if and only if β and τ are left (respectively right) non-singular.*

Proof. Since τ satisfies (A.1)–(A.4), its equivalent (1), which is $\operatorname{Im} \tau_\ell \subseteq H^{op\,o}$ and $\tau_\ell : K \longrightarrow H^{op\,o}$ is a bialgebra map, holds. Likewise, since β satisfies (B.1)–(B.4), by Lemma 4.1 it follows that $\operatorname{Im} \beta_\ell \subseteq R^{\underline{op}\,o}$ and $\beta_\ell : T \longrightarrow R^{\underline{op}\,o}$ is a map of algebras and a map of coalgebras. Now β_ℓ is τ_ℓ-linear and colinear since τ and β satisfy (C.1)–(C.2) by Proposition 4.2. Therefore $\beta_\ell\#\tau_\ell : T\#K \longrightarrow (R^{\underline{op}})^o\#(H^{op})^o$ is a bialgebra map by Proposition 5.1. At this point part a) follows by Propositions 6.1 and 7.1.

Part b) is a direct consequence of definitions. As for part c) we first note that the tensor product of two linear maps is one-one if and only if each tensorand is. Since φ^{-1} and $(\varphi^{-1})^*$ are linear isomorphisms, part c) now follows from part b). $\qquad\square$

Apropos of the theorem, requiring that $\beta\#\tau$ be left or right non-singular seems to be a rather stringent condition. We will find it very natural, and desirable, for β to be non-singular in connection with representation theory. See Section 9.

We shall call a tuple $(K, H, \tau, T, R, \beta)$ which satisfies the hypothesis of the preceding theorem *twist datum*. In applications morphisms of algebras of the type $((T\#K)\otimes(R\#H))_\sigma$, where σ is the two-cocycle given by $\beta\#\tau$, will be of interest to us. As an immediate consequence of Theorem 8.1 and Proposition 5.1:

Corollary 8.2. *Let K and H be Hopf algebras with bijective antipodes and let $(K, H, \tau, T, R, \beta)$ and $(\overline{K}, \overline{H}, \overline{\tau}, \overline{T}, \overline{R}, \overline{\beta})$ be twist data. Suppose that $\varphi : K \longrightarrow \overline{K}$ and $\nu : H \longrightarrow \overline{H}$ are bialgebra maps, $f : T \longrightarrow \overline{T}$ and $g : R \longrightarrow \overline{R}$ are algebra and coalgebra maps, where f is φ-linear and colinear and g is ν-linear and colinear. Assume further that $\overline{\tau}\circ(\varphi\otimes\nu) = \tau$ and $\overline{\beta}\circ(f\otimes g) = \beta$. Then $(\overline{\beta}\#\overline{\tau})\circ((f\#\varphi)\otimes(g\#\nu)) = \beta\#\tau$. In particular*

$$(f\#\varphi)\otimes(g\#\nu) : \big((T\#K)\otimes(R\#H) \big)_\sigma \longrightarrow \big((\overline{T}\#\overline{K})\otimes(\overline{R}\#\overline{H}) \big)_{\overline{\sigma}}$$

is a bialgebra map. $\qquad\square$

8.2. The fundamental case $U = \mathfrak{B}(W)\#K$ and $A = \mathfrak{B}(V)\#H$

We specialize the results of the preceding section to the case of most interest to us: when $U = \mathfrak{B}(W)\#K$ and $A = \mathfrak{B}(V)\#H$ are biproducts of Nichols algebras with Hopf algebras having bijective antipodes. We are able to express assumptions involving $\mathfrak{B}(W)$ and $\mathfrak{B}(V)$ in terms of W and V.

Theorem 8.3. *Let K and H be Hopf algebras with bijective antipodes and let $\tau : K\otimes H \longrightarrow k$ be a linear form which satisfies (A.1)–(A.4). Suppose $W \in {}^K_K\mathcal{YD}$, $V \in {}^H_H\mathcal{YD}$, $\beta : W\otimes V \longrightarrow k$ is a linear form, and τ and β satisfy (C.1)–(C.2). Then:*

a) *$(K, H, \tau, \mathfrak{B}(W), \mathfrak{B}(V), \mathfrak{B}(\beta))$ is a twist datum.*
b) *The linear form $\mathfrak{B}(\beta)\#\tau : (\mathfrak{B}(W)\#K)\otimes(\mathfrak{B}(V)\#H) \longrightarrow k$ satisfies (A.1)–(A.4).*

Proof. The form $\mathfrak{B}(\beta) : \mathfrak{B}(W)\otimes\mathfrak{B}(V) \longrightarrow k$ of Corollary 4.3 satisfies the hypothesis of Theorem 8.1. $\qquad\square$

We shall call a tuple $(K, H, \tau, W, V, \beta)$ which satisfies the hypothesis of the preceding theorem *Yetter-Drinfel'd twist datum*. By the preceding theorem if $(K, H, \tau, W, V, \beta)$ is a Yetter-Drinfel'd twist datum then

$$(K, H, \tau, \mathfrak{B}(W), \mathfrak{B}(V), \mathfrak{B}(\beta))$$

is a twist datum. We now turn our attention to morphisms.

Proposition 8.4. *Let K, H be Hopf algebras with bijective antipodes and*

$$(K, H, \tau, W, V, \beta) \quad and \quad (\overline{K}, \overline{H}, \overline{\tau}, \overline{W}, \overline{V}, \overline{\beta})$$

be Yetter-Drinfel'd twist data. Suppose that $\varphi : K \longrightarrow \overline{K}$ and $\nu : H \longrightarrow \overline{H}$ are bialgebra maps which satisfy $\overline{\tau} \circ (\varphi \otimes \nu) = \tau$. Let $f : W \longrightarrow \overline{W}$ be φ-linear and colinear, let $g : V \longrightarrow \overline{V}$ be ν-linear and colinear, and suppose that $\overline{\beta} \circ (f \otimes g) = \beta$. Then

$$F : \big((\mathfrak{B}(W) \# K) \otimes (\mathfrak{B}(V) \# H) \big)_\sigma \longrightarrow \big((\mathfrak{B}(\overline{W}) \# \overline{K}) \otimes (\mathfrak{B}(\overline{V}) \# \overline{H}) \big)_{\overline{\sigma}}$$

is a bialgebra map, where $F = (\mathfrak{B}(f) \# \varphi) \otimes (\mathfrak{B}(g) \# \nu)$.

Proof. First of all we observe that

$$(K, H, \tau, \mathfrak{B}(W), \mathfrak{B}(V), \mathfrak{B}(\beta)) \quad and \quad (\overline{K}, \overline{H}, \overline{\tau}, \mathfrak{B}(\overline{W}), \mathfrak{B}(\overline{V}), \mathfrak{B}(\overline{\beta}))$$

are twist data by part a) of Theorem 8.3. By assumption $\overline{\tau} \circ (\varphi \otimes \nu) = \tau$. Since $\overline{\beta} \circ (f \otimes g) = \beta$, it follows by Corollary 4.3 that $\mathfrak{B}(\overline{\beta}) \circ (\mathfrak{B}(f) \otimes \mathfrak{B}(g)) = \mathfrak{B}(\beta)$. At this point we apply Corollary 8.2 to complete the proof. $\square$

Let $(K, H, \tau, W, V, \beta)$ be a Yetter-Drinfel'd twist datum. Axiom (C.1) implies that $V^\perp$ is a left K-submodule of W and axiom (C.2) implies that $W^\perp$ is a left H-submodule of V. Thus the projections $\pi_W : W \longrightarrow W/V^\perp$ and $\pi_V : V \longrightarrow V/W^\perp$ are module maps. If $V^\perp$ is a left K-subcomodule of W then π_W is also a left K-comodule map and likewise if $W^\perp$ is a left H-subcomodule of V then π_V is also a left H-comodule map.

Suppose that $V^\perp$ and $W^\perp$ are subcomodules. By part b) of Corollary 1.3 note that $\mathfrak{B}(\pi_W) : \mathfrak{B}(W) \longrightarrow \mathfrak{B}(W/V^\perp)$ and $\mathfrak{B}(\pi_V) : \mathfrak{B}(V) \longrightarrow \mathfrak{B}(V/W^\perp)$ are onto since π_W and π_V are. Recall that the linear form $\overline{\beta} : W/V^\perp \otimes V/W^\perp \longrightarrow k$ determined by $\overline{\beta} \circ (\pi_W \otimes \pi_V) = \beta$ is non-singular. With φ, ν the identity, $f = \pi_W$, and $g = \pi_V$, the preceding proposition gives:

Corollary 8.5. *Let K, H be Hopf algebras with bijective antipodes and let $(K, H, \tau, W, V, \beta)$ be a Yetter-Drinfel'd twist datum. Suppose that $V^\perp$ and $W^\perp$ are subcomodules of W and V respectively. Let $\pi_W : W \longrightarrow W/V^\perp$ and $\pi_V : V \longrightarrow V/W^\perp$ be the projections. Then $(K, H, \tau, W/V^\perp, V/W^\perp, \overline{\beta})$ is a Yetter-Drinfel'd twist datum and*

$$\big((\mathfrak{B}(W) \# K) \otimes (\mathfrak{B}(V) \# H) \big)_\sigma \xrightarrow{\ F\ } \big((\mathfrak{B}(W/V^\perp) \# K) \otimes (\mathfrak{B}(V/W^\perp) \# H) \big)_{\overline{\sigma}}$$

is a surjective bialgebra map, where $F = (\mathfrak{B}(\pi_W) \# I_K) \otimes (\mathfrak{B}(\pi_V) \# I_H)$. $\square$

The preceding corollary can be used in our study of a class of irreducible representations of an extensive class of examples of two-cocycle twists in [11]. We will be able to replace the domain of F by its image and thus assume that β is non-singular. The non-singularity of β has very interesting consequences.

9. The case when K and H are group algebras of abelian groups

Here we specialize the results of Section 8.2 to a very typical case. Let Γ be an abelian group. We set $_{k[\Gamma]}^{k[\Gamma]}\mathcal{Y}D = {}_{\Gamma}^{\Gamma}\mathcal{Y}D$. Suppose that $V \in {}_{\Gamma}^{\Gamma}\mathcal{Y}D$, $g \in \Gamma$, and $\chi \in \widehat{\Gamma}$. We set

$$V_g = \{v \in V \mid \delta(v) = g \otimes v\}$$

and

$$V_g^\chi = \{v \in V_g \mid h \cdot v = \chi(h)v \text{ for all } h \in \Gamma\}.$$

A Yetter-Drinfel'd module in $_{\Gamma}^{\Gamma}\mathcal{Y}D$ can be described as a Γ-graded vector space which is a Γ-module such that all g-homogeneous components $g \in \Gamma$ are stable under the Γ-action.

In this section we fix abelian groups Λ and Γ, positive integers n and m, elements $z_1, \ldots, z_m \in \Lambda$ and $g_1, \ldots, g_n \in \Gamma$, and nontrivial characters $\eta_1, \ldots, \eta_m \in \widehat{\Lambda}$ and $\chi_1, \ldots, \chi_n \in \widehat{\Gamma}$. Suppose $W \in {}_{\Lambda}^{\Lambda}\mathcal{Y}D$ has basis $u_i \in W_{z_i}^{\eta_i}, 1 \leq i \leq m$, and that $V \in {}_{\Gamma}^{\Gamma}\mathcal{Y}D$ has basis $a_j \in V_{g_j}^{\chi_j}, 1 \leq j \leq n$.

For any character $\Psi \in \widehat{\Gamma}$ we define the algebra map

$$\widetilde{\Psi} : \mathfrak{B}(V)\#k[\Gamma] \to k$$

by $\widetilde{\Psi}(a_j\#1) = 0, \widetilde{\Psi}(1\#g) = \Psi(g)$ for all $1 \leq j \leq n$ and $g \in \Gamma$.

The following corollaries will play a useful role in [11, Lemma 3.1].

Corollary 9.1. *In addition to the above, let* $\varphi : \Lambda \longrightarrow \widehat{\Gamma}$ *be a group homomorphism,* $l_1, \ldots, l_m \in k$, *and* $s : \{1, \ldots, m\} \longrightarrow \{1, \ldots, n\}$ *be a function.*

Let $\tau : k[\Lambda] \otimes k[\Gamma] \longrightarrow k$ *and* $\beta : W \otimes V \longrightarrow k$ *be the linear forms defined by*

$$\tau(z \otimes g) = \varphi(z)(g) \text{ and } \beta(u_i \otimes a_j) = l_i \delta_{s(i),j} \tag{17}$$

for all $z \in \Lambda, g \in \Gamma, 1 \leq i \leq m$ *and* $1 \leq j \leq n$. *Assume further that for all* $1 \leq i \leq m$ *with* $\lambda_i \neq 0$ *and* $z \in \Lambda$

$$\varphi(z_i) = \chi_{s(i)}^{-1} \text{ and } \eta_i(z) = \varphi(z)(g_{s(i)}). \tag{18}$$

Then $(k[\Lambda], k[\Gamma], \tau, W, V, \beta)$ *is a Yetter-Drinfel'd twist datum, and the corresponding Hopf algebra map*

$$\Phi = (\mathfrak{B}(\beta)\#\tau)_\ell : \mathfrak{B}(W)\#k[\Lambda] \longrightarrow (\mathfrak{B}(V)\#k[\Gamma])^{o\,cop}$$

can be described as follows:

For all $1 \leq i \leq m$ *let* $\gamma_i = \widetilde{\varphi(z_i)}$, *and let* $\delta_i : \mathfrak{B}(V)\#k[\Gamma] \longrightarrow k$ *be the unique* (ε, γ_i)-*derivation with* $\delta_i(a_j\#1) = \beta(u_i \otimes a_j), \delta_i(1\#g) = 0$ *for all* $1 \leq j \leq n$ *and* $g \in \Gamma$. *The algebra map* Φ *is determined by*

$$\Phi(1\#z) = \widetilde{\varphi(z)}, \quad \Phi(u_i\#1) = \delta_i$$

for all $z \in \Lambda$ *and* $1 \leq i \leq m$.

Proof. We first show that $(k[\Lambda], k[\Gamma], \tau, W, V, \beta)$ is a Yetter-Drinfel'd twist datum. Since φ is a group homomorphism τ satisfies (1), an equivalent of (A.1)–(A.4). Note that (C.2) holds for τ and β if and only if for all $1 \le i \le m$ and $1 \le j \le n$ the equation $\beta(u_i \leftarrow \tau_r(g), a_j) = \beta(u_i, S^{-1}(g) \cdot a_j)$ holds for all $g \in \Gamma$. The latter is $\tau_r(g)(z_i)\beta(u_i, a_j) = \beta(u_i, \chi_j(g^{-1})a_j)$, or $\tau_\ell(z_i)(g)l_i\delta_{s(i),j} = \chi_j(g^{-1})l_i\delta_{s(i),j}$, an equivalent of the first equation of (18). Next we note that (C.1) is equivalent to $\beta(z \cdot u_i, a_j) = \beta(u_i, a_j \leftarrow \tau_\ell(z))$ which is the same as $\beta(\eta_i(z)u_i, a_j) = \beta(u_i, \tau_\ell(z)(g_j)a_j)$, or $\eta_i(z)l_i\delta_{s(i),j} = \varphi(z)(g_j)l_i\delta_{s(i),j}$, for all $z \in \Lambda$, $1 \le i \le m$, and $1 \le j \le n$. The latter is equivalent to the second equation of (18).

We have shown that $(k[\Lambda], k[\Gamma], \tau, W, V, \beta)$ is a Yetter-Drinfel'd twist datum, and the corollary follows by Theorem 8.3 and Theorem 8.1. $\square$

Corollary 9.2. *Assume the situation of Corollary 9.1. Let $I' = \{1 \le i \le m \mid l_i \ne 0\}$, and assume that the restriction of s to I' is injective. Let $V' \subseteq V$ and $W' \subseteq W$ be the Yetter-Drinfel'd submodules with bases $a_{s(i)}, i \in I'$, and $u_i, i \in I'$.*

   Then

a) *$(k[\Lambda], k[\Gamma], \tau, W', V', \beta')$ is a Yetter-Drinfel'd twist datum. $V^\perp \subseteq W$ and $W^\perp \subseteq V$ are Yetter-Drinfel'd submodules, and the inclusion maps $W' \subseteq W$ and $V' \subseteq V$ define isomorphisms $W' \cong W/V^\perp$ and $V' \cong V/W^\perp$. The restriction*

$$\beta' : W' \otimes V' \to k$$

   of β is nondegenerate.

b) *The projections $\pi_W : W \to W'$, $\pi_V : V \to V'$ define a surjective bialgebra map*

$$\big((\mathfrak{B}(W) \# k[\Lambda]) \otimes (\mathfrak{B}(V) \# k[\Gamma]) \big)_\sigma \xrightarrow{F} \big((\mathfrak{B}(W') \# k[\Lambda]) \otimes (\mathfrak{B}(V') \# k[\Gamma]) \big)_{\sigma'}$$

   given by

$$F = (\mathfrak{B}(\pi_W) \# \mathrm{id}) \otimes (\mathfrak{B}(\pi_V) \# \mathrm{id}).$$

Proof. By Corollary 9.1 $(k[\Lambda], k[\Gamma], \tau, W, V, \beta)$ is a Yetter-Drinfel'd twist datum. Thus $(k[\Lambda], k[\Gamma], \tau, W', V', \beta')$ is a Yetter-Drinfel'd twist datum as well.

To show that β' is non-singular we compute $W^\perp$ and $V^\perp$. Suppose that $a \in V$ and write $a = \sum_{j=1}^n x_j a_j$, where $x_1, \ldots, x_n \in k$. Then $a \in W^\perp$ if and only if $\beta(u_i, \sum_{j=1}^n x_j a_j) = 0$, or $\sum_{j=1}^n x_j l_i \delta_{s(i),j} = 0$, for all $1 \le i \le m$. Therefore $W^\perp$ has basis $\{a_j \mid j \in \{1, \ldots, n\} \setminus s(I')\}$. We have shown that $W^\perp \oplus V' = V$.

Let $u \in W$ and write $u = \sum_{i=1}^m y_i u_i$ where $y_1, \ldots, y_m \in k$. Then $u \in V^\perp$ if and only if $\beta(\sum_{i=1}^m y_i u_i, a_j) = 0$, or $\sum_{i=1}^m y_i l_i \delta_{s(i),j} = 0$, for all $1 \le j \le n$. Since s is one-one we conclude that $u \in V^\perp$ if and only if $y_i l_i = 0$ for all $i \in I'$. Thus $V^\perp$ has basis $\{u_i \mid i \in \{1, \ldots, m\} \setminus I'\}$. Thus β' is non-singular, and part a) is established.

Note that the maps $\pi_W : W \longrightarrow W'$ and $\pi_V : V \longrightarrow V'$ of Yetter-Drinfel'd modules can be identified with the projections $W \longrightarrow W/V^\perp$ and $V \longrightarrow V/W^\perp$ respectively. At this point we apply Corollary 8.5 to complete the proof. $\square$

References

[1] E. Abe, "Hopf Algebras," **74**, *Cambridge Tracts in Mathematics*, Cambridge University Press, Cambridge, UK, 1980.

[2] Nicolás Andruskiewitsch and Hans-Jürgen Schneider,Finite quantum groups over abelian groups of prime exponent, *Ann. Sci. École Norm. Sup.* (4) **35** (2002), 1–26.

[3] Nicolás Andruskiewitsch and Hans-Jürgen Schneider, Pointed Hopf algebras. New directions in Hopf algebras, 1–68, *Math. Sci. Res. Inst. Publ.* **43**, Cambridge Univ. Press, Cambridge, 2002.

[4] M. Beattie, Duals of pointed Hopf algebras, *J. of Algebra* **262** (2003), 54–76.

[5] Yukio Doi and Mitsuhiro Takeuchi, Multiplication alteration by two-cocycles – the quantum version, *Comm. Algebra* **22** (1994), 5715–5732.

[6] Larry A. Lambe and David E. Radford, Introduction to the quantum Yang-Baxter equation and quantum groups: an algebraic approach. Mathematics and its Applications **423**, Kluwer Academic Publishers, Dordrecht (1997), xx+293 pp.

[7] Shahn Majid, Algebras and Hopf Algebras in Braided Categories. "Advances in Hopf Algebras", Lecture Notes in Pure and Applied Mathematics **158**, 55–105, Marcel Dekker, 1994.

[8] S. Montgomery, "Hopf Algebras and their actions on rings," **82**, *Regional Conference Series in Mathematics*, AMS, Providence, RI, 1993.

[9] David E. Radford, Coreflexive coalgebras, *J. of Algebra* **26** (1973), 512–535.

[10] D.E. Radford, The structure of Hopf algebras with a projection, *J. Algebra* **92**(1985), 322–347.

[11] David E. Radford and Hans-Jürgen Schneider, On the Simple Representations of Generalized Quantum Groups and Quantum Doubles, to appear in J. Algebra.

[12] Moss E. Sweedler, Hopf algebras. Mathematics Lecture Note Series W. A. Benjamin, Inc., New York (1969), vii+336 pp.

David E. Radford
University of Illinois at Chicago
Department of Mathematics, Statistics and Computer Science
(m/c 240), 801 South Morgan Street
Chicago, IL 60608-7045, USA
e-mail: radford@uic.edu

Hans Jürgen Schneider
Mathematisches Institut
Ludwig-Maximilians-Universität München
Theresienstr. 39
D-80333 München, Germany
e-mail: Hans-Juergen.Schneider@mathematik.uni-muenchen.de